TO: Jerry, Ann and Eve

PRACTICAL FLUORESCENCE

Theory, Methods, and Techniques

by GEORGE G. GUILBAULT

Department of Chemistry
Louisiana State University in New Orleans
New Orleans, Louisiana

with contributions from
R. F. CHEN; GOVINDJEE,
G. PAPAGEORGIOU, and
E. RABINOWITCH; E. L. WEHRY

MARCEL DEKKER, INC.   NEW YORK AND BASEL

MARCEL DEKKER, INC.

270 Madison Avenue, New York, New York 10016

LIBRARY OF CONGRESS CATALOG CARD NUMBER: 72-90964

ISBN: 0-8247-1263-3

Current printing (last digit):
10 9 8 7 6 5 4

PRINTED IN THE UNITED STATES OF AMERICA

PREFACE

Luminescence is one of the oldest and most established analytical
techniques, having been first observed by Monardes in 1565 from the extract
Ligirium Nephiticiem. Sir David Brewster noted the red emission from
chlorophyll in 1833, and Sir G. G. Stokes described the mechanism of the
absorption and emission process in 1852. Stokes also named fluorescence
after the mineral fluorspar (Latin fluo = to flow + spar = a rock) which
exhibits a blue-white fluorescence.

Phosphorescence dates back to the early 1500's, and was so named after
the Greek "light bearing." In fact, the element phosphorus was named from
this same Greek word in 1669, since it was found to produce a bright light in
a dark room.

Luminescence is one of the most active research fields in science today,
as evidenced by the increasing number of papers, reviews, and monographs
published each year. Fluorescence, phosphorescence, chemiluminescence, and
atomic fluorescence provide some of the most sensitive and selective methods
of chemical analysis.

In this book, attempts are made to introduce the reader to the entire
field of luminescence spectroscopy. This volume starts with an introduction
to luminescence and then proceeds to a discussion of luminescent instrumentation, structural and environmental effects on luminescence, and phosphorescence. Finally, the use of luminescence in determination of inorganic ions
and organic compounds, and the use of fluorescence in enzymology, in photosynthesis and for the assay of proteins are discussed. Special chapters on
chemiluminescence, atomic fluorescence, solid surface monitoring,
fluorescence indicators, and forensic and environmental analysis complete
the presentation.

Throughout the book, emphasis is placed upon fundamental principles and actual analytical applications. It is felt that complete absorption of the material presented in the book should result in a broad understanding of fluorescence and phosphorescence spectrometry which should prove very beneficial to the beginner and also to the experienced worker in the field of luminescence.

I would like to thank Mrs. Mercedes Weiser and Mrs. Judy Snowden for typing the entire book in final form for direct reproduction. I am especially grateful to my wife, Palma Covington Guilbault, for her constant encouragement throughout the preparation of this book.

New Orleans, Louisiana   G. G. Guilbault
January, 1973

CONTENTS

PRACTICAL FLUORESCENCE

Theory, Methods, and Techniques

Chapter 1

INTRODUCTION TO LUMINESCENCE

I. INTRODUCTORY REMARKS

A. History of Luminescence

Luminescence is one of the oldest and most established analytical techniques, having been first observed by Monardes in 1565 from an extract of <u>Ligirium nephiticiem</u>. Sir David Brewster noted the red emission from chlorophyll in

1

1833, and Sir G. G. Stokes described the mechanism of the absorption and emission process in 1852. Stokes also named fluorescence after the mineral fluorspar (Latin fluo = to flow + spar = a rock), which exhibits a blue-white fluorescence.

Phosphorescence dates back to the early 1500s, being so named after the Greek work for "light bearing." In fact the element phosphorus was named from this same Greek work in 1669 since it was found to produce a bright light in a dark room.

Luminescence is one of the most active research fields in science today, as evidenced by the increasing number of papers, reviews, and monographs published each year. Fluorescence, phosphorescence, chemiluminescence, and atomic fluorescence provide some of the most sensitive and selective methods of chemical analysis.

Some of the better general references in luminescence spectroscopy are listed at the end of this chapter. Books by Guilbault [1], Hercules [2], Passwater [3], Phillips and Elevitch [4], Udenfriend [5,6], White and Argauer [7], Konstantinova-Shlesinger [8], and Pringsheim [9] are worth reading. Excellent chapters by Weissler and White on fluorescence appear in the Handbook of Analytical Chemistry [10] and in Scott's Standard Methods of Chemical Analysis, [11]. Workers in fluorescence should also consult the reviews [12] by White in Analytical Chemistry every two years and receive the free pamphlets published monthly by the American Instrument Company [13] and G. K. Turner Associates [14].

B. Light and Its Interaction with Matter

Light is a form of electromagnetic radiation, the propagation of which is regarded as a wave phenomenon. Light is characterized by a wavelength λ and a frequency ν interrelated by the equation

$$\nu = \frac{c}{\lambda} , \tag{1}$$

where c is the velocity of light, 3×10^{10} cm/sec.

When light impinges upon matter, two things can happen: it can pass through the matter with no absorption taking place, or it can be absorbed either entirely or in part. In the latter case energy is transferred to the molecule in the absorption process.

Absorption of energy must occur in integral units, called quanta. The quanta-energy relationship can be expressed by the equation

$$E = h\nu = \frac{hc}{\lambda} , \tag{2}$$

where E is the energy, h is Planck's constant (6.62 X 10^{-27} erg-sec). Note that energy E is inversely related to wavelength λ.

Every molecule possesses a series of closely spaced energy levels and can go from a lower to a higher energy level by the absorption of a discrete quantum of light equal in energy to the difference between the two energy states (Fig. 1). Only a few molecules are raised to this higher excited state and hence capable of exhibiting luminescence. Between each main electronic state are the various vibrational levels of the molecule. In Fig. 1, which illustrates the various potential energy levels of a diatomic molecule, are indicated the various vibrational levels, represented as 0, 1, 2, 3, and 4, of each curve. The ground state is indicated by G, the first excited singlet electronic state by S^*, and the first excited triplet by T^*. Differences in the singlet and triplet are differences in the spin of the electron, S. All electrons have a spin S equal to

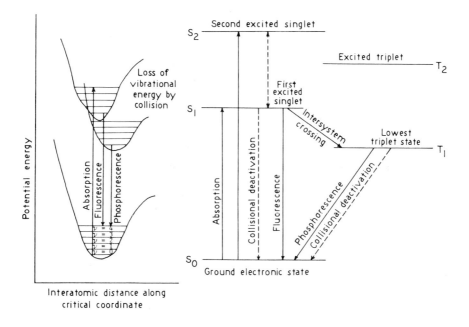

FIG. 1.   Schematic energy-level diagram for a diatomic molecule.

$\pm$ 1/2.   The arrow designations often used to denote spin ($\uparrow\downarrow$) arise from the right-hand rule. If one curls the fingers of the right hand in the direction of the spin, the thumb points in the direction of the arrow, and vice versa. The spin, S, can be denoted as + 1/2 or -1/2.

A normal polyatomic molecule in the ground state G usually has an even

number of electrons with paired spins. Thus there is one electron with
$S = +1/2$ and one with $S = -1/2$. Multiplicity is a term used to express the
orbital angular momentum of a given state and is related to the spin by the
equation

$$M = 2S + 1. \tag{3}$$

Thus when all electrons are paired, $S = 0$ ($+1/2$ $-1/2 = 0$), and the multiplicity
equals 1. This is called a singlet electronic state. When the spin of a single
electron is reversed, the molecule finds itself with two unpaired electrons and
$S = 1$ ($+ 1/2 + 1/2 = 1$) and the multiplicity is 3$[2(1) + 1 = 3]$. This electronic
state is called a triplet.

The relationships between wavelength, energy, and color are shown in
Table 1. Photons in the ultraviolet and visible regions of the electromagnetic
spectrum have energies of 35 to 145 kcal/mole and promote electronic transitions. More energetic photons cause photodecomposition, rather than
electronic transitions. Less energetic photons have only enough energy to
cause a vibrational or rotational transition. The less energetic vibrational
transitions are often superimposed on the electronic transition and are observed
as fine structure.

When a quantum of light impinges on a molecule, it is absorbed in about
10^{-15} sec, and a transition to a higher electronic state takes place (Fig. 1-1).
This absorption of radiation is highly specific, and radiation of a particular
energy is absorbed only by a characteristic structure. The electron is raised
to an upper excited singlet state, S_1, S_2, etc. These ground-to-singlet
transitions are responsible for the visible- and ultraviolet-absorption spectra
observed for molecules. The absorption transitions usually originate in the
lowest vibrational level of the ground electronic state.

During the time the molecule can spend in the excited state, 10^{-4} sec,
some energy in excess of the lowest vibrational energy level is rapidly dissipated. The lowest vibrational level ($\nu = 0$) of the excited singlet state S is
attained. If all the excess energy is not further dissipated by collisions with
other molecules, the electron returns to the ground electronic state, with the
emission of energy. This phenomenon is called fluorescence. Because some
energy is lost in the brief period before emission can occur, the emitted energy
(fluorescence) is of longer wavelength than the energy that was absorbed.

TABLE 1

Relationship of Wavelength to Energy and Color

Region	Wavelength (nm)	E (kcal/mole)	Color	Complement
Ultraviolet:				
Far	200	143. 0	--	--
Near	250	114. 5	--	--
	300	95. 4	--	--
	350	82. 0	--	--
Visible	380	75. 5	Violet	Yellow-green
	400	71. 5	Violet	Yellow-green
	450	63. 8	Blue	Yellow
	500	57. 3	Blue-green	Red
	550	52. 1	Green	Purple
	600	47. 7	Orange	Green-blue
	650	44. 1	Red	Blue-green
	700	40. 9	Red	Blue-green
	750	38. 2	Red	Blue-green
Near infrared	780	36. 7	--	--
	800	35. 8	--	--

The phenomenon of phosphorescence involves an intersystem crossing, or transition, from the singlet to the triplet state. A triplet state results when the spin of one electron changes so that the spins are the same, or unpaired. The transition from the ground state to the triplet excited state is a forbidden (highly improbable) transition. Internal conversion from the singlet to the triplet (electron-spin reversal) is more probable since the energy of the lowest vibrational level of T^* is lower than that of S^*. Molecules in T^* can then return to the ground state G directly, since a return via S^* could result only by acquiring energy from the environment (this sometimes occurs and is called delayed fluorescence, as we shall see). Transition times of 10^{-4} to 10 sec are observed in phosphorescence. Hence a characteristic feature of phosphores-

cence is an afterglow, that is, emission that continues after the exciting source
is removed. Because of the relatively long lifetime of the triplet state,
molecules in this state are much more susceptible to radiationless deactivation
processes, and only substances dissolved in a rigid medium phosphoresce.

II. TYPES OF LUMINESCENCE

The various types of luminescence can be classified according to the means
by which energy is supplied to excite the luminescent molecule.

When molecules are excited by interaction with photons of electromagnetic
radiation, the form of luminescence is called photoluminescence. If the release
of electromagnetic energy is immediate or from the singlet state, the process
is called fluorescence, whereas phosphorescence is a delayed release of energy
from the triplet state. Some molecules exhibit a delayed fluorescence that
might incorrectly be assumed to be phosphorescence. This results from two
intersystem crossings, first from the singlet to the triplet, then from the
triplet to the singlet.

If the excitation energy is obtained from the chemical energy of reaction,
the process is chemiluminescence. In bioluminescence the electromagnetic
energy is released by organisms.

Triboluminescence (Greek tribo, to rub) is produced as a release of energy
when certain crystals, such as sugar, are broken. The energy stored on
crystal formation is released in the breaking of the crystal.

Other types of luminescence--cathodoluminescence resulting from a
release of energy produced by exposure to cathode rays, or thermoluminescence
which occurs when a material existing in high vibrational energy levels emits
energy at a temperature below red heat, after being exposed to small amounts
of thermal energy--are much less commonly encountered.

III. TYPES OF FLUORESCENCE AND EMISSION PROCESSES

The fluorescence normally observed in solutions is called Stokes
fluorescence. This is the reemission of less energetic photons, which have a
longer wavelength (lower frequency) than the absorbed photons.

If thermal energy is added to an excited state or a compound has many
highly populated vibrational energy levels, emission at shorter wavelengths
than those of absorption occurs. This is anti-Stokes fluorescence, often

observed in dilute gases at high temperatures. A common example is the green
emission from copper-activated cadmium sulfide excited by red light.

Resonance fluorescence is the reemission of photons possessing the same
energy as the absorbed photons. This type of fluorescence is never observed in

solution because of solvent interactions, but it does occur in gases and crystals. It is also the basis of atomic fluorescence. Atomic fluorescence spectroscopy is an excellent technique for the assay of many elements and will be discussed later in this volume.

If an electron is excited by an absorbed photon of energy to a higher vib-rational level with no electronic transition, energy is entirely conserved and a photon of the same energy is reemitted within 10^{-15} sec as the electron returns to its original state. The emitted light has the same wavelength as the exciting light since the absorbed and emitted photons are of the same energy. The emitted light is referred to as Rayleigh scattering and occurs at all wavelengths. Its intensity, however, varies as the fourth power of the wavelength, so its effect can be minimized by working at longer wavelengths. It is a problem when the intensity of fluorescence is low in comparison with the exciting radiation and when the absorption and fluorescence spectra of a substance are close together.

Another form of scattering emission related to Rayleigh scattering is the Raman effect. Raman scatter appears in fluorescence spectra at higher and lower wavelengths (the former being more common) than the Rayleigh-scatter peak, and these Raman bands are satellites of the Rayleigh-scatter peak with a constant frequency difference from the exciting radiation. These bands are due to vibrational energy being added to, or subtracted from, this excitation photon. The Raman bands are much weaker than the Rayleigh-scatter peak but become significant when high-intensity sources are used. The relationship between the fluorescence band, Rayleigh scatter, and Raman scatter is shown in Fig. 2.

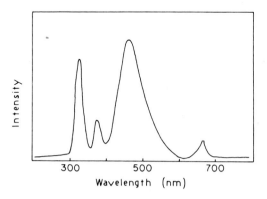

FIG. 2. Fluorescence spectra of quinine sulfate in 0.1 N sulfuric acid (λ_{ex} = 320 nm). Peaks: 320 nm, Rayleigh scatter; 360 nm, Raman scatter of water; 450 nm, quinine fluorescence; 640 nm, second-order Rayleigh scatter; 720 nm, second-order Raman scatter.

IV. EXCITATION SPECTRUM

Any fluorescent molecule has two characteristic spectra: the excitation spectrum (the relative efficiency of different wavelengths of exciting radiation to cause fluorescence) and the emission spectrum (the relative intensity of radiation emitted at various wavelengths).

The shape of the excitation spectrum should be identical with that of the absorption spectrum of the molecule and independent of the wavelength at which fluorescence is measured. This is seldom the case, however, the differences being due to instrumental artifacts. Examination of the excitation spectrum indicates the positions of the absorption spectrum that give rise to fluorescence emission; for example, the excitation spectrum of the aluminum chelate of acid Alizarin Garnet R (Fig. 3) indicates peaks at 350, 430, and 470 nm. The absorption spectrum (run on a spectrophotometer) exhibits peaks at 270, 350, and 480 nm. The two spectra do not agree because (a) photomultiplier sensitivity changes, (b) the bandwidth of the monochromator changes, and (c) the slits remain constant in fluorescence. To obtain the true, or "corrected," spectra of the compound the apparent excitation curve would have to be corrected for these factors and then the absorption spectrum should be obtained.

A general rule of thumb is that the longest wavelength peak in the excitation spectrum is chosen for excitation of the sample. This minimizes possible decomposition caused by the shorter wavelength, higher energy, radiation.

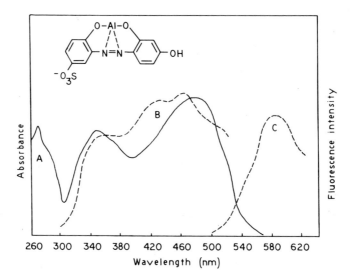

FIG. 3. Absorption and fluorescence spectra of the aluminum complex with acid Alizarin Garnet R (0.008%): curve A, the absorption spectrum; curve B, the fluorescence-excitation spectrum; curve C, the fluorescence-emission spectrum.

Finally, it should be pointed out that the excited state of a molecule differs from the ground state in chemical and physical properties. The excited state possesses a different geometry with different interatomic distances and a different dipole moment. Even stereo effects like cis and trans isomers can be introduced. Chemically, vast changes in acidity result in going from the ground to the excited state; for example, phenol has a pK_a of 10.0 in the ground state, 4.0 in the singlet excited state, and 8.5 in the triplet excited state. Since a change in pH or dissociation can vastly affect the fluorescence of a compound, this factor becomes very important in the measurement of luminescence.

Another fact of concern is changes in chemical structure that result in the photoexcitation of a molecule. The molecule 9,10-dihydroxy-anthracene loses a proton when excited to the singlet; the resulting ion is then easily oxidized to the aldehyde. This chemical change interferes with fluorometric measurement.

V. EMISSION SPECTRUM

The emission, or fluorescence, spectrum of a compound results from the reemission of radiation absorbed by that molecule. The quantum efficiency and the shape of the emission spectrum are independent of the wavelength of the exciting radiation. If the exciting radiation is at a wavelength that differs from the wavelength of the absorption peak, less radiant energy will be absorbed and hence less will be emitted. The emission spectrum of the aluminum-acid Alizarin Garnet R complex indicates a fluorescence peak at 580 nm (curve C, Fig. 3).

Each absorption band to the first electronic state will have a corresponding emission, or fluorescence, band. These two bands, or spectra, will be approximately mirror images of each other. In fact this mirror-image principle is useful in distinguishing whether an absorption band is another vibrational band in the first excited state or a higher electronic level. Fluorescence peaks other than the mirror image of the absorption spectrum indicate scatter or the presence of impurities. Rayleigh and Tyndall scatter can be observed in the emission spectrum at the same wavelength as the excitation wavelength and also at twice this value (second-order grating effect). In very dilute solutions one may also observe Raman scatter. The wider the fluorescence band, the more complex and less symmetrical the compound.

Figure 4 shows the absorption and emission spectra of anthracene and quinine. Four major absorption peaks are observed in the anthracene spectrum; all correspond to transitions from S_0 to S_1^* but denote transitions to different vibrational levels. Four major emission peaks, each a mirror image of the peaks in the absorption spectrum, are likewise observed. For quinine two excitation peaks are observed, one at 250 nm corresponding to an $S_0 \longrightarrow S_2^*$

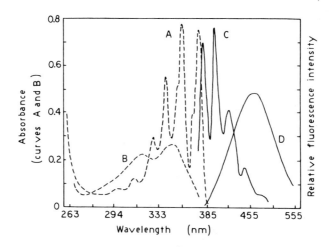

FIG. 4. Absorption and fluorescence spectra of anthracene (in ethanol) and quinine (in 0.1 N sulfuric acid): curve A, anthracene absorption; curve B, quinine absorption; curve C, anthracene fluorescence; curve D, quinine fluorescence.

transition and a second at 350 nm corresponding to an $S_0 \longrightarrow S_1^*$ transition. Only one emission peak, corresponding to the $S_1 \longrightarrow S_0$ transition, is observed.

The fact that some compounds possess several excitation and/or emission peaks is of analytical usefulness. If two compounds have overlapping excitation bands, as in the case of anthracene and quinine, both could be excited together and then differentiated by their emission spectra. Quinine could be measured at a λ_{em} of 450 nm, whereas anthracene could be monitored at a λ_{em} of 400 nm. Similarly, if two compounds emit radiation at the same wavelength, they can still be measured together in the same solution if they have different, nonoverlapping, excitation peaks. This, in fact, is one of the major advantages that fluorescence spectroscopy has over absorption spectroscopy.

Any portion of the spectrum where absorption occurs can produce fluorescence since emission almost always takes place from the lowest vibrational level of the first excited singlet state in solution regardless of the vibrational level or the state to which the molecule is originally excited. The fluorescence peak will be at the same wavelength regardless of the excitation wavelength; however, the intensity of the fluorescence will vary with the relative strength of the absorption (or the sum total of all the absorptions).

A physical constant that is characteristic of luminescent molecules is the difference between the wavelengths of the excitation and emission maxima. This constant is called the Stokes shift and indicates the energy dissipated during the lifetime of the excited state before return to the ground state:

$$\text{Stokes shift} = 10^7 \left(\frac{1}{\lambda_{ex}} - \frac{1}{\lambda_{em}} \right), \tag{4}$$

where λ_{ex} and λ_{em} are the corrected maximum wavelengths for excitation and emission, and are expressed in nanometers. The Stokes shift is of interest to analytical chemists since the emission wavelength can be greatly shifted by varying the form of the molecule being excited. The fluorescence-maximum shift of 5-hydroxyindole from 330 nm at pH 7 to 550 nm in strong acid occurs with no change in the excitation peak (295 nm) and is due to excited-state protonation.

The Stokes shifts for various molecular species of 3-hydroxypyridine are listed in Table 2. The Stokes shifts for the cations undergoing excited-state ionizations are much higher than those of the neutral ionic species, indicating that processes other than light absorption and emission are involved, and that energy is dissipated in bringing about ionization in the excited state.

TABLE 2

Stokes Shifts for 3-Hydroxypyridine [a]

Species	pH optimum	Stokes shift (cm^{-1})
Normal cation	8 N HCl	5750
Excited-state cation	3 N HCl	9830
Dipolar ion	pH 6	6240

[a]Reprinted from Ref. [15], by courtesy of publisher.

VI. FLUORESCENCE QUANTUM EFFICIENCY

Every molecule possesses a characteristic property that is described by a number called the quantum yield, or quantum efficiency, Φ. This is the ratio of the total energy emitted per quantum of energy absorbed:

$$\Phi = \frac{\text{number of quanta emitted}}{\text{number of quanta absorbed}} = \text{quantum yield} \tag{5}$$

The higher the value of Φ, the greater the fluorescence of a compound. A non-fluorescent molecule is one whose quantum efficiency is zero or so close to zero that the fluorescence is not measurable. All energy absorbed by such a molecule is rapidly lost by collisional deactivation. Some typical quantum yields of various substances are shown in Table 3.

The value of Φ can be determined by measuring the fluorescence of a dilute solution (F_1) of a substance, such as quinine sulfate, whose quantum efficiency, Φ_1, is known. The fluorescence determined is then measured, and the quantum efficiency is calculated as follows (use $\Phi_1 = 0.55$ for quinine sulfate):

TABLE 3

Quantum Yields of Fluorescence, Φ_F

Compound	Solvent	Φ_F	Ref.
9-Aminoacridine	Water	0.98	16
	Ethanol	0.99	16
2-Methoxy-6-chloro-9-N-glycylacridine	Water	0.74	16
Acriflavin	Water	0.54	16
Anthracene	Benzene	0.29	16
	Hexane	0.33	17
	Ethanol	0.29	17
	Methyl methacrylate plastic	0.24	18
9,10-Dichloroanthracene	Hexane	0.54	17
	Ethanol	0.58	17
9,10-Diphenylanthracene	Methyl methacrylate plastic	0.83	18
Chlorophyll a	Benzene, ether, dioxane	0.32	16
	Acetone, cyclohexanol	0.30	16
	Ethanol, methanol	0.23	16
	Benzene	0.18	16
Chlorophyll b	Benzene	0.11	16
	Ether	0.12	16
	Acetone	0.09	16
	Methanol	0.10	16
Eosin	0.1 N NaOH	0.19	16
Fluorene	Ethanol	0.53	16
	Hexane	0.54	16
Fluorescein	0.1 N NaOH	0.92	16
Indole	Water, pH 7	0.65	16
	Water	0.45	16

Compound	Solvent	Φ_F	Ref.
Naphthalene	Alcohol	0.12	16
	Hexane	0.10	16
1-Aminonaphthalene-3,6,8-sulfonate	Water	0.15	16
1-Dimethylaminonaphthalene-4-sulfonate	Water	0.48	16
1-Dimethylaminonaphthalene-7-sulfonate	Water	0.75	16
2-Naphthol	0.05 M Borate, pH 10	0.21	18
Phenanthrene	Alcohol	0.10	16
Phenol	Water	0.22	16
Pheophytin a	Benzene	0.18	16
Proflavine	0.05 M Acetate, pH 4	0.34*	18
Pyrene	Methyl methacrylate plastic	0.61	18
Pyridoxal	0.05 M Phosphate, pH 7	0.048	19
Pyridoxamine	0.05 M Phosphate, pH 7	0.11	19
Quinine Sulfate	1 N H_2SO_4	0.55*	18
Rhodamine B	Ethanol	0.97	16
	Ethanol	0.69	20
Riboflavin	Water, pH 7	0.26	16
Skatole	Water	0.42	16
Sodium salicylate	Water	0.28	16
Sodium sulfanilate	Water	0.07	16
Sodium toluene-4-sulfonate	Water	0.05	16
Uranyl acetate	Water	0.04	16

Note: Most solutions are at 10^{-3} M and temperatures of 21 to 25°C. Those marked with an asterisk are extrapolated to infinite dilution. All values are referred to either quinine sulfate (0.51 at 10^{-3} M or 0.55 at infinite dilution) or anthracene in ethanol (Φ_F = 0.28).

$$\Phi_{unk} = \Phi_{std} \frac{F_{unk}}{F_{std}} \cdot \frac{q_{std}}{q_{unk}} \cdot \frac{A_{std}}{A_{unk}} , \qquad (6)$$

where F is the relative fluorescence determined by integrating the area beneath the corrected fluorescence spectrum, q is the relative photon output of the source at the excitation wavelength (taken directly from the curve), and A is the absorbance.

The quantum yield of many compounds has been found to be dependent on the wavelength used for excitation. Chen [21] showed that the emission spectrum of quinine sulfate (used as a standard in ϕ calculations) shifts from 458 to 466 nm when it is excited at 390 nm. Chen showed that this shift was not the result of excited-state ionization, however. It was suggested that quinine fluorescence arises simultaneously from two singlet states, and hence the observed quantum yield of emission varies with λ_{ex}. The variation of the quantum yield of quinine with λ_{ex} is shown in Table 4.

Chen [21] also found that the quantum efficiency is temperature dependent, as might be expected. Figure 5 shows the temperature dependence of the quantum yields of vitamin B_6 compounds: pyridoxal, pyridoxamine, and pyridoxamine phosphate.

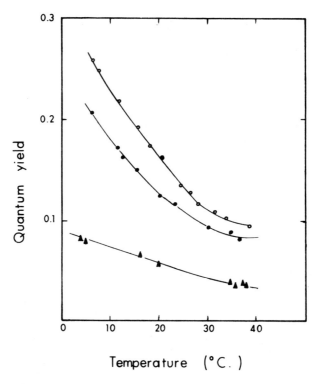

FIG. 5.   Temperature dependence of the quantum yields of vitamin B_6 compounds: ▲, pyridoxal; ●, pyridoxamine; O, pyridoxamine phosphate.

VII. LIFETIME OF THE EXCITED STATE

The fluorescence lifetime of most organic molecules is in the nanosecond region. The fluorescence lifetime τ refers to the mean lifetime of the excited state; the probability of finding a given molecule that has been excited still in the excited state after time t is $e^{-t/\tau}$. The general equation relating the fluorescence intensity I and the lifetime τ is

$$I = I_0 e^{-t/\tau} , \tag{7}$$

where I is the fluorescence intensity at time t, I_0 is the maximum fluorescence intensity during excitation, t is the time after removing source of excitation, and τ is the average lifetime of the excited state.

The average lifetimes of the excited states of some typical compounds are listed in Table 5.

Precise measurement of the observed lifetime is important since it can be used to calculate the natural lifetime τ_0 or the absolute quantum efficiency Φ_0, if one or the other is known:

$$\tau = \Phi_0 \tau_0. \tag{8}$$

The availability of nanosecond-flash fluorometers has provided apparatus for direct measurements of τ that are more precise than any made previously.

TABLE 4

Change of Φ of Quinine with Excitation Wavelength [a]

λ_{ex} (nm)	Relative Φ
250	1.02
313	1.00
345	0.98
348	0.99
366	1.09
380	1.20
390	1.23

[a] From Chen [21]. Measurements in 0.1 N H_2SO_4 at $25^{\circ}C$. Quantum yield at 313 nm taken as 1.00.

Some workers, such as Ware and Baldwin [22], have reported τ values (Table 5) determined with their own laboratory-built equipment. Others, such as Chen, Vurek, and Alexander [23], have used a TRW nanosecond-flash fluorometer to determine the decay times of a large number of substances of biological importance (Table 5).

TABLE 5

Fluorescence-Decay Times τ

Compound	Solvent	τ (nsec)	Ref.
Acridine orange	Water	2.0	23
9-Aminoacridine	100% EtOH	15.2	22
Acridone	Water	15.2	23
Albumin:	100% EtOH	12.5	22
Bovine serum	Water	4.6	23
Egg	Water	4.5	23
Human serum	Water	4.5	23
Anthracene	Benzene	4.26	22
	Hexane	5.75	24
9,10-Dichloroanthracene	Benzene	9.98	22
	Hexane	7.4	24
9,10-Diphenylanthracene	Benzene	7.37	22
Anthranilic acid	Water	8.4	23
3-Hydroxyanthranilic acid	Water	10.9	23
Apomyoglobin:			
Horse heart	Water	3.7	23
Sperm whale	Water	3.0	23
α-Chymotrypsin	Water	3.4	23
Chymotrypsinogen	Water	2.9	23
Coproporphyrin	0.01 N HCl	7.1	23
Eosin	0.1 N NaOH	1.7	23
Flavine-adenine dinucleotide	Water	4.9	23
Fluorescein	0.1 N NaOH	4.6	22
Flavine-adenine mononucleotide	Water	5.6	23
γ-Globulin, human	Water	3.2	23
Glutamic dehydrogenase	Water	4.1	23

Compound	Solvent	τ (nsec)	Ref.
Indole	Water	2.7	23
Indoleacetic acid	Water	2.6	23
Lysozyme, chicken	Water	2.0	23
1-Aminonaphthalene-4-sulfonic acid	Water	11.5	23
1-Aminonaphthalene-5-sulfonic acid	Water	5.5	23
NADH	0.1 N $NaHCO_3$	4.5	23
NADPH	0.1 N $NaHCO_3$	4.3	23
Nicotinamide mono-nucleotide, reduced	0.1 N $NaHCO_3$	3.8	23
Proflavin	Water	4.5	23
Protoporphyrin I	0.01 N HCl	7.2	23
Pyridoxal	Water	4.2	23
Pyridoxamine-5-phosphate	Water	4.3	23
Quinine	0.1 N H_2SO_4	19.0	23
	1 N H_2SO_4	15.2	22
Quinacrine	Water	4.0	23
6-Methoxyquinoline	0.1 N H_2SO_4	22.8	23
Resorcinol	Water	1.7	23
Rhodamine 3GO	Water	3.9	23
Rhodamine 6GO	Water	5.8	23
Riboflavin	Water	4.2	23
Salicylic acid	Water	3.9	23
Serotonin	Water	2.7	23
Skatole	Water	6.4	23
Sulfanilic acid	Water	2.5	23
Trypsin	Water	2.0	23
Tryptophan	Water	2.6	23
Tyrosine	Water	2.6	23

Note: Solutions reported in Ref. [23] contained 0.01 M Tris-HCl, pH 7.0, and were generally 10^{-5} M, except for proteins, which were 1 mg/ml.

VIII. RELATION BETWEEN FLUORESCENCE INTENSITY AND CONCENTRATION

The basic equation defining the relationship of fluorescence to concentration is

$$F = \Phi I_0 (1 - e^{-\epsilon bc}), \tag{9}$$

where Φ is the quantum efficiency, I_0 is the incident radiant power, ϵ is the molar absorptivity, b is the path length of the cell, and c is the molar concentration.

The basic fluorescence intensity-concentration equation indicates that there are three major factors other than concentration that affect the fluorescence intensity:

1. The quantum efficiency Φ. The greater the value of Φ, the greater will be the fluorescence, as already discussed.

2. The intensity of incident radiation, I_0. Theoretically, the more intense source will yield the greater fluorescence. In actual practice a very intense source can cause photodecomposition of the sample. Hence one compromises on a source of moderate intensity (i. e. , a mercury or xenon lamp is used).

3. The molar absorptivity of the compound ϵ. In order to emit radiation a molecule must first absorb radiation. Hence, the higher the molar absorptivity, the better will be the fluorescence intensity of the compound. It is for this reason that saturated nonaromatic compounds are nonfluorescent.

For very dilute solutions the equation reduces to one comparable to Beer's law in spectrophotometry,

$$F = K \, \Phi I_0 \epsilon \, bc. \tag{10}$$

Thus a plot of fluorescence versus concentration should be linear at low concentrations and reach a maximum at higher concentrations (Fig. 6).   At high

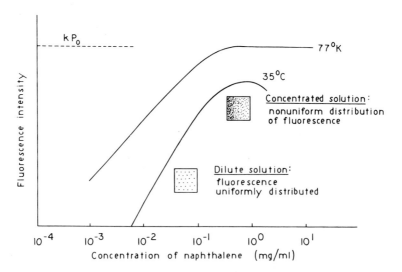

FIG. 6.   Dependence of fluorescence on the concentration of fluorophor and temperature.

concentrations quenching becomes so great that the fluorescence intensity decreases (inner-cell effect). The linearity of fluorescence as a function of concentration holds over a very wide range of concentration (Fig. 6). Measurements down to $10^{-5} \mu g/ml$ are feasible, and linearity extends up to 100 $\mu g/ml$ or higher.

Generally a linear response will be obtained until the concentration of the fluorescent species is large enough to absorb significant amounts of exciting light. For a linear response to be obtained the solution must absorb less than 5% of the exciting radiation. At higher concentrations light scattering as well as the inner-cell effect become important. In the concentration regions where fluorescence is proportional to concentration, fluorescence is measured in the absence of significant radiation. In this region the energy available for excitation is uniformly distributed through the solution.

IX. INTRODUCTION TO EXPERIMENTATION

The fundamental principles of fluorescence measurement are illustrated by Fig. 7, a simplified schematic representation of a filter fluorometer. The desired narrow band of wavelengths of exciting radiation is selected by a filter (called the primary filter) placed between the radiation source and the sample. The wavelength of fluorescence radiant energy to be measured is selected by a second optical filter (called the secondary filter) placed between the sample and a photodetector located at a $90°$ angle from the incident optical path. The output of the photodetector, a current that is proportional to the intensity of the fluorescent energy, is amplified to give a reading on a meter or a recorder. In a spectrofluorometer the filters are replaced by prism or grating monochromators, and an x-y recorder is used to display the excitation and emission spectra.

The geometry and nature of the fluorometric measurement accounts for its excellent sensitivity, which surpasses absorption methods by three to four

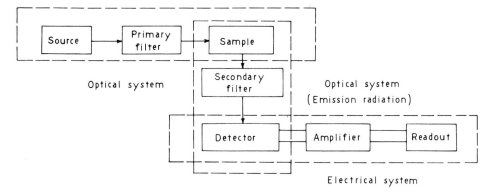

FIG. 7. Schematic diagram of the optical components of a typical filter fluorometer.

orders of magnitude. The fluorometer is capable of measuring low concentrations of substances with good reliability, whereas the spectrophotometer loses accuracy as the signal from the sample approaches that of the reference.

Because of the 180° geometry of the spectrophotometer, the detector constantly views the source, which is very intense. This results in a large electronic signal. In fluorometry it is not necessary to measure small differences between the sample and blank; rather, the detector senses only the sample fluorescence since it is placed at a right angle. Hence the background is zero, and the sample reading can be adjusted to 100% by controlling the intensity of the source by varying the monochromator slits or by controlling the detector amplification. Thus the accuracy is almost completely independent of concentration.

X. USE OF LUMINESCENCE

Fluorescence, phosphorescence, and chemiluminescence provide some of the most sensitive and selective methods of analysis for many compounds. Some typical examples of analysis in clinical pathology, inorganic analysis, agricultural chemistry, and public health are listed in Table 6.

TABLE 6

Applications of Luminescence

Field	Type of analysis	Compounds or procedures
Clinical pathology	Electrolytes	Calcium, magnesium, inorganic sulfate, inorganic phosphate
	Steroids	Corticosteroids, estrogens, progesterone, androgens, testosterone, bile acids
	Lipids	Lipoproteins, phospholipids, cholesterol, triglycerides
	Proteins	Serum albumin, protein electrophoresis
	Amino acids and metabolites	Tryptophan, serotonin, phenylalanine, tyrosine, catecholamines, 3-O-methylcatecholamines, homovanillic acid, DOPA, tyramine and 3-methoxy-tyramine, histidine and histamine, creatine, kynurenic acid, xanthurenic acid

Field	Type of analysis	Compounds or procedures
	Immunology	Fluorescent antibodies, fluorescent antigens, blood typing
	Enzymes	Dehydrogenases, transaminases, phosphatases, lipases, creatine kinase, LDH-isoenzymes, peroxidases
	Drugs	Barbiturates, salicylates, quinidine, LSD, tetracyclines
	Metabolites	Blood glucose, porphyrins, carboxylic acids and ketones
	Other	Blood urea nitrogen, ammonia, hippuric acid, hematin iron
Inorganic analysis	Anions	Cyanide, fluoride, sulfate, silicate, iodide
	Cations	Aluminum, arsenic, beryllium, boron, cadmium, cerium, calcium, gallium, iron, lithium, magnesium, rare earths, selenium, silicon, tin, tungsten, uranium, zinc, zirconium
Agricultural chemistry	Inorganic	As noted above, especially selenium, magnesium, boron, fluorides, aluminum, tin
	Tracing techniques	Insecticide and pesticide spray-coverage studies; residue evaluations
	Natural products	Gibberellic acid, chlorophylls, pigments
	Vitamins	A, B_1, B_2, B_6, C, D, and E
	Proteins	Protein in milk
Public health	Pollution control	Insecticide-aerial-drift studies, water-pollution studies, spent sulfite liquor
	Bacteriology	Identification and counting of bacteria
	Metal poisioning	Beryllium, boron, lead, uranium, cadmium
	Immunology	Fluorescent-antibody control
	Screening programs	Phenylketonuria, histidemia

XI. PRACTICAL CONSIDERATIONS

A. Advantages of Fluorescence

Molecular emission (fluorescence and phosphorescence) is a particularly important analytical technique because of its extreme sensitivity and good specificity. Fluorometric methods can detect concentrations of substances as low as one part in ten billion, a sensitivity 1000 times greater than that of most spectrophotometric methods. The main reason for this increased sensitivity is that in fluorescence the emitted radiation is measured <u>directly</u> and can be increased or decreased by altering the intensity of the exciting radiant energy. An increase in signal over a zero background signal is measured in fluorometric methods. In spectrophotometric methods the analogous quantity, absorbed radiation, is measured indirectly as the difference between the incident and the transmitted beams. This small decrease in the intensity of a very large signal is measured in spectrophotometry with a correspondingly large loss in sensitivity.

The specificity of fluorescence is the result of two main factors: (a) there are fewer fluorescent compounds than absorbing ones because all fluorescent compounds must necessarily absorb radiation, but not all compounds that absorb radiation emit; (b) two wavelengths are used in fluorometry, but only one in spectrophotometry. Two compounds that absorb radiation at the same wavelength will probably not emit at the same wavelength. The difference between the excitation and emission peaks ranges from 10 to 280 nm (emission at the same wavelength as excitation is scatter, as we have seen).

Materials that possess native fluorescence, those that can be converted to fluorescent compounds (fluorophors), and those that extinguish the fluorescence of other compounds can all be determined quantitatively by fluorometry.

B. Limitations of Fluorescence

The principal disadvantage of fluorescence as an analytical tool is its serious dependence on the environment (temperature, pH, ionic strength, etc.).

1. Photochemical Decomposition

The ultraviolet bright light used for excitation may cause photochemical changes in, or destruction of, the fluorescent compound, giving a gradual decrease in the intensity reading (Table 7). In a practical sense one can take three measures to avoid photochemical decomposition: (a) always use the longest wavelength radiation for excitation; (b) measure the fluorescence of the sample immediately after excitation--**not** allow the exciting radiation to strike the sample for long periods; and (c) protect photochemically unstable standard solutions, such as quinine sulfate, from sunlight and ultraviolet laboratory lights by storing in a black bottle.

TABLE 7

Effect of Ultraviolet Irradiation on a 0.01-μg/ml Quinine Sulfate Solution

Irradiation time (min)	Fluorescence (% of initial)
0	100
2	99.4
21	92.9
38	88.1
86	74.3
97	69.2
100	68.0

2. Viscosity

The fluorescence of a compound is affected by the viscosity of the medium, increasing with increasing viscosity. Energy transfer is reduced by a reduction in the number of molecular collisions.

Thus the fluorescence of most compounds can be increased by using a more viscous solvent, such as glycerol or gelatin.

3. Quenching

Quenching, the reduction of fluorescence by a competing deactivating process resulting from the specific interaction between a fluorophor and another substance present in the system, is also frequently a problem.

The general mechanism for the quenching process can be denoted as follows:

$$M + h\nu \longrightarrow M^* \quad \text{(light absorption)}, \tag{11}$$
$$M^* \longrightarrow M + h\nu \quad \text{(fluorescence emission)}, \tag{12}$$
$$M^* + Q \longrightarrow Q^* + M \quad \text{(quenching)}, \tag{13}$$
$$Q^* \longrightarrow Q + \text{energy}. \tag{14}$$

Four common types of quenching are observed in luminescence processes: temperature, oxygen, concentration and impurity quenching. One of the most notorious quenchers is dissolved oxygen, which causes a reduction in fluorescence intensity and a complete destruction of phosphorescent intensity. Small amounts of iodide and nitrogen oxides are very effective quenchers and interfere.

Small amounts of highly absorbing substances like dichromate interfere by robbing the fluorescent species of the light available for excitation. For this reason most workers prefer not to wash their cuvettes with dichromate cleaning solution.

a. Temperature Quenching. As the temperature is increased, the fluorescence decreases. This is illustrated in Fig. 8, which shows the effect of temperature on the fluorescence of four common substances. The degree of temperature dependence varies from compound to compound. Tryptophan, quinine, and indoleacetic acid are compounds whose fluorescence varies greatly with temperature.

Temperature effects on luminescence are a type of excited-state quenching by encounter. The fluorescence changes are nearly those of molecular activity with temperature, which suggests that increasing temperature increases molecular motion and collisions, and hence robs the molecule of energy.

The change in fluorescence is normally 1 % per 1°C; however, in some compounds, such as tryptophan or Rhodamine B, it can be as high as 5 %.

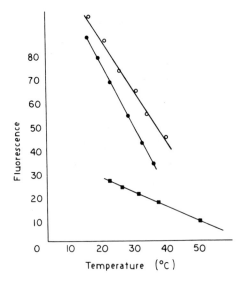

FIG. 8. Variations in the fluorescence intensity of several compounds as a function of temperature. All compounds were dissolved in 0. 1 M phosphate buffer, pH 7. 0, except quinine. ●, Tryptophan or indoleacetic acid; O, indole-acetic acid in buffer saturated with benzene; ■ , quinine in 0. 1 N sulfuric acid.

In a practical sense, temperature control should be exercised for maximum precision and accuracy. For some instruments, such as the Aminco-Bowman spectrophotofluorometer, temperature-control accessories are available. In cases where the sample must be stored under refrigeration, care must be exercised to bring the sample back to room temperature before measurement. This may pose problems of photodecomposition or change in sample concentration due to evaporation of the solvent that should be considered.

Temperature exerts other effects than quenching on fluorescence. The total fluorescence at the normal excitation, or fluorescence maxima, is lowered due to a dissipation of energy by vibrational energy transitions. Also higher temperatures produce more band maxima due to the increased population of higher energy vibertional levels. For best results all standards and samples should be kept at the same temperature during measurements of fluorescence.

b. Oxygen Quenching. Oxygen, present in solutions at a concentration of 10^{-3} M, normally reduces the fluorescence of a typical compound by 20 %. Unsubstituted aromatics are still more severely affected by oxygen. Oxygen must be completely removed for phosphorimetry.

We shall investigate the causes of oxygen interference in Chapter 3. Let us say now that oxygen quenching is a type of excited-state quenching, and it is possible to measure the dissolved-oxygen content by its quenching. The ratio of the observed fluorescence in aerated and non-aerated solutions is given by the ratio L_0/L. A few of these ratios are presented in Table 8. A value of 1 means that the fluorescence in a deaerated solution, L_0, is the same as that in an aerated solution, L, and hence oxygen has no effect on the luminescence.

The analytical sensitivity can be increased by oxygen removal. This can be accomplished by bubbling an inert gas, such as nitrogen, through the

TABLE 8

Ratio L_0/L for Several Organic Compounds

Compound	Ratio L_0/L
Anthracene	1.25
Benzene	2.4
Carbazole	1.9
Naphthalene	6.5
Pyrene	1.0
Toluene	3.0

solution for 5 to 10 min or, better, by a freeze-thaw cycle. It is important that all solvents for use in phosphorimetry be purchased degassed and stored under nitrogen.

c. Concentration Quenching. Absorption causes many problems during a fluorometric assay, just as fluorescence causes a problem when the absorbance of a solution is measured.

In order for fluorescence to be observed absorption must occur. As we have already seen, the fluorescence intensity is proportional to the molar absorptivity: the more highly absorbing the substance, the greater its fluorescence. But when the absorption is too large, no light can pass through to cause excitation. Thus, at low concentrations, when the absorbance is less than about 0.05, there is a linear relationship between fluorescence and concentration (Fig. 9). At intermediate concentrations the light is not evenly distributed along the path of light. The portion of the solution nearest the light source absorbs so much radiation that less and less is available for the rest of the solution. As a result, considerable excitation occurs at the front of the solution (Fig. 10), but less and less occurs throughout the rest of the cell. This type of concentration quenching causes a fluorescence loss that is called the inner-cell effect.

When fluorescence is measured at the surface (Fig. 10), the fluorescence increases linearly and then levels off as predicted by Eq. (9). This is explained by the fact that even in concentrated solutions only the surface is observed, and hence quenching is not important. However, in solution, quenching does become important; hence the fluorescence decreases at high concentrations.

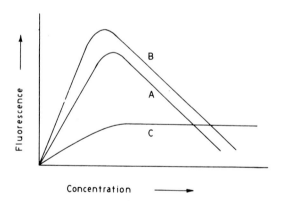

FIG. 9. Relationship between fluorescence and concentration.

A: End-on detector

B: Right angle detector

C: Surface detector

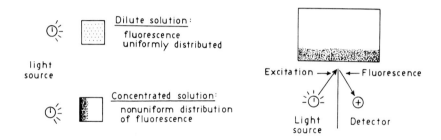

FIG. 10.   Effect of concentration on fluorescence.

Hence an unknown sample should always be tested for concentration quenching.   Consider curve A in Fig. 9.   There are two possible concentrations for each relative fluorescence value, one on each side of the maximum.   To know which is correct, the analyst must dilute his sample and read again.   If the fluorescence increases, then the previous reading was on the negative slope, and an incorrect value was obtained.   In this case the solution should be diluted and read again.   If the fluorescence decreases, then a correct value is obtained.   All fluorescence methods should incorporate a linear curve with checks for concentration quenching.

Concentration quenching can be reduced instrumentally by using wide slits so that little sample remains between the source, the fluorescing region of the sample and the detector.   A special cell with thick walls is also helpful.

Concentration quenching can also take the form of energy transfer between molecules of the sample.   This probability increases with the sixth power of the concentration and becomes important at concentrations above 10^{-3} M.

An important form of concentration quenching involves dimer or polymer formation, called excimer quenching.   The excimer has a different electron orientation and a longer emission wavelength than the monomer.   Hence if an instrument monitors the fluorescence at one wavelength, the emission at the longer wavelength will go undetected.   So the observed fluorescence will decrease with increasing concentration.

d. Impurity Quenching.   Many researchers have become completely disenchanted with fluorescence as an analytical tool because of the phenomenon of impurity quenching.   Fluorometry is generally considered a specific and sensitive tool, not subject to chemical interference, since most measurements are made in dilute solutions that contain only trace amounts of impurities.   When these impurities are present at moderate concentrations, interferences result.   This interference can be in the form of the inner-cell effect, collisional quenching, energy transfer, charge transfer, or the heavy-atom effect.   To get around this problem the analyst could try to reduce the interference by dilution or by

separation techniques. If an interference is known to be present, it should be
added to the standards used in preparing the calibration curve.

REFERENCES

1. G. G. Guilbault, ed. , Fluorescence. Theory, Instrumentation, and Practice,
    Dekker, New York, 1967.

2. D. D. Hercules, ed. , Fluorescence and Phosphorescence Analysis,
    Interscience, New York, 1966.

3. R. A. Passwater, Guide to Fluorescence Literature, Plenum, New York,
    1967.

4. R. E. Phillips and F. R. Elevitch, Fluorometric Techniques in Clinical
    Pathology in Progress in Clinical Pathology, Grune and Stratton, New York,
    1966, Chapter 4.

5. S. Udenfriend, Fluorescent Assay in Biology and Medicine, Vol. I, Academic
    Press, New York, 1966.

6. S. Udenfriend, Fluorescent Assay in Biology and Medicine, Vol. II,
    Academic Press, New York, 1970.

7. C. E. White and R. Argauer, Fluorescence Analysis, a Practical Approach,
    Dekker, New York, 1970.

8. M. A. Konstantinova-Shlesinger, Fluorometric Analysis, N. Kamer,
    transl. , Davey, New York, 1965.

9. P. Pringsheim, Fluorescence and Phosphorescence, Interscience, New
    York, 1949.

10. A. Weissler and C. E. White, "Fluorescence Analysis," in Handbook of
    Analytical Chemistry, L. Meites, ed. , McGraw-Hill, New York, 1963,
    Chapter 6.

11. A. Weissler and C. E. White, "Fluorometric Analysis," in Standard
    Methods of Clinical Analysis, Vol. 3A, F. W. Welcher, ed. , D. Van
    Nostrand, Princeton, N. J. , 1966, Chapter 5.

12. C. E. White, "Fluorometric Analysis, Fundamental Reviews,"Anal. Chem.,
    21, 104 (1949); 22, 69 (1950); 24, 85 (1952); 26, 129 (1954); 28, 621 (1956);
    30, 729 (1958); 32, 47R (1960); 34, 82R (1962); 36, 116R (1964); 38, 115R
    (1966); 40, 114R (1968); 42, 57R (1970).

13. Fluorescence News Monthly, American Instrument Co. , Silver Spring, Md.

14. Traces, Monthly, G. K. Turner Associates, Palo Alto, Calif.

15. J. W. Bridges, D. S. Davies, and R. T. Williams, Biochem. J. , 98, 451
    (1966).

16. G. Weber and F. W. Teale, Trans. Faraday Soc. , 53, 646 (1957).

17. E. J. Bowen and D. Seaman, in Luminescence in Organic and Inorganic
    Materials, H. Kallmann and G. Spruch, eds. , Wiley, New York, 1962,
    p. 153.

18. W. H. Melhuish, J. Opt. Soc. Amer. , 54, 183 (1964).

19. R. F. Chen, Science, 150, 1593 (1965).

20. C. A. Parker and W. T. Reese, Analyst, 85, 587 (1960).

21. R. F. Chen, Anal. Biochem. , 19, 374 (1967).

22. W. R. Ware and B. A. Baldwin, J. Chem. Phys. , 40, 1703 (1964).

23. R. F. Chen, G. G. Vurek, and N. Alexander, Science, 156, 949 (1967).

24. W. R. Ware and B. A. Baldwin, J. Chem. Phys. , 43, 1194 (1965).

Chapter 2

INSTRUMENTATION

I. COMPONENT PARTS

The basic components of any instrument designed to measure luminescence are shown in Fig. 11. They are the light source, wavelength selectors, sample compartment, and detector system. The instrumentation is the same as that used in spectrophotometry, with two exceptions: (a) the detector is rotated 90° to the incident-light path and (b) a second wavelength selector is placed in front of the detector. Thus any good spectrophotometer can be adopted to fluorescence work at a small additional cost.

In fluorescence the detector is placed at 90° to the path of incident light so that little of the incident radiation will strike the detector. Since only light emitted from the sample reaches the detector, this device will register zero signal when no luminescence occurs. Thus an increase in signal indicates an emission from the sample. This is the major reason for the greater sensitivity of luminescence over spectrophotometric methods.

The second wavelength selector is placed before the detector to remove all radiation except that emitted by the sample. This provides another degree of specificity to the analysis.

A discussion of each of the components used in fluorescence instrumentation

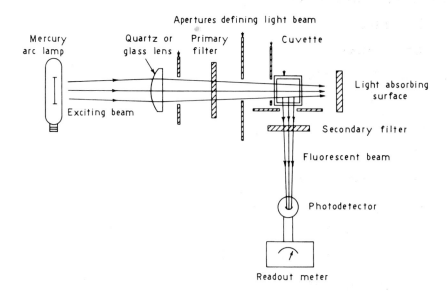

FIG. 11. Schematic diagram of the optical components of a typical fluorometer.

will help the reader to be aware of how these parameters can be optimized for maximum performance.

There are two principal types of instrument: filter fluorometers and spectrofluorometers. The filter instrument uses filters to select the desired excitation and emission wavelengths by absorbing or reflecting unwanted radiation. We shall discuss the various types of instruments available commercially in each of these classes later in this chapter.

II. LIGHT SOURCES

Several types of light source are available for spectroscopic studies:
tungsten incandescent, mercury, xenon, hydrogen, and deuterium lamps. The
usefulness of a lamp can be evaluated by plotting its relative intensity versus
wavelength. Since the total fluorescence observed is proportional to the intensity
of the source of excitation, it is necessary to have a significant amount of energy
available in the absorption region of the sample to be determined.

<u>A. Incandescent Lamps</u>

Incandescent lamps act as blackbodies. As the temperature and current are
increased, the total radiation of the lamp is increased, and the wavelength of
maximum emission is shifted to the ultraviolet. The output of a typical incandescent lamp is pictured in Fig. 12. As is evident, very little energy is released in
the ultraviolet region; the lamp yields greatest energy in the red region. The
tungsten lamp has a maximum emission at 580 nm at 3000°K and no output below
350 nm. Since many compounds are excited by ultraviolet-near visible radiation,
the incandescent lamp has found very little usefulness in fluorescence
instrumentation.

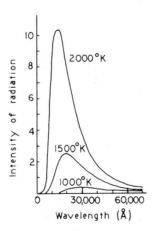

FIG. 12. Emission of radiation from a black body at various temperatures.
The area under the curve between specified wavelengths, divided by 10^4, gives
the energy in calories per second radiated from 1 cm^2 of a blackbody, in the
range of wavelengths.

B. Gas-Discharge Lamps

1. General Remarks

Gas-discharge lamps provide a major source of ultraviolet radiation. On application of a voltage to the lamp, the small number of electrons present in the gas are accelerated by the electric field until their energies become large enough to ionize the remaining gas by collisions with neutral particles. These electrons then act as new current carriers; the ionization rate increases in a cascade-type effect. The electric current keeps the electrons, ions, and neutral atoms in continuous motion, the particles become excited to a higher state, and then radiate photons on returning to the ground state.

Every gas radiates only certain characteristic wavelengths, depending on its nature. At high pressures the radiation becomes dominated not by the resonance lines, but by other spectral lines corresponding to transitions between higher energy levels.

The spectral lines radiated by a gas are not monochromatic except at very low pressures. As the pressure in the lamp is raised, the emission band broadens out.

The most common sources used in luminescence instrumentation are the gas-discharge lamps: mercury and xenon arc. The spectral characteristics of the 150-W xenon-arc lamp and the Pyrex-jacketed 100-W H-4 mercury lamp are shown in Fig. 13.

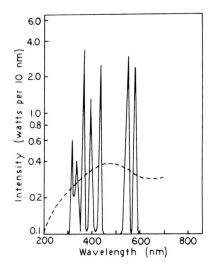

FIG. 13.   Spectral characteristics of the 150-W xenon-arc lamp (broken curve) and of the Pyrex-jacketed 100-W H-4 mercury lamp (solid curve).

2. Mercury-Vapor Lamp

The mercury-vapor lamp is widely used in filter fluorometers because of its intense light emission and stability. Since an elaborate power supply is not needed, the lamp is economical. The most commonly used wavelength in the mercury-vapor lamp is the resonance lines at 365 to 366 nm, although other useful lines are present at 254, 302, 313, 405, 436, 546, 577, and 579 nm. A list of the relative intensities of lines of the mercury-vapor lamp is presented in Table 9.

TABLE 9

Relative Intensity of Spectral Lines in a Mercury-Vapor Lamp

Line (nm)	Relative intensity	Line (nm)	Relative intensity
253.7	1	366.3	1.4×10^{-3}
296.5	6.0×10^{-3}	404.7	8.9×10^{-3}
302.2	1.1×10^{-2}	435.8	1.7×10^{-2}
312.2	7.1×10^{-3}	546.1	1.2×10^{-2}
313.2	1.1×10^{-2}	577.0	1.7×10^{-3}
365.0	8.9×10^{-3}	579.0	1.8×10^{-3}
365.5	2.1×10^{-3}		

A low-pressure mercury-vapor lamp can be modified to provide energy at wavelengths other than the resonance lines. The inner surface of the lamp is painted with a thin layer of crystalline phosphors that absorb the mercury-vapor resonance radiation and generate a broad band at longer wavelengths. The Aminco and Turner filter fluorometers, for example, use a 4-W low-pressure mercury-phosphor lamp with a broad emission band having its peak at 360 to 365 nm. A green lamp with a 525-nm peak, a blue lamp with 405- and 436-nm peaks (Fig. 14), and a blue lamp with 405-, 436-, 546-, and 577-nm peaks (Fig. 15) are available.

3. Xenon Lamp

Most grating instruments, such as the Aminco Bowman SPF, use a high-pressure xenon lamp (commonly 150-W). The xenon lamp has a good continuum (Fig. 16), better in the ultraviolet than that of the tungsten lamp, but it does not have the intensity that the mercury lamp has at its resonance

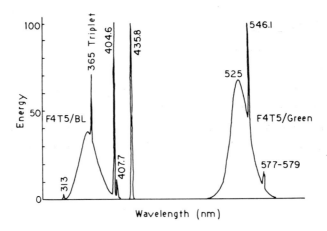

FIG. 14. Energy spectra of fluorescent lamps.

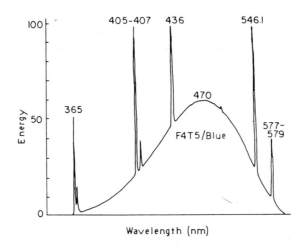

FIG. 15. Energy spectrum of the F4T5 blue-coated mercury lamp.

lines (at those lines the mercury lamp has twice the intensity of the xenon lamp).
Moreover, it is necessary to use an expensive dc converter and stabilizer with
this lamp. A mercury lamp cannot be used with a scanning spectrofluorometer,
however, because the excitation spectrum obtained would simply be that of the
mercury resonance lines superimposed on the sample's excitation spectrum.
A lamp with a smooth continuum is needed in the ultraviolet region also.

The usefulness of the xenon lamp is limited by the regulation of the power
supply, which directly affects the stability and life of the lamp. Any fluctuations
in the power supply will appear in the lamp. "Arc wander," the variation in the
position of the arc passing between the electrodes of the xenon lamp, is
frequently a problem. Better lamp construction and greater stability in the
power supply have helped to minimize arc wander. Arc wander will cause a slight
shift in the excitation spectrum and hence must be eliminated.

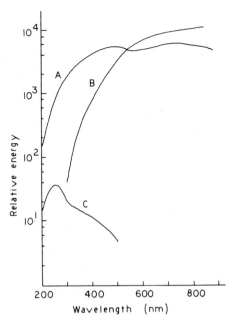

FIG. 16. Relative output of various lamps.
A -Xenon   B -Tungsten   C -Deuterium

The ellipsoidal condensing system accessory to the Aminco Bowman SPF is much more sensitive to arc instability. But the excitation wavelength remains constant due to the entrance -slit arrangement.

4. Other Lamps

Xenon and mercury gases have been combined to yield a high-pressure mercury lamp that can be used with the same power supply as that used for the xenon lamp. The xenon is added to facilitate arcing. The spectral output from a xenon -mercury lamp is shown in Fig. 17.

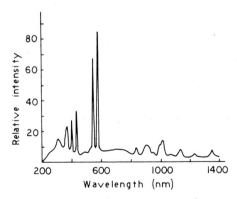

FIG. 17. Typical relative emission spectrum from xenon -mercury lamps. Note that the xenon only contributes an insignificant 5 % to the background continuum. The xenon, however, facilitates lamp starting, adds stability, and prolongs the life of the lamp.

Hydrogen and deuterium gas-discharge lamps provide a continuum in the ultraviolet region. However, the output above 300 nm is relatively low in the hydrogen lamp. The deuterium lamp has about three times the energy of a comparable hydrogen lamp and is useful up to about 500 nm (Fig. 16).

The life expectancy of ionized-gas lamps ranges from 200 to 900 hours, depending on various conditions of wattage and use. Any lamp should be allowed to stabilize for 15 to 30 min before any readings are taken.

5. Hazards of Gas-Vapor Lamps

Gas-ionization lamps are filled with high-pressure gases, 5 atm at room temperature or 20 atm at operating temperatures. Since the lamp can easily explode on receiving a shock, several layers of cloth should be wrapped around it when handled. Furthermore, the gas-vapor lamps produce ultraviolet radiation, which can severely burn the retina of the eye. A pair of ultraviolet filter glasses should be worn when working with these lamps, and one should never look directly at a gas-ionization lamp.

Gas-ionization lamps also convert air into ozone and nitrogen oxides, which are toxic. Ozone is produced by ultraviolet radiation of 180 to 210-nm wavelength; in fact the production of ozone is a good measure of the ultraviolet output of the xenon lamp. Ozone is easily detected by its characteristic odor.

The fluorometer lamp should be vented to a hood or exhaust to remove ozone. If venting is impossible, the ozone should be catalytically converted to molecular oxygen by passing the exhaust through a molecular sieve (10X) or platinum dust.

III. MONOCHROMATORS

The purpose of the monochromator is to isolate a narrow band of electromagnetic radiation from the source. It is a wavelength selector. Two types of monochromator are used in luminescence equipment: filters and gratings. In fact instruments are classified as filter fluorometers (nonscanning) or grating spectrofluorometers (scanning). Prisms are not used in luminescence instruments because they give their greatest dispersion in the ultraviolet, not the visible where most measurements are made (Fig. 18), and to obtain adequate sensitivity a large prism would be needed. This is expensive.

A. Filters

1. General Remarks

Tinted-glass, gelatin, and liquid filters selectively absorb unwanted wavelengths; interference filters depend on constructive interference of light

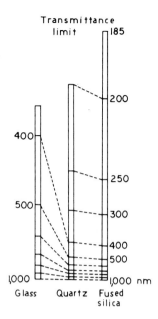

FIG. 18. Approximate range and dispersion of prism materials.

rays for their transmission characteristics. In selecting a filter one
compromises between peak transmittance and bandpass width. The transmittance should be high, the bandpass as narrow as possible.

The tinted-glass filter is a solid sheet of glass that has been colored by a
pigment dissolved or dispersed in the glass. The gelatin (Wratten) filter consists
of a layer of gelatin impregnated with suitable organic dyes and placed between
two glass sheets. Liquid filters are solutions of absorbing substances.

There are three types of filter:

1. Neutral tint. These give a nearly constant transmission over a wide
range. They are designed to decrease the intensity of the fluorescence signal
uniformly and are used with strongly fluorescing compounds.

2. Cutoff filters. This type of filter is used to cut off stray or unwanted
radiation since it produces a sharp cutoff in the spectrum with complete
transmission on one side of the cutoff with little or no transmission on the other.
Some typical cutoff filters are shown in Figs. 19 and 20.

3. Bandpass filters. These are composite filters constructed from
sets of cutoff filters. One part consists of a long-wavelength sharp-cutoff
filter (blue and green series), and the other is a short-wavelength cutoff filter
(red and yellow series). Figure 21 illustrates the use of two such filters in
composing a glass absorption filter with a nominal wavelength of 590 nm. Other
such filters are available for wavelengths of 360 to 740 nm, as illustrated in

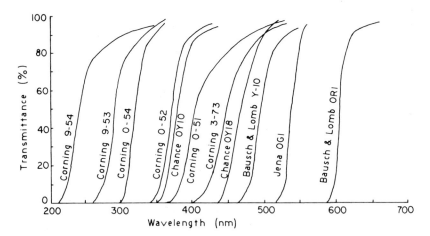

FIG. 19. Transmittance characteristics of secondary filters made by
the Corning Glass Works (U.S.A.), the Chance Pilkington Optical Works
(England), Bausch & Lomb (U.S.A.), and Jena Glaswerke (Germany).

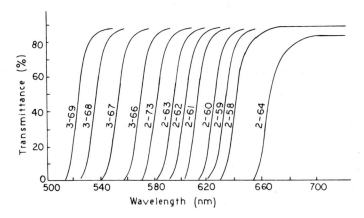

FIG. 20. Sharp-cutoff filters.

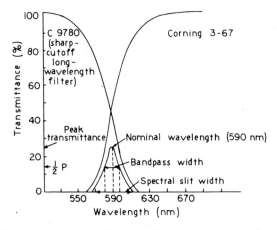

FIG. 21. Spectral transmittance characteristics of a composite glass
absorption filter and its components.

Fig. 22. Normally these filters have relatively wide bandpass widths (width of
the band at one-half the peak transmittante) with peak transmittances of about
25%.

There are several typical filters that are useful in conjunction with the
mercury-vapor lamp (Table 10). Many other filters are available for isolating
almost any spectral region, and the reader is advised to consult Eastman Kodak,
Turner, or Aminco catalogs for the wide selection available.

TABLE 10

Wratten Filters Useful with Mercury-Vapor Lamps

Filter No.	Principal lines transmitted (nm)
18 A	365
22	577 and 579
50	436
74	546

2. Interference Filters

Narrower bandwidths are obtained with interference filters, which are
evaporated coatings of a transparent dielectric spacer of low refractive index
sandwiched between semitransparent silver films (Fig. 23). Magnesium
fluoride (n_D = 1.38) is usually used as the dielectric. Light incident on the face
of the filter (usually at $90°$) is reflected back and forth between the metal films.
Constructive interference between different pairs of superimposed light rays
occurs only when the path difference is exactly one wavelength or some multiple
thereof. These filters have a bandwidth of 10 to 17 nm and peak transmittances
of 40 to 60%.

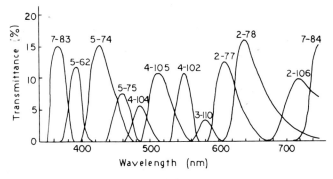

FIG. 22. Transmittance spectra of some glass filters (Corning Glass
Works).

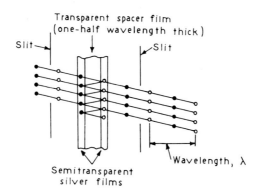

FIG. 23. Path of light rays through an interference filter.

Multilayer interference filters are prepared by successive evaporations of
5 to 25 layers of high- and low-refractive-index dielectric in alternating layers.
Multilayer filters are characterized by a bandpass width of 8 nm and a peak
transmittance of 60 to 95% (see Fig. 24). The transmittance characteristics of
some typical commercial interference filters are shown in Fig. 25.

3. Primary and Secondary Filters

The primary filter is used in the filter fluorometer to select the excitation
wavelength. A Wratten filter is generally not used as a primary filter because
of its instability. Some typical primary filters are listed in Table 11. Secondary
filters are used to prevent unwanted wavelengths from reaching the detector in
filter fluorometers and to remove Rayleigh, Raman, and Tyndall scattering in
spectrofluorometers. The secondary filter must be chosen so as to be
"compatible" with the primary filter. This means that the transmission of the
primary and secondary filters must not overlap significantly. Otherwise,
scattered radiation from even slightly turbid samples will reach the detector and
be measured as apparent fluorescence. Table 12 lists a number of secondary
filters and their compatability with primary filters.

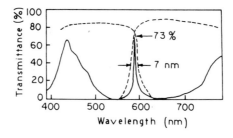

FIG. 24. Wavelength-versus-transmittance curve for a multilayer interference filter peaked at 591. 7 nm. The broken lines are transmittance curves
of glass (blocking) filters to cut out the side bands.

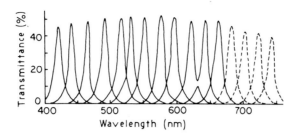

FIG. 25. Transmittance characteristics of interference filters available from the Photovolt Corporation.

The choice of the proper filter is empirical. For practical purposes, for example, excitation at the strongest mercury line at 365 nm will give the greatest sensitivity, if the compound absorbs there, even though this might not be the wavelength of maximum absorption. Hence a filter peaking at 365 nm, rather than one peaking at, say, 390 nm, might be used, even though the λ_{max} were 390 nm.

Another potential source of error is the fluorescence of filters themselves. A fluorescent secondary filter, such as the Corning 22, will cause a positive error in a turbid solution. And very little turbidity can cause quite large errors when improper filters are used.

TABLE 11

Primary (Activation) Filters

Wavelength (nm)	Filter No.	Comments
254	110-810 (7-54)	Passes also the 313-, 365-, and part of the 405-nm lines, both of which are relatively minor in the required 110-851 lamp; quartz cuvettes (110-802) must be used
254	110-810 (7-54) + 110-815	Provides pure 254-nm activation when measurement in the near ultraviolet is required; quartz cuvettes (110-802) and 110-851 lamp required

Table 11 (Continued)

Wavelength (nm)	Filter No.	Comments
325	110-810 (7-54) + Wratten 34A (unmounted) 110-836	Gelatin held in place by the Corning filter; used with the standard (110-850) lamp to provide a narrow band of activating light peaking at 325 nm; with the 110-851 lamp it isolates the 313-nm line. In neither case is the activating light very intense, the former being generally best, but it is suitable for a great deal of work. Pyrex cuvettes are generally satis- factory
365	110-811 (7-60) or 110-834 (7-37)	The general-purpose primary filter supplied with the Turner fluorometers; normally used with the standard (110-850) lamp and Pyrex cuvettes; 110-834 recommended for solid samples or paper-chromatogram door
405	110-812 (405)	Should be used with the dark Corning 7-51 glass away from the light as the Wratten 2C is slightly fluorescent and must be blocked; used with either 110-850 or 110-851 lamp (about two-fold gain in sensitivity with latter) and Pyrex cuvettes
405 + 436	110-813 (47B)	Seldom used alone, but for some appli- cations where either 405 or 436 may be used provides increased sensitivity by providing both; like the 405 filter, it is normally used with 110-850 lamp, but the 110-851 lamp provides increased sensitivity; used with Pyrex cuvettes
436	110-816 (2A) + 110-813 (47B)	The 2A is placed nearest the lamp and eliminates the 405-nm line. Other comments as for the 405
470	110-827 (3) + 110-831 (48)	Used only with the 110-853 blue lamp
546	110-814 (1-60) + 110-822 (58) or 110-832 (546)	The 1-60 is placed nearest the lamp. Other comments as for the 405. The 110-832 is recommended for tracer work with Rhodamine B or Pontacyl Brilliant Pink B or for best results with paper-chromatogram door

TABLE 12

Compatibilities of Filters [a]

Secondary filter	Primary filter									
	7-54 254 nm	7-54 + 815 pure 254 nm	7-54 + 34A 325 nm	7-60 365 nm	7-37 360 nm	405 405 nm	47B + 2A 436 nm	3 + 48 470 nm [b]	3 + 48 + 48 470 nm [b]	1-60 + 58 546 nm
			Sharp-Cutoff Series							
110-816 (2A) 415 nm		X	X[c]	X	X	X				
110-827 (3) 455 nm	X	X	X	X	X	X				
110-828 (4) 465 nm	X	X	X	X	X	X				
110-817 (8) 486 nm	X	X	X	X	X	X				
110-818 (2A-12) 510 nm	X	X	X	X	X	X	X		X[c]	
110-826 (2A-15) 520 nm	X	X	X	X	X	X	X	X[c]	X	
110-829 (16) 535 nm	X	X	X	X	X	X	X	X	X	
110-819 (22)[d] 560 nm	X	X	X	X	X	X	X	X	X	X[c]
110-824 (23A) 570 nm	X	X	X	X	X	X	X	X	X	X

Primary filter

Secondary filter	7-54 254 nm	7-54 + 815 pure 254 nm	7-54 + 34A 325 nm	7-60 365 nm	7-37 360 nm	405 405 nm	47B + 2A 436 nm	3 + 48 470 nm[b]	3 + 48 + 48 470 nm[b]	1-60 + 58 546 nm
110-820 (25) 595 nm	X	X	X	X	X	X	X	X	X	X
Sharp-Cutoff Series										
Narrow-Bandpass Series										
110-810 (7-54) + Wratten 34A (110-836) 325 nm		X								
110-817 (7-60) 360 nm		X								
110-812 405 nm		X	X[c]	X[c]	X					
110-813 (47B) 430 nm		X	X[c]							
110-813 (47B) + 110-815 435 nm	X[c]	X	X[c]	X	X	X[c]				
110-816 (2A) + 110-831 (48) 360 nm	X[c]	X	X[c]	X	X	X[c]				
110-821 (75) 485 nm	X	X	X	X	X	X				
110-825 (65A) 495 nm	X	X	X	X	X	X				

Primary filter

Narrow-Bandpass Series--Continued

Secondary filter	7-54 254 nm	7-54 + 815 pure 254 nm	7-54 + 34A 325 nm	7-60 365 nm	7-37 360 nm	405 405 nm	47B + 2A 436 nm	3 + 48 470 nm[b]	3 + 48 + 48 470 nm[b]	1-60 + 58 546 nm
110-825 (65A) + 110-817 (8) 510 nm	X	X	X	X	X	X	X[c]			
110-822 (58) 525 nm	X	X	X	X	X	X	X			
110-822 (58) + 110-818 (2A-12) 535 nm	X	X	X	X	X	X	X		X[c]	
110-822 (58) + 110-826 (2A-15) 540 nm	X	X	X	X	X	X	X	X	X	
110-822 (58) + 110-814 (1-60) 545 nm	X	X	X	X	X	X	X	X	X	
110-833 590 nm	X	X	X	X	X	X	X	X	X	X

[a] X indicates compatibility.

[b] See Table 11.

[c] Use with care, as slight overlap exists.

B. Gratings

1. General Remarks

A grating consists of a large number of parallel lines or grooves ruled at extremely close intervals (e. g. , 30,000 lines per inch) on a highly polished surface, such as aluminum. A master grating is used as a mold in the production of replica gratings. A film of parting compound is applied to the master, the film is aluminized, the crevices are filled with epoxy resin, and an optical flat is bonded by the epoxy to the aluminum replica of the master-grating pattern. When the epoxy hardens, the replica grating, completely anodized, is separated from the master.

When a beam of monochromatic light is focused on a transmission grating, each line acts like a source of this radiation. At certain points on the opposite side of the grating the wavelengths reinforce each other. At other points there is destructive interference and darkness. The result is a series of bright lines with dark regions between them. Since each wavelength has its own diffraction angle through the grating, the polychromatic radiation passing through is separated into a spectrum.

Figure 26 shows the diagram of a typical diffraction grating. Two factors

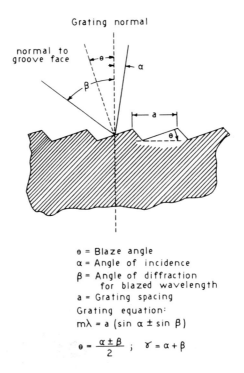

θ = Blaze angle
α = Angle of incidence
β = Angle of diffraction
      for blazed wavelength
a = Grating spacing
Grating equation:
$$m\lambda = a (\sin \alpha \pm \sin \beta)$$

$$\theta = \frac{\alpha \pm \beta}{2} \quad ; \quad \gamma = \alpha + \beta$$

FIG. 26. Cross-sectional diagram of diffraction grating. The "angles" of a single groove that are shown are microscopic on an actual grating.

are important in discussing the characteristics of a grating: blaze angle and
the number of lines or grooves in the grating. The basic equation for a grating
is

$$n\lambda = 2d \sin \theta \tag{15}$$

where n is the order of diffraction, d is the distance between adjacent grooves,
θ is the angle of reflectance, and λ is the wavelength of radiation.

When radiant energy strikes the grating so that the angles of incidence and
diffraction are equal but opposite in sign, then $n\lambda = 0$. This is the zero order,
which corresponds to spectral reflection in a reflection grating. When n = 1,
the diffraction is of the first order; when n = 2, it is of the second order, etc.
(Fig. 27).

The resolving power R of a grating is given by the formula

$$R = \frac{\lambda}{\Delta\lambda} = mN, \tag{16}$$

where $\Delta\lambda$ is the wavelength difference between two lines that are just barely
distinguishable, λ is their average wavelength, N is the number of lines or
grooves in the grating, and m is the length of the grating in centimeters.

For example, a grating with 600 lines per millimeter and 15 cm long would
have a theoretical resolving power of 90,000 (15 cm X 6,000 lines per centimeter).
At 5400 Å this grating would yield a theoretical $\Delta\lambda$ of 0. 06 Å:

$$\Delta\lambda = \frac{\lambda}{R} = \frac{5400}{90,000} = 0.06 \text{ Å}.$$

Gratings with resolving powers of 500,000 have been ruled. The greater the
number of lines, the greater the resolution. The more lines per unit length, the

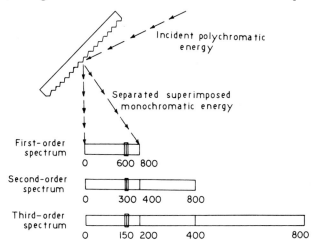

FIG. 27. Overlap of first-, second-, and third-order spectra from a
reflection grating.

greater the dispersion in the first order. The Aminco SPF-1000 has a grating of 30,000 in the first order, or a theoretical resolution of 0.02 nm at the 577-nm line of mercury. The observed resolution is only about 0.7 nm, however, due to the influence of other factors in the design of the monochromator.

The blaze is the second concept that is important to understand in selecting a grating. The blaze is the wavelength at which the maximum output of the grating is concentrated, and the blaze angle is the angle at which the grooves are ruled in the grating (Fig. 26). If a grating is blazed at 500 nm, its maximum output is 500 nm. The efficiency of a grating drops off rapidly as the wavelength differs from the wavelength for which the grating is blazed. The first-order grating efficiency drops off more rapidly on the short-wavelength side than on the longer wavelength side. The bandpass of the efficiency curve for a grating extends from two-thirds of the blaze wavelength and extends to twice the blaze wavelength.

In the Aminco Bowman SPF, for example, the excitation grating is blazed at 300 nm and the emission grating is blazed at 500 nm. Thus the former can be used from 200 to 600 nm, the latter from 335 to 1,000 nm. For maximum efficiency, however, a grating blazed at 750 nm should be used for scanning the near-infrared region. The typical efficiencies and bandpass widths of gratings used in spectrofluorometry are listed in Table 13.

2. Advantages and Disadvantages of Gratings

The advantages of gratings are the following:

1. Gratings have uniform resolution and linear dispersion at all wavelengths.
2. Up to 80 % of the incident radiation can be directed into the first order by blazing the grating.
3. Gratings are less expensive than prisms.
4. All wavelengths can be dispersed.

The major disadvantage of the grating is that several orders of spectra are passed. This can be offset by using filters in the optical path. Thus to observe the 600-nm spectral line without interference from the second-order 300-nm spectral line, a filter cutting off all radiation below 400 nm should be used.

IV. CELL COMPARTMENT

Pyrex cells are useful for measurements above 320 nm (which compose 95 % of all common analyses). Only below 320 nm are quartz or fused-silica cells required. Hence for all practical purposes the large additional cost of quartz or fused silica (Supersil, etc.) is not justified.

Below 320 nm, where quartz is necessary, the researcher should consider

TABLE 13

Typical Efficiencies and Bandpass Widths of Gratings
Used in Spectrofluorometry

Parameter	Grating blaze (nm)[a]		
	300	500	750
First-order blaze (nm)	300	500	750
Second-order blaze (nm)	150	250	375
First-order efficiency--first-order blaze (%)	85	90	90
Second-order efficiency-- second-order blaze (%)	60	80	75
First-order bandpass (nm)	200-600	335-1000	500-1500
Second-order bandpass (nm)	100-255	170-750	250-1125

[a]All gratings with 600 lines per millimeter.

the properties of the different kinds of quartz. Corning quartz possesses a lower native fluorescence than do other varieties. However, even this quartz possesses sufficient fluorescence to be detected on the spectrophotofluorometer at high sensitivity (λ_{ex} = 265 and 330 nm; λ_{em} = 500 nm). Fused silica (e. g. , Supersil) is preferred over fused quartz. Care should also be given to the selection of sample cuvettes, especially with respect to scratches and surface flaws.

V. CELL CONFIGURATION

A few words on cell configuration are desirable at this time. One could use any cell configuration from 0 to 180° for measurement of fluorescence. A 30 or 45° configuration works as well as a 90° one. However, the lowest backgrounds from incident radiation are obtained at 90°, and, for this reason, this configuration is most commonly used in all instruments.

When concentration quenching is encountered in the conventional 90° configuration and the sample cannot be diluted, the worker can try one of the following: (a) use a front-surface configuration with a solid-sample accessory or (b) use a microcell to eliminate all regions of the sample solution that are not both fluorescing and observed by the detector.

VI. SLITS

The slit width is the most important parameter in determining the resolution of an instrument. The distribution of energy as a function of wavelength for the light passing through the exit slit of a monochromator can be represented as an isosceles triangle if the entrance and exit slits are of equal width (Fig. 28). The middle wavelength (peak transmittance) is called the nominal wavelength and is the value read on the dial of the instrument. The bandpass is the bandwidth at one-half the peak transmittance and is essentially the width of the exit slit. The spectral slit width is twice the bandpass and is the total width of the base line. Within the bandpass width is contained three-quarters of the transmitted radiant energy. In a grating instrument the bandpass for a given slit is constant through the spectrum and depends on the ruling of the grating.

Figure 29 illustrates the effect of slit width on the spectral isolation of the monochromatic 546-nm mercury line in the Beckman DU instrument. At a slit opening of 0.1 mm the bandpass is 3.4 nm. This means that two peaks closer than approximately 6 nm could not be resolved. At a slit width of 0.02 mm the bandpass is 1.3 nm, but at 0.01 mm it is only 1.0 nm. Hence there is no linear relationship between bandpass and slit width, but it can be said that better resolution can be obtained by decreasing the slit width. However, decreasing the slit width decreases sensitivity since the intensity of light emerging from the monochromator decreases. This is why most good fluorometers use a photo-multiplier, which is a more sensitive device for measuring the emergent radiation.

There are three types of slit: fixed, unilateral, and bilateral. Fixed slits are slots cut in an opaque material. Some instruments, such as the Aminco Bowman SPF, use a series of interchangeable fixed slits to provide reproducible settings. Unilateral slits are made from two beveled blades or jaws, with one jaw movable, and allow continuous variation through a limited range. The

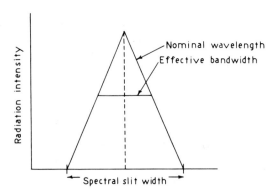

FIG. 28. Distribution of radiant energy emerging from a slit as a function of wavelength.

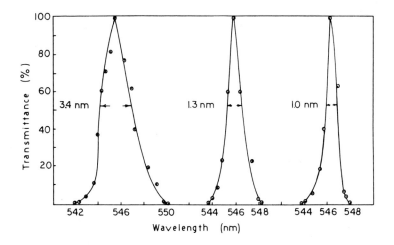

FIG. 29. Spectral isolation with various slit widths. Beckman Model DU spectrophotometer. After Cary and Beckman [1].

disadvantage of the unilateral slit is that the center of the spectral line is shifted as the slit width is altered. Bilateral slits have two blades that move symmetrically to maintain a constant center line. This is the type used in the Perkin-Elmer MPF-2A. Slits should be cleaned periodically with cellophane or similar material.

Resolution is normally expressed in spectral bandpass, since this takes into account the spread of wavelength leaving the exit slit. The bandpass is related to the dispersion D, the focal length F, and the slit width W by the equation

$$BP = \frac{W}{FD} \tag{17}$$

In the Aminco Bowman SPF which has a 250-nm focal length, the equation results in

$$BP = 5.5W$$

or each millimeter of slit width yields 5.5 nm of bandpass, down to the maximum resolution of the monochromator, which is about 0.7 nm. The shorter-focal-length Aminco SPF-125, which has a 125-mm focal length, has a 11-nm bandpass for every 1 mm of slit width.

VII. DETECTORS

Most of the fluorometers developed over five years ago used barrier-layer cells for detection in the visible region of the spectrum. This detector relied heavily on external electrical circuitry for amplification of the very weak signal. Recent advances, however, have made it possible to use photomultiplier tubes as the detecting elements in fluorometers. The photomultiplier is much more sensitive.

A. Barrier-Layer Cells

The construction of a typical barrier-layer cell is shown in Fig. 30. The cell consists of a plate of a metal on which has been deposited a thin layer of a semiconductor, frequently selenium on an iron base. A thin layer of silver is placed on the semiconductor to act as the collector electrode. The metal base acts as the electrode.

Energy falling on the surface of the semiconductor excites electrons at the silver-selenium interface; these are released and passed to the collector electrode. The cell generates its own emf; no external power supply is needed to observe a photocurrent. Usually the cell is connected to a galvanometer to record current.

The spectral response of the selenium cell adequately covers the visible region with greatest sensitivity at 500 to 600 nm (Fig. 31).

The barrier-layer cell is simple but has several disadvantages: it fatigues rapidly, and the signal is hard to amplify; hence it is used only where a large energy is involved. It is used in the Klett and Lumetron instruments.

B. Photomultiplier Tubes

Photomultiplier tubes are used exclusively in newer instruments (within the last five years) because of their extreme sensitivity and fast response. The photomultiplier combines photocathode emission with multiple cascade stages of electron amplification to achieve a large amplification of primary photo-current within the envelope of the phototube itself, with retention of linear response. As shown in Fig. 32, the photomultiplier is constructed so that the primary photoelectrons from the cathode are attracted and accelerated to the first dynode with considerable energy. Each dynode consists of a plate coated with a substance having a small force of attraction for the escaping electrons. The impinging high-energy electrons strike with enough energy to dislodge and

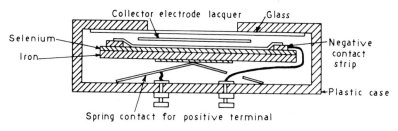

FIG. 30. Construction of a barrier-layer cell. (Courtesy of the General Electric Company.)

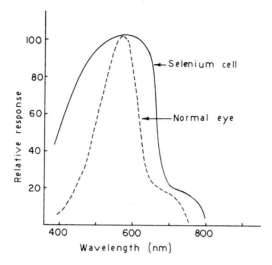

FIG. 31. Spectral response of a selenium barrier-layer cell with protective glass cover. Response of human eye is shown for comparison.

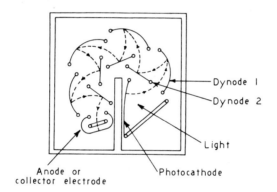

FIG. 32. Schematic diagram of a photomultiplier tube. Broken lines are the paths traveled by the secondary electrons as they are focused by each succeeding dynode's field in turn.

eject two to five secondary electrons. These electrons are accelerated to the second dynode by an additional positive potential, and so on. The process can be repeated 10 to 20 times in stages of potential from 20 to 100 V. This requires a high-voltage, low-current power supply that is very stable.

The photomultiplier surface contains a thin layer of one or more elements possessing a low ionization potential. As a result, valence electrons are easily released when struck by photons. The alkali metals are commonly used and are plated on an Ag/Ag_2O cathode. Photomultiplier tubes can be made to respond to different wavelengths by varying the elements of the photosensitive surface. The tubes are classified by "S" ratings from S-1 to S-22 (see Fig. 33). The response of the human eye most closely resembles the S-4 or S-11 curve.

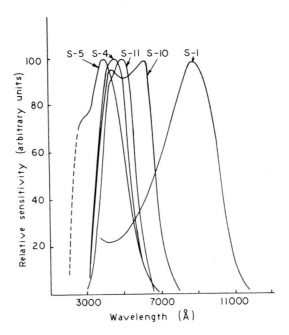

FIG. 33. Spectral response curves of some commercial photoemissive surfaces. All curves have been adjusted so that the wavelength of maximum sensitivity is 100 ordinate units.

Because of manufacturing problems, it is difficult to get two photomultiplier tubes with the same response characteristics. The researcher is advised to select carefully and to buy tubes with known sensitivity and spectral response.

There are several sources of noise in a photomultiplier tube, two of which are shot noise, which arises from extraneous electron emission, and dark-current noise, which arises from thermionic emission in the dynodes.

For best operation and low noise, it is recommended that the photocathode be kept in the dark at all times. Sensitivity can be increased by lowering the noise. The signal-to-noise ratio (S/N) should be at least 2, but preferably 10. Thus lowering the noise by cooling the photomultiplier could increase sensitivity.

VIII. BASIC INSTRUMENTS

As we have already mentioned, all instruments available for the measurement of luminescence fall into two classes: filter instruments and grating instruments.

A. Filter Instruments

1. General Remarks

Filter fluorometers are inexpensive, very sensitive, and simple in design.

The instrument uses a mercury lamp, a primary filter to pass only certain wavelengths to excite the sample, a secondary filter to pass only the fluorescence emission, and not the incident energy, strong radiation, and light scatter, to the detector, which is usually a photomultiplier. The filter fluorometer is more sensitive than the grating instrument and can do almost anything the latter can do, except scan. Table 14 lists the characteristics of some commonly available filter fluorometers.

To use a filter instrument one must first determine the wavelengths of excitation and emission in order to choose the filters to use for maximum sensitivity. To do this, one first runs the absorption spectrum of the compound on any spectrophotometer. Since the excitation spectrum should match the absorption spectrum, one need simply pick a primary filter to peak around the λ_{max} of the compound. Then one places a dilute solution (10^{-5} M) of the compound to be assayed into the fluorometer and tries several secondary filters (realizing that the emission peak is usually separated from the excitation peak by 20 to 150 nm) until a maximum signal is obtained.

2. Commercial Instruments

Commercial instruments differ from each other in the component parts and in the manner of assembly. Some instruments like the Coleman Model 12C, the Photovolt Model 540, the Farrand, and the Aminco possess a single-beam circuit, whereas the Klett, the Lumetron, the Turner, the Beckman, and the Hilger Spekker instruments utilize some sort of a null-indicating circuit for eliminating lamp fluctuations (a type of double-beam circuit).

A photograph of the Klett fluorometer is shown in Fig. 34. The

Fig. 34. The Klett fluorometer.

TABLE 14

Comparison of Common Filter Fluorometers

Fluorometer[a]	Lamp	Lowest detectable concentration (at $S/N = 1$) of quinine sulfate (μg/ml)	Cost	Remarks
Aminco	Hg	0.0002	$1150	Photomultiplier detector; single beam; temperature control; seven scales; quartz optics
Baird-Atomic Fluorimet	Hg	0.001	$670	Solid state
Beckman ratio	Hg (phosphor-coated sleeve)	0.0010	$1075	Photomultiplier detector; double beam, ratio recording; Vycor optics
Coleman 12C	Hg arc	0.003	$480	Phototube detector; single beam; one scale; glass optics
Farrand ratio	Hg	0.0001	$1695	1P28 Photomultiplier, optical balance
Technicon	Hg	0.001	$2755	Photomultiplier detector; double beam
Turner 110, 111	Hg	0.003	$1195 (110) $1685 (111)	Photomultiplier detector; double beam; optical balance; temperature control; quartz optics; four scales; 110 = null balance; 111 = recording

a All instruments listed use a glass-filter monochromator.

Klett is one of the few instruments still utilizing a barrier-layer-cell detector.

G. K. Turner Associates markets two good filter fluorometers that are quite widely used. The Turner Model 110 is pictured in Fig. 35. It is a null-balance instrument. The Model 111 has direct output to a recorder and is more versatile, yet more expensive. Figure 36 shows the optical system for the Model 110. The Turner has a calibrated rear light path to compensate for fluctuations in the light source.

A true ratio fluorometer, such as the Beckman shown in Fig. 37, places the reference solution in one beam while the sample solution is irradiated by the second beam. This makes the fluorometer insensitive to the temperature changes that affect other instruments. A special lamp irradiates each solution alternately, and a discriminator circuit presents the ratio signal to the meter.

The new Technicon continuous-flow fluorometer is shown in Fig. 38. It is used in conjunction with Technicon automated clinical procedures.

The Coleman Model 12C fluorometer has been used in a number of laboratories for about 30 years. It has a blue-sensitive phototube as a detector. The use of a phototube limits its sensitivity, however.

Aminco offers a good filter fluorometer seen pictured in Fig. 39. It is available in solid state and has seven linearly calibrated scales.

The Photovolt Model 540 fluorometer is pictured in Fig. 40. Both low-pressure mercury and phosphor-coated lamps can be used with the instrument. A sensitive 1P21 photomultiplier tube is used in the instrument.

The Lumetron Model 402-EF fluorometer is also manufactured by the Photovolt Corporation. It is a versatile instrument but not as sensitive as the Model 540 since it uses a barrier-layer cell as the detector. It does have a split-beam circuit with two photocells to balance out lamp fluctuations. A diagram of the Model 402-EF is shown in Fig. 41.

3. Fluorescence Accessories for Spectrophotometers

Fluorescence attachments were developed several years ago for both the Beckman DU and DK spectrophotometers. Both attachments use a mercury-vapor lamp, a filter for selecting the wavelength of excitation radiation, a sample holder, and the optics to reflect the emitted fluorescence into the DU or DK monochromator, which then converts this into a fluorescence spectrum (Fig. 42). The photomultiplier attachment to each instrument must be used to obtain good sensitivity.

A fluorescence accessory is also available for the Cary Model 10 and 11 spectrophotometers as well as for the Zeiss spectrophotometer (Fig. 43).

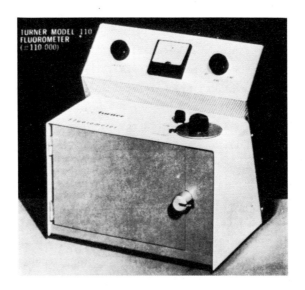

FIG. 35.   The Turner Model 110 fluorometer.

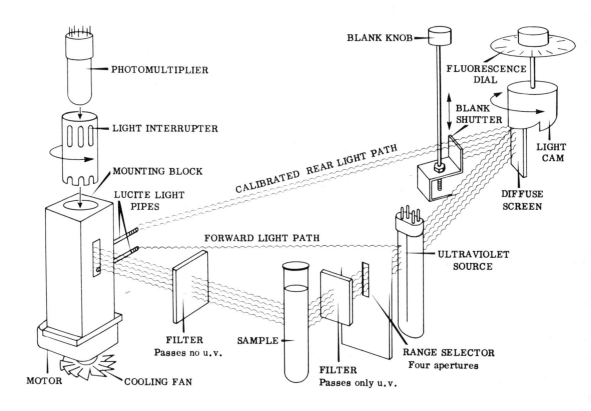

FIG. 36.   Optical system of the Turner Model 110 fluorometer.

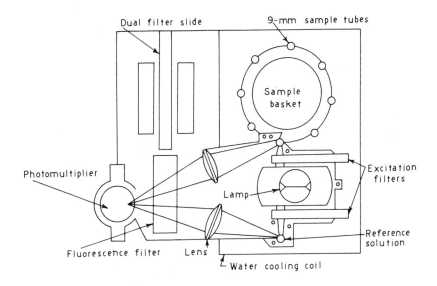

FIG. 37. Optical diagram of the Beckman ratio fluorometer. The lamp
alternately illuminates the reference and sample solutions.

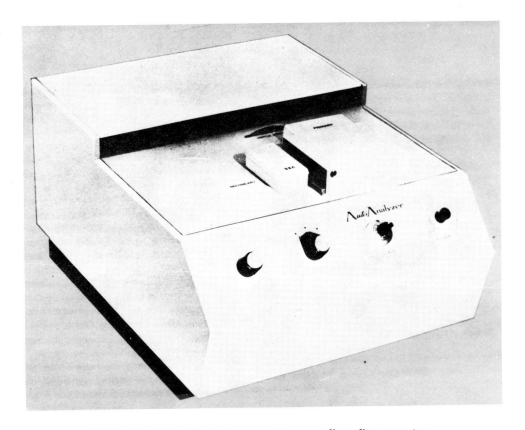

FIG. 38. New Technicon continuous-flow fluorometer.

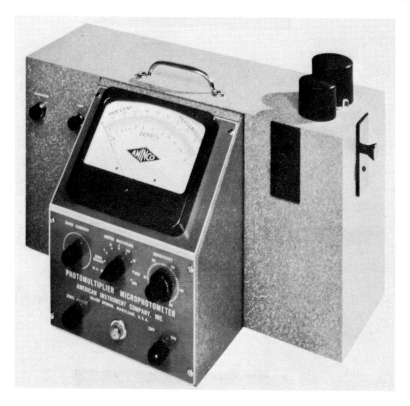

FIG. 39. Aminco Filter Fluorometer.

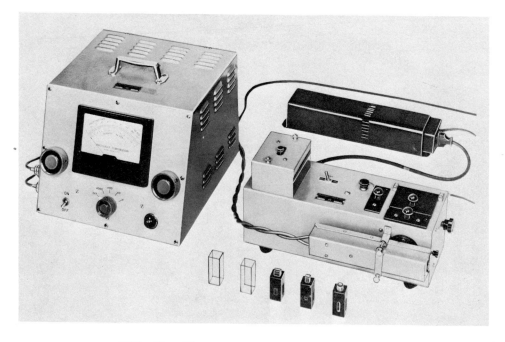

FIG. 40. Photovolt Model 540 fluorometer.

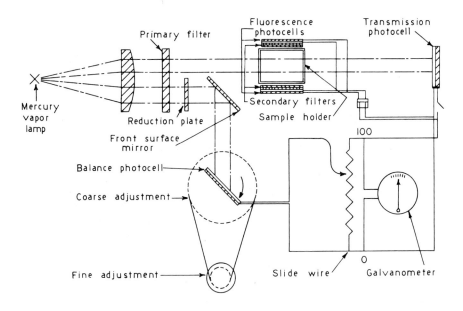

FIG. 41. Diagram of the Lumetron Model 402-EF fluorometer.

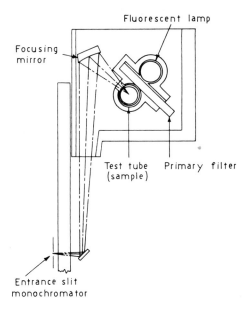

FIG. 42. Fluorescence attachment for the Beckman DU or DK spectrometers.

FIG. 43.   Fluorescence attachment ZFM4 for the Zeiss Model PMQ 11
spectrophotometer.

B. Grating Instruments

1. General Remarks

As grating instrument generally uses a xenon lamp as the source of
radiation, two gratings to disperse and select the desired excitation and
fluorescence energy, and a photomultiplier as the detector.   Various slits in
the focal planes of the gratings determine the bandpass and the intensity of the
energy striking the sample or the detector.

The grating spectrofluorometer is more versatile than the filter
instrument and can be used for luminescence research projects.   A comparison
of some spectrofluorometers is presented in Table 15.   In this table are
compared the various characteristics of each instrument: lamp, monochromator,
slit widths, lines on the grating, lowest detectable concentration of quinine
sulfate, cost, resolution, and various characteristics of each.   This table is
intended to give an unbiased comparison according to manufacturer's
specifications.

2. Commercial Instruments

One of the most popular spectrofluorometers is the Aminco-Bowman
spectrophotofluorometer (SPF), an adaptation of an instrument developed by
Dr. Robert L. Bowman.   It is shown in Fig. 44.   The components of the
spectrofluorometer are diagrammed in Fig. 45.   The xenon source covers the
range of excitation wavelengths from 200 to 1500 nm.   A blue- or red-sensitive

TABLE 15

Comparison of Spectrofluorometers

Instrument	Lamp	Monochromator (slits)	Lowest detectable concentration (at S/N = 1) of quinine sulfate (ppb)	Cost	Resolution (nm)	Remarks
Aminco:						
Aminco-Bowman	Xenon (150 W)	Grating (1-30 nm) 15,000 lines per inch; (30,000 optional)	0.2	$4850[a]	1[b]	1P21 Photomultiplier; seven scales; temperature control
SPF-1000	Xenon	Grating 30,000 lines per inch (0.1-4 nm)	0.5	$2500	0.2	Photomultiplier; corrected spectra instrument $25000
SPF-125	Hg	Grating 15,000 lines per inch (1.5-44 nm)	0.005	$3950	1.5	1P21 Photomultiplier; seven scales; solid state
Baird-Atomic:						
Fluorispec SF/1	Xenon (150 W)	Dual grating (2-32 nm)	0.1	$5775	1.6	1P28 Photomultiplier or 7102 for IR; four scales at 1 and 1/10 sec time constants
Fluoripoint	Hg or Xe	Grating (10 or 20 nm)	0.05	$3000 + light source	10	EM 1 9771B Photomultiplier; seven scales
Farrand:						
MK-1	Xenon (150 W)	Grating (0.5-20 nm)	0.1	$5300	0.5	1P28 Photomultiplier; ten scanning speeds

Instrument	Lamp	Monochromator (slits)	Lowest detectable concentration (at S/N = 1) of quinine sulfate (ppb)[b]	Cost	Resolution (nm)	Remarks
Perkin-Elmer:						
MPF-2A	Xenon (150 W)	Grating (1-40 nm) 14,400 lines per inch	0.05[c]	$7450	1	R-106 Photomultiplier; ratio recording; temperature control; three scan speeds; variable slits
MPF-3	Xenon (150 W)	Grating (1-40 nm) 14,400 lines per inch	0.005	$8950	1	Same as 2A except six scan speeds; $13800 with constant quanta accessory
204	Xenon (150 W)	Grating (10 nm) 14,400 lines per inch	0.05	$3750	10	Meter readout with thirty-six scales; $4600 with recorder
Turner:						
210	Xenon (75 W)	Grating (0.5-25 nm) 30,000 lines per inch	2[d]	$18750	0.1[e]	Corrected spectra; double beam; temperature control; can be used as spectrophotometer
430	Xenon (150 W)	Grating (15 nm)	0.005	$3890	15	R-136 Photomultiplier

a Cost of corrected-spectra instrument, $8575; ratio-recording attachment, $2700.
b Resolution of 0.5 nm optional.
c At 10-nm bandpass.
d At 15-nm bandpass.
e Readability.

FIG. 44. Aminco-Bowman spectrophotofluorometer.

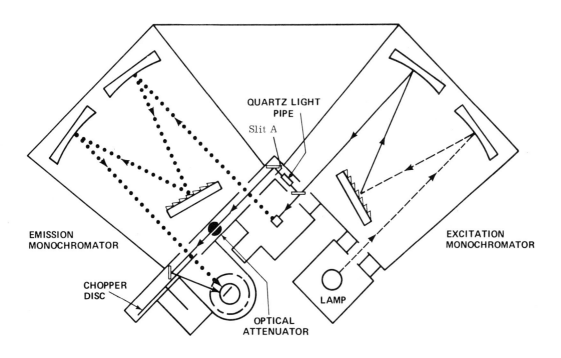

FIG. 45. Optical system of Aminco-Bowman SPF

photomultiplier tube is used to measure the spectral signal. Two gratings are
used for monochromaticity, and the resolution of the instrument is quite good
($<$ 1 nm).

The Baird-Atomic, Inc., spectrofluorometer is a two-grating monochromator based on a design originally suggested by Bowman, Caulfield, and
Udenfriend [2]. The use of two excitation and two emission gratings (Fig. 46)
gives the instrument good resolution and low background scatter. The
instrument is shown in Fig. 47.

The Baird-Atomic Fluoripoint spectrofluorometer is pictured in Fig. 48.
This is a low cost instrument that has two grating monochromators, a Hg or Xe

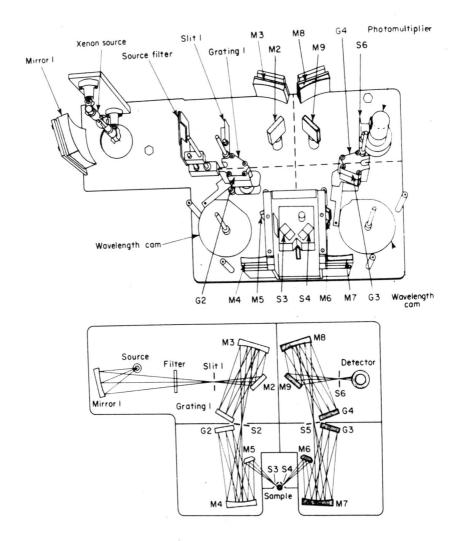

FIG. 46. Schematic optical diagram of a spectrophotofluorometer with
two double gratings (the Baird-Atomic Fluorispec Model SF/1). The sequence
of numbers denotes the path followed by the incident and fluorescent light.
Symbols: S, slit; G, grating; M, mirror.

FIG. 47.   Baird-Atomic, Inc., spectrofluorometer.

source, a sample compartment with four position sample turret, and a solid state
photomultiplier power supply and amplifier.

The Farrand Model MK-1 spectrofluorometer is also based on a design
originally suggested by Bowman et al. [2].   A schematic diagram of the Farrand
is shown in Fig. 49.   Two grating monochromators (220 to 650 nm) are used
along with a 1P21 photomultiplier tube.   An attachment that uses a lead sulfide
phototube is available for measurements in the infrared region.

The Perkin-Elmer MPF-3 instrument is shown in Fig. 50.   It has two
monochromators with good resolution and also slits that are variably adjustable
from 1 to 40 nm.   The instrument has a ratio-recording system to compensate
for drifts in the light source.

Turner Instrument Company now markets a new, lost cost ($3890) grating
instrument, the Model 430 pictured in Fig. 51.   The instrument has a sensitivity
of 5 parts per trillion with a 15-nm bandwidth.   A jacketed cell holder provides
temperature control, and a range selector has precision ranges of 1, 3, 10, 30,
100, 300 and 1,000.

FIG. 48. Baird-Atomic Fluoripoint spectrofluorometer.

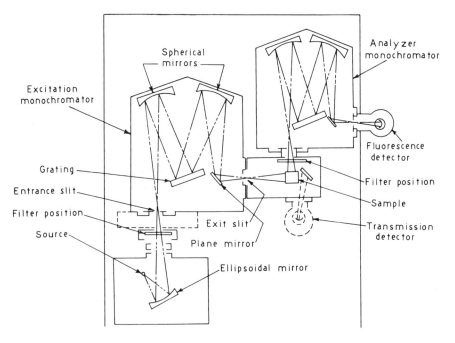

FIG. 49. Diagram of the Farrand Model MK-1 spectrofluorometer.

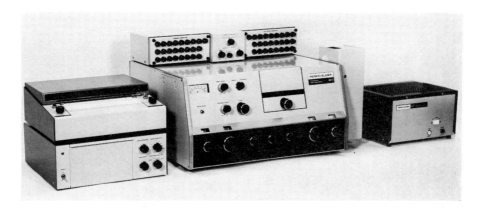

FIG. 50. Perkin-Elmer Model MPF-3 spectrofluorometer.

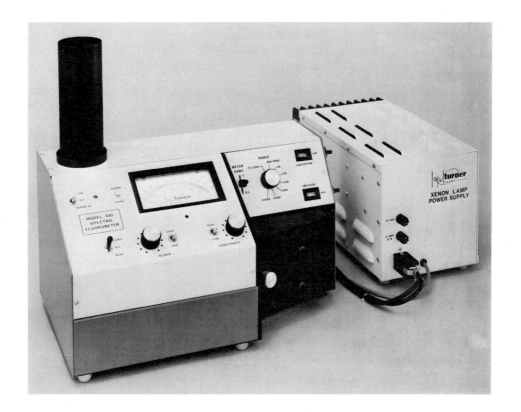

FIG. 51. Turner Model 430 spectrofluorometer.

3. Corrected-Spectra Instruments

The American Instrument Company has a corrected-spectra attachment for the Aminco-Bowman spectrophotofluorometer (Fig. 52). For correction of the

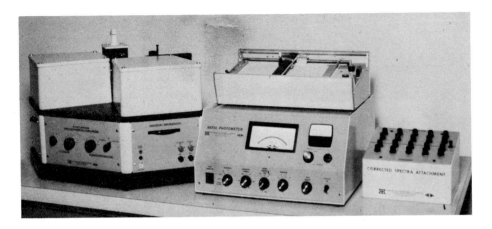

FIG. 52.   Corrected spectra attachment for Aminco-Bowman Ratio Photometer.

excitation spectra a beam splitter focuses a part of the excitation source onto a thermopile whose voltage output is independent of the spectral character of the excitation radiation.   The ratio of the signal from the sample to the output of the thermopile is measured.   Emission spectra are corrected by a potentiometer generating a predetermined correction program of voltage versus wavelength. The potentiometer is calibrated to compensate for instrumental variables, such as mirrors, monochromators, and detectors.   The voltage output of the photo-multiplier tube is automatically multiplied by the voltage output of the correction potentiometer.

A corrected-spectra accessory is available for the Perkin-Elmer Model MPF-2A and 3.   The ratio-recording capability of this model is coupled with a quantum counter consisting of a solution of Rhodamine B in ethylene glycol.   This solution maintains a constant ratio of quanta absorbed at 200 to 600 nm to quanta emitted at 630 nm.   Both the excitation and emission spectra are corrected.   A block diagram of the corrected-spectra accessory is shown in Fig. 53.

The Turner Model 210 Spectro (Fig. 54) can function as an ordinary highly sensitive spectrofluorometer or as an absolute instrument that provides corrected excitation and emission spectra.   In the latter mode the photomultiplier alternately receives two beams of light, one the emission from the sample, the second that from a fixed-wavelength reference lamp whose intensity is auto-matically adjusted to match the energy of the exciting light reaching the sample. With the emission monochromator fixed and the excitation monochromator scanning, an excitation spectrum corrected to constant energy is obtained.   With the excitation monochromator fixed and the emission monochromator scanning, one obtains an emission spectrum that can be calibrated in terms of quanta per unit bandwidth.

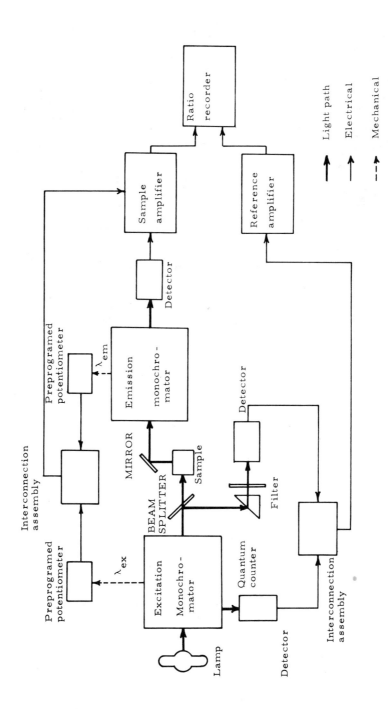

FIG. 53. System block diagram of the corrected-spectra accessory for the Perkin-Elmer Model MPF-2A spectrofluorometer.

FIG. 54.   Turner Model 210 Spectro.

A schematic diagram of the Model 210 is shown in Fig. 55. Emission spectra are corrected up to 690 nm, although the instrument can be used to 1100 nm.

C.   Instruments for Measuring Decay Times

The introduction of nanosecond-flash lamps has made it possible to measure fluorescence lifetimes by direct observation of the decay. Bennett [3] and Ware and Baldwin [4] have designed nanosecond-flash instruments to measure fluorescence lifetimes. Steingraber and Berlman [5] designed and built a flash apparatus for measuring decay times that contains many of the features used in a subsequent commercial instrument. Mackey et al [6] designed an intense light source of nanosecond duration that could be operated at 5000 pulses per second. This became the basis for the TRW Model 31B nanosecond spectral source system, which is the first commercial instrument for measuring fluorescence-decay times. The instrument is shown in Fig. 56, and a block diagram of the instrument is presented in Fig. 57. The instrument can be used to measure τ from 200 to 600 nm.

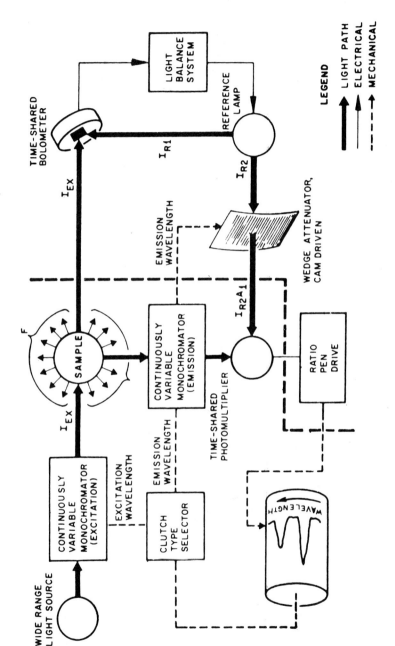

FIG. 55. Schematic representation of the Turner Model 210 Spectro.
Units to the left of the heavy broken line comprise those found in the standard
uncorrected fluorescence spectrometer; those on the right comprise the
correcting systems.

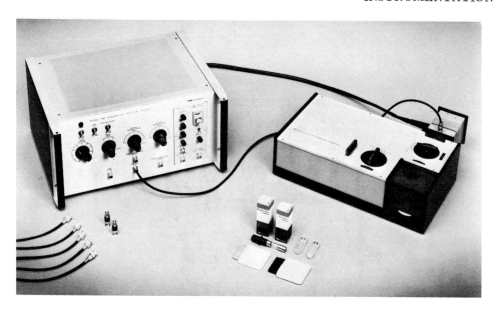

FIG. 56.   TRW Model 31 B nanosecond spectral source system.

D.   Modification of the Cell Compartment for Increased Sensitivity

In an ordinary fluorescence cell light comes in on one side of the cell and goes out at a 90° angle. The other two sides of the cell are usually backed with a black absorbing surface. Researchers at the American Instrument Company found that, if these two black surfaces were replaced with mirrors (Fig. 58), a large increase in sensitivity could be obtained (Fig. 59). This is due to the fact that the species emits at all angles, not only 90°. The mirrors serve to focus and amplify this emitted radiation so it is measured at 90° by the detector.

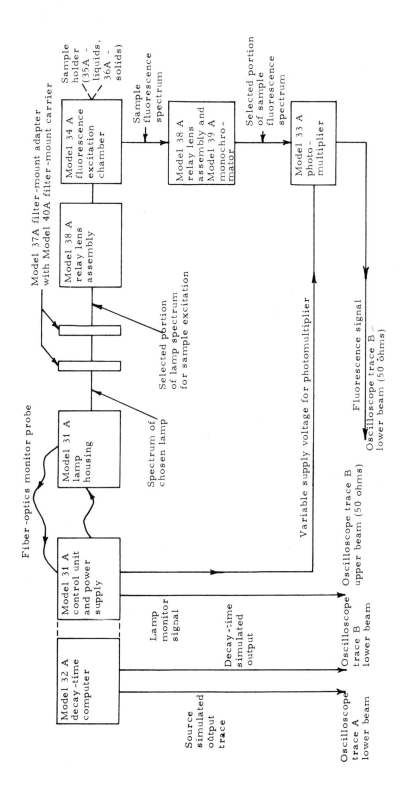

FIG. 57. Block diagram of TRW 31B nanosecond spectral source system as used for measuring fluorescence-decay times shorter than 1 μsec.

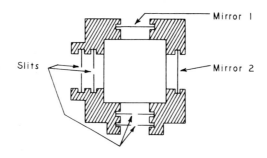

FIG. 58. Arrangement of mirrors in the Aminco-Bowman spectrophoto-fluorometer cell compartment.

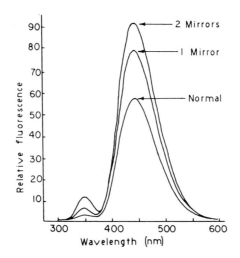

FIG. 59. Increase in spectrophotofluorometer sensitivity obtained by the use of mirrors, arranged as in Fig. 58. A 49% net increase is obtained with any one mirror in either of the two positions. Sample solution: quinine sulfate (0.1 μg) in 0.1 N sulfuric acid.

REFERENCES

1. H. H. Cary and A. O. Beckman, J. Opt. Soc. Amer., 31, 682 (1941).

2. R. L. Bowman, P. A. Caulfield, and S. Udenfriend, Science, 122, 32 (1955).

3. R. G. Bennett, Rev. Sci. Instr., 31, 1275 (1960).

4. W. R. Ware and B. A. Baldwin, J. Chem. Phys., 40, 1703 (1964).

5. O. J. Steingraber and I. B. Berlman, Rev. Sci. Instr., 34, 524 (1963).

6. R. C. Mackey, S. A. Pollack, and R. S. Witte, Rev. Sci. Instr., 36, 1715 (1965).

Chapter 3

EFFECTS OF MOLECULAR STRUCTURE AND
MOLECULAR ENVIRONMENT ON FLUORESCENCE

E. L. Wehry
Department of Chemistry
University of Tennessee
Knoxville, Tennessee

I. INTRODUCTION

In order to effectively utilize luminescence as an analytical tool it is
necessary that the researcher know the basic effects of structure and the
environment on the emission process. The fluorescence of a molecule
depends on its structure and on the environment in which the luminescence is

measured. The researcher may be able to convert a nonfluorescent molecule into a fluorescent species in some cases and will have available a wide choice of media in which to study the fluorescence of his samples. Under structural effects we shall briefly consider what types of compounds fluoresce and how we might increase the total emission by changes in structure. Under environmental effects we shall study how pH, the solvent, oxygen, metal ions, and other variables affect the luminescence characteristics of compounds.

II. STRUCTURAL EFFECTS

A. General Considerations

Fluorescence phenomena are not sensitive to the finer details of molecular structure; fluorescence is not generally useful as a "fingerprinting" technique. Of the huge number of known organic and inorganic compounds, only a small fraction exhibits intense luminescence. In order to understand how molecular structure affects fluorescence, one must realize that fluorescence always competes with a variety of other processes. When a molecule is promoted to an electronically excited state, it may divest itself of its excess energy in a number of different ways the principal decay processes being (a) fluorescence (b) nonradiative decay (internal conversion or intersystem crossing), and (c) photochemical reaction. Which of these three processes dominates depends entirely on their relative rates. Thus, for fluorescence to dominate, one desires that the rate constant for radiative transitions be large relative to those for nonradiative decay or photodecomposition.

In general, therefore, strongly fluorescent molecules possess the following characteristics:

1. The spin-allowed electronic absorption transition of lowest energy is very intense (i. e. , has a large ϵ_{max}). The intensity of absorption is directly proportional to the rate constant for the radiative transition. Because fluorescence is simply the reverse of absorption, it follows that, the more probable the absorption transition, the more probable (hence more rapid) will be the reverse (fluorescence) transition. Therefore, in order to predict whether or not a given molecule will fluoresce, an examination of its absorption spectrum is of considerable assistance.

2. The energy of the lowest spin-allowed absorption transition should be reasonably low. The greater the energy of excitation, the more probable the occurrence of photodissociation.

3. The electron that is promoted to a higher level in the absorption transition should be located in an orbital not strongly involved in bonding.

Otherwise, bond dissociation may accompany excitation, and fluorescence is unlikely to be observed.

4. The molecule should not contain structural features or functional groups that enhance the rates of radiationless transitions. Although the theory of nonradiative processes is still in the process of development, we observe that certain structural features greatly increase the rates of radiationless processes and therefore adversely affect fluorescence intensities.

On the basis of these few simple considerations we can easily understand why aromatic hydrocarbons are usually very intensely fluorescent. In these systems π electrons, which are less strongly held than σ electrons, can be promoted to π^* antibonding orbitals by absorption of electromagnetic radiation of fairly low energy without extensive disruption of bonding (Fig. 60). Furthermore, $\pi \longrightarrow \pi^*$ transitions in most aromatic hydrocarbons are strongly allowed ($\epsilon_{max} \sim 10^4$). The combination of these two factors signifies that aromatic compounds possessing low-lying (π, π^*) singlet states usually fluoresce strongly.

The situation becomes more complicated in molecules containing carbonyl groups (aldehydes, ketones, carboxylic acids) and in heterocyclic compounds. In such molecules nonbonding (n) electrons are available for excitation into π^* orbitals [1]. As noted in Fig. 60, the energy of $n \longrightarrow \pi^*$ absorption will usually,

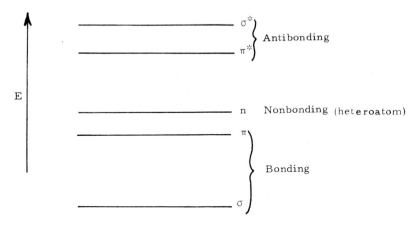

Transition		Band	Molar absorptivity	Comments
n $\longrightarrow$ π^*		R	100	Observed in fluorometry
π $\longrightarrow$ π^*		K	12,000	Observed in fluorometry
σ $\longrightarrow$ π^*		E	200	Not observed--high energy (low λ)

FIG. 60. Transitions involved in the absorption process.

though not invariably, be lower than that of the lowest $\pi \rightarrow \pi^*$ transition. This fact is important because $n \rightarrow \pi^*$ transitions are much less intense ($\epsilon_{max} \sim 10^2$) than $\pi \rightarrow \pi^*$ transitions. Therefore, even if there were no other factors to consider, we would expect aromatic carbonyl compounds and heteroaromatics to be less strongly fluorescent than aromatic hydrocarbons. This conclusion is reinforced by some additional factors, which we shall consider, later.

In saturated hydrocarbons there are no π-bonding or nonbonding electrons; thus all electronic transitions involve σ-bonding electrons. As noted in Fig. 60, we expect transitions involving σ electrons to occur at very high energies and, in addition, to significantly disrupt bonding in the molecule. In fact saturated hydrocarbons do fluoresce [2], though the fluorescence is very weak (fluorescence quantum efficiencies are on the order of 10^{-3}) and occurs in the 140- to 170-nm region (vacuum ultraviolet). In aliphatic carbonyl compounds $n \rightarrow \pi^*$ transitions can occur, and these compounds consequently can exhibit fluorescence in the "normal" ultraviolet or even the visible region, though the quantum efficiencies are likewise very small. Some nonaromatic, but highly conjugated, compounds, such as ß-carotene (1), vitamin A (2), and vitamin A aldehyde (3), are fluorescent, due to the occurrence of $\pi \rightarrow \pi^*$ transitions. In general, however, the vast majority of intensely fluorescing organic compounds are aromatic, and it is with such systems that we shall be mainly concerned.

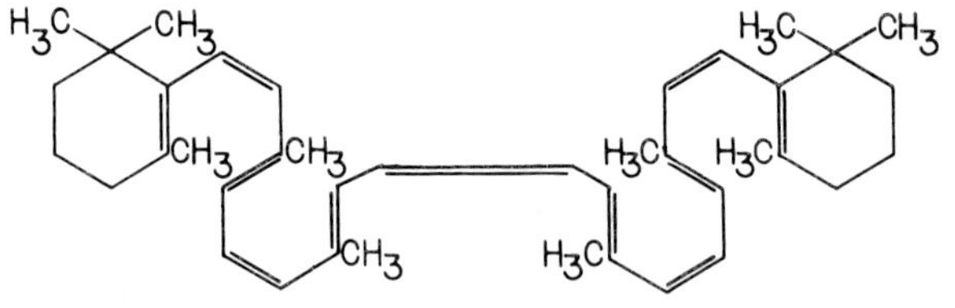

(1)

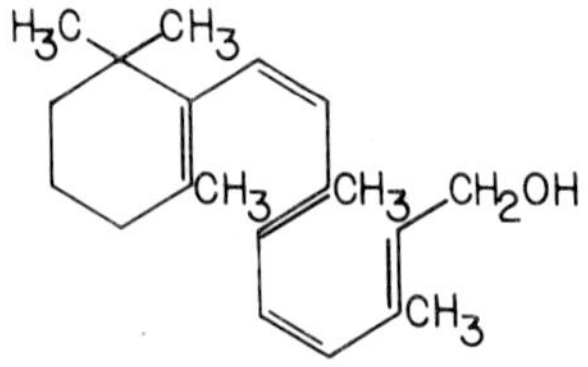

(2)

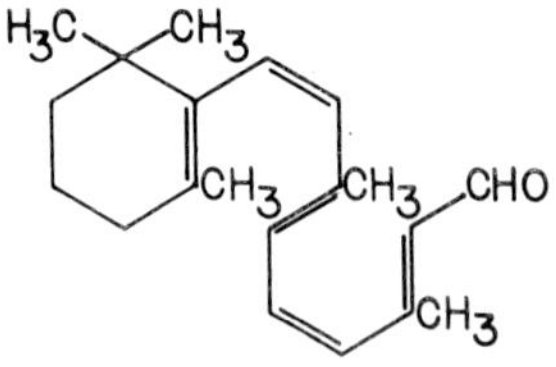

(3)

B. Aromatic Hydrocarbons

Luminescence phenomena in aromatic hydrocarbons have been discussed in great detail in a recent monograph by Birks [3], and we shall therefore consider only some of the more important and interesting fluorescence properties of these compounds.

Most unsubstituted aromatic compounds exhibit an intense fluorescence in the ultraviolet or visible region. As the degree of conjugation increases, the intensity of fluorescence often increases and a bathochromic shift (shift to longer wavelengths) is observed (Table 16). Thus benzene (**4**) and naphthalene

TABLE 16

Luminescence of Condensed Linear Aromatics in EPA[a] Glass at 77°K

Compound	Φ_F	λ_{ex}(nm)	λ_{em}(nm)
Benzene[b]	0.11	205	278
Naphthalene[b]	0.29	286	321
Anthracene[c]	0.46	365	400
Tetracene[d]	0.60	390	480
Pentacene[e]	0.52	·580	640

[a] A mixture of diethyl ether, isopentane, and ethanol, 5:5:2 (v/v/v).
[b] Fluoresces in the ultraviolet.
[c] Fluoresces in the blue.
[d] Fluoresces in the green.
[e] Fluoresces in the red.

(**5**) fluoresce in the ultraviolet, anthracene (**6**) in the blue, tetracene (**7**) in the green, and pentacene (**8**) in the red. For a given number of aromatic rings it is nearly always observed that linear ring systems fluoresce at longer wavelength than nonlinear systems. Thus λ_{em} is at 400 nm for anthracene (**6**) and at 350 nm for phenanthrene (**9**); similarly, λ_{em} is at 480 for tetracene (**7**) and at 380 nm for benz[a]anthracene (**10**).

  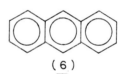

(**4**)   (**5**)   (**6**)

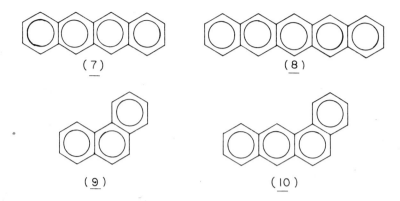

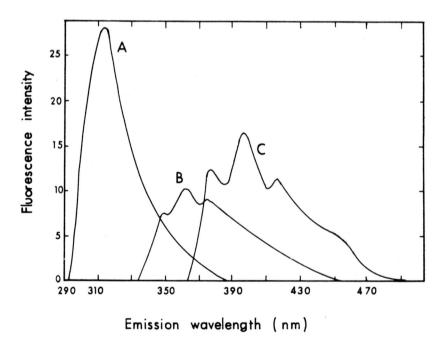

The fluorescence spectra of aromatic hydrocarbons usually are mirror images of their lowest singlet ⟶ singlet absorption bands. The fluorescence bands often possess considerable fine structure, even in room-temperature solutions. Figure 61 shows the fluorescence spectra in room-temperature liquid solution of three hydrocarbons (fluorene, phenanthrene, and anthracene); note the resolved fine structure in the spectra of the latter two compounds. Note also that the fluorescence spectra (as well as the excitation spectra) of these three compounds are sufficiently free from overlap, so that, by careful

FIG. 61. Fluorescence spectra in room-temperature liquid solution of three hydrocarbons [4]. Curve A, fluorene spectrum, λ_{ex} 265 nm; curve B, phenanthrene spectrum, λ_{ex} 265 nm; Curve C, anthracene spectrum, λ_{ex} 365 nm.

choice of excitation and emission wavelengths, one may perform simultaneous analyses of these three compounds in mixtures without prior separations [4] (provided that the concentration of each hydrocarbon in solution is about 10^{-3} M or less, so that quenching of the fluorescence of one solute by another will in many cases be negligible [5]. Polycyclic aromatic hydrocarbons are important air pollutants; Sawicki [6] has given a number of examples in which selection of excitation and emission wavelengths has enabled assay of hydrocarbon mixtures without prior separation.

Most fluorescent aromatic hydrocarbons also phosphoresce, and the quantum yields for fluorescence and phosphorescence are usually comparable in magnitude. It seems generally true that $S_1^* \rightarrow S_0$ internal conversion is an unimportant process in aromatic hydrocarbons; hence the principal process competing with fluorescence is $S_1^* \rightarrow T_1^*$ intersystem crossing. Usually the sum of the quantum yields for these two processes is unity [7]. This fact does not necessarily mean that the sum of the fluorescence and phosphorescence yields is unity, since $T_1^* \rightarrow S_0$ intersystem crossing or other decay processes may compete with phosphorescence in the deactivation of the triplet.

Some especially interesting luminescence phenomena occur in nonrigid aromatic hydrocarbons consisting of two or more distinct chromophores. For example, the wavelength maximum and general appearance of the fluorescence spectrum of 2-phenylnaphthalene (11) are observed to be dependent on the exciting wavelength, in violation of the general rule that fluorescence wave-

(11)

lengths and quantum yields are independent of exciting wavelength. In the case of 2-phenylnaphthalene the unusual wavelength dependence of the fluorescence spectrum has been attributed to the presence of a distribution of ground-state rotational conformers [8]. Hence there is a spectrum of ground-state energies, such that the energy required to populate any given electronically excited state is different for different conformers. Similar observations [9] have been described for a number of substituted aromatics containing two chromophores that are not rigidly fixed in a single relative geometrical configuration. Berlman [10] has described the use of fluorescence spectra to provide qualitative evidence concerning the degree of planarity in the ground and excited states of such molecules.

Molecules consisting of more than one aromatic ring system separated by alkyl chains or C-C single bonds frequently exhibit unusual fluorescence spectra. In simple cases the fluorescence spectrum may simply resemble that of two or more essentially noninteracting aromatic compounds; for example, 1,1-binaphthyl fluorescence is almost indistinguishable, particularly at low temperatures, from that of naphthalene. For some related systems, however, intramolecular energy transfer takes place. A particularly interesting system is

$$n = 1 - 3$$

The absorption spectra of the naphthalene and anthracene moieties are sufficiently different to permit reasonably selective excitation of the naphthalene entity by proper choice of exciting frequency. When that is done, however, only anthracene fluorescence is observed [11], even when n = 3. The fact that the absorption spectra of these "double molecules" are essentially superimpositions of the absorption spectra of 1-methylnaphthalene and 9-methylanthracene suggests that conjugation through the alkyl chain is insignificant in both ground and excited states (though one must be cautious in inferring excited-state conjugated from absorption spectra). Presumably the actual act of intramolecular energy transfer occurs via Forster resonance coupling, with the nonrigid structure facilitating close approach of the naphthalene and anthracene moieties during the excited-state lifetime.

In extreme cases the approach may be so facile in excited states that a nonrigid molecule may form an intramolecular dimer. For example, in di-ß-naphthylalkanes (12) only one fluorescence band is noted for most values of n. However, when n = 3, two fluorescence bands are observed, one of which occurs at a very low energy and is devoid of fine structure [12].

$$(12)$$

Such behavior is frequently observed in liquid solutions of "simple" aromatic hydrocarbons at high concentrations, wherein a low-frequency, broad fluorescence band is attributed to the formation of excimers,

$$A + {}^1A^* \longrightarrow (A - A)^* \longrightarrow A + A + h. \qquad (18)$$

Excimers are transient dimers, stable only in an excited state, formed by
interaction between a ground-state and an excited singlet-state molecule [3].
The similarity between fluorescence behavior characteristic of "normal"
intermolecular excimer formation and that observed for 1, 3-di-ß-naphthyl-
propane suggests that the latter compound has an excimer excited state that is,
in essence, an intramolecular excimer formed between the two phenyl groups.
One would expect intramolecular excimer formation to be highly sensitive to the
alkyl-chain length, as is observed [12]. It is further noted that intramolecular
excimer formation is exhibited by 1, 3-di-α-naphthylpropane, but not by the
asymmetrical compound 1-α, 3-ß-dinaphthylpropane, indicating that the stable
excimer configuration is a symmetrical "sandwich" arrangement of the two
naphthalene rings.

Aromatic compounds consisting of two aryl groups separated by an alkene
chain e. g. , stilbene (13) exhibit cis-trans isomerism. In such systems it is
usually observed that the planar trans isomer (13a) is intensely fluorescent,
whereas the cis isomer (13b) is nonfluorescent, or at best very weakly fluorescent, in room-temperature solution. The fluorescence yields of the sterically

(13a)   (13b)

hindered, nonplanar cis isomers generally increase as the temperature of the
solution is decreased and the viscosity is increased. As the temperature is
decreased, the extent of molecular vibrations in the cis compounds is decreased,
resulting in a decrease in the efficiency of internal conversion within the singlet
manifold. A detailed discussion of these processes in terms of specific vibrational modes in cis-stilbene has been presented [13].

C. Substituted Aromatics

The nature of substituent groups (especially chromophoric ones) plays an
important role in the nature and extent of a molecule's fluorescence. Fluorescence yields (intensities) and energies of aromatic and heterocyclic hydrocarbons
are usually altered by ring substitution. Unfortunately we must be careful in
making broad generalizations. Substituent effects on the chemical and physical
properties of organic molecules in their ground electronic states constitute a

lively area of investigation at present. Furthermore only little is known about the influence of substituents on the behavior of excited states. Both effects must be understood before generalizations concerning the effect of various substituent groups can be made.

A simple generalization is that ortho-para-directing substituents often enhance fluorescence, whereas meta-directing groups repress it. Many of the common meta-directing substituents possess low-lying (n, π^*) singlets. The -NO_2 group is especially notorious for repressing fluorescence. The low-lying (n, π^*) singlet increases the extent of singlet $\longrightarrow$ triplet intersystem crossing relative to that in the parent hydrocarbon; for example, in both nitrobenzene and 1-nitronaphthalene the quantum efficiency of S_1^* $\longrightarrow$ T_1^* intersystem crossing is about 0.6 [14]. Surprisingly, many nitroaromatics exhibit only very weak phosphorescence (Φ_p for nitrobenzene is less than 10^{-3}) despite the high efficiency of triplet formation; apparently T_1^* $\longrightarrow$ S_0 intersystem crossing and photochemistry (hydrogen-atom abstraction) occur much more rapidly than phosphorescence [14]. If the intersystem-crossing efficiency of nitrobenzene is 0.6, then the S_1^* $\longrightarrow$ S_0 internal conversion efficiency must be approximately 0.4, because nitrobenzene is nonfluorescent.

In very acidic glassy solvents at 77°K many nitroaromatics fluoresce but exhibit no phosphorescence. This effect is attributed [15] to protonation of the nitro group, which has relatively little effect on the energy of the lowest (n, π^*) singlet but significantly decreases the energy of the lowest (π, π^*) singlet.

In a similar manner carbonyl substituents (ketone, aldehyde, ester, carboxylic acid), which are meta-directing, repress fluorescence because carbonyl-substituted aromatics possess low-lying (n, π^*) singlets. For example, the intersystem-crossing yield in benzophenone is virtually unity [16], and benzophenone accordingly exhibits intense phosphorescence (especially in nonprotic media) but no fluorescence. In contrast, the -CN substituent is meta-directing, yet cyanosubstituted aromatics often fluoresce more intensely than the parent hydrocarbon. Evidently (n, π^*) singlet states in cyanoaromatics are sufficiently higher in energy than the lowest (π, π^*) singlet for the former to have no significant perturbing effect [17].

Some ortho-para-directing substituents (e. g., -OH, -NH_2, -OCH_3) tend to enhance the fluorescence of aromatic compounds. Great care must be exercised in discussing the effect of these substituents because they have a strong tendency to hydrogen-bond with the solvent or occasionally with other solutes. For example, the dissociated -OH group (-O⁻) is a strongly ortho-para-directing group, yet most (but not all) phenolates are less intensely fluorescent than their conjugate acids. In most cases ionized phenols interact very strongly with the solvent, increasing the efficiency of S_1^* $\longrightarrow$ S_0 internal conversion. In other words, for aromatic compounds with acidic or basic functional groups it is inherently impossible to separate "structural" from "environmental" effects on their luminescence behavior.

As one traverses the substituent series F, Cl, Br, I, phosphorescence is usually increasingly favored relative to fluorescence. This effect is illustrated for emission of substituted naphthalenes in Fig. 62 and Table 17. As noted in

TABLE 17

Substituent Effects on Naphthalene Luminescence

Compound	Φ_P/Φ_F	v_F (cm^{-1})	v_P (cm^{-1})	τ_P (sec)
Naphthalene	0.093	31,750	21,250	2.5
1-Methylnaphthalene	0.053	31,450	21,000	2.5
1-Fluoronaphthalene	0.068	31,600	21,150	1.4
1-Chloronaphthalene	5.2	31,360	20,700	0.23
1-Bromonaphthalene	6.4	31,280	20,650	0.014
1-Iodinaphthalene	> 1000	Not observed	20,500	0.0023
1-Nitronaphthalene	> 1000	Not observed	19,750	0.049

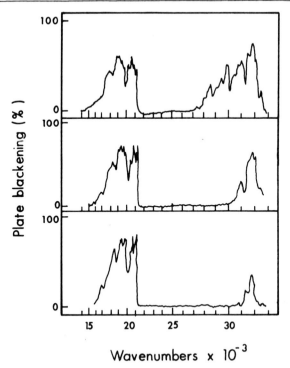

FIG. 62. The heavy-atom effect in intersystem crossing. Total emission spectra of halogenated naphthalenes in EPA at 77°K. Top, 2-chloronaphthalene; center, 2-bromonaphthalene; bottom, 2-iodinaphthalene. Fluorescence is at higher frequencies. After Kasha [18].

the table, the ratio ϕ_P/ϕ_F increases and ϕ_P (the phosphorescence lifetime) decreases in going from F to I. The trends for luminescence yields and life-times in haloaromatics are usually rationalized by postulating that heavy-halogen substitution increases the rates of $S_1^* \longrightarrow T_1^*$ intersystem crossing and $T_1^* \longrightarrow S_0$ phosphorescence [19]. This occurrence is commonly termed the "heavy-atom effect." As we shall subsequently note, the "heavy atom" need not even be a constituent of the luminescent solute to effect the perturbation.

It is worth noting that, though there is considerable evidence of the operation of heavy-atom effects in some systems, the effect cannot be considered a panacea. Fluorine-substituted aromatics, for example, usually fluoresce less efficiently than the parent compounds, yet the fluorine substituent often has little or no effect on the intersystem-crossing rate [20]. Instead, the main effect of fluorosubstitution seems to be to enhance $S_1^* \longrightarrow S_0$ internal conversion. It is also instructive to consider the luminescence of substituted fluorescein dyes; a summary of the results is given in Table 18.

TABLE 18

Luminescence of the Dianions of Fluorescein Dyes at $77^{\circ}K$[a]

Substituent and number	ϕ_F	ϕ_P/ϕ_F	τ_P (msec)
None	0.83	0	--
Cl, 2	0.79	0	--
Br, 1	0.60	0.13	50
Br, 2	0.29	0.21	44
Br, 4	0.40	0.082	9.4
I, 1	0.15	0.67	15.8
I, 2	0.054	1.05	10.4
I, 3	0.061	0.71	5.1
I, 4	0.066	0.40	1.3

[a]Data from Ref. [21].

Fluorescein and chlorofluoresceins do not exhibit measurable phosphorescence, whereas the bromine- and iodine-substituted derivatives do phosphoresce. Successive substitutions of bromine or iodine effect a decrease in the phosphorescence lifetime and in the fluorescence quantum yield, but no substantial increase in ϕ_P/ϕ_F (compare the data for halonaphthalenes, Table 17). It is difficult to infer from these data that heavy-halogen substitution produces an increase in $S_1^* \longrightarrow T_1^*$ intersystem crossing. In fact the decrease in ϕ_F, without a corresponding increase in ϕ_P, suggests that heavy-halogen substitution effects an increase in the rate of $S_1 \longrightarrow S_0$ intersystem crossings, which effectively cancel, with no significant increase in phosphorescence intensity.

Heavy-atom effects are by no means limited to compounds containing halogen substituents; note the luminescence properties of caffeine and 6-thiocaffeine in Table 19. Finally, we note that there is now considerable

TABLE 19

Luminescence of Caffeine and Thiocaffeine[a]

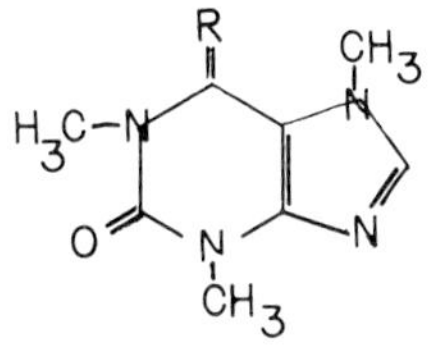

Compound	R	ϕ_F	ϕ_P	τ_P (sec)
Caffeine	O	0.22	0.14	1.9
Thiocaffeine	S	10^{-5}	0.43	0.024

[a]Data from Ref. [22].

evidence indicating that the relative importance of resonance to inductive substituent effects is much greater in excited states (especially singlet states) than in the ground state [23,24]. We summarize our consideration of substituent effects in Table 20, with the warning that this table must be treated only as a very general guide. Numerous exceptions to almost every entry in the table have been documented, particularly in the case of substituent groups that can interact strongly with the solvent.

D.   Heteroaromatics

Heteroatom substitution greatly affects the luminescence of aromatic compounds. The major reason for this is that most heteroatoms possess at least one "lone" pair of nonbonding (n) electrons. Absorption of radiation results in an $n \longrightarrow \pi^*$ transition; this $n \longrightarrow \pi^*$ transition in heterocyclic

TABLE 20

Effects of Substituents on the Fluorescence of Aromatics

Substituent	Effect on frequency of emission	Effect on intensity
Alkyl	None	Very slight increase or decrease
OH, OCH_3, OC_2H_5	Decrease	Increase
CO_2H	Decrease	Large decrease
NH_2, NHR, NR_2	Decrease	Increase
NO_2, NO	Large decrease	Large decrease
CN	None	Increase
SH	Decrease	Decrease
F Cl Br I $\downarrow$	Decrease	Decrease
SO_3H	None	None

compounds is responsible for many of the differences between their luminescence properties and those of aromatic hydrocarbons. Often the lowest energy
excited-singlet state in a heteroaromatic molecule is an (n, π^*) singlet. There
are several important properties of n $\longrightarrow$ π^* transitions and (n, π^*) excited
states that must be considered. First, in many aromatic molecules, excitation
of the first (π, π^*) singlet is strongly allowed ($\epsilon > 10^4$), whereas population of
(n, π^*) singlets is formally forbidden ($\epsilon \sim 10^2$). Consequently the radiative
lifetimes of (π, π^*) singlets tend to be considerably shorter than those of
(n, π^*) singlets, so that processes competing with fluorescence (radiationless
transitions, photochemistry, etc.) are likely to be of greater relative significance for molecules whose lowest singlets are (n, π^*). Second, energy
differences between the first excited singlet and the lowest triplet are frequently
much larger for (π, π^*) than for (n, π^*) excited states. The latter fact has two
important consequences. First, $S_1^* \longrightarrow T_1^*$ intersystem-crossing
probabilities tend to be inversely dependent on the energy separation between
the initial and final states. Therefore population of the triplet state by intersystem crossing from the singlet is more probable for heteroaromatic
molecules than for their hydrocarbon analogs, with the result that usually
greatly enhanced phosphorescence-to-fluorescence ratios are observed as a
result of the heteroatom substitution. Second, provided that the energy gap

between the lowest (n, π^*) and (π, π^*) singlets is not extremely large, a heteroaromatic whose lowest singlet is (n, π^*) will frequently have a (π, π^*) lowest triplet, phosphorescence from which may be quite similar in energy distribution and polarization to that observed in smaller yield from the parent hydrocarbon.

The arguments presented here, though oversimplified, have in the past been sufficient to enable rationalization of most of the available facts concerning heterocycle luminescence. Recent experiments and theoretical studies suggest, however, that the situation may be somewhat more complicated than indicated above. Virtually all meaningful experimental data presently available are concerned with nitrogen heterocyclics, and we accordingly restrict the discussion to this class of compounds. Many nitrogen heterocyclics are nonfluorescent, even for cases in which the fluorescence yields from the parent hydrocarbons are quite large. One often attempts to rationalize the lack of fluorescence in heterocyclics by invoking the argument that the radiative lifetime for $S_1^* \longrightarrow S_0$ radiative processes is so long that competing processes dissipate the energy before emission can occur. Although partially valid, this explanation neglects the fact that fluorescence is often noted from aromatic hydrocarbons, even when the radiative lifetime of S_1^* is relatively long. The usual nonfluorescence of nitrogen heterocyclics must therefore be due to greater efficiencies of radiationless processes (especially intersystem crossing) from (n, π^*) singlets than from (π, π^*) singlets, which would lead us to predict a greater efficiency of triplet-state formation in nitrogen heterocyclics.

Figure 63 shows the luminescence spectra of quinoline and naphthalene. The peaks at the left of each spectrum are phosphorescence; those at the right, fluorescence. For quinoline the phosphorescence is more efficient ($S_1^* \longrightarrow T_1^* \longrightarrow S_0$), whereas for naphthalene the fluorescence is more intense ($S_1^* \longrightarrow S_0$).

Theoretical treatments, however, suggest that knowledge of the energy gap between (π, π^*) or (n, π^*) singlets and triplets is insufficient to enable prediction of intersystem-crossing efficiencies. A detailed theoretical treatment of intersystem crossing in nitrogen heterocyclics has been reported by El-Sayed [25], who showed that direct spin-orbit coupling between singlet and triplet (n, π^*) or singlet and triplet (π, π^*) states, or vice versa, is much more efficient. El-Sayed's treatment therefore suggests that a prerequisite for efficient $S_1^* \longrightarrow T_1^*$ intersystem crossing in nitrogen heterocyclics is the availability of a triplet (π, π^*) state lower in energy than the lowest excited (n, π^*) singlet (case I, Fig. 64); in cases in which the lowest (π, π^*) triplet lies

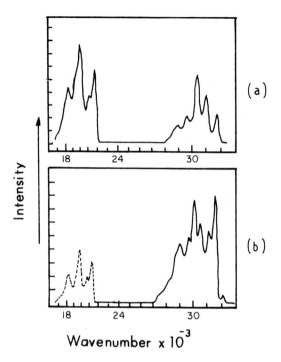

FIG. 63. Luminescence spectra of quinoline (top) and naphthalene (bottom). Reprinted from Ref. [25] by courtesy of the American Institute of Physics.

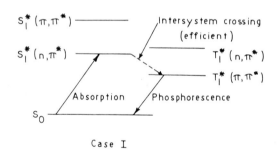

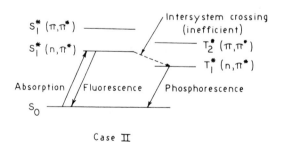

FIG. 64. Energy levels for a heteroaromatic molecule with a low-lying (n, π*) excited singlet. Depending on the relative energies of the lowest ^{1}(n, π*) and ^{3}(π, π*) excited states, π* ⟶ π phosphorescence (case I) or π* ⟶ n fluorescence (case II) may be observed.

above the first (n, π^*) singlet (case II, Fig. 64) intersystem crossing should be inefficient, and we may reasonably expect to observe $\pi^* \longrightarrow$ n fluorescence. In order to predict the emission behavior of a nitrogen heterocyclic, then, one wishes to know the relative energies of the lowest (π, π^*) triplet, which can in some cases be determined by the careful examination of phosphorescence excitation spectra [26]. It is also worth noting that, contrary to theoretical predictions [25], the rate constant for S_1^* (π, π^*) $\longrightarrow$ T_1^* (π, π^*) was slower by only a factor of about 3 than the supposedly strongly preferred S_1^* (n, π^*) $\longrightarrow$ T_1^* (π, π^*) intersystem-crossing process, in the specific case of 9,10-diazaphenanthrene as compared with the parent hydrocarbon [27].

That this viewpoint possesses some validity is demonstrated by the observation of $\pi^* \longrightarrow$ n fluorescence in several nitrogen heterocyclics, most notably the polyazines (heterocyclic analogs of benzene containing two or more nitrogen atoms). The isomeric diazines (14a) exhibit weak emission that has been characterized [28, 29] as $\pi^* \longrightarrow$ n fluorescence, and 1,2,4,5-tetrazine (14b) exhibits a rather intense $\pi^* \longrightarrow$ n fluorescence [30].

$$(14a) \qquad\qquad (14b)$$

Similar emission is also observed from a more complet nitrogen heterocyclic, 9,10-diazaphenanthrene. Some of the polyazines also exhibit $\pi^* \longrightarrow$ n phosphorescence, whose efficiency varies considerably from one compound to another [30].

The luminescence behavior of nitrogen heteroaromatics is significantly more complex than that of aromatic hydrocarbons and does not always follow the same general "rules." A good case in point is phthalazine (15), whose phosphorescence yield is strongly dependent on the wavelength of excitation.

$$(15)$$

This dependence is explicable in terms of rapid intersystem crossing from higher singlet states (S_2^* or S_3^*) to the triplet manifold, in violation of the cherished generalization that internal conversion from higher singlets to S_1^* is almost invariably more rapid than any competing process. Indeed Li and Lim [31] contend that intersystem crossing from higher singlets to the triplet

manifold may be a rather common phenomenon in nitrogen heterocyclics. It is
also worth noting that, in phthalazine, the phosphorescence quantum yields are
substantially smaller than the intersystem-crossing yields, implying that
$T_1^* \longrightarrow S_0$ intersystem crossing competes successfully with emission from T_1^*
[32].

It is further of interest that photochemistry plays an important role in the
deactivation of photoexcited nitrogen heteroaromatics. For example, pyridine
isomerizes to a transient "Dewar structure" on ultraviolet irradiation [33]:

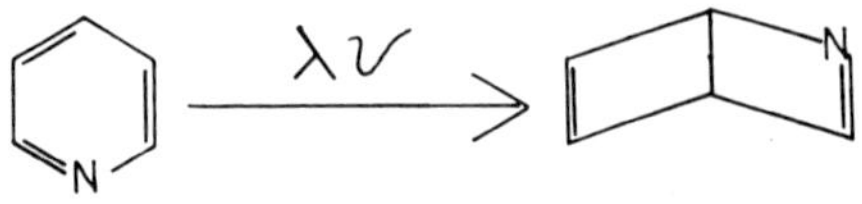

Similar behavior may occur in other heterocyclics and may therefore compete
rather effectively with luminescence in specific cases. Finally, it should be
obvious that, because heterocyclics are usually very basic, their fluorescence
and phosphorescence will in general be extraordinarily dependent on the solvent
(especially its acidity or hydrogen-bonding ability).

Generally there has been a lack of meaningful quantitative experimental
information on heteroaromatic compounds other than those containing nitrogen.
Fluorescence spectra have been reported for many heteroaromatic molecules,
but compound purity is often, at best, suspect, and corrected spectra or
quantum yields are scarce. Progress in the analytical utilization of heterocycle
fluorescence and phosphorescence will continue to be slow until a more complete
understanding of basic processes is attained, and this requires acquisition and
careful reporting of quantitative emission spectra.

E.   Aromatic Carbonyl Compounds

Most aromatic carbonyl compounds possess lowest-excited-singlet states
of (n, π^*) character, as do heterocyclics; intersystem crossing to the triplet
manifold is usually very efficient. For example, the rate constant for intersystem crossing in acetophenone and benzophenone is so large (1.5×10^{11} sec^{-1}
[34]) that the quantum yield for the formation of T_1^* is essentially unity [16].
Table 21 lists some intersystem-crossing effiencies in aromatic ketones.
Because of the efficiency of intersystem crossing, a large number of aromatic
aldehydes and ketones exhibit fairly intense phosphorescence, but no fluorescence. There are a few exceptions (e. g. , 9-fluorenone) in which fluorescence
has been observed [35], presumably because the lowest singlet is (π, π^*). Also,
in carefully deoxygenated solutions in aprotic solvents, some aromatic ketones

TABLE 21

Intersystem-Crossing Efficiencies in Aromatic Ketones

Compound	Φ_{isc}
Acetophenone	0.99
Benzophenone	1.00
Michler's ketone	1.01
2-Acetonaphthone	0.84
9-Fluorenone	0.93
Benzil	0.87
9,10-Anthraquinone	0.87

exhibit both fluorescence and thermally activated delayed fluorescence [36] in liquid solution [37].

Phosphorescence in aromatic carbonyl compounds may originate from either (π, π^*) or (n, π^*) triplets, whichever lies lower in energy. It is often relatively easy to ascertain which type of triplet is responsible for the observed phosphorescence of a carbonyl compound, for the intensities of phosphorescence from (n, π^*) triplets are enhanced slightly, if at all, by heavy-atom substitution in the molecule. In contrast, $\pi^* \longrightarrow \pi$ phosphorescence is subject to strong heavy-atom enhancement [38]. It has also been shown to be relatively easy to distinguish $\pi \longrightarrow \pi^*$ from $n \longrightarrow \pi^*$ singlet $\longrightarrow$ singlet transitions in the absorption spectra of aryl ketones, by noting the effect of oxime formation on the spectrum [35].

Aromatic ketones exhibit some interesting examples of intramolecular energy transfer. A classic case is 4-phenylbenzophenone ($\underline{16}$).

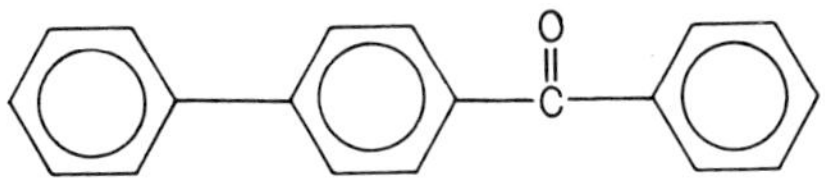

($\underline{16}$)

The absorption spectrum of this molecule very closely resembles that of benzophenone, but its phosphorescence bears a striking resemblance to that of biphenyl, rather than that of benzophenone [39]. Presumably this behavior is a manifestation of intermolecular energy transfer from the carbonyl (n, π^*) triplet to the lower biphenyl-like (π, π^*) triplet. Similar effects may be anticipated whenever a carbonyl group is conjugated with an aromatic system whose (π, π^*) triplet is appreciably lower in energy than the carbonyl (n, π^*) triplet.

A very striking example of intramolecular-energy-transfer processes is afforded by the compound series

$$n = 1-3$$

in which the benzophenone is separated from the alkylnaphthalene by an aliphatic chain. The absorption spectra of these compounds are essentially super-impositions of 4-methylbenzophenone and 1-methylnaphthalene absorptions. The absorptions do not completely overlap, and hence the naphthalene moiety can be selectively excited by properly chosen incident wavelength. When this is done, efficient intramolecular energy transfer occurs from the naphthalene (π, π^*) singlet to the lowerlying carbonyl (n, π^*) singlet, which then undergoes efficient intersystem crossing to the benzophenone triplet. But the naphthalene (π, π^*) triplet lies below the benzophenone (n, π^*) triplet; hence triplet energy is transferred back to the naphthalene system, from which characteristic 1-alkyl-naphthalene phosphorescence is observed [40], with an appreciably greater quantum yield than that from 1-methylnaphthalene itself. Such back-and-forth energy flow is made possible by two important factors. First, as for hetero-cyclics, intersystem crossing between an (n, π^*) singlet and a (π, π^*) triplet, or vice versa, is much more efficient than that between two (n, π^*) or (π, π^*) states [25]. Second, energy differences between (π, π^*) singlets and triplets are much larger than those for (n, π^*) singlets and triplets. Hence the naphthalene singlet lies above the lowest benzophenone (n, π^*) singlet, but the naphthalene triplet lies well below the benzophenone (n, π^*) triplet.

We have already mentioned that aliphatic aldehydes and ketones frequently fluoresce, albeit weakly, in solution. This is probably the only general class of organic compounds in which aliphatic members fluoresce more strongly than a majority of their aromatic analogs. This fact is not difficult to understand, for the lowest (π, π^*) triplet in, for example, acetone lies much higher in energy than the (n, π^*) singlet, so that intersystem crossing can occur only to the (n, π^*) triplet, a process that is, as we have noted, relatively inefficient. Hence weak $\pi^* \longrightarrow n$ fluorescence can be observed from many aliphatic aldehydes and ketones; in fact the principal process competing with fluorescence is usually photochemical reaction, rather than intersystem crossing. Some aliphatic carbonyl compounds, especially biacetyl, $CH_3COCOCH_3$, exhibit the unusual property of phosphorescence in liquid solution. However, as the chain lengths of alkyl substituents on diketones is increased, phosphorescence yields in solution rapidly decrease, presumably due to increased vibrational dissipation of singlet and triplet energy [41].

The luminescence properties of carboxylic acids are also worth briefly noting. It has been observed that both the -COOH and -COOR (ester) substituents repress fluorescence from aromatic compounds, whereas the dissociated carboxyl group, $-COO^-$, has little effect on fluorescence [42]. The presence of the -COOH or -COOR group evidently increases the rate of singlet → triplet intersystem crossing in aromatic systems; for example, for toluene, $\phi_p / \phi_F = 0.96$, but for toluenecarboxylic acids, $\phi_p / \phi_F = 2$ [42]. Rather interestingly, in carboxylate salts the luminescence behavior appears to be dependent on the cation [43], indicating that there is considerable mixing between ring and $-COO^-$ (π, π^*) excited states.

III. ENVIRONMENTAL EFFECTS

Environmental factors can strongly influence the fluorescence of polyatomic molecules. The molecular environment constitutes an important parameter that can be used by the analyst to increase the sensitivity and selectivity of fluorometry. A large number of environmental effects are of importance. We shall only discuss a few of these: nature of solvent, pH, heavy atoms, metal ions, oxygen, temperature, and presence of other solutes. It cannot be too strongly emphasized that pure solvents must be used in fluorometry to obtain good results. It is not generally sufficient to demonstrate that the solvent does not itself fluoresce since nonluminescent impurities can act as quenchers.

A. Solvent Effects

Electronic transitions occur at rates rapid relative to the rates of inter-nuclear motion in molecules. Hence, during an electronic transition (absorption or emission), the nuclei remain essentially stationary (the Franck-Condon principle). Accordingly, when a molecule in its ground state absorbs a photon, it finds itself in a metastable excited state ("Franck-Condon" excited state), in which the molecular geometry and solvent configuration are those characteristic of the ground state. Solvent reorientation then occurs approximately 10^{-11} to 10^{-12} sec after excitation, producing an "equilibrium" excited state, in which the solvent configuration is optimal for the geometry and electron distribution of the molecule. Emission occurs from the equilibrium excited state, forming a Franck-Condon ground state. Solvent relaxation then occurs, forming the equilibrium ground state. This process is schematically represented by Eqs. (19), and the energy relationships of the various equilibrium and Franck-Condon states are represented in Fig. 65. It is worth noting that, because electronic excitation can produce drastic changes in both molecular geometry and electronic charge distribution, the energy differences between equilibrium and Franck-Condon excited states can be quite large in some cases.

$$(A)_{eq} + h\nu \longrightarrow (A^*)_{FC}, \tag{19a}$$

$$(A^*)_{FC} \longrightarrow (A^*)_{eq}, \tag{19b}$$

$$(A^*)_{eq} \longrightarrow (A)_{FC} + h\nu, \tag{19c}$$

$$(A)_{FC} \longrightarrow (A)_{eq}. \tag{19d}$$

Because solvent-relaxation phenomena occur so rapidly, they have, until very recently, not been studied directly. The advent of picosecond-duration laser-flash spectroscopy systems [44] means that such rapid phenomena are now amenable to experimental observation. The laser-flash technique has been applied to excited-state solvent-relaxation phenomena by Ware and co-workers [45] in a series of elegant experiments; we shall refer to the results of some of these studies below.

It is evident that the ground and excited states involved in absorption and fluorescence are "different". Accordingly there is no reason to expect precise

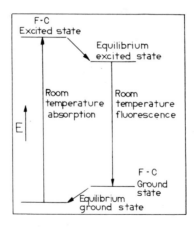

FIG. 65. Schematic representation of equilibrium and Franck-Condon (F-C) electronic states.

correspondence between absorption and fluorescence solvent effects. Any single solution absorption or fluorescence spectrum involves either a ground or an excited state that is not even approximately in equilibrium with its solvent cage. Accordingly great care must be exercised in the interpretation of "single" spectra, either absorption or emission [46, 47].

In most polar molecules the excited state is more polar than the ground state. Hence an increase in the polarity of the solvent produces a greater stabilization of the excited state than of the ground state. Consequently a shift in both absorbance and fluorescence spectra to lower energy, or longer wave-length (red shift), is usually observed as the dielectric constant of the solvent increases (Fig. 66 and Table 22).

Note in Table 22 that both the absorption and fluorescence spectra shift to lower frequencies as the solvent dielectric constant increases, but that the magnitude of the shift is substantially larger in fluorescence than in absorption. Note also that such correlations are not infallible, as indicated by the anomalous position of methanol in Table 22.

The direction of the red shift is general, but its magnitude depends on the specific nature of the solute-solvent interaction. Several important types are the following:

1. Dipole-dipole interactions between solute and solvent, both being polar.

2. Interactions between a solute possessing a permanent dipole and dipoles induced in the solvent, in the case where the former is polar but the latter is not.

3. Interactions between solvents that possess a permanent dipole and

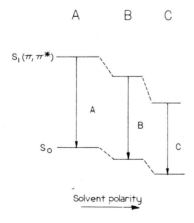

FIG. 66. Schematic representation of the red shift in $\pi^* \longrightarrow \pi$ fluorescence spectra as the polarity of the solvent is increased. Solvent A is less polar than solvent B, which is in turn less polar than solvent C.

TABLE 22

Electrostatic Solvent Effects on Anthracene Fluorescence

Solvent	Dielectric constant at 25° C	$\bar{\nu}_F$ (cm^{-1})	λ_F (nm)	$\bar{\nu}_F - \bar{\nu}_A$ (cm^{-1})[a]
(Vapor)	--	27380	365	180
Hexane	1.9	26520	377	156
Methanol	32.6	26460	378	177
Dioxane	2.2	26220	381	280
Toluene	2.4	26170	382	228
Benzene	2.3	26120	383	262
Chlorobenzene	5.7	26060	384	279
Acetonitrile	38.0	25970	385	285
Formamide	109.5	25510	392	347
N-Methylformamide	182.4	25110	398	368

[a] $\bar{\nu}_F$ is the low-frequency maximum in the anthracene fluorescence spectrum; $\bar{\nu}_A$ is the low-frequency maximum in the absorption spectrum.

dipoles thereby induced in the solute, in the case where the solvent is polar but the solute is not.

4. Interactions between the transition dipole of the solute and the dipole induced thereby in the solvent--when neither possesses a permanent dipole.

The fluorescence spectrum will be influenced by one or more of these categories; the general effect is termed an "electrostatic solvent effect".

In the case of solute-solvent pairs in which neither is appreciably polar, a red shift from the vapor spectrum is still observed. This shift is caused by a dispersive interaction resulting from the fact that electronic transitions produce changes in the electron densities of solute molecules. Even if both solvent and solute are nonpolar in the initial and final states, the occurrence of an electronic transition in a solute requires a finite transition dipole which polarizes the surrounding solvent shell. This "polarization shift" governs the magnitude of the red shift observed in going from gas-phase to solution spectra [48].

The fluorescence intensities of aromatic compounds can also be affected by electrostatic solvent effects. These effects are generally insignificant if both solute and solvent are nonpolar. For polar solute-solvent pairs electrostatic intensity perturbations are minor relative to those produced by specific short-range interactions (complex formation, hydrogen bonding, etc.).

B. Effect of Heavy-Atom Solvents

We have already noted that heavy-atom substituents tend to enhance phosphorescence at the expense of fluorescence in aromatic molecules; this effect is often observed even if the heavy atom is not a constituent of the luminescent molecule. For example, the value of ϕ_P / ϕ_F for naphthalene in an alcoholic glass containing 10% n-propyl bromide is larger by a factor of 48 than the value in the pure alcoholic matrix [49]. Heavy-atom solvents usually induce a significant decrease in ϕ_F and a concomitant increase in the efficiency of $S_1^* \rightarrow T_1^*$ intersystem crossing [50]. The heavy-atom effect increases the rates of both $T_1^* \rightarrow S_0$ intersystem crossing and $T_1^* \rightarrow S_0$ phosphorescence, but the effect on the latter is usually greater [51], so that the net effect is often an increase in ϕ_P.

There is some uncertainty regarding the actual processes responsible for external-heavy-atom effects [52]. In the case of halogen-containing perturbers (such as haloalkanes [53] or alkali halide salts [54], there is strong evidence that a 1:1 complex is formed between an excited state of the solute and the heavy-atom species. The extent of spin-orbit coupling is much greater in this "exciplex" than in the unperturbed solute. The exciplexes appear to be strongly of a charge-transfer nature, and we shall see that excited-state-complex formation is increasingly being recognized as a general fluorescence-quenching process.

One cannot understand all heavy-atom effects in terms of exciplex formation. For example, the emission spectra of naphthalene and phenanthrene, as determined in solid rare-gas matrices (argon, krypton, and xenon), all exhibited an increase in ϕ_P/ϕ_F and a shortening of the phosphorescence-decay time as the atomic number of the matrix species increased [55]. These observations cannot logically be interpreted as due to excited-state-complex formation; such phenomena have instead been discussed in terms of exchange interactions between the solute and the matrix [56].

There exist several unusual types of heavy-atom perturbations that cannot now be satisfactorily rationalized. For example, it is noted that the value of ϕ_P / ϕ_F for naphthalene in cyclohexane glasses at 77°K increases as ethyl iodide is added to the glass, so long as the quantity of ethyl iodide added is fairly small. When the glass contains more than about 10% ethyl iodide, however, further addition of ethyl iodide produces a sharp decrease in ϕ_P/ϕ_F [57]. There is also some evidence that heavy atoms occasionally decrease the rates for intersystem-crossing processes in organic molecules [58]. These phenomena that we categorize as "heavy-atom effects" are complex and comprise several fundamentally different types of molecular interactions.

The influence of ethyl iodide on the phosphorescence intensities of several solutes is shown in Table 23, from which it is obvious that, in some cases, the heavy-atom effect produces a striking increase in phosphorescence efficiency, which can be exploited in analytical applications of phosphorometry [59]. Heavy-atom solvents must, of course, usually be avoided in high-sensitivity fluorometry. One example in which the external heavy-atom effect was useful in fluorometry is the analysis of a mixture of polycyclic hydrocarbons (3,4-benzpyrene, tetracene, and perylene). Fluorescence of the first two compounds, but not of perylene, is quenched by CH_3I. Thus in a mixture of these three similar compounds the interference of 3,4-benzpyrene and tetracene in the fluorometric determination of pyrene can be suppressed by adding CH_3I to the sample [60].

TABLE 23

External-Heavy-Atom Effect on Phosphorescence Intensity[a]

Compound	Concentration (μg/ml)	(P/P_0)[b]
Naphthalene	13	0.1
Anthracene	18	0.9
Triphenylene	2.3	0.05
Naphthacene	6.0	4.6
1,2-Benzanthracene	23	1.4
3,4-Benzpyrene	25	3.5
1,2-Benzfluorene	22	13
2,3-Benzfluorene	22	25

[a] From Ref. [59].

[b] Ratio of phosphorescence intensity in ethanol-ethyl iodide (5:1, v/v) to that in ethanol.

C. Effects of pH

The fluorescence spectra of most aromatic compounds containing acidic or basic functional groups are very sensitive to the pH and hydrogen-bonding ability of the solvent. Most proton-transfer reactions in polar solvents are very fast, such that Brønsted acid-base reactions can occur during the lifetime of an excited-singlet state of an aromatic molecule. The acid-base properties of 2-naphthol serve as an excellent example of excited-state acid-base chemistry. The 2-naphthol molecule exhibits a single broad fluorescence peak in aqueous solution at 359 nm, whereas the 2-naphtholate anion exhibits a fluorescence peak at 429 nm. The large energy separation of these fluorescence spectra makes it easy to determine the occurrence of acid-base reactions in the excited state. The pK_a of the ground state of 2-naphthol is about 9.5. Let us therefore assume that we are measuring the fluorescence of 2-naphthol in a medium of pH $<$ 9.5, such that the neutral molecular form predominates in the ground state. We will observe the fluorescence of the anionic form of 2-naphthol at pH values much smaller than 9.5, indicating that the electronically excited 2-naphthol molecule is a substantially stronger acid in its first excited singlet state than it is in the ground state. The observation of fluorescence from 2-naphtholate in solutions of pH $<$ 9.5 also indicates that the excited-state proton-transfer reaction is rapid relative to the rate of decay of the 2-naphthol singlet.

In the pH range in which neutral 2-naphthol is the predominant protolytic form in the ground state, we may represent the overall excitation-reaction sequence as follows:

$$C_{10}H_7OH + h\nu \xrightarrow{1} C_{10}H_7OH^* \xrightarrow[+H_2O]{4} C_{10}H_7O^{-*} + H_3O^+ \qquad (20)$$

The steps in reaction sequence (20) can be explained as follows:

1. Absorption of radiant energy to produce the excited molecule. Remember the pH is less than 9.5, and so neutral 2-naphthol predominates.

2. Deactivation of the excited 2-naphthol molecule by molecular fluorescence (359 nm).

3. Radiationless deactivation of the excited neutral molecule.

4. Dissociation of the excited molecule, producing a proton and an excited anion.

5. Deactivation of the naphtholate anion by fluorescence (429 nm).

6. Radiationless deactivation of the excited anion.

By measuring the relative fluorescence intensities for the neutral molecule and the anion as a function of pH, it can be determined that the pK_a for

2-naphthol is about 3.1; that is, the excited singlet exhibits an acid strength that is more than 10^6 times greater than that of the ground state. Techniques for measuring pK_a values for electronically excited molecules have been described and compared by Vander Donckt [61] and by Schulman [62]. In the case of 2-naphthol it is clear that the pH of the solution must be less than 3.5 before the molecular form predominates over the anionic form in the fluorescence spectrum. Since the fluorescence quantum efficiency of neutral 2-naphthol is substantially larger than that of the naphtholate anion, it is clear that, to obtain maximum sensitivity in fluorometric analyses of 2-naphthol, the pH must be not less than 3.5. Knowledge of the equilibrium constants for excited-state protolysis, coupled with knowledge of the relative fluorescence efficiencies of the two protic forms, can thus be of great value in enhancing the sensitivity of fluorometric analyses of solutes containing dissociable functional groups [62].

Brønsted acidity differences between the ground and the lowest excited singlet states of organic molecules are large, commonly ranging from 4 to 9 pK units. Some compound classes, especially phenols, thiols, and aromatic amines, become much stronger acids on excitation, whereas others (nitrogen and sulfur heterocyclics, carboxylic acids, aldehydes, and ketones with lowest (π, π^*) singlets become much more basic (Table 24). For most organic

TABLE 24

Excited-State Acidities for Some Aromatic Compounds[a]

Compound	pK_a		
	Ground	Singlet[b]	Triplet[c]
Phenol	10.00	4.0	8.5
4-Methoxyphenol	10.21	5.6	8.6
2-Naphthol	9.5	3.1	7.7
1-Naphthoic acid	3.7	~11	4.6
2-Naphthoic acid	4.2	~11	4.2
Quinolinium ion	5.1	10.5	5.8
Acridinium ion	5.5	10.6	5.6
2-Naphthylammonium ion	4.1	~ -2	3.1

[a] Data from Jackson and Porter [63], Weller [64], and Wehry and Rogers [65].

[b] First excited-singlet state.

[c] Lowest triplet state.

compounds changes in phosphorescence spectra with pH rather closely resemble those observed in absorption, since differences between ground-state and lowest-triplet acidities are usually less than 2 pK units [63].

The nature of excited-state acid-base chemistry in compounds containing more than one acidic or basic functional group is especially interesting. Consider, for example, methyl salicylate (17).

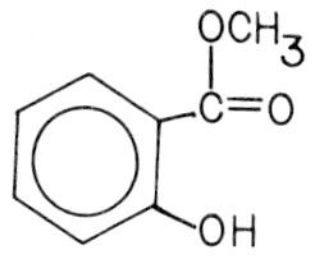

(17)

In the lowest-excited-singlet state of this compound the hydroxyl group greatly increases in acidity, and the ester carbonyl greatly increases in basicity, relative to the ground state. Accordingly, during the lifetime of the excited state, an intramolecular acid-base equilibrium is established [64]:

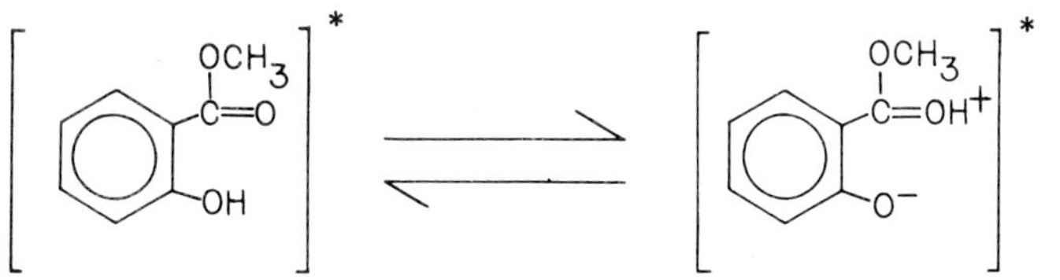

Rather surprisingly, this is one of the very few intramolecular excited-state proton transfers that has ever been firmly authenticated. It has been shown that in the closely related 3-hydroxy-2-naphthoic acid series an intramolecular proton-transfer equilibrium is not established during the excited-singlet lifetime [66]. There have also been a number of fluorometric studies of bifunctional nitrogen heterocyclics (e. g. , hydroxyquinolines [67,68], aminoquinolines [69]) and other compounds (e. g. , aminonaphthols [70]) in which there is some evidence for intramolecular proton transfers, but the problem is complicated by strong hydrogen-bonding interactions with the solvent (see Section III. D).

One may use pH as a parameter in fluorometric analysis to reduce interference by extraneous solutes in a mixture or to obtain the most strongly fluorescent species for analysis. Since excited-state pK_a values and proton-transfer rates are, in some cases, rather sensitive to changes in molecular structure [58,59,61] analytical utilization of excited-state proton-transfer reactions may assist in increasing the selectivity and versatility of analytical procedures.

D. Effects of Hydrogen Bonding

Hydrogen-bonding interactions of substituted aromatic molecules with the solvent or with other solutes can greatly affect their fluorescence behavior. The effects of hydrogen bonding on the fluorescence of organic molecules have recently been reviewed [71]; we consider some of the more important aspects here.

We assume that a solute molecule A can hydrogen-bond with a molecule or solvent (or other solute) B. Formation of an excited-state hydrogen-bonded complex between the two can occur by excitation of such a species already present in the ground state:

$$A\text{-}B \xrightarrow{\ h\nu\ } (A\text{-}B)^{*}. \tag{21}$$

Alternatively, during the lifetime of an uncomplexed, excited A^{*} molecule hydrogen bonding with B can occur:

$$A \xrightarrow{\ h\nu\ } A^{*}; A^{*} + B \longrightarrow (A\text{-}B)^{*} \tag{22}$$

In the former case it is obvious that both the absorption and fluorescence spectra of A will be affected by hydrogen bonding with B. However, when hydrogen bonding takes place only after excitation, only the fluorescence spectrum of A will be perturbed by the interaction. Thus, by carefully comparing the effect of B on the absorption and fluorescence spectra of A, one can compute the relative abilities of A and A^{*} to hydrogen-bond with B. In the specific case of 3- and 4-aminophthalimide in n-propanol (solvent) formation of the hydrogen-bonded "exciplex" $(A\text{-}B)^{*}$ occurred so rapidly that the process could not be followed by nanosecond flash spectroscopy [45].

It is often (but not always) observed that excited-state hydrogen bonding reduces the ϕ_F of A. An interesting example of this effect is 5-hydroxyquinoline. In hydrogen-bond-accepting solvents (acetonitrile, dioxane, DMSO, ether, etc.) it is found [67] that, as the enthalpy of formation of hydrogen bonds between a phenol and the solvent increases, ϕ_F decreases in room-temperature solutions (Table 25).

Unlike most nitrogen heterocyclics, 5-hydroxyquinoline exhibits very inefficient $S_1^{*} \longrightarrow T_1^{*}$ intersystem crossing in any solvent. The influence of solvent on ϕ_F for this compound must therefore be attributed to increased $S_1^{*} \longrightarrow S_0$ internal conversion induced by hydrogen bonding with the solvent [67]. Such observations have been noted in other hydrogen-bonding systems [73, 74]. It should be noted that enhanced internal conversion, effected by excited-state hydrogen bonding, reduces both ϕ_F and ϕ_P. Thus, as a general

TABLE 25

Fluorescence Yield for 5-Hydroxyquinoline in
Various Solvents[a]

Solvent	$-\Delta H$[b] (kcal/mole)	ϕ_F at 298°K	ϕ_F at 77°K
Isopentane	--	0.30	0.27
Acetonitrile	3.5	0.24	--
Sulfolane	3.5	0.21	--
Dioxane	4.4	0.19	--
Diethyl ether	5.1	0.12	0.24
Dimethylformamide	5.3	0.09	--
Tetrahydrofuran	5.5	0.09	--
Dimethyl sulfoxide	6.4	0.07	0.20

[a]From Ref. [67].

[b]Enthalpy of hydrogen-bond formation between phenol and solvent; data from Ref. [72].

rule, the analyst performing fluorometric or phosphorometric analyses of molecules containing such functional groups as $-OH$, $-CO_2H$, $-NR_2$, or $-SH$ should, when feasible, choose solvents that will not hydrogen-bond strongly with the substituent.

The general tendency for intermolecular hydrogen bonding to reduce fluorescence intensities can sometimes be put to use. For example, it is well known that hydrogen bonding of phenols with carboxylate anions (e.g., acetate) causes total attenuation of phenol fluorescence. As we have already noted, formation of a hydrogen-bonded complex between excited phenol and carboxylate can occur either before or after excitation of the phenol. In the case where the complex is formed only after the phenol is excited, the effect of carboxylate on the fluorescence of phenol is equivalent to collisional fluorescence quenching. As such, it should be accurately described, at least at moderate concentrations of "quencher" (carboxylate), by the Stern-Volmer quenching equation. This equation will not be valid if the carboxylate hydrogen-bonds with the ground state of phenol ("static quenching"). Thus, if one chooses a suitable non-hydrogen-bonding solvent and measures the effect of carboxylate concentration on the fluorescence efficiency of phenol, the extent to which the actual data deviate from ideal Stern-Volmer behavior can be employed to compute the equilibrium constant for the hydrogen bonding of carboxylate ions with the ground state of phenol [75].

The influence of hydrogen bonding on the fluorescence of organic molecules is not limited to interactions with the solvent. As already noted, the fluorescence yield of 5-hydroxyquinoline can be correlated with the ability of the solvent to act as a hydrogen-bond acceptor. It therefore might seem reasonable to expect similar correlations to obtain for 8-hydroxyquinoline, particularly since the absorption spectra of 5- and 8-hydroxyquinoline are quite similar.

However, as noted in Table 26, such a correlation is actually not observed for 8-hydroxyquinoline [67].

TABLE 26

Fluorescence Yield for 8-Hydroxyquinoline in
Various Solvents[a]

Solvent	$-\Delta H$[b] (kcal/mole)	ϕ_F at $298^\circ K$	ϕ_F at $77^\circ K$
Isopentane	--	0.0002	0.0002
Acetonitrile	3.5	0.003	--
Sulfolane	3.5	0.005	--
Dioxane	4.4	0.001	--
Diethyl ether	5.1	0.0002	0.07
Dimethylformamide	5.3	0.003	--
Tetrahydrofuran	5.5	0.002	--
Dimethyl sulfoxide	6.4	0.002	0.11

[a]From Ref. [67].

[b]Enthalpy of hydrogen-bond formation between phenol and solvent; data from Ref. [72].

We note first that ϕ_F for 8-hydroxyquinoline is generally about 100 times smaller than that for 5-hydroxyquinoline in the same solvent, despite the fact that the absorption spectra of the two compounds are almost identical. Second, there is no obvious correlation between the hydrogen-bond-accepting ability of a solvent and the fluorescence yield of 8-hydroxyquinoline in that solvent. To understand these differences we must note that 8-hydroxyquinoline can engage in intramolecular and intermolecular hydrogen bonding, whereas only the latter is possible for 5-hydroxyquinoline. One may conclude [67] that both intermolecular and intramolecular hydrogen bonding affects the fluorescence of 8-hydroxyquinoline by increasing the rate constant for S_1^* $\longrightarrow$ S_0 internal conversion; this effect has also been noted in other systems exhibiting intra-molecular hydrogen bonding [76, 77]. In a molecule like 8-hydroxyquinoline competition between intramolecular and intermolecular hydrogen bonding can have important effects. In a solvent that is sufficiently effective as a hydrogen bonder to disrupt the intramolecular hydrogen bond, but does not form an extremely strong intermolecular hydrogen bond with the phenolic hydroxyl, the fluorescence efficiency of a molecule like 8-hydroxyquinoline should be maximal. Thus, whenever fluorescence or phosphorescence assays are designed for a molecule capable of intramolecular hydrogen bonding, such competition should be taken into consideration in defining solvent and tempera-ture conditions. It is also important to note that proton-transfer reactions (either intermolecular or intramolecular) can also occur in excited states of bifunctional compounds like 8-hydroxyquinoline [68].

The general effects of excited-state hydrogen bonding on the luminescence of nitrogen heterocyclics are dependent on whether the lowest excited singlet is (n, π^*) or (π, π^*). It is normally observed that the excited (π, π^*) singlets of nitrogen heterocyclics are much more strongly basic than the ground state; one therefore expects the excited state to hydrogen-bond more strongly with protic solvents than the ground state. Thus for a nitrogen heterocyclic with a low-lying (π, π^*) excited singlet the frequencies of both absorption and fluorescence should shift to lower frequency as the hydrogen-bond-donating ability of the solvent increases, but the magnitude of the fluorescence shifts should be greater than those in absorption. In contrast (n, π^*) excited singlets of nitrogen heterocyclics are obviously much less basic than the ground state and should therefore be less susceptible to hydrogen-bonding interactions. Thus, while energies of n $\longrightarrow$ π^* absorption spectra should increase dramatically with increasing hydrogen-bond-donor power of the solvent, π^* $\longrightarrow$ n fluorescence spectra should be virtually insensitive to this property of the solvent (Fig. 67). A very interesting demonstration of the insensitivity of π^* $\longrightarrow$ n fluorescence to hydrogen-bonding effects has been presented for various diazines (see Table 27). These solvent shifts are the basis for the classical test for distinguishing π $\longrightarrow$ π^* from n $\longrightarrow$ π^* transitions in hetero-cyclics and carbonyl compounds [1].

It is often observed that aromatic carbonyl compounds and nitrogen heterocyclics fluoresce very weakly, or not at all, in nonpolar, aprotic sol-vents, but that their fluorescence yields increase sharply on the addition of hydrogen-bonding solvents. In such compounds the lowest excited state is usually (n, π^*) in aprotic solvents, but the lowest (π, π^*) singlet often is not

A   B   C

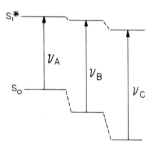

Increasing solvent hydrogen-bond donation

FIG. 67. Schematic representation of the blue shift in n $\longrightarrow$ π^* absorption spectrum as the hydrogen-bonding power of the solvent is increased. Hydrogen-donating power increases in the order A$<$B$<$C.

very much more energetic than the (n, π^*) singlet. On increasing the "proticity" of the solvent, the (π, π^*) states shift to lower energies, relative to the ground state, while the energies of (n, π^*) singlets increase as noted in the preceding paragraph. Therefore it is possible for the opposing energy shifts (n, π^*) and (π, π^*) singlets , on addition of polar, hydrogen-bonding solvents,

to be sufficiently large for a molecule whose lowest singlet is (n, π^*) in nonpolar media to be (π, π^*) in hydroxylic solvents (see Fig. 68). Since fluorescence from (π, π^*) singlets is usually more efficient than that from (n, π^*) singlets, addition of hydroxylic solvents often produces increased fluorescence from hydrocarbon solutions of aldehydes, ketones, and heterocyclics [78, 80]. For example, addition of n-propanol to glassy solutions of 3, 4-benzquinoline enhances ϕ_F by a factor of as much as 35 while simultaneously effecting a significant decrease in ϕ_P and an increase in the phosphorescence radiative lifetime [79]. Such observations are often noted for heterocyclics in which the lowest (n, π^*) and (π, π^*) singlets do not interchange as the proticity of the solvent is increased; the origins of these observations have been discussed in detail by Lim and Yu [81].

E.   Effects of Other Solutes

Fluorescence is not usually considered applicable to the analysis of very complex samples unless the sample constituents are separated first. The fluorescence of an organic molecule can be affected by interactions not only with the solvent but also with other solutes; we consider some important

TABLE 27

Solvent Effects on the Absorption[a] and Fluorescence[b]
Spectra of Diazines[c]

	Solvent				
Compound	Isooctane	Ether	Acetonitrile	Methanol	Water
Pyridazine:					
ν_A	29,740	30,150	31,080	32,360	33,580
$(\Delta \nu)_A$	--	410	1,340	2,620	3,840
ν_F	23,530	23,810	24,070	24,130	24,240
$(\Delta \nu)_F$	--	280	540	600	710
Pyrimidine:					
ν_A	34,250	34,420	34,840	35,710	36,900
$(\Delta \nu)_A$	--	170	590	1,460	2,650
ν_F	26,950	26,950	26,850	26,810	26,320
$(\Delta \nu)_F$	--	0	- 100	- 140	- 630
Pyrazine:					
ν_A	31,620	31,620	31,750	32,260	33,170
$(\Delta \nu)_A$	--	0	130	640	1,550
ν_F	29,110	29,070	28,940	28,820	28,780
$(\Delta \nu)_F$	--	- 40	- 170	- 290	- 330

[a]Absorption frequency ν_A (cm^{-1}) and frequency shift $(\Delta \nu)_A$ (cm^{-1}) relative to isooctane.

[b]Fluorescence frequency ν_F (cm^{-1}) and frequency shift $(\Delta \nu)_F$ (cm^{-1}) relative to isooctane.

[c]Data from Ref. [78].

solute-solute interactions (most of which bring about fluorescence quenching) in this section.

1. Fluorescence Quenching by Electronic Energy Transfer

If a fluorescent solute A is present in a sample containing one or more other compounds, Q, having excited-singlet states lower in energy than the first excited singlet of A, electronic energy transfer to that solute may occur:

$$^1A^* + Q \longrightarrow A + {}^1Q^*. \tag{23}$$

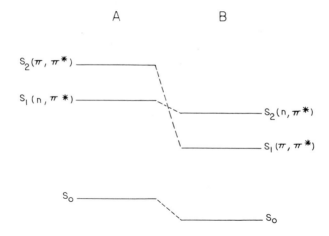

FIG. 68. Schematic representation of the interchange of lowest (n, π*) and (π, π*) excited-singlet states in going from a hydrocarbon (A) to a hydroxylic (B) solvent.

Clearly this process quenches the fluorescence of A. At least three distinct mechanisms for singlet-singlet energy transfer can be delineated [82].

a. Absorption of $^1A^*$ Fluorescence by Q ("Inner-Filter" Effect). This process is analytically significant only if (a) the absorption spectrum of Q overlaps the fluorescence spectrum of A; and (b) the concentration of Q is sufficiently large for its absorbance to be at least 0.05 in the region of spectral overlap. Assuming that A and Q are present in the analytical sample in comparable quantities, the apparent quenching by Q can be suppressed simply by so diluting the solution that the absorbance of Q is very small. If A is a minor constituent of the sample relative to Q, there is usually no alternative to separation of A from Q prior to measurement of fluorescence.

b. Collisional Nonradiative Transfer. In some cases nonradiative energy transfer from $^1A^*$ to Q requires collisional interaction of the two species. Such processes usually proceed at or near the diffusion-controlled rate, which depends strongly on the viscosity of the medium [5]. Collisional quenching usually obeys the Stern-Volmer equation,

$$\frac{\phi_F^o}{\phi_F} - 1 = k_Q \, \tau_F^o [Q], \tag{24}$$

where ϕ_F^o is the fluorescence quantum yield in the absence of Q; ϕ_F is the fluorescence quantum yield in the presence of a specified concentration [Q] of quencher; k_Q is the bimolecular quenching-rate constant (normally about $10^8 \, M^{-1} \, sec^{-1}$); and τ_F^o is the fluorescence lifetime for $^1A^*$ in the absence

of Q. Clearly diffusion-controlled quenching can be alleviated by increasing the
solvent viscosity or by decreasing [Q]. Collisional quenching can usually be
rendered negligible by so diluting the solution that no solute is present at a
concentration greater than 10^{-3} M [5]. Again, only if Q is present in a
significantly greater quantity than A in the original sample will this strategy
fail.

    c. Long-Range Nonradiative Transfer. Singlet-singlet energy transfer,
Eq. (23), does not require collisional interaction of $^1A^*$ with Q, provided that
two conditions are fulfilled: (a) the absorption spectrum of Q must partially
overlap the fluorescence spectrum of $^1A^*$ and (b) the absorption transition
$Q \longrightarrow {}^1Q^*$ in the quencher must be intense [82]. If these conditions are
satisfied, energy transfer from $^1A^*$ to Q may cause significant quenching of the
former over average separations of the two species of 50 $\overset{\circ}{A}$ ·or more [83].
This form of energy-transfer quenching is a significant analytical nuisance
because it is difficult to eliminate simply by diluting the sample solution. It
may in fact be necessary to reduce the concentration of Q to 10^{-4} M or less
before its quenching action ceases to be significant [5]. Because the conditions
for long-range energy transfer are often satisfied in complex mixtures of
fluorescent solutes, it is often necessary to perform rather extensive separations prior to fluorescence measurement. The analysis of complex mixtures of
polynuclear aromatic hydrocarbons adsorbed on airborne particulates [6] is an
excellent example of an analytical system in which energy-transfer quenching
can be a serious problem, often necessitating chemical separation prior to
analysis.

    Although the usual effect of energy transfer is analytically detrimental,
there are situations in which it can be exploited for analytical benefit. Clearly
energy transfer provides a technique for forming $^1Q^*$ without direct excitation.
In some cases an analyte Q either absorbs very weakly or else absorbs in an
inconvenient spectral region. For example, many rare-earth ions luminesce in
solution. However, their absorption bands are weak and very narrow.
Efficient direct excitation is therefore difficult to accomplish. However,
excited-states of rare-earth ions can be produced by energy transfer from
excited triplet states of aromatic carbonyl compounds (A) [84]:

$$^3A^* + {}^5Tb^{3+} \longrightarrow {}^1A + {}^7Tb^{3+}{}_* .\tag{25}$$

Note that no change in overall multiplicity is produced by this energy-transfer
process. Provided that the excited rare-earth ion luminesces or forms a
luminescent excited state by radiationless decay from the state initially

populated, the sensitized emission can be much more intense than that produced by direct excitation of the ion. There are two reasons for this: (a) aromatic carbonyl compounds absorb much more strongly than rare-earth ions; and (b) the efficiency of intersystem crossing in aromatic carbonyls is, as already noted, very high. Hence the efficiency of formation of the $^3A^*$ precursor is very large.

Another intriguing application of energy transfer has been described by Abbott and Hercules [85]. If one wishes to carry out luminescence assays of a compound (such as a carbonyl or a heterocyclic) in which the efficiency of singlet $\longrightarrow$ triplet intersystem crossing is large, he must ordinarily resort to phosphorescence analyses. The experimental difficulties associated with phosphorescence can sometimes be circumvented by employing triplet $\longrightarrow$ singlet energy transfer:

$$A + h\nu \longrightarrow {}^1A^* \xrightarrow[\text{intersystem crossing}]{} {}^3A^*, \quad (26a)$$

$$^3A^* + Q \longrightarrow A + {}^1Q^*, \quad (26b)$$

$$^1Q^* \longrightarrow Q + h\nu'. \quad (26c)$$

Under the proper circumstances [85] the fluorescence intensity observed for quencher Q in liquid solution will depend directly on the original concentration of analyte A. This analytical approach has been described for the specific case of A = aromatic carbonyl compounds and Q = polycyclic hydrocarbons [85]. Since reaction (26b) is a long-range transfer process, it can be efficient even when the original concentration of A is quite small, so long as triplet quenchers other than Q are not present in high concentration.

2. Fluorescence Quenching Not Involving Energy Transfer

Many substances are capable of efficiently quenching the excited states of fluorescent solutes even if the required conditions for electronic energy transfer are not fulfilled. Solutes that are strong electron acceptors often are especially effective quenching agents. A photoexcited molecule of a fluorescent substance is both a stronger electron acceptor and a stronger electron donor than the same molecule in its ground state; therefore the excited molecule is more likely than the ground state to engage in charge-transfer interaction with other solutes. We have already noted excited-state charge-transfer complexation as one possible mechanism for the external-heavy-atom effect.

Charge-transfer fluorescence quenching of aromatic hydrocarbons by electron acceptors (such as N, N-diethylaniline) has been carefully studied by Weller et al. [86]. In charge-transfer quenching the excited singlet state of the

fluorescent solute A forms an "encounter complex," and then an actual excited
charge-transfer complex, with the quencher Q:

$$^1A^* + Q \rightleftharpoons {}^1A^*-Q \longrightarrow {}^1(A^+-Q^-)^* \longrightarrow A + Q + h\nu'$$
$$\downarrow$$
$$A + Q + kT \tag{27}$$

In nonpolar solvents (dielectric constant $\langle 10 \rangle$) fluorescence from the excited
charge-transfer complex $^1(A^+-Q^-)^*$ can be observed; the fluorescence is red-
shifted relative to that of $^1A^*$ and is devoid of fine structure. During the forward
and reverse electron-transfer processes the electronic energy of $^1A^*$ may be
dissipated as vibrational energy to the solvent [kT in Eq. (27)]. Thus fluorescence yields for excited charge-transfer complexes are often rather low, and
the net effect of excited-state charge-transfer interaction in nonpolar media
usually is a significant decrease in sensitivity for the analysis of A.

In more polar solvents quenching of fluorescence of $^1A^*$ by Q is not
usually accompanied by fluorescence from an excited charge-transfer complex.
Instead the encounter complex appears to be converted into solvated ions:

$$^1A^* + Q \rightleftharpoons {}^1A^*-Q \longrightarrow A_s^+ + Q_s^- . \tag{28}$$

In flash experiments both A^+ and $^3A^*$ are detected as products of the interaction
of $^1A^*$ with Q [86, 87]. The origin of triplets of A is not now clear. The
triplets may be produced via enhanced intersystem crossing in the encounter
complex, and perhaps also by ion-recombination reactions:

$$A^+ + Q^- \longrightarrow {}^3A^* + Q. \tag{29}$$

In any case these processes lead to very efficient collisional fluorescence
quenching in polar solvents. Their occurrence is widespread; Weller et al. [86]
have provided an excellent bibliography of charge-transfer quenching.

It has also been demonstrated that the fluorescence of "electron-deficient"
organic compounds (e.g., the methylacridinium cation, riboflavin) can be
quenched by a variety of nucleophiles, including such diverse species as diethyl
ether, ethanol, the sulfate ion, water, pyridine, and the chloride ion, [88]. The
details of excited-state interactions involved in these quenching processes are
not now clear, but it is possible that they are formally analogous to reaction (27)
except that Q acts as an electron donor.

We must also note that collisional fluorescence quenching can be significant
even when the quenching solute does not, at least superficially, appear to be one
that would complex with $^1A*$. For example, conjugated dienes (e.g., 4-methyl-
1,3-pentadiene) effectively quench the fluorescence of aromatic hydrocarbons

like naphthalene or anthracene. Since the energy of S_1^* for the diene is much
higher than that of the aromatic hydrocarbon, electronic energy transfer is not
a viable quenching mechanism. It has been noted [89] that the quenching of
aromatic hydrocarbon fluorescence by conjugated dienes is sharply reduced by
the presence of alkyl substituents in the fluorescent hydrocarbon. It is also
noted that small quantities of photoproducts having the composition of aromatic
hydrocarbon-diene adducts are produced. It is therefore inferred [89] that
excited-state collisional interaction between aromatic hydrocarbon and diene is
responsible for the quenching, which presumably proceeds via enhanced $S_1^* \rightarrow S_0$
internal conversion in the aromatic hydrocarbon. On energetic grounds it is
probably unreasonable to postulate excited-state charge-transfer complex
formation [as Eq. (27)] between excited aromatic hydrocarbons and conjugated
dienes; nonetheless the experimental data leave little doubt that a strong
collisional quenching interaction does occur in this system. The commonness of
collisional quenching of fluorescent organic molecules by extraneous solutes is
thus inferred to be a serious problem that, as already noted, can be alleviated
only by diluting the sample solution, increasing the viscosity of the medium, or
separating the fluorescent solute from potential quenchers prior to measurement.
It is also absolutely necessary to use solvents that are highly pure. For
example, it has been observed that nitrogen heterocyclics form fluorescent
complexes with peroxide impurities in ether solvents, [90], even when the
peroxide concentration is very small. Solvents such as ethers, which are
difficult to purify and cannot be stored for long periods without decomposition,
should be avoided whenever possible in fluorescence analysis.

3. Fluorescence Quenching by Oxygen

Perhaps the most ubiquitous quencher of fluorescence and phosphorescence
is molecular oxygen. In liquid solutions quenching of excited-singlet states of
organic molecules by dissolved O_2 molecule has a very large diffusion coefficient, especially in polar solvents, fluorescence quenching by O_2 can be a
very serious problem.

It has long been known that some organic molecules are much more susceptible than others to fluorescence quenching by O_2. While the ground state of O_2
is a triplet, it is no longer believed that "paramagnetic" effects play a generally
important role in O_2 quenching. Among the mechanisms that have recently
been proposed to account for the quenching of excited-singlet states by O_2 are
the following [92]:

1. Chemical oxidation of excited singlet: $^1A* + O_2 \longrightarrow A^+ + O_2^-$.   (30a)

2. Energy transfer from $^1A*$ to O_2: $\quad ^1A* + {}^3O_2 \longrightarrow {}^3A* + {}^1O_2^*,$ (30b)

$$^1A* + {}^3O_2 \longrightarrow {}^1A + {}^1O_2^*.$$ (30c)

3. Enhanced intersystem crossing in $^1A*$: $\quad ^1A* + {}^3O_2 \longrightarrow {}^3A* + {}^3O_2.$ (30d)

4. Enhanced internal conversion in $^1A*$: $\quad ^1A* + {}^3O_2 \longrightarrow {}^1A + {}^3O_2.$ (30e)

5. Formation of a complex between O_2 and the ground state ^1A.

Which (if any) of these processes pertains to O_2 quenching of a particular fluorescent solute depends on the properties of both the solute and the medium. Thus fluorescence quenching by ground-state complex formation (mechanism 5) is almost never significant in liquid solution but can be important for certain solutes in rigid media [93]. Likewise excited-state redox (process 1) occurs only for a few very strongly reducing solute species.

There has recently been substantial interest in the chemistry of singlet oxygen [92], and it has been thought possible that excited-singlet states of O_2 might be produced by energy transfer from electronically excited organic molecules in solution. It appears, however, that neither reaction (30b) nor reaction (30c) can serve as a general description of O_2 quenching of molecular fluorescence. If reaction (30b) were a general quenching process, one would predict that fluorescent molecules in which the energy gap between S_1^* and T_1^* was smaller then that required for excitation of $^3O_2$ to $^1O_2^*$ would not be efficiently quenched by O_2. In fact, however, aromatic hydrocarbons with small singlet-triplet separations are quenched by O_2 just as efficiently as those with very large singlet-triplet energy splittings [94]. Likewise, reaction (30c) has been shown to be inconsistent with photochemical data, as is mechanism 5 (oxygen-enhanced $S_1^* \longrightarrow S_O$ internal conversion) [95]. By a process of elimination, enhanced $S_1^* \longrightarrow T_1^*$ intersystem crossing in the presence of oxygen remains the only plausible general oxygen-quenching process [94].

Affirmative evidence for the operation of process 4 has recently been acquired by laser-flash spectroscopy [87]. When oxygen-containing solutions of pyrene are flashed, the yield of pyrene triplet states is significantly larger than it is when O_2 is absent. It is inferred [87] that the basic process is charge transfer in nature (see Section III. E. 2), with the efficiency of intersystem crossing in $^1A*$ being greatly enhanced when it forms an encounter complex with O_2. Because the actual quenching act is exceedingly rapid (2×10^{-10} sec), no direct evidence of charge-transfer quenching (such as the observation of an exciplex) has been obtained [96]. It is especially suggestive that compounds most susceptible to O_2 quenching tend to have high ionization potentials [97].

That the presence of O_2 appears to enhance the efficiency of $S_1^* \longrightarrow T_1^*$ intersystem crossing in many fluorescent molecules does not mean that dissolved oxygen can enhance the sensitivity of phosphorometry, for in fact O_2 is also a very effective triplet quencher [92]. Because the triplet has a much longer lifetime than an excited singlet, it is much more susceptible to collisional quenching processes involving impurities such as O_2 (or impurities produced by photodecomposition of the solute). Consequently phosphorescence is rarely observed in liquid solution unless great care is exercised in the purification of solutes and solvent, and then only when exhaustive vacuum-degassing procedures are followed [98]. Triplet states may be quenched in fluid media not only by O_2 or other impurities, but also by other triplets, or the ground state, of the solute of interest [99]:

$$^3A* + \, ^3A* \longrightarrow \, ^1A* + A, \tag{31a}$$

$$^3A* + A \longrightarrow \, ^3(A^*A) \longrightarrow A + A + kT. \tag{31b}$$

Reaction (31a), triplet-triplet annihilation, may lead to delayed fluorescence [100]. In reaction (31b) the triplet excimer, $^3(A^*A)$, whose existence is still conjectural is thought to decay nonradiatively to the ground state [99].

Clearly in most real analytical samples the degree of solvent and solute purity required to observe phosphorescence in liquid solution is impractical. One must therefore resort to rigid media, in which diffusion rates are very much lower. However, as we shall note in Section III. G, the "rigid" media commonly used in phosphorometry give rise to a new set of practical problems.

It should be noted that the ability of O_2 to quench the photoluminescence of organic molecules can be exploited for the determination of O_2 in solution [101]. Also, the rate of O_2 diffusion into bulk polymers can be measured via the quenching of phosphorescence from organic molecules dissolved in the polymer [102].

4. Fluorescence Quenching by Metal Ions

The presence of metal ions can influence the luminescence characteristics of organic molecules, even when the ions do not form stable complexes with the ground state of the solute of interest. Paramagnetic transition-metal ions generally produce the largest effects, suggesting that the paramagnetic species increases the rate constants for intersystem crossing in the organic molecule. Diamagnetic, non-transition-metal ions are usually poor quenchers. It has, however, become evident that, for a given fluorescent solute, it is not generally possible to correlate the extent of quenching produced by metal ions with their paramagnetic moments. For example, Mn^{2+}, which possesses five unpaired electrons, is generally an inefficient quencher [103, 104]. It is clear that one cannot rationalize all available information simply by assuming that metals quench excited states by paramagnetic enhancement of spin-orbit coupling.

At least two additional processes by which metal ions can quench fluorescence or organic molecules can be suggested:

1. Excited-state-complex formation, which would be at least superficially analogous to the charge-transfer quenching processes described in Section III. E.
2. Generally metal ions that are poor quenchers are (a) very weakly complexing ions (e. g. , Zn^{2+}, Cd^{2+}); (b) lanthanides (in which the unpaired electrons are contained in well-submerged f orbitals); or (c) ions that are very difficult to reduce (e. g. , Mn^{2+}). In some cases, therefore, it may be reasonable to postulate excited-state-complex formation between a metal ion and an excited singlet of an organic molecule, in which the former acts as electron donor [103]. Insufficient data are available to assess the general validity of this suggestion. It is worth noting that, in the quenching of riboflavin fluorescence by metal ions, the quenching efficiencies correlate very poorly with the ionization potentials of the ions [104], in spite of the knowledge that singlet riboflavin is susceptible to quenching by electron donors [88].

2. In some specific instances electronic energy transfer from excited-singlet states of organic molecules to metal ions may occur. For example, the first excited-singlet state of riboflavin sensitizes the photooxidation of Fe^{2+} as well as the photoaquation of $Cr(NH_3)_5NCS^{2+}$ [104]. The indication is that these sensitization processes involve energy transfer from singlet riboflavin to spin-allowed excited states of the metal ions, though this interpretation is not unequivocal.

In general it appears unlikely that a single mechanism will be able to rationalize the various observations of metal-ion quenching. Since this is an area in which little work has yet been done, analysts encountering metal-containing systems should perform fluorescence measurements on them with great care, the more so as many metal ions undergo photochemical oxidation-reduction reactions in the visible region [105]. Metal ions are also effective triplet-state quenchers [106], but these processes are normally of little analytical significance because most phosphorescence measurements are performed in rigid media, in which collisional quenching is minimized.

The analytical implications of metal-ion quenching are evident, and often unfortunate. Most transition-metal complexes, even of aromatic ligands, are nonfluorescent [107], and uncomplexed heavy metals frequently quench the fluorescence of other species. The extent of such quenching can in principle be utilized to determine metal ions in solution. However, the only reported applications of fluorescence quenching to the determination of metals involve "static quenching", that is, the reaction of a metal with the ground state of a fluorescent species to produce nonfluorescent products [108]. There is hope for a relatively new technique, in which one attempts to surmount metal-ion quenching by forming an anionic complex of the metal ion of interest; this complex is then allowed to form an ion pair with a fluorescent organic cation. The ion-pair formation often effects significant fluorescence-frequency shifts without causing serious quenching, at least for diamagnetic metal complexes [109].

5. Effect of Solute Concentration on Fluorescence

It is usually found that the fluorescence intensity of a given solute increases linearly with increasing concentration at relatively low concentrations. At higher concentrations the fluorescence intensity may reach a limiting value and even decrease with further increases in concentration. Several processes are responsible for these "concentration-quenching" effects:

1. Because the low-frequency tail of the absorption spectrum of a solute often overlaps the high-frequency end of its fluorescence spectrum, fluorescence from a $^1A^*$ molecule can be reabsorbed by a ground-state molecule of the same solute. The probability of such an event increases with increasing solute concentration. "Self-absorption" distorts the shape of the fluorescence spectrum, since only the higher frequencies in the spectrum are reabsorbed. Self-absorption ultimately reduces the fluorescence intensity, unless ϕ_F for $^1A^*$ is equal to unity. The importance of self-absorption errors can often be reduced by using "front-surface" excitation [110].

2. Many aromatic molecules (especially those with functional groups capable of hydrogen bonding) form dimers or higher aggregates in solution, particularly in nonpolar and non-hydrogen-bonding solvents (hydrocarbons, CCl_4, CS_2. etc.). This tendency will of course be greater at high solute concentrations. Often the dimers are less strongly fluorescent than the monomer. Also, since the energy of S_1^* for the dimer is invariably lower than that for the monomer, the dimer can quench monomer emission by radiative, or long-range nonradiative, energy transfer [111].

3. We have already referred to the propensity of excited singlets $^1A^*$ to form excimers with ground-state solute molecules [3, 112]:

$$^1A^* + A \longrightarrow {}^1(A^*A) \tag{32}$$

The excimer has its own characteristic emission spectrum, which is red-shifted relative to the monomer spectrum. Many aromatic compounds show excimer emission at concentrations well below those required for the formation of ground-state dimers. For example, aromatic hydrocarbons form excimers at concentrations on the order of 10^{-3} M.

In principle concentration quenching is not an important analytical difficulty since it can easily be remedied by diluting the solution. However, numerous published studies show that in practice concentration quenching occurs and is not properly compensated for.

F.   Effects of Temperature

Temperature per se is a relatively unimportant variable in analytical fluorometry. The rates of radiative and radiationless decay of electronically excited organic molecules in liquid solution are usually dependent of temperature. The following represent some of the more important ways in which temperature can affect fluorescence in liquid solution:

1.   As already noted in Section III. E, the rate of collisional quenching decreases as the viscosity of the medium is increased. Thus collisional quenching of fluorescence in liquid media tends to become somewhat less serious as the temperature is lowered.

2.   Other "chemical" processes involving excited-singlet states (e. g. , intermolecular hydrogen bonding) tend to occur more slowly as the temperature decreases. The positions of excited-state equilibria also shift slightly with temperature.

3.   The generalities expressed in the first paragraph of this section are untrue if the solute in question has two or more excited states that are very slightly different in energy. Consider, for example, the energy-level diagram in Fig. 69, in which T_1^* lies just below S_1^*. At sufficiently high temperatures intersystem crossing from S_1^* to T_1^* may be followed by thermal excitation of the triplet back to S_1^*. In that case the fluorescence of the molecule will exhibit two spectrally identical components, one of which has a "normal" decay time while the other has a lifetime slightly shorter than that of T_1^*. The fluorescence intensity will increase with temperature. Such "delayed fluorescence," thermally activated, is exhibited by a number of molecules (e. g. , anthraquinone [113]).

Fluorescence yields for a number of substituted anthracenes decrease very significantly with increasing temperature. These observations are explained in terms of the energy-level scheme in Fig. 70, wherein a second triplet state lies slightly higher in energy than S_1^*. Hence in this case $S_1^* \longrightarrow T_2^*$ system crossing is thermally activated and competes increasingly effectively with fluorescence as the temperature increases. The relative positions of S_1^* and T_2^* in anthracenes are very dependent on the identity, number, and position of substituent groups as well as on the solvent.

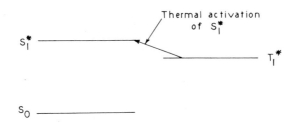

FIG. 69. Energy-level diagram for a molecule that exhibits thermally activated delayed fluorescence.

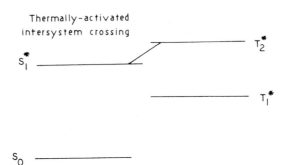

FIG. 70. Energy-level diagram for substituted anthracenes in which Φ_F decreases with increasing temperature. Note that T_2^* lies slightly above S_1^*, so that $S_1^* \longrightarrow T_2^*$ intersystem crossing is thermally activated.

G. Fluorescence and Phosphorescence in Rigid Media

There are two principal advantages to measuring luminescence spectra in rigid media:

1. Bimolecular collisional quenching processes are much less significant in rigid than in fluid media, owing to the much greater viscosity of the former. Consequently fluorescence yields often are larger in rigid than in fluid media. We have already discussed the general impossibility of measuring phosphorescence in liquid media.

2. Luminescence spectra obtained in rigid media at low temperatures often contain well-resolved vibrational fine structure not observed in liquid solution [115] (compare the aromatic hydrocarbon total emission spectra at $77^\circ K$ in Fig. 71 with the fluorescence spectra in room-temperature solution in Fig. 61).

It is usually observed that the fluorescence spectrum of a compound is blue-shifted in a rigid medium relative to the spectrum in liquid solution. This shift results from the fact that solvent reorientation after excitation is considerably less facile in a rigid medium than it is in solution. Hence the Franck-Condon excited state is effectively "frozen in" at low temperatures (Fig. 72 and Table

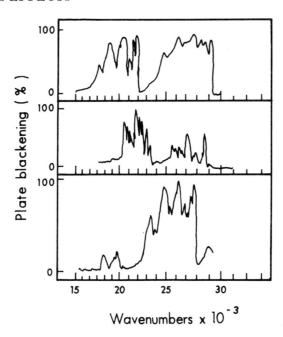

FIG. 71. Total emission spectra of three aromatic hydrocarbons in EPA glass at 77°K. Top, phenanthrene; center, triphenylene; bottom, chrysene. Fluorescence is at higher frequencies, phosphorescence at lower frequencies. From Kasha [18].

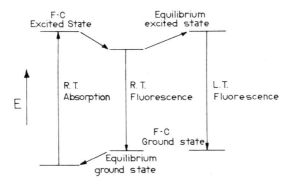

FIG. 72. Schematic representation of equilibrium and Franck-Condon (F-C) electronic states at room temperature (R. T.) and low temperatures (L. T.).

28). That the "low-temperature blue shift" is actually an effect of viscosity, rather than temperature, is indicated by the observation of very similar blue shifts (relative to solution spectra) in the room-temperature fluorescence of aromatic compounds embedded in plastic matrices [116]. Interestingly, Lippert, Lüder, and Moll [117] have noted that the fluorescence of aromatic compounds in solution also undergoes a blue shift with increasing temperature. At high temperatures the thermal motion of solvent molecules acts to inhibit

TABLE 28

Temperature Effect on Fluorescence Maxima
of Hydroxynaphthalenes[a]

| Compound | Form | $\bar{\nu}$ (cm^{-1}) | | Difference |
		Room Temp.	77°K	
1-Naphthol	Molecular	27,400	29,000	1600
	Ionic	20,600	23,900	3300
2-Naphthol	Molecular	27,700	28,200	500
	Ionic	23,000	25,300	2300
1,3-Naphthalenediol	Molecular	26,400	26,850	450
	Hemi-ionic	23,000	24,100	1100
1,6-Naphthalenediol	Molecular	27,350	27,550	200
	Hemi-ionic	23,000	25,500	2500
2,3-Naphthalenediol	Molecular	29,300	29,600	100
	Hemi-ionic	25,050	27,250	2200

formation of tight solvent "cages" around the solute, so that the difference in structure between equilibrium ground- and excited-state solvent shells are, in effect, blurred out by thermal motion.

The rigidity of a low-temperature glass can have interesting effects on the fluorescence of nonrigid aromatic molecules [118]. For example, in liquid solution the fluorescence of anthracene-9-carboxylic acid is red-shifted relative to that of the parent hydrocarbon. Also, the spectrum of anthracene in room-temperature solution is structured (Fig. 61), whereas that of the carboxylic acid is structureless. These differences are attributed [119] to rotation of the carboxyl group in the excited state of the acid. In the ground state the carboxyl group is not coplanar with the aromatic ring system; in the excited state it may rotate into a position approximately coplanar with the ring. In the "rotated" position the carboxyl group may engage in intramolecular hydrogen bonding with a peri hydrogen in the aromatic ring. In a rigid glass, such as methanol-ethanol at 77°K, the carboxyl is "locked" in its ground-state position. The intramolecular hydrogen bonding therefore does not occur, so that, in the rigid glass, the fluorescence spectrum of anthracene-9-carboxylic acid rather closely resembles that of unsubstituted anthracene [119].

The rigid matrices usually used in phosphorometry are optically transparent glasses formed by freezing pure organic solvents, or mixtures thereof, at liquid-nitrogen temperature (77°K) [120, 121]. Although the use of these glassy solvents is by now a matter of experimental routine, a number of problems associated with their use often escape consideration. The most obvious, and yet most often overlooked, fact is that the solubilities of most organic compounds decrease with decreasing temperature. In the act of freezing a solution it is therefore possible that the solute will crystallize out before the solvent has solidified. In that case the solute will not be uniformly dispersed through the matrix, but will instead be concentrated in microcrystalline aggregates. These aggregates may exhibit very much lower emission yields than the isolated molecules [122] (and therefore decrease sensitivity) or may exhibit a different spectrum than the isolated molecules [123]. These effects should be minimized by (a) not employing excessively high solute concentrations, (b) by choosing solvents in which the solutes are very soluble at room temperature, and (c) by freezing the sample as rapidly as possible (so as to minimize the segregation of solute and solvent molecules).

It must also be noted that at 77°K many organic glasses are far from "rigid." For example, it has recently been reported that the phosphorescence of toluene in 2-methylbutane glasses at 77°K is quenched by cis-1, 3-pentadiene and that the quenching data satisfy the Stern-Volmer equation [124]. It is believed that bimolecular collisional quenching can occur in this relatively nonviscous glass. The apparent rigidity of organic solvent glasses may change with time, especially if the glass undergoes a phase transition at a temperature approximately equal to that of the coolant [125]. Glassy solvents can undergo configurational relaxation requiring periods of as much as several hours, which may have pronounced effects on phosphorescence spectra and decay times. The extent of such configurational changes may depend on the purity of the solvent, the rate at which the solvent was frozen, and even the geometry of the sample tube. Great care must therefore be exercised in glass preparation if reproducible spectra and emission yields are to be obtained. Some pure solvents that can be used for phosphorometry at 77°K are listed in Table 29.

The situation becomes especially complex in glasses consisting of mixtures of two or more solvents. For example, in a study of the effect of excited-state hydrogen bonding on the luminescence of nitrogen heterocyclics in 3-methylpentane glasses containing varying quantities of 1-propanol, it was noted that a sudden change in luminescence characteristics occurred at about 1% propanol. This change was attributed to a phase transition in the glass [126], rather than to any fundamental change in solute-solvent interactions. Another interesting example is the luminescence of xanthone, an aromatic ketone, in ether-isopentane-ethanol (EPA), a very common glassy solvent mixture. Two distinct phosphorescence spectra, with different excitation

TABLE 29

Pure Solvents Which Can Be Used for Phosphorometry at 77°K

n-pentane

3-methylpentane

Diethyl ether

Ethanol

1-Propanol

2-Methyltetrahydrofuran

Sulfolane

2-Bromobutane

spectra, are observed for xanthone in EPA. In view of the sensitivity of ketone luminescence to the hydrogen-bonding ability of the solvent (see Section III. D) it is likely that the two-component phosphorescence spectrum arises from solute molecules in different molecular environments [127]. An excited xanthone molecule in an "isopentane" region of the glass would exhibit a different spectrum than one located in an "ethanol" region. Indeed it is not inconceivable that, in a mixed glass like EPA different ketone molecules may have either (n, π^*) or (π, π^*) lowest triplet states, depending on their local solvent environments. It appears that glassy solvent mixtures may be far from homogeneous, especially if they are frozen slowly. It is also apparent that diffusion rates in glasses may be slow enough for different molecular environments not to be averaged out during the lifetime of an excited-triplet state. For this reason it is the author's belief that the use of EPA and similar mixtures should be avoided whenever possible in phosphorometry.

There are very few quantitative data (especially quantum efficiencies) for the luminescence of organic molecules in glassy media. Experimental problems in, and procedures for, luminescence-quantum-yield measurements in glassy solvents have been discussed recently by Demas and Crosby [128].

Little attention has been devoted to the use of solid rare gases and other very weakly interacting rigid media in analytical luminescence spectroscopy. The principal advantage of rare-gas matrices over organic solvent glasses is that the former interact only very weakly with solute molecules; hence luminescence spectra obtained in rare-gas matrices are predominantly characteristic of the solute, with environmental perturbations being minor. (It should, however, be noted that heavy-atom effects are observed in rare-gas matrices [55]; see Section III. B.) Meyer [129] has reviewed the fluorescence and phosphorescence or organic molecules in very-low-temperature inert matrices.

The experimental complexities associated with low temperatures can be avoided by measuring phosphorescence spectra in rigid plastic matrices at room temperature. Three basic techniques for preparing solid solutions of phosphorescent solutes in plastics have been used [130]:

1. Preparing a solution of the solute in a liquid monomer (e.g., methyl methacrylate is a liquid at room temperature) and then thermally polymerizing the monomer.

2. Dissolving both the polymer (e.g., polyethylene) and the sample in a liquid solvent and then evaporating the solvent. The solute will be more or less uniformly dispersed in the polymer.

3. Mixing a solid polymer with a sample and then thermally melting the polymer. The sample is allowed to dissolve in the molten plastic, after which the plastic is refrozen.

Polymers formed in any of these ways are rather rigid (probably more viscous than most glassy organic solvents, even at temperatures as high as $100^{\circ}C$). Large aromatic molecules are effectively immobilized in plastics, though smaller molecules (O_2; free radicals produced by photodegradation of the polymer) can diffuse at sufficiently high rates to quench phosphorescent triplets [131]. Thus polymer samples must normally be vacuum-degassed to obtain high phosphorescence yields; in some cases it may be necessary to operate at sub-ambient temperatures to reduce collisional quenching [132].

The most important advantage of plastics over low-temperature glasses is the evident lack of solute aggregation in the former. For example, cellulose acetate samples containing pyrene concentrations greater than 0.07 M exhibited no excimer fluorescence [133], whereas low-temperature glasses containing such high solute concentrations are very prone to exhibit excimer emission and microcrystallite formation. Solvent shifts of fluorescence and phosphorescence spectra as a function of polarity or hydrogen-bonding ability of plastic matrices are analogous to those observed in liquid media [134].

In spite of the fact that phosphorescence is easily observable from solutes dispersed in plastic matrices at room temperature, these media have not become popular in analytical phosphorometry for a number of reasons:

1. It is very difficult to remove the last traces of residual monomer from a plastic. If any liquid solvents were employed in the sample preparation, these also can be removed from the plastic matrix only with great effort. Many monomers (e.g., methyl methacrylate and acrylonitrile) photochemically decompose to produce free radicals that can react with or quench phosphorescent triplet states. Methyl methacrylate and some other monomers exhibit their own luminescence as well, so that poly(methyl methacrylate) that has not been care-

fully purged of monomer exhibits its own background luminescence. If the polymer was prepared in the presence of a liquid solvent, the phosphorescence spectra and decay times of solutes in the matrix are dependent on the amount and distribution of residual solvent therein [134]. Residual monomer and solvent can be removed by prolonged vacuum degassing (accomapnied by heating, if neither the matrix nor the sample is heat-sensitive), but the procedure is quite time-consuming [133].

2. The internal structure of most polymers is not understood in detail, and plastics are susceptible to many of the odd effects previously enumerated for low-temperature glasses. The interior structure of most polymers is not uniform. Different solute molecules can therefore find themselves in different microenvironments, which, due to the rigidity of the plastic, are not averaged out over excited-state lifetimes. Hence phosphorescence decay times for solutes in plastic matrices are often nonexponential, and the spectra may consist of several slightly different, overlapping subspectra. Also, even though rather rigid, plastics do undergo various mechanical, thermal, and dielectric relaxation processes that can affect phosphorescence spectra and decay times [135]. Reproducible preparation of samples in polymeric matrices is very difficult.

3. Phosphorescence self-quenching (see below) and solute environment inhomogeneities can be very severe unless thin polymer films are used. Preparation of reproducible thin, polished polymer films is difficult and time-consuming [132].

4. The fluorescence and phosphorescence spectra of aromatic molecules in plastics contain much less fine structure than those obtained in low-temperature glasses [134].

For these reasons the use of plastics in phosphorometry offers few advantages (and several disadvantages) over the common low-temperature glassy matrices. An important exception to this statement is the case in which preparation of a low-temperature glass is accompanied by solute aggregation [122]; in that situation (which is more common than has generally been realized) plastic matrices may be preferable for high-sensitivity phosphorometry.

Finally it should be noted that, irrespective of the type of rigid matrix used, self-quenching of phosphorescence can be a serious problem [136]. A photon emitted by an excited-triplet state has little chance of being absorbed by a ground-state molecule, because singlet $\longrightarrow$ triplet transitions are forbidden. However, triplet $\longrightarrow$ triplet absorption is allowed and may be as intense as $S_0 \longrightarrow S_1^*$ absorption in aromatic molecules (i. e. , ϵ_{max} approaching 10^5). Thus a triplet-state molecule may be promoted to a higher triplet level by absorbing a photon emitted by another excited triplet [136]. This problem (absorption of luminescence of one molecule by another molecule already in an excited state) is normally unimportant in fluorescence, due to the short lifetimes of excited-singlet states, but it may be very important for the longer lived triplet state. The extent of phosphorescence reabsorption will depend on the solute concentration, the triplet lifetime, the homogeneity with which the solute is dispersed in the matrix, the intensity of the exciting light, and the thickness of the sample. For this reason front-surface excitation [110] of very thin samples should be used in high-sensitivity phosphorometry, particularly if the lifetime of the emitting triplet is long and there is reason to believe that the distribution of solute in the matrix may be markedly inhomogeneous.

REFERENCES

1. M. Kasha, in Light and Life (W. D. McElroy and B. Glass, eds.), Johns Hopkins University Press, Baltimore, Md. , 1961, p. 31.

2. F. Hirayama and S. Lipsky, J. Chem. Phys., 51, 3616 (1969).

3. J. B. Birks, Photophysics of Aromatic Molecules, Wiley, New York, 1970.

4. G. A. Thommes and E. Leininger, Anal. Chem., 30, 1361 (1958).

5. C. A. Parker, Photoluminescence of Solutions, Elsevier, Amsterdam, 1968, p. 77.

6. E. Sawicki, Talanta, 16, 1231 (1969).

7. A. R. Horrocks and F. Wilknson, Proc. Roy. Soc. (London), A306, 257 (1968).

8. E. Hughes, Jr. , J. H. Wharton, and R. V. Nauman, J. Phys. Chem., 75, 3097 (1971).

9. A. N. Fletcher, J. Phys. Chem., 72, 2742 (1968).

10. I. B. Berlman, J. Phys. Chem., 74, 3085 (1970).

11. O. Schnepp and M. Levy, J. Amer. Chem. Soc., 84, 172 (1962).

12. E. A. Chandross and C. J. Dempster, J. Amer. Chem. Soc., 92, 3586 (1970).

13. S. Sharafy and K. A. Muszkat, J. Amer. Chem. Soc., 93, 4119 (1971).

14. R. Hurley and A. C. Testa, J. Amer. Chem. Soc., 90, 1949 (1968).

15. O. S. Khalil, H. G. Bach, and S. P. McGlynn, J. Mol. Spectry., 35, 455 (1970).

16. A. A. Lamola and G. S. Hammond, J. Chem. Phys., 43, 2129 (1965).

17.   C. A. Parker, Photoluminescence of Solutions, Elsevier, Amsterdam, p. 435.

18.   M. Kasha, Radiation Res., Suppl. 2, 243 (1960).

19.   S. P. McGlynn, T. Azumi, and M. Kinoshita, Molecular Spectroscopy of the Triplet State, Prentice-Hall, Englewood Cliffs, N. J., 1969, pp. 40-43.

20.   H. M. Rosenberg and S. D. Carson, J. Phys. Chem., 72, 3531 (1968).

21.   L. S. Forster and D. Dudley, J. Phys. Chem., 66, 838 (1962).

22.   G. Lancelot and C. Helene, Chem. Phys. Letters, 9, 327 (1971).

23.   H. H. Jaffe and H. L. Jones, J. Org. Chem., 30, 964 (1965).

24.   E. L. Wehry, J. Amer. Chem. Soc., 89, 41 (1967).

25.   M. A. El-Sayed, J. Chem. Phys., 38, 2834 (1963).

26.   W. A. Case and D. R. Kearns, J. Chem. Phys., 52, 2175 (1970).

27.   H. Dewey and S. G. Hadley, Chem. Phys. Letters, 12, 57 (1971).

28.   B. J. Cohen, H. Baba, and L. Goodman, J. Chem. Phys., 43, 2902 (1965).

29.   L. M. Logan and I. G. Ross, J. Chem. Phys., 43, 2903 (1965).

30.   M. Chowdhury and L. Goodman, J. Chem. Phys., 38, 2979 (1963).

31.   Y. H. Li and E. C. Lim, J. Chem. Phys., 56, 1004 (1972).

32.   H. Baba, I. Yamazaki, and T. Takamura, Spectrochim. Acta, 27A, 1271 (1971).

33.   K. E. Wilzbach and D. J. Rausch, J. Amer. Chem. Soc., 92, 2178 (1970).

34.   S. Dym and R. M. Hochstrasser, J. Chem. Phys., 51, 2458 (1969).

35.   K. Yoshihara and D. R. Kearns, J. Chem. Phys., 45, 1991 (1966).

36.   E. L. Wehry and L. B. Rogers, in Fluorescence and Phosphorescence Analysis (D. M. Hercules, ed.), Interscience, New York, 1966, pp. 118-125.

37.   J. Saltiel, H. C. Curtis, L. Metts, J. W. Miley, J. Winterle, and M. Wrighton, J. Amer. Chem. Soc., 92, 410 (1970).

38.   D. R. Kearns and W. A. Case, J. Amer. Chem. Soc., 88, 5087 (1966).

39.   V. L. Ermolaev and A. Terenin, Sov. Phys. Usp., 3, 423 (1960).

40.   A. A. Lamola, P. A. Leermakers, G. W. Byers, and G. S. Hammond, J. Amer. Chem. Soc., 87, 2322 (1965).

41.   H. H. Richtol and F. H. Klappmeier, J. Chem. Phys., 44, 1519 (1966).

42. J. Tournon and M. A. El-Bayoumi, J. Amer. Chem. Soc., 93, 6396 (1971).

43. H. J. Maria and S. P. McGlynn, J. Chem. Phys., 52, 3399 (1970).

44. P. M. Rentzepis and C. J. Mitschele, Anal. Chem., 42 (14), 20A (1970).

45. W. R. Ware, S. K. Lee, G. J. Brant, and P. P. Chow, J. Chem. Phys., 54, 4729 (1971).

46. E. L. Wehry and L. B. Rogers, Spectrochim. Acta, 21, 1976 (1965).

47. S. G. Schulman, P. T. Tidwell, J. J. Cetorelli, and J. D. Winefordner, J. Amer. Chem. Soc., 93, 3179 (1971).

48. P. Suppan, J. Chem. Soc., A, 3125 (1968).

49. S. P. McGlynn, J. Daigre, and F. J. Smith, J. Chem. Phys., 39, 675 (1963).

50. T. Medinger and F. Wilkinson, Trans. Faraday Soc., 61, 620 (1965).

51. G. G. Giachino and D. R. Kearns, J. Chem. Phys., 52, 2964 (1970); 54, 3248 (1971).

52. S. P. McGlynn, T. Azumi, and M. Kinoshita, Molecular Spectroscopy of the Triplet State, Prentice-Hall, Englewood Cliffs, N. J., 1969, pp. 261-280.

53. V. Ramakrishnan, R. Sunseri, and S. P. McGlynn, J. Chem. Phys., 45, 1365 (1966).

54. R. Sahai, R. H. Hofeldt, and S. H. Lin, Trans. Faraday Soc., 67, 1690 (1971).

55. J. L. Metzger, B. E. Smith, and B. Meyer, Spectrochim. Acta, 25A, 1177 (1969).

56. G. W. Robinson, J. Mol. Spectry., 6, 58 (1961).

57. T. H. Bolotnikova and O. N. Sichkar, Opt. Spectry., 28, 43 (1970).

58. N. J. Turro, G. Kavarnos, V. Fung, A. L. Lyons, Jr., and T. Cole, Jr., J. Amer. Chem. Soc., 94, 1394 (1972).

59. L. V. S. Hood and J. D. Winefordner, Anal. Chem., 38, 1922 (1966).

60. M. Zander, Z. Anal. Chem., 229, 352 (1967).

61. E. Vander Donckt, Progr. React. Kinet., 5, 273 (1970).

62. S. G. Schulman, Crit. Rev. Anal. Chem., 2, 85 (1971).

63. G. Jackson and G. Porter, Proc. Roy. Soc. (London), A260, 13 (1961).

64. A. Weller, Progr. React. Kinet., 1, 189 (1961).

65. E. L. Wehry and L. B. Rogers, J. Amer. Chem. Soc., 87, 4234 (1965).

66. W. R. Ware, P. R. Shukla, P. J. Sullivan, and R. V. Bremphis, J. Chem. Phys. , 55, 4048 (1971).

67. M. Goldman and E. L. Wehry, Anal. Chem. , 42, 1178 (1970).

68. S. G. Schulman, Anal. Chem. , 43, 285 (1971).

69. S. G. Schulman and L. B. Sanders, Anal. Chim. Acta, 56, 91 (1971).

70. D. W. Ellis and L. B. Rogers, Spectrochim. Acta, 20, 1709 (1964).

71. E. L. Wehry, Fluorescence News, 6(1), 1 (1971).

72. D. P. Eyman and R. S. Drago, J. Amer. Chem. Soc. , 88, 1617 (1966).

73. J. W. Eastman and E. J. Rosa, Photochem. Photobiol. , 7, 189 (1968).

74. T. Foerster and K. Rokos, Chem. Phys. Letters, 1, 279 (1967).

75. D. K. Kunimitsu, A. Y. Woody, E. R. Stimson, and H. A. Scheraga, J. Phys. Chem. , 72, 856 (1968).

76. A. A. Lamola and L. J. Sharp, J. Phys. Chem. , 70, 2634 (1966).

77. J. R. Merrill and R. G. Bennett, J. Chem. Phys. , 43, 1410 (1965).

78. H. Baba, L. Goodman, and P. C. Valenti, J. Amer. Chem. Soc. , 88, 5410 (1966).

79. J. L. Kropp and J. J. Lou, J. Phys. Chem. , 75, 2690 (1971).

80. R. Rusakowitz, G. W. Byers, and P. A. Leermakers, J. Amer. Chem. Soc. , 93, 3263 (1971).

81. E. C. Lim and J. M. H. Yu, J. Chem. Phys. , 47, 3270 (1967).

82. A. A. Lamola and N. J. Turro, Energy Transfer and Organic Photochemistry, Interscience, New York, 1969, pp. 18-43.

83. E. J. Bowen and B. Brocklehurst, Trans. Faraday. Soc. , 51, 774 (1955).

84. W. J. McCarthy and J. D. Winefordner, Anal. Chem., 38, 848 (1966).

85. S. R. Abbott and D. M. Hercules, Anal. Chem. , 42, 171 (1970).

86. K. H. Grellmann, A. R. Watkins, and A. Weller, J. Phys. Chem. , 76, 469 (1972).

87. C. R. Goldschmidt, R. Potashnik, and M. Ottolenghi, J. Phys. Chem. , 75, 1025 (1971).

88. D. G. Whitten, J. W. Happ, G. L. B. Carlson, and M. T. McCall, J. Amer. Chem. Soc. , 92, 3499 (1970).

89. L. M. Stephenson and G. S. Hammond, Pure Appl. Chem. , 16, 125 (1968).

90. S. J. Ladner and R. S. Becker, J. Phys. Chem. , 67, 2481 (1963).

91.   W. R. Ware, J. Phys. Chem. , 66, 455 (1962).

92.   D. R. Kearns, Chem. Rev. , 71, 395 (1971).

93.   J. L. Rosenberg and F. S. Humphries, J. Phys. Chem. , 71, 330 (1967).

94.   C. S. Parmenter and J. D. Rau, J. Chem. Phys. , 51, 2242 (1969).

95.   B. Stevens and B. E. Algar , Ann. N. Y. Acad. Sci. , 171, 50 (1970).

96.   I. B. Berlman, C. R. Goldschmidt, G. Stein, Y. Tomkiewicz, and
      A. Weinreb, Chem. Phys. Letters, 4, 338 (1969).

97.   T. Brewer, J. Amer. Chem. Soc. , 93, 775 (1971).

98.   S. C. Tsai and G. W. Robinson, J. Chem. Phys. , 49, 3184 (1968).

99.   J. Langelaar, G. Jansen, R. P. H. Rettschnick, and G. J. Hoytink,
      Chem. Phys. Letters, 12, 86 (1971).

100. E. L. Wehry and L. B. Rogers, in Fluorescence and Phosphorescence
      Analysis (D. M. Hercules, ed.), Interscience, New York, 1966,
      pp. 122 and 123.

101. L. J. Tolmach, Arch. Biochem. Biophys. , 33, 120 (1951).

102. E. I. Hormats and F. C. Unterleitner, J. Phys. Chem. , 69, 3677 (1965).

103. H. Linschitz and L. Pekkarinen, J. Amer. Chem. Soc. , 82, 2411 (1960).

104. A. W. Varnes, R. B. Dodson, and E. L. Wehry, J. Amer. Chem. Soc. ,
      94, 946 (1972).

105. E. L. Wehry, in Analytical Photochemistry and Photochemical Analysis
      (J. M. Fitzgerald, ed.), Dekker, New York, 1971, Chapter 6.

106. C. O. Hill and S. H. Lin, J. Chem. Phys. , 53, 608 (1970).

107. F. E. Lytle, Appl. Spectry. , 24, 319 (1970).

108. S. B. Zamochnick and G. A. Rechnitz, Z. Anal. Chem. , 199, 424
      (1964).

109. N. R. Andersen and D. M. Hercules, Anal. Chem. , 36, 2138 (1964).

110. C. A. Parker, Photoluminescence of Solutions, Elsevier, Amsterdam,
      1968, p. 226.

111. K. K. Rohatgi and G. S. Singhal, Indian J. Chem. , 7, 1020 (1970).

112. E. L. Wehry and L. B. Rogers, in Fluorescence and Phosphorescence
      Analysis (D. M. Hercules, ed.), Interscience, New York, 1966,
      pp. 113-118.

113. S. A. Carlson and D. M. Hercules, J. Amer. Chem. Soc. , 93, 5611
      (1971).

114. R. G. Bennett and P. J. McCartin, J. Chem. Phys. , 44, 1969 (1966).

115. R. A. Passwater, Fluorescence News, 5(5), 4 (1971).

116. M. L. Bhaumik and R. Hardwick, J. Chem. Phys., 39, 1595 (1963).

117. E. Lippert, W. Lüder, and F. Möll, Spectrochim Acta, 15, 858 (1959).

118. D. M. Hercules and L. B. Rogers, J. Phys. Chem., 64, 397 (1960).

119. T. C. Werner and D. M. Hercules, J. Phys. Chem., 73, 2005 (1969).

120. F. J. Smith, J. K. Smith, and S. P. McGlynn, Rev. Sci. Instr., 33, 1367 (1962).

121. J. D. Winefordner and P. A. St. John, Anal. Chem., 35, 2211 (1963).

122. R. A. Keller and D. E. Breen, J. Chem. Phys., 43, 2562 (1965).

123. E. Loewenthal, J. Chem. Phys., 48, 2819 (1968).

124. P. Froehlich and H. Morrison, Chem. Commun., 184 (1972).

125. T. E. Martin and A. H. Kalantar, J. Phys. Chem., 74, 2030 (1970).

126. J. L. Kropp and J. J. Lou, J. Phys. Chem., 74, 3953 (1970).

127. H. J. Pownall and J. R. Huber, J. Amer. Chem. Soc., 93, 6429 (1971).

128. J. N. Demas and G. A. Crosby, J. Amer. Chem. Soc., 92, 7262 (1970).

129. B. Meyer, Low-Temperature Spectroscopy, Elsevier, Amsterdam, 1971, Chapter 5.

130. G. Oster, N. Geacintov, and A. V. Khan, Nature, 196, 1089 (1962).

131. B. A. Baldwin and H. W. Offen, J. Chem. Phys., 49, 2933 (1968).

132. P. F. Jones and S. Siegel, J. Chem. Phys., 50, 1134 (1969).

133. R. E. Kellogg and R. P. Schwenker, J. Chem. Phys., 41, 2860 (1964).

134. E. H. Park, A. H. Kadhim, and H. W. Offen, Photochem. Photobiol., 8, 261 (1968).

135. S. Rodriguez and H. W. Offen, J. Chem. Phys., 52, 586 (1970).

136. J. S. Brinen and W. G. Hodgson, J. Chem. Phys., 47, 2946 (1967).

Chapter 4

PRACTICAL ASPECTS OF MEASUREMENT

I. FILTER VERSUS GRATING INSTRUMENTS

In Chapter 2 we discussed the differences between filter and grating instruments, and the features of various instruments.

The choice of a filter or grating instrument will mainly depend on the budget of one's institution. A good filter instrument costs about $1000; a good grating instrument, about $5000. More money than this will buy various attachments, such as corrected-spectra, temperature control, and variable slits.

Actually a good filter instrument will enable one to do anything a grating instrument will do, except scan and resolve closely absorbing or emitting species. It is durable, efficient, and more sensitive than the grating instrument.

The grating instrument will allow one to achieve good resolution and to choose the wavelengths of maximum excitation and emission for analysis. The spectra obtained will not be the true spectra, however, because of differences in the intensity of the spectral source, the blaze of the monochromator, and differences in the detector response with differences in wavelength. This will be discussed later.

Filters are an essential part of the filter fluorometer, but they can be quite useful in grating instruments. The proper filter placed between the emission monochromator and the phototube can often eliminate many difficulties caused by scatter and second-order spectra.

II. CARE OF FILTERS

Filters are very delicate and should be handled with care. Handle filters only by the edges or extreme corners. Keep filters flat and dry--moisture tends to cloud them. Never wash the filter with water under any circumstances. If water should come in contact with the gelatin at the edges of the filter, this will cause the filter to swell. The glasses will separate so air can enter the filter.

If the filter becomes so dirty that it cannot be cleaned by simply rubbing, a piece of lens cleaning paper or soft cloth moistened with a solvent should be used. Be careful, however, that the cloth and glass is free of grit, which might scratch the glass, and be sure the solvent does not touch the cemented edges of the filter. Dry, cool storage is desirable for filters; a desiccated sealed container is desirable. Filters should never be used at temperatures higher than 120 to 130° F.

III. SELECTION OF THE EXCITATION AND EMISSION WAVELENGTHS

In choosing the best wavelength for excitation of the molecule to be studied, several factors should be considered. The wavelength should be near a strong intensity point of the instrument's radiation source and also at a strong absorption band of the compound. The excitation wavelength should also be reasonably separated from the emission band. Since scatter is usually observed, there should be at least 20 to 30 nm separating the two peaks, and preferably 50 nm. If the compound possesses two or more peaks in the excitation spectrum, all of good intensity, the one at longest wavelengths should, generally, be chosen. This will ensure minimum photodecomposition of the compound.

If absorbing interferences are present in the sample, one might wish to choose a point on the excitation spectrum other than the maximum wavelength. Consider, for example, the absorption spectrum of serum pictured in Fig. 73a. A broad band is observed with λ_{max} at 430 nm. If one were to attempt to use naphthol AS-BI phosphate for assaying alkaline phosphatase in serum, one would encounter difficulty in attempting to monitor the naphthol AS-BI liberated at the 405-nm peak (Fig. 73b) with a 405 filter (Fig. 73c). However, a 7-54 filter can be used to pass the 318-nm peak, thus eliminating most of the interferences from bilirubin and other absorbing substances in blood.

If a grating instrument is available, the emission peaks are found by running

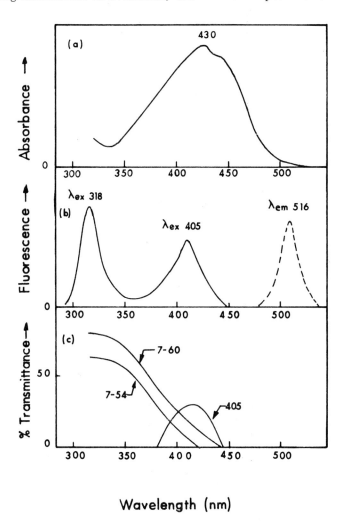

FIG. 73. (a) Absorption spectrum of serum at pH 9.8. (b) Excitation and emission spectra of naphthol AS-BI. (c) Transmittance characteristics of Wratten filters 7-54, 7-60, and 405.

the emission spectrum at the wavelength of maximum excitation. The latter is determined by setting the emission monochromator at various wavelengths from 350 to 600 nm in 50-nm increments and manually scanning the excitation monochromator. When a maximum is observed in the excitation spectrum, the emission monochromator is scanned until a maximum is observed. A permanent excitation spectrum, uncorrected of course, is then obtained by using an x-y recorder at the wavelength of maximum emission. Then the emission spectrum is obtained by recording at the wavelength of maximum excitation.

If only a filter fluorometer is available, the absorption spectrum of the compound is obtained with any good spectrophotometer. The λ_{max} should be the λ_{ex} of the compound. One then chooses a filter that best fits the λ_{ex} of the compound and one of the lines of the mercury lamp that is used in the filter fluorometer. Since the lamp will have a maximum intensity at these spectral lines, it is often best to pick a filter to match this line rather than the λ_{ex} of the compound, provided the compound exhibits appreciable absorption at this wavelength. The 254- or 365-nm lines of the lamp would be best for the excitation of, for example, quinine sulfate (λ_{ex} 250 and 350 nm).

Then one chooses a filter for selection of the emission wavelength by trying different secondary filters at 50-nm intervals, starting at λ_{ex} + 50 nm. Care should be taken to choose a secondary filter that is compatible with the primary filter (see Table 12 in Chapter 2). When one has found a filter that gives a good signal, one could try other filters that peak in the same area to attempt to optimize the signal.

Next let us consider the emission spectrum of the compound. The wavelength chosen to measure emission should be separated by at least 20 to 50 nm from the excitation peak to eliminate scatter. The fluorescence spectrum should closely resemble a mirror image of the absorption spectrum. In many cases frequencies of vibration excitation in the lowest excited singlet are within 10 to 20% of those in the ground state, which is the factor responsible for the mirror-image relationship between absorption and fluorescence spectra. The absence of such a relationship between the two spectra may thus be taken to indicate a large change in the configuration of the molecule caused by photoexcitation.

Fewer peaks are sometimes observed in the emission spectra than in the excitation spectra; this is because emission frequently occurs from the first excited state, whereas absorption can take place to the first, second and sometimes the third excited singlet. Thus quinine sulfate has peaks at 250 and 350 nm but only one emission peak at 450 nm. If the intensity of the emission peak is less than that of the excitation peak, photodecomposition should be suspected. Any other peaks in the emission spectrum are either scatter or Raman peaks. The

Raman peaks in water appear at 248, 313, 365, and 436 nm. In the spectrum of quinine sulfate scatter would be observed at either 250 or 350 nm, depending on the wavelength used for excitation. Second-order scatter would appear at 500 and 700 nm.

IV. SOURCES AND DETECTORS

Xenon lamps are used in all commercial grating instruments, mercury lamps in all filter instruments. Safety precautions that should be followed in using these lamps were presented in Chapter 2.

Most fluorometers and spectrophotofluorometers today use a photomultiplier tube as the detector. The most common photomultiplier tube is the 1P21 (300 to 600 nm), but investigators should be familiar with other tubes that can be used alternatively, such as the 1P28 (200 to 600 nm), or the 7102 (500 to 1000 nm) for infrared studies. As we mentioned in Chapter 2, no two photomultiplier tubes, even from the same manufacturer, will have the same characteristics. Care in ordering and evaluating photomultipliers should be exercised. In no case should the response characteristics of a photomultiplier tube given by a manufacturer be accepted and used in correcting spectra. Photodetectors should be kept in the dark at all times to obtain low noise and good operation.

V. TEMPERATURE

Fluorescence is quite sensitive to changes in temperature, as we have already pointed out. Some compounds, like indoleacetic acid, undergo a 50% change in fluorescence in the 20 to 30°C temperature region.

Heat radiating from the light source used for excitation is the major cause of temperature instability. In some instruments the temperature of the solution in the cell compartment will rise as much as 6 to 10° in a few minutes. Under these circumstances it would be impossible to measure the fluorescence of a compound like indoleacetic acid, which has a high temperature coefficient.

The cell compartment should be kept as close to room temperature as possible. Some instruments, such as the Aminco-Bowman, have cooling attachments to maintain a constant temperature. Circulating water or air also help to maintain room temperature in the sample cell. A constant-temperature room does not always help. Also, samples that must be carried through a procedure involving heating or cooling should be allowed to come to room temperature before assay.

VI. CUVETTES

A. Emission from Cuvettes

Parker [1] has shown that various types of quartz cuvettes emit an appreciable amount of fluorescence (Fig. 74). Fused quartz has an extremely high background fluorescence; synthetic silica has a much lower background and is preferred for studies in the ultraviolet. As we have already mentioned, for studies in the visible region, which constitute about 80% of all measurements, Pyrex or glass cells are sufficient for good results, and the large additional cost of quartz or silica is not necessary.

Fused-quartz cuvettes also emit in some regions of the visible. For example, Parker and Joyce [2] noted phosphorescence in the 370- to 430-nm region. Synthetic silica cells show a much lower background.

B. Care of Cells

1. Causes of Cell Deterioration

Cell deterioration is caused by films, etching, and contamination.

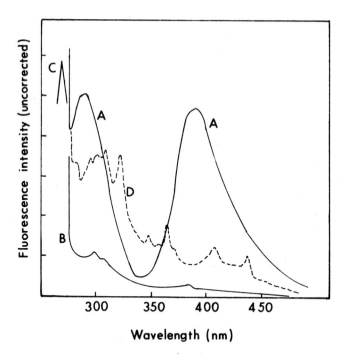

FIG. 74. Interference from scattered light and fluorescence of cuvettes. Specpure cyclohexane excited by light from a Bausch & Lomb monochromator set at 250 nm. Curve A, in fused-quartz cuvette, with filters (sensitivity 1000); curve B, in synthetic silica cuvette, with filters (sensitivity 1000); curve C, as for curve B, but sensitivity 500 to show intensity of Raman band; curve D, in synthetic silica cuvette, without filters (sensitivity 1000). Reprinted from Ref. [1] by courtesy of C. A. Parker.

Films are deposited by solvent evaporation, strong wetting agents, or inadequate cleaning.

Etching is due to continued use of strong alkalies or concentrated mineral acids, either in the sample or in the cleaning solution. It is also due to weak alkaline solutions left in the cell for long periods.

Contamination is caused by solutions being allowed to evaporate in the cell, leaving deposits of salts, organic material, etc.

2. Effects of Cell Deterioration

The principal effects of cell deterioration are reduced transmission and light scattering caused by etching of the cell windows. Films and particulate matter also decrease transmission and may contaminate the sample.

3. Detection

Cell deterioration may be evident from poor base lines. The extent of deterioration may be judged by the following specifications: at 220, 240, and 270 nm (silica cells) or 320 nm (Vycor or Pyrex cells) the transmittance of a new cell filled with distilled water will be at least 70% of transmittance with air alone in the cell space.

4. Prevention

The following procedures should be observed for proper cell maintenance:

1. Always start cleaning by rinsing the cell thoroughly with distilled water (aqueous samples) or a suitable nonfluorescent, purified organic solvent.

2. Clean the cell with a mild agent as soon as possible after each use. Mild inorganic detergents (Calgonite) may be used if it is certain that they produce true solutions and do not contain particulate matter.

3. For hard-to-remove deposits use a solution of 50% 3 N hydrochloric acid and 50% ethanol.

4. Whenever possible, rinse the cell with the sample solution before filling.

5. Remember that if a reagent is not of spectrograde purity, it may leave a deposit on the cell window after evaporation.

6. Never blow the cell dry with air. It is better to speed evaporation of the solvent with the aid of vacuum.

7. Never use any brush or instrument that might scratch the sides of the cell.

8. Never use alkalies, abrasives, etching materials, or hot concentrated acids.

9. Never use ultrasonic devices to clean cells.

VII. SOLVENTS

It cannot be too strongly emphasized that pure solvents must be used in fluorometric work in order to obtain meaningful results. It is not sufficient merely to demonstrate that the solvent or reagent is not fluorescent in itself. Nonluminescent impurities can act as quenchers. Traces of peroxides in diethyl ether can significantly reduce the luminescence of nitrogen heterocyclics due to a charge-transfer reaction [3]. This occurs even when the peroxide level is low. The same type of excited-state reaction occurs between nitrogen hetero- cyclics and aliphatic alcohols, which are frequently present as solvent contaminants [4].

Generally, most commercial solvents must be purified before use in fluorescence. Methyl and ethyl alcohols and dimethylformamide should be distilled before use. Distillation of absolute ethanol from potassium hydroxide has been found to be satisfactory. Figure 75 shows the spectra obtained from commercial absolute ethanol (curve 1) and ethanol purified by distillation (curve 2). The anthracene present as impurity (curve 3) is removed by distillation.

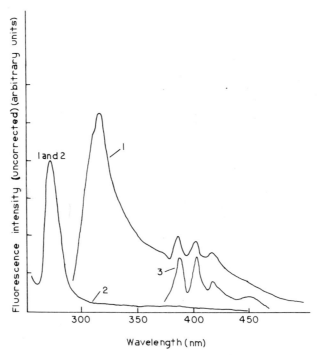

FIG. 75. Fluorescence of trace impurities in ethanol excited by light of 250-nm wavelength: curve 1, absolute ethanol as received; curve 2, ethanol purified by distillation; curve 3, purified ethanol with added anthracene (0.0003 µg/ml). The peak at 270 nm in curves 1 and 2 is the Raman emission. Reprinted from Ref. [1] by courtesy of C. A. Parker.

N-Butanol, benzene, heptane, ethylene dichloride, ether, and acetone have been purified for fluorescence assay by shaking with 0. 1 M hydrochloric acid, 0. 1 M sodium hydroxide, and then water [5]. Purification can also be effected by passing solvents over silica gel.

Harleco reagents for fluorescence studies have been marketed by the Hartmann-Leddon Company (Philadelphia) for about 10 years. Alcohols, water, chlorinated solvents, hydrocarbons, and acids with very low background fluorescences are available. Good solvents for fluorescence are those marketed as "gas-chromatography grade." These solvents have a very low concentration of impurities because of the sensitivities of gas-chromatography methods.

Frequently contamination not originally present in the solvent can result from packaging and from contact with rubber, glass, and cork stoppers, grease, and filter papers, which contain fluorescent contaminants or quenchers. Boron can be extracted from Pyrex glass, and aluminum, calcium, and silica from glass. Many reagents (e. g. , hydrochloric acid) contain aluminum in appreciable concentrations.

Many synthetic detergents are fluorescent. The luminescent properties of a detergent should be checked before use. Inorganic cleaning agents, such as Calgonite, are good for fluorometry. Dichromate cleaning solutions should never be used because even traces of it absorb significant radiation and will interfere with an assay. When necessary, hot nitric acid can be used to clean cuvettes.

VIII. LIGHT SCATTERING

The term "scattered light" refers to light which is of the same or longer wavelength as that of the exciting beam and which emerges from the cuvette. The scatter that occurs at the wavelength of excitation is composed of Rayleigh scattering from the solvent, Tyndall scattering from colloidal particles, and scatter from the surface of the container. There is another type of scatter, called Raman scatter, which occurs at wavelengths longer than the exciting radiation. It is a physical property of the pure solvent and differs from λ_{ex} by a constant frequency. The Raman peaks of several common solvents are listed in Table 30. At high sensitivities all types of scatter appear, but Rayleigh and Tyndall scatter are more intense than Raman scatter. Hence, in water solution, excitation at 313 nm will produce Rayleigh peaks at 313 and 626 nm (second order) and a Raman peak at 350 nm. When the fluorescence and excitation wavelengths are close together, the distortion due to scattering severely limits sensitivity.

Scatter can be diminished by the use of a secondary filter. With radiation at

TABLE 30

Raman Bands for Several Solvents Corresponding to the
Various Mercury Lines[a]

Solvent	Wavelength (nm) of Raman band produced by excitation at				
	248	313	365	405	436
Carbon tetrachloride	--	320	375	418	450
Chloroform	--	346	410	461	502
Cyclohexane	267	344	408	458	499
Ethanol	267	344	409	459	500
Water	271	350	416	469	511

[a] After Parker [6].

365 nm one would use a secondary cutoff filter that has zero transmission below 420 nm. This would eliminate both Rayleigh and Raman scatter.

Price, Kaihara, and Howerton [7] showed that the scatter components, Raman and Rayleigh, differ from fluorescence in their degree of polarization and can be removed by polarized filters. However, such filters also reduce signal intensity and limit sensitivity. Chen and Bowman [8] used the quartz Polarcoat polarizing filter to eliminate scattered light. Such filters have a higher transmission in the ultraviolet [9]. Chen [10] pointed out that scatter produced by the ultraviolet source can be decreased by a single polarized filter in the excitation beam. He found that the use of two polarizers in the excitation beam was not more effective.

IX. PHOTODECOMPOSITION

Photodecomposition was discussed in Chapter 1. It becomes an extremely important problem in dilute solutions; in concentrated solutions the amount of substance decomposed is minor in comparison with the total amount. To minimize photodecomposition it is important that the fluorometer have a shutter, so that the solution is only irradiated during the short period of measurement. The measurement should be made as rapidly as possible. If labile compounds are under study, a weaker source and a more sensitive photodetector should be used; for example, proteins which are photosensitive [11], should be handled in this manner.

Photodecomposition occurs locally in a 1-cm cuvette, and often diffusivity allows the solution to recover and reequilibrate. Reequilibration could also be accomplished by stirring or inverting the cuvette. Photodecomposition becomes more of a problem with microcuvettes.

X. EFFECTS OF CONCENTRATION

As predicted by basic equations, fluorescence is linearly related to concentration at low concentrations (10^{-9} to 10^{-6} M), reaches a limiting maximum, and then decreases at high concentration due to the inner-cell effect and excimer formation.

In attempts to lower the range of concentrations that could be determined by fluorometry, Lowry et al. [12] designed a microcuvette that permits measurement on as little as 50 μl of solution. For some instruments, such as the Aminco, one can obtain adapters that permit measurements on 100 μl of solution. With the small volumes, many compounds can be measured in the nanogram and picogram regions. A microfluorometer for measuring fluorescence within living cells was reported by Chance et al. [13-15]. Wied et al. [16] modified a Zeiss photomicroscope for fluorescence studies. Cytochemical studies have been made on endometrial and cervial materials with greater reliability than was possible with other methods.

Let us consider also the opposite case, wherein high concentrations of a fluorophor must be used and dilution is not possible.

One of the problems in measuring high concentrations is the large width of the cell, generally 1 cm; the excitation beam must pass through 40 mm of solution needlessly. Likewise, the emission beam must pass unnecessarily through a large volume of solution. Hence a significant amount of fluorescent radiation is reabsorbed, and the signal decreases. Chen and Hayes [17] minimized the effects of light absorption on both excitation and emission by modifying the cell compartment as shown in Fig. 76. Two sides of the cell compartment are removed and spacers are added. Now the beam passes through only a small portion of the cell in excitation and emission. With this arrangement sensitivity is the same and absorption effects are minimized. In measuring the fluorescence of NADPH, Chen and Hayes [17] found that linearity could be obtained to 10^{-4} M and that the useful range of measurement could be extended to almost 10^{-3} M. This may be compared with the 5×10^{-5} M obtainable with the conventional 1-cm cell arrangement (Fig. 77).

Chen and Hayes [17] also pointed out that greater linearity could be obtained by exciting NADPH fluorescence at a point on the absorption curve other than at the excitation maximum. Longer wavelengths were preferred to shorter ones because the interference from other biological substances is

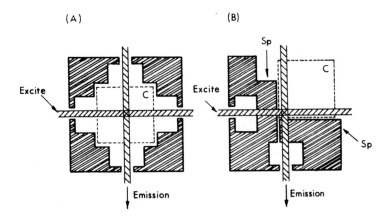

FIG. 76. Cuvette placement in the Aminco-Bowman spectrophotofluorometer. Cell holders are shown diagrammatically as seen from above. (a)
Usual placement of a 1 x 1-cm cuvette illuminated and observed through 1-mm
slits. (b) Modified cell holder that has had parts of two sides removed to enable
the cell (C) to be placed, with the help of spacers (Sp) in such a way that only a
corner of the cell is both illuminated and observed. The spacers depicted are
3 mm thick. Reprinted from Ref. [17] by courtesy of R. A. Chen and J. E.
Hayes.

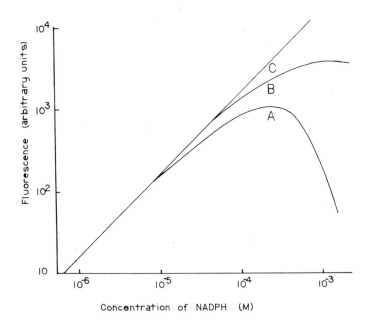

FIG. 77. Fluorescence-versus-concentration curves for NADPH, excited at
350 nm. Bottom curve (A) obtained with a 1 x 1-cm cell centrally placed as
shown in Fig. 75a. Middle curve (B) obtained with an eccentrically placed
1 x 1-cm cell as shown in Fig. 75b, but with 2-mm spacers. The top curve (C)
is a straight line indicating the ideal response. Log-log plot. Reprinted from
Ref. [17] by courtesy of R. A. Chen and J. E. Hayes.

minimized. By using microcells and a 395-nm excitation, Chen and Hayes were able to measure NADPH up to 10^{-3} M.

Alternatively, corrections for self-absorption can be applied since absorption by the fluorophor reduces fluorescence to the same extent that it diminishes transmission. Correction factors are derived for correcting to 100% transmission. Parker and associates [18,19] were the first to describe such methods of correcting for self-absorption.

XI. STANDARDIZATION, CALIBRATION, AND CORRECTION

A. Standardization

Because of differences in lamp intensity and photomultiplier sensitivity, as well as other instrumental variables, it is impossible to obtain exactly the same reading each and every day on a fluorometer. Hence the instrument should be standardized and set to a constant sensitivity each day. Generally a solution of a stable fluorophor of definite, known concentration is used to set the instrument to a given level. One of the most widely accepted standards is quinine sulfate, which has a λ_{ex} of 350 nm and a λ_{em} of 450 nm. This compound is particularly good for assays that use the 365-nm mercury line. A solution of quinine sulfate, say 10^{-5} M, is prepared and is used to set the instrument to a preset level (e. g. , 80 on a particular scale) every day.

Quinine sulfate in 0.1 N sulfuric acid, although a good standard, has certain limitations:

1. It is useful as a standard only for substances that are close to it in fluorescent properties (350 and 450 nm). In other regions other standards, such as indole, phenols, and fluorescein, have been used.

2. It is light sensitive, particularly at concentrations less than 1 μg/ml, and should be kept refrigerated in a black bottle.

3. It will appear to decrease in concentration at low concentrations (<50 μg/ml) due to adsorption onto glass surfaces [20]. For best results a more concentrated stock solution of quinine sulfate should be stored and dilutions made when necessary.

A set of six fluorescence standards useful in standardization of any fluorometer is sold by Perkin-Elmer (Norwalk, Conn.). The set consists of six fluorescent compounds dissolved in a plastic matrix: (a) a mixture of anthracene and naphthalene, (b) orilene, (c) p-terphenyl, (d) tetraphenylbutadiene, (e) a proprietary mixture, and (f) Rhodamine B. By using any one of these six standards, any region can be standardized. Since the samples are solid (they look like plastic blocks), they are stable and can be used indefinitely without

special storage. The set is recommended for anyone who does routine fluorescence measurements.

Alternatively, the Raman spectrum of the solvent could be used to standard-
ize instrument sensitivity daily [6]. Since the Raman peak is always located near
the excitation maximum, this is a more versatile method of standardization. The
Raman peak appears on the spectrum as the internal standard of sensitivity.

B. Calibration of Wavelength

Original calibration is performed at the factory, but every spectrofluorometer must be calibrated periodically to ascertain that the reading on the wavelength dial is the true wavelength. Differences could cause loss in sensitivity
and error in an assay.

In calibration the difference between the apparent wavelength and the known
wavelength for a series of mercury emission lines is used as a test of wavelength
accuracy. A low-pressure mercury-arc lamp (Pen Ray quartz lamp, Ultra-
Violet Products, Inc., San Gabriel, Calif., or equivalent) is placed in the
sample-cell holder of the instrument. The 12 mercury lines at 253.65, 296.73,
302.15, 313.16, 334.15, 366.33, 404.66, 407.78, 435.84, 546.07, 576.96, and
597.07 nm are used. The second-order lines at 626.32, 668.3, 732.66, and
809.32 nm can also be used for calibration if desired. For each line the position
of the wavelength dial is adjusted to give maximum signal, and the wavelength
reading is recorded. The difference between the observed reading and the true
reading represents the correction factor that must be applied to the reading on
the dial to give the true value. To compensate for dial backlash, always adjust
the dial to the peak reading from the same side. In scanning instruments turn
the dial to the peak in the same direction as the dial is turned by the scan motor.

To adjust the excitation monochromator once the emission monochromator
has been adjusted, place in the cuvette a suspension of Ludox (an aqueous suspension of colloidal silica, E. I. du Pont de Nemours, Wilmington, Del.).
Place the Pen Ray lamp in the standard position and adjust the excitation-
wavelength dial setting until a maximum meter signal is produced at each of the
previously determined wavelength settings for the emission monochromator.
The correction factors for each of the excitation wavelengths are determined for
future use.

In many cases the errors obtained in wavelength reading are sizable. For
example, Hercules [21] found an error of 15 to 20 nm in the fluorescence maxima
of 1-naphthol. He pointed out that the pen-and-ink recorders can also be
responsible for spectral errors unless they are checked at various recording
speeds.

Some practical aspects of the calibration of a fluorescence spectrometer (the Aminco -Bowman) have been described by Chen [22].

C.   Correction of Spectra

1.   General Remarks

The excitation and emission curves recorded from most spectrofluoro- meters are only approximate curves and do not represent the true spectra. If one simply wishes to report the maximum wavelengths of excitation and emission for a compound in describing an analytical procedure, he need only calibrate his instrument as described in the preceding section and need not worry about what the true spectrum looks like. If, however, one wants to obtain the true spectrum of a compound for publication, he must correct the spectrum. Table 31 summarizes the various methods that are available for correcting the excitation and emission sources.

Of course instruments are available for the direct recording of the excitation and emission spectra (see Chapter 2).

2.   Excitation Spectrum

The excitation spectrum obtained for any compound should agree with its absorption spectrum. In reality, it does not, as we discussed in Chapter 1. The true spectrum of quinine sulfate (curve Ac, Fig. 78) shows that the main excitation peak is at 250 nm, and not 350 nm, as indicated by the apparent spectrum (curve A, Fig. 78). Two of the main reasons for the differences are the following:

1.   The blaze of the grating. A typical grating blazed at 300 nm gives only 50% of this radiation at 200 nm and 80% at 400 nm (Fig. 79).

2.   The radiation emitted from the source varies as a function of wavelength.

The chemical actinometer method [23, 24] is inexpensive, simple, and accurate from 250 to 350 nm, but it is time consuming and tedious. After a solution of potassium iron(III) oxalate is placed in the solution cell of the spectrofluorometer and exposed to radiant energy for a set time, the amount of iron(II) that is produced is titrated:

$$2K_3Fe(C_2O_4)_3 + h\nu \longrightarrow 2FeC_2O_4 + 3K_2C_2O_4 + 2CO_2.$$

Alternatively, the intensity of the radiation source can be determined by using a photographic film that has uniform response over the spectral range measured [23].

A standardized calibrated photomultiplier tube can be purchased from the

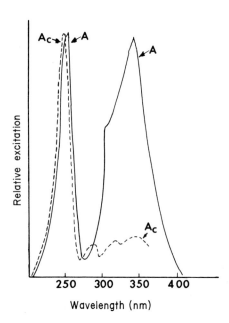

FIG. 78. Fluorescence excitation of quinine sulfate, 0.25 μg/ml in 0.05 M H$_2$SO$_4$: curve A, as recorded with a spectrofluorometer; curve Ac, corrected for quantum distribution of the excitation source.

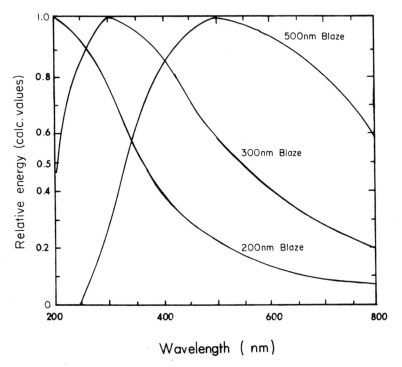

FIG. 79. Relative energy response (calculated values) from diffraction gratings blazed for maxima at 2000, 3000, and 5000 Å. Reprinted by courtesy of Bausch & Lomb.

manufacturer along with its response curve. The RCA 7200 tube, for example, covers the range 250 to 580 nm. The response curve is recorded, and the readings are corrected at 5-nm intervals from the curve furnished by the manufacturer (see Ref. 23 for more details).

A temperature-compensated thermopile (Kipp and Zonen No. E20-8170 [31]) can be used over the range 200 to 600 nm with good results. The method is simple and efficient [25], but it requires an electronic galvanometer together with the thermopile.

Two fluorescent methods have been described for the evaluation of the excitation source. Method 1 uses the fluorescence emission from the front surface of a concentrated fluorescent solution [26-28]. Rhodamine B is used. Method 2 uses a combination of the absorption curve and the excitation curve of the aluminum chelate of Pontachrome Blue Black R (PBBR) [25]. Figure 80 shows the absorption and excitation spectra of the Al-PBBR chelate.

Figure 81 compares the relative spectral energy distribution of the excitation source and monochromator for the thermopile, the RCA 7200 phototube, and the Al-PBBR chelate methods of correcting the excitation spectrum.

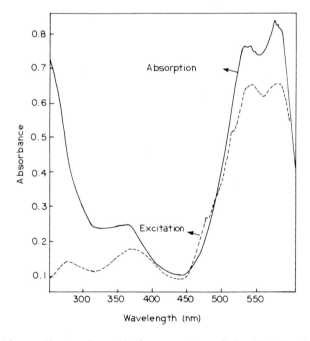

FIG. 80. Absorption and excitation spectra of the PBBR-aluminum chelate in 95% ethanol. Absorption concentrations: PBBR, 3.84×10^{-5} M; $AlCl_3 \cdot 6H_2O$, 3.3×10^{-3} M. Excitation concentrations: above solution diluted 2:25 with 95% ethanol. Reprinted from Ref. [25] by courtesy of the American Chemical Society.

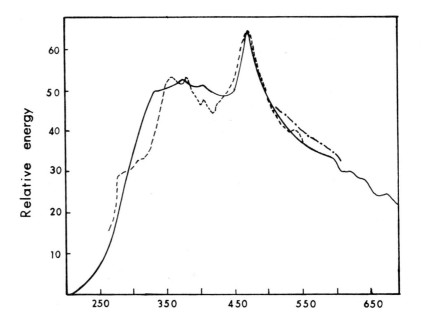

Wavelength (nm)

FIG. 81.   Relative spectral energy distribution of the excitation source (Osram xenon lamp): (——) thermopile, (•——•) phototube, and (- - -) PBBR-aluminum chelate methods.   An RCA 7200 photomultiplier tube, using the manufacturer's data as to spectral response.   Reprinted from Ref. [25] by courtesy of the American Chemical Society.

3.   Emission Spectrum

Like the excitation spectrum, the emission spectrum must be corrected to yield a true spectrum.   The emission curve is affected by both the blazing of the emission monochromator and the response of the detector phototube.   Both of these change with wavelength, and hence a correction factor must be applied. The emission grating is usually blazed at 500 nm and has a reduced output of radiant energy at 400 and 600 nm.   The photomultiplier tube that is most commonly used in commercial spectrofluorometers is the 1P28; it has a maximum response at 350 nm, reduced to 10% at 600 nm (Fig. 82).   The 1P21 photomultiplier tube has a maximum response at 400 nm, reduced to 10% of this value at 600 nm.

Correction factors for the emission spectra can be obtained by one of three methods (see Table 31):

1.   Standard lamp method.   In this method a calibrated tungsten lamp, set about 3 ft from the fluorometer, is used.   Its radiation is decreased by passage

TABLE 31

Methods of Correcting the Excitation Source and
Calibrating the Emission Unit

Method	Description	Refs.
Correction of the Excitation Source		
Chemical means:		
Chemical actinometer		23, 24
Photographic plate		23
Thermopile with galvanometer	Kipp thermopile with response independent of wavelength (e. g. , Kipp No. 4-8170), attached to an electronic galvanometer (e. g. , Kintel Model 204A), with a recorder attachment	25
Calibrated photomultiplier	A calibrated photomultiplier (e. g. , RCA 7200) can be purchased with its characteristic response curve	23
Fluorescent solution	1. A Rhodamine B solution (3 μg/l in ethylene glycol) is placed so that the emission may be recorded from the front surface	26-28
	2. The excitation curve of a fluorescent solution like the aluminum chelate of Pontachrome Blue Black R in ethanol is compared with its absorption curve	25
Calibration of the Emission Unit		
Standard-lamp method	A calibration tungsten lamp is set an appropriate distance from the fluorometer, and the light, with its intensity decreased by passage though a blackened copper-wire mesh, is sent directly through the slit of the emission unit; the intensity is recorded with the phototube which is to be used for future work	23, 25

Method	Description	Refs.
	Calibration of the Emission Unit--Continued	
Fluorescent-solution method	Certain compounds have been found to serve as satisfactory emission standards; these may be used in dilute solutions with the sample cell in its usual position	25, 29
Reflection of the calibrated source into the emission unit		30

through a blackened copper-wire mesh and then is sent through the slit of the emission unit. The intensity is recorded with the phototube to be used. This method is described in detail in Refs. 23 and 25.

2. Fluorescent-solution method. Argauer and White [25] have described a relatively simple method for determining the correction factors for the emission unit by comparing the emission curve obtained on an instrument with that obtained with certain fluorescent dyes on a standardized instrument. Five standard fluorescent compounds were found to be useful over the range 450 to

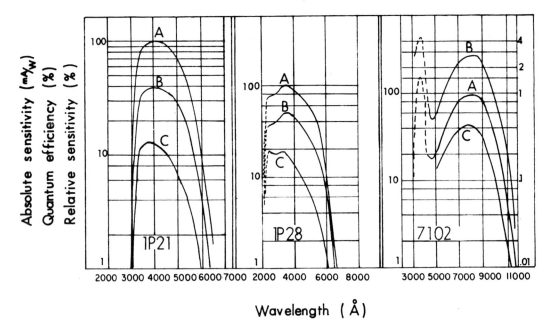

FIG. 82. Average relative responses of 1P21, 1P28, and 7102 photo-multiplier tubes. Reprinted by courtesy of the Radio Corporation of America.

750 nm: quinine sulfate, 3-aminophthalimide, m-nitrodimethylaniline, the
aluminum chelate of PBBR and 4-dimethylamino-4'-nitrostilbene (Fig. 83).

3.   Reflection of the excitation source.   Once the corrected spectrum source
has been found, the emission spectrum can be easily corrected by reflection of
the exciting radiation into the emission slit.   Eastman [30] found that a solution
of polystyrene (Styron 666, Dow Chemical Company) in methyl ethyl ketone
scatters a reproducible fraction of radiation from the excitation source into the
emission unit.   Borresen [31] used both Rhodamine B and fluorescein as
excitation monitors with a magnesium reflector to obtain the corrected emission
spectra in the ultraviolet and visible regions.

D.   Reporting Fluorescence Spectra

In 1963 a committee of international scientists made suggestions on the
uniform reporting of fluorescent spectra [32].   In reporting a routine method of
analysis the spectra need not be corrected since only the wavelengths of
maximum excitation and emission need be reported.   However, if one wants to
report the spectrum of a substance in a publication it is highly desirable, if his
data are to be meaningful to others, that he follow the following recommenda-
tions:

1.   The spectrum should be corrected for variations in photomultiplier
sensitivity, spectrometer dispersion, and light losses.   An instrument-
calibration curve should be prepared by using the standard procedures described
in the preceding sections.   If the correction of the spectrum is not vital to the
study being reported, reference should be made to the correction curve of the
instrument used so that others may apply the correction themselves.

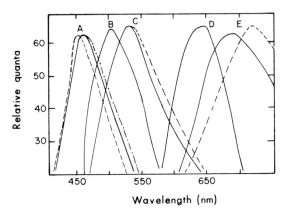

FIG. 83.   Fluorescence-emission standards: curve A, quinine sulfate;
curve B, 3-aminophthalimide; curve C, m-nitrodimethylaniline; curve D, PBBR-
aluminum chelate; curve E, 4-dimethylamino-4'-nitrostilbene.   Reprinted from
Ref. [25] by courtesy of the American Chemical Society.

2. The corrected spectrum should be plotted as relative quanta per unit frequency interval on the vertical scale versus wavelength in reciprocal microns (μ^{-1}) or centimeters (cm^{-1}) on the horizontal scale. When possible, the absolute frequency efficiency should be stated or quantitatively compared with a suitable standard solution (e. g. , quinine sulfate in 0. 1 N H_2SO_4 at stated concentration and absorbance per centimeter).

3. The spectrum should be corrected for "background" (fluorescence from solvent or cuvette, Raman emission, scattered radiation, etc.). Alternatively, the total "background" should be recorded on the graph by taking measurements on the pure solvent.

4. The following experimental details should be stated: (a) the spectrum of the exciting light--if a monochromator is used, the light source and the band-width; if filters are used, the transmission curves and the nature of the light source; (b) the geometrical arrangement of the cuvette, the path lengths of the exciting and emitting radiations through the solution, the absorbance per centimeter of solution for the frequency of exciting light, and an estimate of the distortion of the spectrum by self-absorption; (c) purity and concentration of the solute, and the nature of the solvent; (d) temperature of the solution; (e) whether the solution is aerated or deaerated; and (f) the type of analyzing monochromator and the bandwidth at an appropriate frequency.

The authors [32] believe that, although these proposals would involve the reporting of much more detail than usual, the information provided would make the spectra of maximum and lasting value to other workers.

REFERENCES

1. C. A. Parker, Proc. Soc. Anal. Chem. , 3, 158 (1966).
2. C. A. Parker and T. A. Joyce, J. Chem. Soc. A, 821 (1966).
3. S. J. Ladner and R. S. Becker, J. Phys. Chem. , 67, 2481 (1963).
4. M. S. Walker, T. W. Bednar, and R. Lumry, J. Chem. Phys., 45, 3455 (1966).
5. B. B. Brodie, S. Udenfriend, and J. E. Baer, J. Biol. Chem., 168, 299 (1947).
6. C. A. Parker, Analyst, 84, 446 (1959).
7. J. M. Price, M. Kaihara, and H. K. Howerton, Appl. Optics , 1, 521 (1962).
8. R. F. Chen and R. L. Bowman, Science, 147 , 729 (1965).
9. M. W. McDermott and R. Novick, J. Opt. Soc. Amer., 51, 1008 (1961).
10. R. F. Chen, Anal. Biochem., 14, 497 (1966).
11. R. F. Chen, Biochem. Biophys. Res. Commun., 17, 141 (1964).

12. O. H. Lowry, N. R. Roberts, K. Y. Leiner, M. L. Wu, A. L. Farr, and R. W. Alberts, J. Biol. Chem., 207, 1 (1954).

13. B. Chance and V. Legallais, Rev. Sci. Instr., 30, 732 (1959).

14. B. Chance, P. Cohen, F. Jobsis, and B. Scoener, Science, 137, 499 (1962).

15. B. Chance, V. Legallais, and B. Schoener, Nature, 195, 1073 (1962).

16. G. L. Wied, A. M. Messina, J. I. Manglano, P. Meier, and R. R. Blough, Acta Cytol., 8, 408 (1964).

17. R. F. Chen and J. E. Hayes, Anal. Biochem., 13, 523 (1965).

18. C. A. Parker and W. J. Barnes, Analyst, 82, 606 (1957).

19. C. A. Parker and W. T. Rees, Analyst, 85, 587 (1960).

20. L. H. Bird, New Zealand J. Sci. Technol., 303, 334 (1949).

21. D. M. Hercules, Science, 125, 1242 (1957).

22. R. F. Chen, Anal. Biochem., 20, 339 (1967).

23. C. E. White, M. Ho, and E. Weimer, Anal. Chem., 32, 438 (1960).

24. C. G. Hatchard and C. A. Parker, Proc. Roy. Soc. (London), A235, 518 (1956).

25. R. Argauer and C. E. White, Anal. Chem., 36, 368 (1964).

26. C. A. Parker, Anal. Chem., 34, 502 (1962).

27. R. F. Chen, Anal. Biochem., 20, 339 (1967).

28. H. C. Borresen and C. A. Parker, Anal. Chem., 38, 1073 (1966).

29. E. Lippert, W. Noegel, and I. Sieboldfalkenstein, Z. Anal. Chem., 170, 1 (1959).

30. J. W. Eastman, Appl. Opt., 5, 1125 (1966).

31. H. C. Borresen, Acta Chem. Scand., 19, 2089 (1965).

32. J. H. Chapman, T. Forster, G. Kortum, C. A. Parker, E. Lippert, W. H. Melhuish, and G. Nebbia, Appl. Spectry., 17, 171 (1963).

Chapter 5

PHOSPHORESCENCE

I. HISTORICAL DEVELOPMENT

In 1944 Lewis and Kasha [1] identified the phenomenon of phosphorescence as a transition from a metastable triplet state and hinted that molecules might be identified by their emission spectra. The design and construction of a phosphorimeter were described by Freed and Salmre [2] in 1958. These authors reported the luminescence characteristics of indole, 5-hydroxytryptamine, tryptophan, and reserpine. They found the sensitivities of analysis for these compounds to be 10 times greater with phosphorimetry than with fluorometry; the sensitivities were limited mainly by the presence of a high solvent background and residual quartz luminescence.

161

   In 1957 Keirs, Britt, and Wentworth [3] critically evaluated a number of
analytical methods based on the phenomenon of phosphorescence. Among the
techniques that were demonstrated to be of analytical importance for improving
the selectivity of analysis were excitation resolution, emission resolution, and
phosphoroscopic resolution. A synthetic mixture of 4-nitrobiphenyl, benzaldehyde, and benzophenone (decay times of 0.08, 0.006, and 0.006 sec, respectively)
was resolved into a single fast-decaying component and two slower decaying
species. Benzaldehyde and benzophenone were resolved by measurement of the
phosphorescence emission at two different wavelengths (emission resolution)
using simultaneous equations. Acetophenone and benzophenone (decay times of
0.008 and 0.006 sec, respectively) were resolved phosphoroscopically with the
aid of two simultaneous equations. Triphenylamine and diphenylamine were
resolved by selective excitation at two different wavelengths (excitation
resolution) using simultaneous equations. Simultaneous equations are necessary
if and only if instrumental resolution is not completely effective. The
instrumentation included a high-pressure mercury arc with filters for excitation
resolution, a Becquerel phosphoroscope with resolution times of 0.001 to 0.02
sec, and either a spectrograph for the photographic recording of spectra or a
photomultiplier for the measurement of total emission during phosphoroscopic
resolution.

   The possibilities of chemical analysis by phosphorimetry were reviewed in
1962 by Parker and Hatchard [4]. A modified spectrofluorometer was used for
the measurement of weak phosphorescence spectra. Two quartz-prism
monochromators, two separate choppers driven by synchronous motors, a
photomultiplier for emission measurement, and a photomultiplier for monitoring
the exciting radiation comprised the instrumental components. Rapidly decaying
phosphors having ratios of phosphorescence intensity to fluorescence intensity as
small as 10^{-5} were studied. Emission spectra were corrected for
monochromator-photodetector characteristics. The areas of the emission
spectra were proportional to the respective fluorescence and phosphorescence
quantum efficiencies. A number of fluorescence and phosphorescence quantum
efficiencies were tabulated for aromatic compounds at 77°K. It was concluded
that all major applications of phosphorimetry would probably require measurements in a rigid solution at low temperatures, which led Parker and Hatchard to
believe that phosphorimetry would be applicable only when fluorometry at room
temperature was insensitive or nonspecific. Parker and Hatchard also
investigated the phosphorescence emission of several compounds at room
temperature and found weak but measurable emission. In spite of this lack of
sensitivity, it was still possible to measure low concentrations of some
compounds.

Freed and Vise [5] constructed a spectrophosphorimeter to determine the phosphorescence excitation and emission spectra of N-acetyl-L-tyrosine ethyl ester, whose spectrum was similar to that of tryptophan. They used a solvent of water-methanol-ethanol in a ratio of 5 : 11 : 4, v/v/v. Chips of fused quartz were added to the sample cells to prevent cracking of the solvent glasses at 77°K. An internal standard of benzyl alcohol was also employed to account for such instrumental instabilities as source drift and sample positioning. Freed and Vise concluded that solvent purity was the chief limitation in phosphorimetry.

In 1963 Winefordner and Latz [6] described the construction of a spectrophosphorimeter consisting of an unfiltered 150-W mercury arc for excitation, a rotating-can phosphoroscope, and a grating monochromator for the measurement of phosphorescence emission spectra. Latz [7] has reported a thorough study of the procedural factors that influence phosphorimetry as a means of chemical analysis.

Phosphorometry has been of limited value in the past because of marginal precision and accuracy, of solvent limitations and of difficulties and time of sampling. In 1968 Hollifield and Winefordner [8] described a rotating-sample-cell method for increasing the precision of low-temperature phosphorescence measurements. In conventional phosphorescence analysis the sample cell is stationary; the achievable relative standard deviation is in the 10 to 20 % range. With a rotating sample cell, the achievable relative standard deviation was about 1 to 2 %, representing an improvement by a factor of 10. Moreover, cracking of the sample on freezing was less of a problem with the rotated cell. This observation was probably one of the most important discoveries in the field, since it changed phosphorescence to a more quantitative technique.

Other significant advances include the use of a more stable source power supply and the use of aqueous solvents. Aqueous solutions (20 µl) can be placed in a quartz capillary tube by capillary action and no cell cracking is observed on cooling to 77°K.

II. THEORETICAL CONSIDERATIONS

A. General Remarks

Each energy level can be occupied by two electrons that must have opposite spins, designated as plus and minus. If all the electrons are paired in this way, the system is in the singlet state. However, if the atom or molecule has two unpaired electrons, both having the same spin, it is in a triplet state. The lowest energy level available, the ground state, is a singlet state. If the absorption of energy causes one of the electrons to be raised to a higher vacant

level, without change in spin, the result is an excited singlet state. If a change
in spin occurs, the result is an excited triplet state.

A molecule can undergo a variety of possible electron-energy transitions,
accompanied by the absorption or emission of light. If a pair of π electrons are
excited to a higher π level, an antibonding state designated as π^*, the resulting
state is a π, π^* singlet if no change in spin has occurred, and a π, π^* triplet if
the spin has flipped over to the opposite sign.

The emission of light from a π, π^* singlet is fluorescence. If the excited
state is a π, π^* triplet, the higher improbability of a spin-flipping transition
back to the ground state ($T_1^* \longrightarrow S_0$) causes the light emission to be greatly
delayed, and the result is phosphorescence. Phosphorescence is the release of
electromagnetic radiation from a photon-excited state of a molecule when it
returns to the ground state via a triplet-to-singlet transition. Because more
energy is lost in the process, the wavelength of phosphorescence is shifted to
longer wavelengths (lower energy) than that of fluorescence (Fig. 83).

The greatest difference between fluorescence and phosphorescence is the
afterglow observed in phosphorescence. When one cuts off the excitation
radiation, fluorescence ceases, but phosphorescence persists for some time.
The duration of the afterglow is expressed as its half-life or mean lifetime.

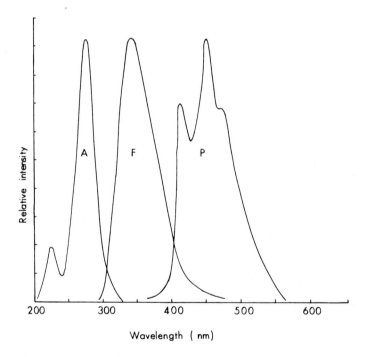

FIG. 84. Absorption (A), fluorescence (F), and phosphorescence (P)
spectra of tryptophan.

A second characteristic of phosphorescence is that the excited state is paramagnetic. This can be shown by magnetic susceptibility measurements [9] and electron-spin-resonance experiments [10].

Another distinction of phosphorescence is that the lifetime of the excited state and the intensity of emission are extremely sensitive to heavy atoms and paramagnetic ions.

B. Structural Effects

The effect of halogen substitution on the luminescence of aromatic hydrocarbons is of considerable importance. One generally observes that as the substituent series fluorine, chlorine, bromine, iodine is traversed, phosphorescence is increasingly favored relative to fluorescence. This effect is illustrated in Table 32, which shows the ratio of the quantum efficiency of phosphorescence to fluorescence, Φ_P/Φ_F, for a number of substituted naphthalenes and fluoresceins. The ratio Φ_P/Φ_F increases across the series F $\longrightarrow$ I.

The trends for yields of luminescence and excited-state lifetimes in haloaromatics can be rationalized only if one postulates that heavy-halogen substitution increases the rate of intersystem crossing from a singlet to a triplet state (a radiationless process) and the subsequent triplet-to-singlet transition (a radiative process). Halogen substitution must therefore increase the extent of spin-orbit coupling in aromatic systems, the increase being larger for heavier halogens. This perturbation is commonly termed the heavy-atom effect. In the case of the substituted naphthalenes the increase in the phosphorescence-to-fluorescence ratio is primarily due to an increased probability of singlet-to-triplet transitions. However, in the fluorescein series the primary effect is the result of an increased nonradiative transition from the excited state back to the ground state.

Since fluorescein and dichlorofluorescein are highly fluorescent, but not phosphorescent, whereas the corresponding bromosubstituted and iodosubstituted derivatives are weakly fluorescent, but are phosphorescent, a study of the luminescence properties of the fluorescein family provides an interesting introduction to some of the structural factors affecting luminescence.

Paramagnetic ions also effect a transition from the singlet to the triplet, with a resulting increase in phosphorescence intensity. The metal chelates of these ions are generally phosphorescent.

TABLE 32

Substituent Effects on the Luminescence of Naphthalenes and
Fluoresceins in EPA[a]

Compound	Φ_P/Φ_F	Φ_F	λ_F (nm)	λ_P (nm)
Naphthalene	0.093			
1-Chloronaphthalene	5.2			
1-Bromonaphthalene	6.4			
1-Iodonaphthalene	>1000			
Fluorescein	0	0.83	527	
2',7'-Dichlorofluorescein	0	0.79	538	
2',5'-Dibromofluorescein	0.21	0.29	540	650
4',5'-Diiodofluorescein	1.05	0.054	544	667

[a] A mixture of diethyl ether, isopentane, and ethanol, 5:5:2 (v/v/v).

C. Choice of Experimental Conditions; Signal-to-Noise Ratio Theory [11]

1. General Considerations [12]

The method of sample illumination and observation and the pertinent
distances are shown in Fig. 85. The assumptions necessary to facilitate
discussion are the following: The excitation and emission monochromator are
identical in all respects; that is, they have the same entrance and exit slit widths
W and slit heights H, the same reciprocal linear dispersion R_d and hence the
same spectral bandwidths s, the same apertures, that is, the same solid angle
Ω of radiation collected, and the same transmittances t of all optical components.
The sample excitation is made at the peak of the uncorrected phosphorescence
excitation spectrum, and the sample phosphorescence is measured at the peak of
the uncorrected phosphorescence emission spectrum. The source area is larger
than the sample-cell area perpendicular to the exciting radiation, and the optical
axis of the emission monochromator is larger than the slit area, WH. A photomultiplier of known characteristics is used to detect the phosphorescence
radiation. It is also assumed that the emission and excitation bands have a
Gaussian profile and half-widths of $\Delta\lambda$ and $\Delta\lambda'$ (in nanometers), respectively.

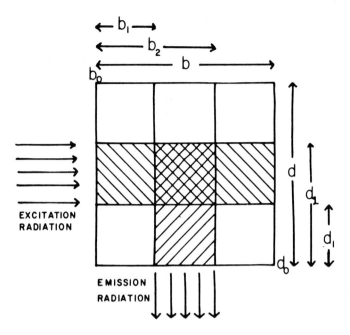

FIG. 85.   Geometry of sample cell and method of illumination and observation.

2.   The Intensity of Phosphorescence [12]

The phosphorescence intensity I_p, in W/cm-ster-nm, is given by

$$I_p = K_0 I_0 f_1 f_2 \Phi_s a' f_3 f_4 f_5. \tag{33}$$

The radiant power from the source radiation passing through the excitation monochromator and incident on the sample-cell surface (at b_0) is given by $K_0 I_0$, where I_0 is the source intensity, in W/(cm^2-ster-nm), and K_0 is the excitation-monochromator factor that accounts for the solid angle of radiation Ω in steradians, collected by the excitation monochromator, the transmittance t of the excitation monochromator and entrance optics for the source, the reciprocal linear dispersion, R_d, in nanometers per centimeter, of the excitation monochromator, and the slit width W and height H, in centimeters, and is given by

$$K_0 = t \Omega H R_d W^2. \tag{34}$$

The fraction of radiant power, f_1, absorbed by impurity species m is given by

$$f_1 = \exp(-2.3 b_1 \sum_m a_m C_m), \tag{35}$$

where b_1 is the distance, in centimeters, given in Fig. 85, a_m and C_m are the molar absorptivity at the excitation wavelength and the molar concentration of impurity species m, and the summation is over all impurities. Thus $K_0 I_0 f_1$ is the radiant power at plane b_1 in Fig. 85, and this power is available for the excitation of the sample s. The fraction of radiant power absorbed by only the sample in the region b_1 to b_2 (this is the only region that is effective in producing a measurable signal) is given by

$$f_2 = \exp(-2.3 b_1 a_s C_s) - \exp(-2.3 b_2 a_s C_s), \tag{36}$$

where a_s and C_s are the molar absorptivity at the excitation wavelength and the molar concentration of sample species s, respectively. Therefore, $K_0 I_0 f_1 f_2$ is simply the radiant power, in watts, absorbed by the sample in the observation region b_1 to b_2 (d centimeters long).

Of course only a fraction of the absorbed power $K_0 I_0 f_1 f_2$ is reemitted as phosphorescence, and this fraction is $\Phi_s a'$, where Φ_s is the power efficiency (watts of radiant power emitted per watt of radiant power absorbed) and a' is the observation-efficiency factor that accounts for light losses in using a rotating shutter. Of course the phosphorescence radiation is emitted isotropically, and so to convert the radiant phosphorescence power $K_0 I_0 f_1 f_2 \Phi_s a'$ to a phosphorescence intensity I_p, we must multiply by f_3, which is the reciprocal of the product of the sample area A_s, the half-width of the emission band, $\Delta \lambda$, and 4π, the number of steradians in a sphere, that is,

$$f_3 = \frac{1}{4\pi A_s \Delta \lambda'}. \tag{37}$$

Of course, $K_0 I_0 f_1 f_2 \Phi_s a' f_3$ is the intensity of phosphorescence if no self-absorption of radiation occurs in the region d_2 to d_1 and if no absorption of fluorescence by sample and impurities occurs in the unexcited region d_1 to d_0. The fraction f_4 accounts for self-absorption, and the fraction f_5 accounts for absorption in the region d_1 to d_0. The parameters f_4 and f_5 are given by

$$f_4 = 1 + \exp[-2.3(d_2 - d_1)\Sigma_j a_j' C_j] \tag{38}$$

and

$$f_5 = \exp(-2.3 d_1 \Sigma_j a_j' C_j), \tag{39}$$

where the d terms are the distances, in centimeters, shown in Fig. 84, the a_j' and C_j are the molar absorptivity at the emission wavelength and the molar concentration, respectively, and the summations are taken over all impurity species and the sample s.

3. The Readout Signal Due to Phosphorescence Radiation Reaching

the Photodetector [12]

The readout voltage signal E_s, in volts, is given by

$$E_s = I_P K_0 \gamma R_L G_e. \tag{40}$$

The radiant power, in watts, of phosphorescence reaching the photocathode of
the photodetector is $I_P K_0$ (K_0 is the same as defined for Eq. (33)). The
photoanodic current, in amperes, due to the radiant power $I_P K_0$ striking the
photoanode is found by multiplying by γ, the photoanodic sensitivity of the
photodetector, in amperes at anode per watt of radiant power striking the
photocathode. The resulting photoanodic current is then dropped across a load
resistor of R_L ohms, resulting in a voltage, which is fed into an electrometer of
overall gain G_e and in an output voltage of E_s volts.

The complete expression for E_s in terms of the above parameters can be
given in detail by simply substituting for the appropriate parameters in Eq. (40).
This will not be done here because several major points can be made by using
the simpler expression resulting after making several analytically useful
assumptions: that no impurities are present in the solution; that the sample in
the entire cell is excited (i. e. , d_2 = d and d_1 = 0); that the sample

phosphorescence in the entire cell is observed (i. e. , b_2 = b and b_1 = 0); and
that the sample does not absorb appreciably at the peak phosphorescence
wavelength (i. e. , a_s' = 0). From the above simplifications and expressions,
E_s is given by

$$E_s = \frac{\gamma R_L G_e I_0 K_0^2 \Phi_s a'}{4\pi A_s \Delta\lambda'} \; [1 - \exp(-2.3a_s bC_s)]. \tag{41}$$

Two very useful limiting cases of Eq. (41) should be noted. If the sample
concentration is low, then $1 - \exp(-2.3a_s bC_s) \tilde{=} 2.3a_s bC_s$, and so

$$E_s = \frac{2.3\gamma R_L G_e I_0 K_0^2 \Phi_s a' a_s bC_s}{4\pi A_s \Delta\lambda'}. \tag{42}$$

Therefore the readout signal is directly proportional to sample concentration.
If the sample concentration is high, as in the case of quantum counters,
$1 - \exp(-2.3a_s bC_s) \tilde{=} 1.0$, and so

$$E_s = \frac{\gamma R_L G_e I_0 K_0^2 \Phi_s a'}{4\pi A_s \Delta\lambda'}. \tag{43}$$

In this case, E_s is independent of sample concentration. However, E_s is still
a linear function of Φ_s, and hence the name "quantum counter. " Therefore
quantum counters are useful for monitoring source intensity.

An analytical curve is a plot of $\log E_s$ versus $\log C_s$. Such a plot should have a slope of unity at low concentrations and a slope of zero at high sample concentrations. Actually at high sample concentrations, severe curvature of analytical curves can occur because of a decrease in Φ_s due to self-quenching and due to the self-absorption factor f_4 and the reabsorption factor f_5. The presence of impurities can also seriously influence the shape of analytical curves.

4. The Readout Noise [12]

The root-mean-square (rms) readout noise signal, $\overline{\Delta E}_s$, in volts, is given by

$$\overline{\Delta E}_s = \overline{\Delta i}_s R_L G_e , \tag{44}$$

where $\overline{\Delta i}_s$ is the total rms photoanodic noise current, in amperes, due to all sources. It is assumed that all noises are random (white) at all frequencies. The total rms noise current $\overline{\Delta i}_s$ in phosphorimetry is given by

$$\overline{\Delta i}_s = (\overline{\Delta i}_p^2 + \overline{\Delta i}_f^2)^{1/2} , \tag{45}$$

where $\overline{\Delta i}_p$ is the rms phototube shot noise, in amperes, and $\overline{\Delta i}_f$ is the rms source flicker noise, in amperes, due to flicker in the light source arising from arc wander or due to flicker in the exciting and luminescent radiation due to turbulence in the thermostating medium surrounding the sample. The rms shot noise $\overline{\Delta i}_p$ is given by

$$\overline{\Delta i}_p = [2eBM\Delta f(i_t + i_b + i_s)]^{1/2} , \tag{46}$$

where the terms in parentheses include all significant contributions to the net phototube current due to background (blank), i_s is the photoanodic current due to the sample, and i_t is the thermionic dark current. The term Δf is the frequency-response bandwidth of the amplifier-readout system with net gain G_e, e is the charge of the electron, in coulombs, and BM is the effective gain of the photo-tube. The term BM is approximately given by

$$BM = G^x + G^{x-1} + \ldots + G + 1 , \tag{47}$$

where G is the gain per stage of the phototube and x is the number of stages (dynodes). The photoanodic current due to background emission, i_b, is a summation of contributions from all different background species. This term is best measured experimentally since the identity and concentration of back-ground species are seldom known. The photoanodic current due to the sample emission, i_s, can be found by dividing E_s by $R_L G_e$. Of course, near the minimum detectable sample concentration, $i_s < (i_t + i_b)$ and can be omitted from Eq. (47). The rms source flicker noise $\overline{\Delta i}_f$ is given approximately by

$$\overline{\Delta i}_f = (\zeta \sqrt{\Delta f} + \epsilon \sqrt{\Delta f}) (i_s + i_b), \tag{48}$$

where ζ is the ratio of the rms fluctuations, in units of $\sec^{1/2}$, in the source intensity (the fluctuations are due to fluctuations in the source itself and due to fluctuations in the thermostating medium, e. g., flicker in incident radiation due to convection of the thermostating medium), ϵ is the ratio of the fluctuation in the emitted luminescence intensity due to fluctuations in the thermostating medium to the emitted intensity in $\sec^{1/2}$, and Δf is the frequency-response bandwidth of the amplifier-readout system. If the decay time of the luminophor is large, then Δf is determined also by the integrating effect of the luminophor; however, in this discussion this effect is neglected. The parameter ζ is related to ϵ by

$$\zeta = (\epsilon^2 + \delta^2)^{1/2}, \tag{49}$$

where δ is the source flicker factor, that is, the ratio of rms fluctuation in the source intensity to the source intensity itself.

The total rms read-out noise voltage $\overline{\Delta E}_s$ is therefore obtained by combining the above expressions, and so

$$\overline{\Delta E}_s = R_L G_e \sqrt{\Delta f} \left[2eBM(i_t + i_b + i_s) + (\delta^2 + 2\epsilon^2 + 2\epsilon\sqrt{\delta^2 + \epsilon^2})(i_s + i_b)^2 \right]^{1/2} \tag{50}$$

5. The Signal-to-Noise Ratio [12, 13]

The signal-to-noise ratio $E_s / \overline{\Delta E}_s$ can be obtained by simply dividing Eq. (41) by Eq. (50). This is not done here, however, but rather it will be left up to the reader to perform this simple operation.

6. Optimization of Experimental Conditions for Maximum Precision [12, 13]

Experimental conditions resulting in the maximum signal-to-noise ratio $E_s/\overline{\Delta E}_s$ also result in the maximum precision of measurement. If systematic errors are present, these same conditions also result in the greatest accuracy of measurement. The optimum value (if one exists) of any variable parameter can theoretically be found in two ways: by differentiation of $E_s / \overline{\Delta E}_s$ with respect to X and maximizing; and graphically, plotting $E_s / \overline{\Delta E}_s$ versus X, this indicates not only whether or not a maximum is present but also the sensitivity of $E_s / \overline{\Delta E}_s$ to variation in the parameter X.

If the differentiation method is used, the optimum value of X, namely, X_0, is obtained by using the expression

$$\left[\frac{\partial (E_s / \overline{\Delta E}_s)}{\partial X} \right]_{Y, Z, \text{ etc.}} = 0, \tag{51}$$

where the partial derivative of $E_s / \overline{\Delta E}_s$ with respect to X is taken, keeping all other parameters constant and then solving for X_0. In most cases such a procedure results in a maximal value of $E_s / \overline{\Delta E}_s$ rather than a minimum. To determine unambiguously that a maximum exists, the second derivative of $E_s / \overline{\Delta E}_s$ with respect to X can be taken, and if a negative value results, a maximum does in fact exist. If X and Y are interrelated, then the optimum values of X and Y can be found by taking the partial derivative with respect to X and solving for X_0. Using this value of X_0, a first approximation of the optimum value of Y_0 can then be found by using an equation similar to Eq. (51) but in terms of Y. Then using this new value of Y_0, X_0 must be redetermined. This reiterative process must be continued until the values of X_0 and Y_0 are consistent.

The above method of maximization of $E_s / \overline{\Delta E}_s$ is not useful for variable parameters having no optimum value. In addition, in cases where an optimum does or does not exist, such a maximization procedure does not give any information concerning variation in $E_s / \overline{\Delta E}_s$ due to variation in the parameter, X.

The graphical method, however, does allow determination not only of the optimum value of X but also of the sensitivity of $E_s / \overline{\Delta E}_s$ to variation in X. The graphical method entails the plotting of $E_s / \overline{\Delta E}_s$ versus X (keeping Y, Z, etc., constant). A maximum in the curve indicates an optimum value of X for the fixed values of Y, Z, etc. If X and Y are interrelated variables, then surface plots must be made; that is, $E_s / \overline{\Delta E}_s$ versus X on one axis and Y on another axis for a constant value of Z.

Figure 86 shows plots of $E_s / \overline{\Delta E}_s$ versus W, the monochromator-slit width, for various sample concentrations of a typical organic phosphor measured by using the Aminco spectrophotofluorometer phosphoroscope attachment. In this case, which is typical of most organic phosphors, $E_s / \overline{\Delta E}_s$ increases with W and in fact never reaches a plateau for low sample concentrations, even when values of W as large as the mechanical limits on the monochromator slits are used. However, at larger sample concentrations a plateau in $E_s / \overline{\Delta E}_s$ values is reached at experimentally attainable slit widths. Therefore, if one were attempting to analyze quantitatively a dilute solution for this species, it would be advantageous to use the smallest slit widt W giving a value of $E_s / \overline{\Delta E}_s$ on the plateau. The spectral bandwidth of the monochromator is $R_d W$, and so the smaller the value of W, the smaller the spectral bandwidth that will result in

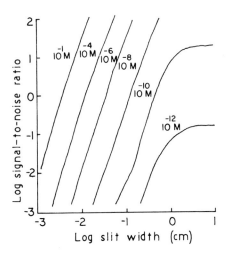

FIG. 86. Plots of signal-to-noise ratio versus monochromator-slit width as a function of solution concentration.

greater selectivity of measurement. Of course it may be impossible to achieve maximum precision, sensitivity, and selectivity because in most cases sensitivity and precision increase as W increases, whereas resolution decreases with increase in W. Therefore a compromise must normally be made between sensitivity, precision, and resolution, depending on the experimental system.

Most instrumental parameters on most commercial luminescence spectrometers do not have to be optimized since they are already fixed (e. g., H, t, Ω, R_d, I_0). However, if the slit height were variable, then a plot similar to the one for W in Fig. 86 would result. Several variables, however, need not be optimized since they either do not influence $E_s/\overline{\Delta E}_s$ significantly or must be adjusted to the largest or smallest value consistent with the type of results desired. A parameter that has no effect on $E_s/\overline{\Delta E}_s$ is the phototube voltage as long as the voltage is below the voltage resulting in a breakdown and above the voltage needed to allow accurate measurement of the signal and the noise. A parameter that will continue to cause a decrease in $E_s/\overline{\Delta E}_s$ as it increases, assuming that there is no drift in the readout signal, is Δf, the frequency-response bandwidth of the amplifier-readout system. Therefore, if no drift is present, then $E_s/\overline{\Delta E}_s$ will increase as $1/\sqrt{\Delta f}$ decreases. The parameter Δf can be made small by using signal-averaging techniques. Of course, $E_s/\overline{\Delta E}_s$ will increase approximately linearly with an increase in the power efficiency, Φ_s.

Finally, a plot of $E_s/\overline{\Delta E}_s$ versus the source intensity I_0 will result in a plot similar to the one for $E_s/\overline{\Delta E}_s$ versus W in Fig. 86. It should be stressed that most sources now in use are relatively-low-intensity sources, and therefore an increase in $E_s/\overline{\Delta E}_s$ will occur if the source intensity I_0 and/or the source area (emitting area) increases.

7. Optimization of Experimental Conditions for Lowest Limits of Detection [13]

The experimental conditions giving greatest precision will also be approximately the conditions giving the maximum sensitivity, that is, lowest limits of detection. The minimum detectable sample concentration C_m is defined as the concentration resulting in a single-to-noise ratio of

$$\frac{E_s}{\overline{\Delta E}_s} = \frac{t\sqrt{2}}{\sqrt{n}}, \tag{52}$$

where t is the Student t, which can be found from statistical tables for a given confidence level (usually 99.5%) and a given number of degrees of freedom (2n - 2), and n is the number of combined sample and blank measurements. To determine the theoretically measurable value of C_m, the previously derived expressions for E_s and $\overline{\Delta E}_s$ can be substituted into Eq. (52), and after substituting in the optimum values of all parameters, C_m can be solved for. If one wishes to use the truly optimum value of all parameters, X, Y, Z, etc., then it is necessary to differentiate C_m with respect to X at constant Y and Z (assuming parameters are independent), minimizing by setting $dC_m/dX = 0$, and solving for X_o. Of course Y_o, Z_o, etc., can be found similarly, and if the parameters are dependent, a reiterative process must be used. Again, it is more convenient to plot C_m versus X at constant X, Y, etc., to determine X_o. However, by the above technique optimum values of all parameters at the minimum detectable sample concentration C_m can be found, and these values result in the lowest values of C_m. Of course, variation of the same parameters that increase $E_s/\overline{\Delta E}_s$ will increase C_m, and so no detailed description of variation in parameters, as done in the preceding subsection, will be given.

Table 33 gives some calculated values of C_m and some corresponding experimentally measured values of C_m for four organic molecules measured by using the Aminco-Bowman spectrophotofluorometer with a phosphoroscope attachment. The excellent agreement between the calculated and the theoretical values establishes the validity of the above equations and demonstrates the use of these expressions.

III. INSTRUMENTATION

A. General Remarks

The instrumentation used to study phosphorescence is very similar to that used for fluorescence. To see phosphorescence in the presence of fluorescence we must take advantage of the time difference involved between the absorption

TABLE 33

Comparison of Calculated and Observed Minimum Detectable
Sample Concentrations C_m for Four Compounds Measured by
Phosphorimetry[a]

Compound[c]	Minimum detectable concentration (moles/l)[b]	
	Calculated[b, d]	Experimental[b]
Retene	2.6×10^{-9}	5×10^{-9}
L-Tyrosine	7.0×10^{-8}	5×10^{-8}
Benzaldehyde	1.2×10^{-8}	7×10^{-9}
Benzyl alcohol	2.7×10^{-8}	2.5×10^{-8}

[a] Reprinted from Ref. 12 by courtesy of the American Chemical Society.

[b] All calculations were performed for the Aminco-Bowman spectrophotofluoro-meter with phosphorescence attachment. All experimental measurements were taken on the same instrument. The calculated and measured values of C_m were found by using Eq. 52.

[c] All experimental measurements were made in ethanol.

[d] All calculated values were obtained by using the average monochromator-slit width of approximately 0.15 cm.

and the emission of radiant energy to phosphorescence; this difference is con-siderably larger than that in fluorescence. This is accomplished with a mechanical device called a phosphoroscope, which is discussed in the next subsection.

At room temperature the energy of the triplet state is readily lost by a collisional deactivation process involving the solvent, and phosphorescence is not observed. At reduced temperatures a solidified sample does not lose energy readily, and phosphorescence can be easily observed. The usual procedure in phosphorescence studies is to place the samples in small quartz tubes, which are then placed in liquid nitrogen (77°K) and held in a quartz Dewar flask. The incident-radiation energy passes through the unsilvered part of the Dewar flask, and luminescence is observed through the same part of the flask at right angles to the incident beam. The sample cell is immersed directly into the collant, which is usually liquid nitrogen.

The solvents used in phosphorimetry at liquid-nitrogen temperature must form clear, rigid glasses, must have good solubility characteristics for the compounds to be studied, must be readily available and inexpensive, and must neither absorb strongly nor luminesce greatly in the spectral regions of interest. The most commonly mixed solvent is EPA, a 5:5:2 (v/v/v) mixture of diethyl ether, isopentane, and ethanol.

Phosphorescence attachments that can be used with a commercial fluorometer are available for the Aminco-Bowman spectrophotofluorometer, the Baird-Atomic Fluirspec, the Farrand MK-1 spectrofluorometer, and the Aminco filter fluorometer. The last is the cheapest and most readily available (accessory D2-63019, Fig. 87). Source light passes through a rotating shutter to excite the sample to phosphoresce; emitted light alternately passes to the detector.

B. The Phosphoroscope

The phosphoroscope is a mechanical device used (a) to modulate the radiation from the light source incident on the sample and (b) to modulate the luminescence radiating from the sample and incident on the photodetector. The modulation is periodic and out of phase, so that no incident exciting or

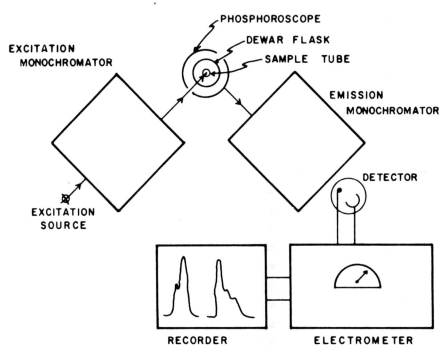

FIG. 87. Schematic diagram of a spectrophosphorimeter.

luminescence radiation reaches the photodetector during one phase, and only long-decaying luminescence (phosphorescence) radiation reaches the photodetector during the second phase. Therefore the main purpose of the phosphoroscope is to allow measurement of phosphorescence in the presence of incident-light scattering and fluorescence.

There are two major types of mechanical phosphoroscopes. The rotating-can phosphoroscope (see Fig. 88) was first described by Lewis and Kasha and is used in the Aminco-Bowman spectrophotofluorometer. It is a hollow cylinder with one or more slits equally spaced in the circumference. As the can is turned by a motor, radiation from the excitation monochromator is alternately allowed to strike the sample, and the light emitted by the sample is alternately, but out of phase with excitation, allowed to reach the emission-monochromator entrance slit. The fluorescence and exciting radiation decay rapidly after the exciting radiation is terminated. Hence only the long-decaying phosphorescence will remain when the phosphoroscope has turned from the point at which excitation is stopped to the point where the luminescence of the sample is measured.

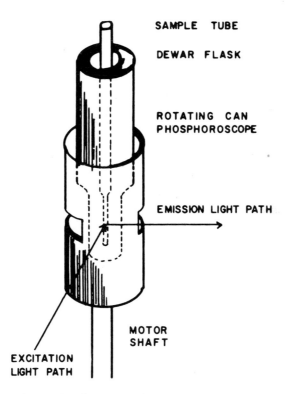

FIG. 88.   Schematic diagram of a rotating-can phosphoroscope.

The other major type of mechanical phosphoroscope (Fig. 89) is the Becquerel, or rotating-disk type. This type of phosphoroscope is currently being used in several of the compensated-luminescence instruments because it is more versatile than the rotating-can type. It consists of two disks with notches cut in them at equal intervals and mounted on a common axis turned by a motor. Once again excitation occurs out of phase with measurement of luminescence.

The operation of the rotating-can and Becquerel phosphoroscopes has been considered by O'Haver and Winefordner [14]. They have derived an expression for the observation-efficiency factor a', which is defined as the ratio of the phosphorescence signal obtained with a phosphoroscope to the phosphorescence signal obtained with continuous excitation and observation. The parameter a' is a function of the phosphorescence decay time of the sample, τ; the time required for one cycle of excitation and observation, and called the cycle time t_C; the time between the termination of excitation and initiation of observation during one cycle, called the delay time t_D; and the time during which the sample is excited during one cycle, called the exposure time t_E.

Some general conclusions can be made from a consideration of the expression for a'. The Becquerel phosphoroscope is the more versatile of the two mechanical phosphoroscopes discussed here. If $\tau > t_C$, then the value of a' is a constant for any phosphoroscope (can or disk) speed. For example, for long-decaying phosphors, a' will be about 0.40 for the phosphoroscope can used on the Aminco phosphorescence attachment. On the other hand, if $\tau < t_C$, then a' decreases rapidly as τ decreases because intensity is lost during the delay time, and the phosphorescence decays significantly during the time of observation. In such a case the value of a' can be increased by increasing the phosphoroscope speed and/or increasing the number of openings in the can or disk. However, for molecules of long τ, the number of openings and the phosphoroscope speed have little effect on the value of a'. O'Haver and Winefordner [14] have also shown that the Aminco phosphoroscope can, at its maximum speed of approximately 15,000 rpm, only be used effectively to

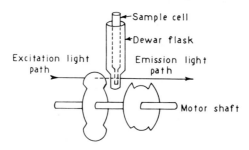

FIG. 89.   Schematic diagram of a Becquerel disk phosphoroscope.

measure the phosphorescence of molecules with τ greater than 1 msec.
Finally, when an analyst is performing a quantitative analysis by means of
phosphorimetry, he should check whether or not a' is constant with
phosphoroscope speed. If.a' is sensitive to phosphoroscope speed, intensity
variations will result from very small fluctuations in the phosphoroscope speed.
Fortunately, for most organic phosphors of interest in biology and medicine a'
will not be a function of phosphoroscope speed.

C. The Sample Cell

In phosphorimetry the sample must be frozen; this limits the internal
dimensions of the sample cell. Too small a cell diameter will prevent the
excitation energy from entering the sample due to reflection; too large a cell
diameter leads to cracking of the frozen solution and a high degree of scattering.
Generally small quartz tubes 1 mm and 3 mm are used. These are placed in
liquid nitrogen (77°K) and held in a quartz Dewar flask. The crystalline nature
of the frozen sample requires precise positioning and repositioning of the cell.
For good precision (1 %) the sample should be rotated as described by Hollifield
and Winefordner [8].

The cells should be cleaned with successive rinses of fuming nitric acid,
distilled water, and finally with the solvent and sample solution to be measured.
The small cells are emptied by means of a polyethylene tube connected to a
water aspirator.

A typical quartz Dewar flask that was used in the phosphorescence attachment to the Aminco-Bowman spectrophotofluorometer is pictured in Fig. 90.
Light passes through the unsilvered part of the Dewar flask, and luminescence
is observed through a similar part of the flask at right angles. The sample cell
is immersed directly into the coolant which is usually liquid nitrogen. This
immersion technique has the advantage of simplicity and speed of operation.
However, it also has several disadvantages. First, the exciting radiation and
the luminescence radiation must pass through three quartz layers, which results
in considerable light loss. Second, the light paths must pass through the
thermostating medium, limiting the choice of collants to transparent,
nonluminescent liquids. Third, as the collant in the sample-viewing area warms
up, convection of the coolant results in a change in its refractive index and a
resultant flicker in the exciting radiation and the luminescence radiation.
Fourth, ice crystals and foreign objects, such as dust, in the flask tend to
produce nucleation sites, causing bubbling of the collant, which again produces
a flicker noise.

These disadvantages could be minimized by using a conduction-cooling

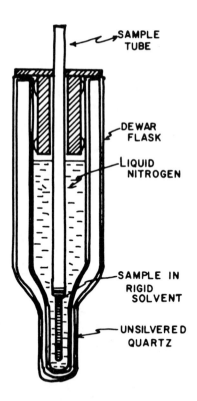

FIG. 90.   Schematic diagram of sampling system used in phosphorimetry.

device.   In this system the sample cell is cooled by contact with a copper rod immersed in some coolant.   Thermal gradients are reduced by the high thermal conductivity of the copper, and quartz surfaces are minimized.   Unfortunately it is difficult to obtain good thermal contact between the sample cell and the copper rod.   Also it is difficult to thermally insulate the cooling apparatus; this results in fogging of the viewing area, making quantitative measurements difficult.   Hoerman and Manciewicz [15] have successfully applied conduction cooling to the study of calcified tissue.   They formed potassium bromide pellets of particles of bone, enamel, and dentin and cooled the pellets to $93^{\circ}K$ by conduction using a liquid-nitrogen reservoir.

At the present time immersion cooling is the only readily available method for quantitative analysis.   Unfortunately the method of positioning the sample cell in the Dewar flask as well as arc wander of the xenon source limit the precision of phosphorimetric measurements to about $\pm$ 5 to 10 % relative standard deviation with the Aminco phosphoroscope attachment.   With rotation of the sample, however, 1 % precision is attainable [8].

D. Solvents

Of all the possible coolants (e. g. , liquid nitrogen, air, oxygen, rare gases, nitrous oxide, and other boiling liquids as well as liquid/solid mixtures) only liquid nitrogen is sufficiently pure to be transparent and nonluminescent at all wavelengths between 200 and 800 nm and safe, convenient, and inexpensive to use.

The solvents used in phosphorimetry at liquid-nitrogen temperature must form clear, rigid glasses, must have good solubility characteristics for the compounds to be studied, must be readily available and inexpensive, and must neither absorb strongly nor luminesce greatly in the spectral regions of interest.

Some of the solvents used for phosphorimetric analysis are listed in Table 34 and are classified as acidic, alcoholic, aqueous, basic, ether, halides, hydrocarbon, and plastics or solid matrices. Ethanol has been combined with a variety of other solvents to form good mixed solvents, the most popular of which is EPA, a 5:5:2 (v/v/v) mixture of ether, isopentane and ethanol (sold by the American Instrument Co. , Silver Spring, Md.).

Solvents, even the so-called spectrograde ones, are generally not sufficiently pure to be used without further purification. Purification in some cases may be as simple as several extractions with dilute acid and base. In any event the procedure with the least number of transfers and least exposure to the atmosphere and to glassware is likely to be best. Absolute ethanol as received from the supplier must be purified. This is accomplished using a 5-ft, vacuum-jacketed distillation column of 1 in. , packed with 3/32-in. glass helices.

Hydrocarbons, such as hexane, isopentane, and heptane, can also be purified by distillation. Alternatively, the hydrocarbons can be dried over anhydrous sodium sulfate or sodium ribbon and then passed through a 2-ft column of 200-mest silica gel activated for 12 hours at $350^{\circ}C$ prior to use [17].

Lukasiewicz, et al. [18, 19, 20] have described another improvement to allow the measurement of aqueous or predominately aqueous solutions at $77^{\circ}K$ by means of phosphorometry: an open quartz capillary cell as the sample cell. Sample solutions (20 μl) fill the cell by capillary action, and the sample cell is then rotated as described in IV A 8. As a result of the open, capillary cell, these cells with predominately (or pure) aqueous solutions do not crack when rapidly cooled to liquid N_2 temperatures or when warmed back to room temperature by use of a heat gun. The major advantages of using aqueous or aqueous-organic mixture solutions in phosphorometry are:

1. Most biochemical species are more soluble in aqueous solutions than in the solvents needed to produce clear, rigid glasses at liquid nitrogen temperatures.

TABLE 34

Solvents for Phosphorimetric Analysis[a]

Type	Solvent and composition (v/v)	Temperature (^{O}K)
Acidic	98% Sulfuric acid	At least to 210
	88% Phosphoric acid	At least to 210
	Acetic anhydride-H_3PO_4, 17:25	At least to 210
	Trifluoroacetic acid-methylamine-trimethylamine, 4:11:5	77
	Ethanol-HCl (conc.), 19:1	77
	Acetic acid	77, 90
	Hydrofluoric acid + BF_3	292
Alcoholic	Ethanol, <5% H_2O	77
	N-Propanol	77
	Ethanol-methanol, 4:1 to 5:2	77
	Isopropanol-isopentane, 3:7	77
	Ethanol-isopentane-diethyl ether, 2:5:5 (EPA)	77
	Isopropanol-isopentane-diethyl ether, 2:5:5	77
	N-Butanol-isopentane, 3:7	77
	Isopropanol-isopentane, 2:8	77
	N-Propanol-isopentane, 2:8	77
	Diethyl ether-ethanol (96%), 1:2	77
	Ethanol-glycerol, 11:1	77
	Ethanol-methanol-diethyl ether, 8:2:1	77
	N-Propanol-diethyl ether, 2:5	77
	N-Butanol-diethyl ether, 2:5	77
	Diethyl ether-isooctane-isopropanol, 3:3:1	77
	Diethyl ether-isooctane-ethanol, 3:3:1	77
	Diethyl ether-isopropanol, 3:1	77
	Isopropanol-methylcyclohexane-isooctane, 1:3:3	77

Type	Solvent and composition (v/v)	Temperature ($^{\circ}$K)
	Glycerol	183-195
	Propylene glycol	183
	Ethyl Cellosolve -N -butanol -N -pentane, 1:2:10	77
Aqueous	Water -propylene glycol, 1:1	183-195
	Water -ethylene glycol, 1:2	123-150
	Propane -1. 2 -diol -phosphate buffer (0. 01 M, pH 7), 1:1	77
Basic	Ethanol -NH_3 (28% aq.), < 20:1	77
	Ethanol -NaOH (0. 5% aq.), < 20:1	77
	Triethylamine -diethyl ether - isopentane, 3:1:3	77
	Triethylamine -diethyl ether - N -pentane, 2:5:5	77
	Trimethylamine -isopentane - diethyl ether, 2:5:5	77
	Triethanolamine	193-213
	Methylhydrazine -methylamine, 2:4	77
	Diethyl ether -ethanol -NH_3 (28% aq.), 10:9:1	77
Ethers	Diethyl ether	77
	2 -Methyltetrahydrofuran	77
	Di -N -propyl ether -isopentane, 3:1	77
	Di -N -propyl ether -methylcyclohexane, 3:1	77
	Di -N -propyl ether -N -pentane, 2:1	77
	Diethyl ether -pentane -2(cis) -pentane - 2(trans), 2:1 (mixture)	77
	Diethyl ether -isopentane, 1:1 to 1:2	77
	N -Butyl ether -isopropyl ether -diethyl ether, 3:5:12	77
	Diethyl ether -isopentane -dimethyl - formamide -ethanol, 12:10:6:1	77
	Diethyl ether -isopentane -dimethyl - formamide -ethanol, 2:4:1:16	77

Type	Solvent and composition (v/v)	Temperature (^{o}K)
Halides	Bromoform	77, 90
	2-Bromobutane	77
	Chloroform	77, 90
	EPA-chloroform, 12:1	77
	Ethyl iodide-isopentane-diethyl ether, 1:2:1	77
	Ethyl bromide-methylcyclohexane-isopentane-methylcyclopentane, 1:4:7:7	77
	Ethanol-methanol-ethyl iodide, 16:4:1	77
	Ethanol-methanol-propyl iodide, 16:4:1	77
	Ethanol-methanol-propyl chloride, 16:4:1	77
	Ethanol-methanol-propyl bromide, 16:4:1	77
	Ethyl iodide-ethanol-methanol, 5:16:4	77
	Ethanol-isopentane-diethyl ether-dry HCl, 1:1:1 (0.01 M)	77
Hydrocarbons	Methane	4.2
	Isopentane	77
	3-Methylpentane	77
	Pentane (Technical grade 1:1 N-pentane, isopentane)	77
	Petroleum ether, 58-60°C fraction	77
	N-Pentane-N-heptane, 1:1	77
	Methylcyclohexane-N-pentane, 4:1 to 3:2	77
	Methylcyclohexane-isopentane, 4:1 to 1:5	77
	Methylcyclohexane-methyl cyclopentane, 1:1	77
	Pentene-2(cis)-pentene-2(trans) (mixed isomers)	77
	Paraffin oil (Nujol)	183-195
	Cyclohexane-decalin, 1:3	20-77

Type	Solvent and composition, (v/v)	Temperature (^{O}K)
Plastics and miscellaneous solvents	Lucite (polymethylmethacrylate)	80-300
	Cellulose acetate	77-300
	Poly(vinyl alcohol)	Room temp.
	Glucose	Room temp.
	Polyacrylonitrile; Orlon A	77-300
	Potassium bromide	93
	Boric acid	Room temp.
	Cellophane	77-300
	Plexiglas	Room temp.
	Sucrose	Room temp.

a Some data from Winefordner [16].

2. Solution conditions, e.g., pH, ionic strength, etc., can be varied for optimal analytical results.

3. Water is easier to obtain in higher phosphorometric purity than most organic solvents.

4. Contamination problems are less when water can be used to clean glassware, cells, and equipment.

E.   Commercial Equipment

Phosphorescence attachments are available to commercial instruments such as the Aminco-Bowman spectrophotofluorometer, the Baird-Atomic Fluorispec, and the Aminco filter fluorometer. The latter is not recommended since it has given very poor results in our laboratory. The attachments to the Aminco-Bowman instrument and the Baird-Atomic Fluorispec consist of a phosphoroscope can, a quartz Dewar flask, and quartz sample tubes. Both have been used in our laboratory and work well. It is recommended that the quartz sample tube be rotated as suggested by Winefordner [8] for optimum precision (1%). It is surprising that no commercial equipment yet uses this simple expedient.

IV. ANALYTICAL APPLICATIONS OF PHOSPHORESCENCE

A. Practical Aspects of Measurement

1. Instrument Calibration

In phosphorescence the instrument sensitivity should be adjusted daily as described for fluorometry in Chapter 4. The sensitivity must be periodically adjusted so that the overall measurement system has the same sensitivity as that used in all previous studies. As an example, Moye [21] used a standard solution of toluene in ethanol as a reference standard to adjust instrument sensitivity.

2. Procedure for Measurement

The excitation and emission monochromators are usually adjusted to the peak excitation and emission wavelengths unless interferences that prevent this are present. The instrument sensitivity is then checked as described above and set to the proper range for measurements to be made. The Dewar flask is cleaned and filled with liquid nitrogen. The sample tube is cleaned and filled with sample, and the sample-tube holder is aligned in the Dewar flask. If the sample solution cracks or forms a snow on cooling, the sample should be discarded and a new sample solution placed in the sample tube. (Note: with the rotating-sample approach of Winefordner [8] cracked glasses and snows can be tolerated). The signal is determined from the readout device, and this reading is used to determine the concentration of the sample by use of an analytical curve of readout signal versus concentration of sample. The standard analytical curve should be prepared by using standard solutions of the sample and measured under experimental conditions identical with those used for the unknown sample. In plotting an analytical curve each point on the curve must be corrected for the solvent blank. It is interesting to note that most phosphorescence analytical curves are linear over a concentration range of 10,000 or more. Negative curvature usually results in the 10^{-4} M and higher concentration range due to self-absorption, molecular aggregation, and concentration quenching. Positive curvature near the minimum detectable sample concentration is generally a result of contaminant luminescence.

3. Sources of Contamination

The sources of phosphorescence background contamination include detergents used to clean glassware; stopcock grease present in the cleanup step; filter paper, ion-exchange resins, and chromatographic materials used in prior separation steps; and plastic containers in which the sample or chemical reagents are stored.

4. Quenching and Enhancement of Phosphorescence

In general, phosphorescence is less suceptible to quenching than is fluorescence. The use of low temperatures and rigid media in phosphorimetry diminishes the probability of nonradiative deactivation of the triplet state by diffusion-controlled processes.

The presence of heavy atoms affects both the radiative and nonradiative triplet-to-singlet transitions. McGlynn, Daigre, and Smith [22] have studied the external-heavy-atom effect and have suggested that it might be a useful means of enhancing the phosphorescence intensity of some species. Hood and Winefordner [23] have studied the influence of ethyl iodide-ethanol as a solvent for enhancing the phosphorescence signal of naphthalene, phenanthrene, and 10 other poly-nuclear aromatic hydrocarbons. They found that a 5:1 (v/v) ethanol-ethyl iodide solvent enhanced the phosphorescence signal of 1,2-benzfluorene and 2,3-benzfluorene by 13 and 25 times, respectively, whereas the same solvent resulted in a diminution of the phosphorescence signal of naphthalene and phenanthrene by 10 and 5 times, respectively. Many of the compounds influenced most significantly by the heavy-atom effect are carcinogens (e. g. , 3,4-benz-pyrene, 1,2,5,6-dibenzanthracene, and 1,2-benzanthracene). Since the heavy-atom effect can enhance or depress the phosphorescence signal, an increase in the selectivity of measurement as well as increased sensitivity can be effected by properly selecting the solvent for a specific system.

It must be kept in mind that, even though the phosphorescence intensity may actually be increased due to the heavy-atom effect, there may still be a decrease in the phosphorescence signal because the heavy atom may result in such a reduction in the decay time τ that there may be a significant decrease in the phosphorescence signal due to phosphoroscope delay time (i. e. , the value of α' decreases considerably).

Paramagnetic ions are also effective quenchers of phosphorescence in many instances. The acetylacetone complexes of paramagnetic transition-metal ions are not phosphorescent, whereas the same complexes of diamagnetic ions are. Oxygen as a quencher of luminescence has been the subject of much debate. However, from an analytical viewpoint, oxygen does not appear to be a major quencher for samples measured in rigid media at low temperatures.

5. Analysis of Mixtures of Phosphorescent Compounds

There are several methods of achieving selectivity in the measurement of mixtures of phosphorescent compounds. These methods are called excitation resolution, emission resolution, phosphoroscopic resolution, and time-resolved phosphorimetry. We shall discuss each of these separately.

a. Excitation Resolution. Excitation resolution is achieved by varying the wavelength of the excitation monochromator to excite each molecule separately. If the phosphorescence of a mixture is excited by a wavelength that is considerably more strongly absorbed by one component than by all the others, then there is obtained predominantly the emission spectrum of this compound. Similarly, if several excitation wavelengths are employed, it becomes possible to obtain the spectra of the individual components more or less undistorted, provided the absorbance spectra of the compounds present in the mixture differ sufficiently. This technique, which has long been used in fluorometry, has been found extremely useful also in phosphorimetry.

Figure 91 gives an example. In it are reproduced the phosphorescence spectra of the three-component mixture of phenanthrene, 1,2-benzpyrene, and peri-(1,8,9)-naphthoxanthene excited by three different wavelengths. For each of these wavelengths a different compound gives an intense absorption maximum, and the others give either minima or weak absorption. Above the spectra of the mixture are shown the spectra of the pure components. As can be seen, in each case only one component appears clearly in the spectrum of the mixture.

b. Emission Resolution. Emission resolution is achieved by varying the wavelength of the emission monochromator in order to measure preferentially the

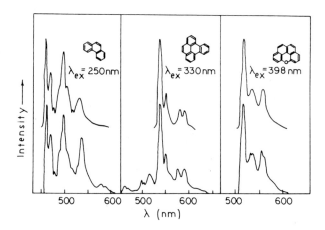

FIG. 91. Phosphorescence spectra of a mixture of 40% phenanthrene, 30% 1,2-benzpyrene, and 30% peri-(1,8,9)-naphthoxanthene on excitation with 250, 330, and 398 nm. The curves were measured on the pure compounds in each case.

luminescence of just one molecule. For example, two pesticides, methoxychlor and Kelthane, can be assayed in this way. As indicated in Fig. 92, the emission monochromator is first adjusted to 525 nm in order to measure Kelthane (B) in the presence of methoxychlor. Then the emission is set at 400 nm in order to measure methoxychlor (A).

c. Phosphoroscopic Resolution. By suitable change in the time of measurement of the luminescence radiation after termination of the exciting radiation by a shutter, the phosphoroscopic resolution of two molecules with sufficiently different decay times (at least a tenfold difference) can be performed. The slow-decaying molecule is measured at a time sufficient for at least 99% of the fast-decaying species to have decayed. Keirs, Britt, and Wentworth [24] were the first to suggest phosphoroscopic resolution based on rotation speed and the geometry of a Becquerel phosphoroscope. If there is a sufficient difference in the decay times of two phosphors, it is possible to vary the phosphoroscope speed in order to measure only the slower decaying species. Once the concentration of the slower decaying species is known, the concentration of the faster decaying species can also be determined.

d. Time-Resolved Phosphorimetry. St. John and Winefordner [25] have developed a new technique called time-resolved phosphorimetry (a type of phosphorescence resolution) to measure two-component mixtures of similar phosphorescent compounds. The method utilizes the difference in the decay times of the phosphors. The exponentially decaying phosphors are resolved by using a logarithmically responding instrument. The concentration of the two components can be determined from the recording of the logarithm of the phosphorescence signal due to the mixture versus time after termination of excitation by a method analogous to that used in radioactive-isotope analysis. Mixtures of tryptophan and tyrosine and of benzoic acid and benzaldehyde were measured by this technique with an overall relative error of about 3%. In

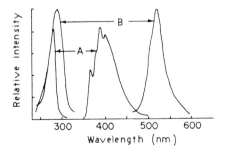

FIG. 92. Resolution of the emission of a mixture of two pesticides: (A) methoxychlor and (B) kelthane.

TABLE 35

Experimentally Measured Limiting Detectable Concentrations

Compound-solvent[a]	λ_{ex} (nm)[b]	λ_{em} (nm)[b]	τ (sec)[c]	Limit of detection (μg/ml)	Refs.
Acenaphthene-E	300	515	--	0.2	23
Acetaldehyde 4-nitrophenylhydrazone-EPA	395	525	0.50	0.06	26
Acetone 4-nitrophenylhydrazone-EPA	392	525	0.48	0.1	26
p-Aminobenzoic acid-E	310	430	3.2	0.04	21, 27
2-Aminofluorene-E	380	590	4.6	0.01	28
2-Amino-5-nitrobiphenyl-EPA	380	520	0.56	0.05	26
2-Amino-6-nitrobenzothiazole-EPA	375	515	0.41	0.08	26
2-Amino-7-nitrofluorene-EPA	340	520	0.38	4.5	26
Anabasine-E	270	390	6.2	0.010	18, 27
Anthracene-E	300	462	--	0.05	23
Apomorphine HCl-E	320	470	3.1	0.001	29
Aramite-E	285	400	3.3	0.034	18, 27
L-Arterenol bitartrate-E	260	455	0.5	1.0	28
Atropine-E	w[d]	410	1.4	0.10	30
Aspirin-EPA	240	380	2.1	0.10	6, 7
Azosulfamide-E	290	440	0.9	N.A.U.[e]	31
Bayer 4646[f]-EPA	290	460	0.6	0.12	21, 27
Bayer 37344[f]-EPA	275	435	< 0.2[g]	0.10	21, 27
Benzaldehyde-E	254	433	3.4	0.004	28
1,2-Benzanthracene-E	310	510	2.2	0.03	23

Compound					Ref.
1,2-Benzfluorene-E	315	502	--	0.2	23
2,3-Benzfluorene-E	325	502	--	0.2	23
Benzocaine-E	310	430	3.4	0.007	23
Benzoic acid-E	240	400	2.4	0.005	30
Benzophenone 4-nitrophenylhydrazone-EPA	365	515	--	2.0	26
4-Benzoylbiphenyl 4-nitrophenylhydrazore-EPA	370	520	0.38	0.4	26
3,4-Benzpyrene-E	325	508	--	3.0	23
Benzyl alcohol-E	219	393	--	0.04	28
Biphenyl-E	270	385	1.0	0.004	32
Brucine-E	305	435	0.9	0.10	29
Butacaine sulfate-E	310	430	5.7	0.05	30
Caffeine-E	285	440	2.0	0.2	30
Caffeine-EPA	w[d]	410	--	1.0	7
Chlorobenzilate-E	275	415	<0.2[g]	0.01	21, 27
4-Chlorophenol[f]-E	290	505	<0.2[g]	0.23	21, 27
Chlorpromazine HCl-E	320	490	0.3	0.03	30
Chlortetracycline-E	280	410	2.7	0.05	30
Cinchophen-E	350	520	0.8	0.02	30
Cocaine HCl-E	240	400	2.7	0.01	30
Codeine-E	275	505	0.3	0.01	29
Co-Ral[f]-E	360	510	<0.2[g]	0.004	21, 27
Cyclaine HCl-E	240	400	2.4	0.006	30
DDD[f]-E	265	415	<0.2[g]	0.01	21, 27
DDE-E	270	425	0.2	0.001	27
DDT-E	270	420	0.2	0.0007	27
Diacetylsulfanilamide-E	280	405	1.3	0.001	31

Compound-solvent[a]	λ_{ex} (nm)[b]	λ_{em} (nm)[b]	τ (sec)[c]	Limit of detection ($\mu g/ml$)	Refs.
Diazinon-E	275	395	5.0	0.03	21, 27
1, 2, 5, 6-Dibenzanthracene-E	340	550	1.3	0.03	23
2, 6-Dichloro-4-nitroaniline-EPA	368	525	0.47	0.04	26
2, 4-Dichlorophenoxyacetic acid-E	290	495	<0.2[g]	0.04	18, 27
2, 6-Diethyl-4-nitroaniline-EPA	388	525	0.66	0.19	26
Dicumarol-E	305	475	0.6	0.001	33
5, 7-Dimethyl 1, 2-benzacridine-E	310	555	0.6	0.03	28
N, N-Dimethyl-4-nitroaniline-EPA	398	525	0.54	0.05	26
Diphenadione-E	260	440	0.6	1.0	33
Dopamine-E	270	420	0.4	1.0	28
Ephedrine-E	225	390	3.6	0.20	30
L-Epinephrine bitartrate-E	270	410	0.4	1.0	28
Guthion-E	325	420	0.6	0.6	21, 27
Hippuric acid-EPA	311	450	4.9	0.004	28
5-Hydroxy indoleacetic acid	w[d]	410	--	0.1	2
DL-5-Hydroxytryptophan-E	315	435	6.3	0.1	28
Imidan[f]-E	305	440	0.8	0.006	21, 27
Indoleacetic Acid-E	300	440	7.1	0.004	28
Indoleacetonitrile-E	285	438	7.1	0.005	28
Indolebutyric Acid-E	295	446	6.4	0.004	28
Indolecarboxylic Acid-E	290	429	5.5	0.0006	28
Indolepropionic Acid-E	292	445	6.5	0.004	28
Isolan[f]-E	285	395	1.6	2.0	21, 27
Kelthane[f]-E	285	515	<0.2[g]	0.006	21, 27

Kepone[f] -E	260	1.2	410	10.0	21,27
Lidocaine-E	265	1.1	400	1.2	30
Mebaral-E	240	2.2	380	0.01	30
Methoxychlor-E	275	0.7	380	0.005	21,27
N-Methyl-4-nitroaniline-EPA	390	0.5	522	0.05	26
Methycaine HCl-E	240	2.7	400	0.006	30
Morphine-E	285	0.3	500	0.01	29
Morphine sulfate-E	265	0.8	460	10.0	29
Naphthacene-E	300	--	518	0.001	23
Naphthalene-EPA	310	1.8	475	0.7	22
Naphthalene-E	290	--	505	0.06	23
ß-Naphthylamine-E	270	2.3	505	0.03	28
α-Naphthol-E	320	1.2	475	0.003	21,27
Narceine-E	290	0.5	440	0.1	29
NIA 10242-E	285	1.6	400	0.007	21,27
Nicotine-E	270	5.2	390	0.01	21,34
5-Nitroacenaphthene-EPA	380	--	540	0.5	26
4-Nitroaniline-EPA	380	0.6	510	0.02	26
9-Nitroanthracene-EPA	248	--	488	0.13	26
1-Nitroanthraquinone-EPA	250	0.28	490	0.25	26
4-Nitrobiphenyl-EPA	330	--	480	0.2	26
2-Nitrofluorene-EPA	340	0.40	517	0.04	26
6-Nitroindole-EPA	372	0.41	520	0.08	26
1-Nitronaphthalene-EPA	340	--	520	1.5	26
2-Nitronaphthalene-EPA	260	0.36	500	0.15	26
4-Nitro-1-naphthylamine-EPA	400	--	578	60	26
3-Nitro-N-ethylcarbaxole-EPA	315	0.37	475	0.01	26
2-Nitro-N-methylcarbazole-EPA	345	--	530	2.8	26

Compound-solvent[a]	λ_{ex} (nm)[b]	λ_{em} (nm)[b]	τ (sec)[c]	Limit of detection (μg/ml)	Refs.
3-Nitro-N-methylcarbazole-EPA	--	--	0.39	0.2	26
4-Nitro-2-toluidine-EPA	375	520	0.53	0.1	26
4-Nitrophenylhydrazine-EPA	390	520	0.48	0.03	26
4-Nitrophenol-E	355	520	<0.2[g]	0.0002	21, 35
DL-Normetanephrine-E	275	440	0.6	0.1	28
Nornicotine-E	270	390	5.3	0.01	28, 34
Orthotran-E	260	395	<0.2[g]	0.02	21, 27
Papaverine HCl-E	260	480	1.5	0.0005	29
Parathion-E	360	515	<0.2[g]	0.08	21, 27
Phenacetin-EPA	w[d]	410	--	0.2	7
Phenanthrene-E	300	499	--	0.002	23
Phenanthrene-EPA	340	465	2.6	1.0	36
Phenindione-E	235	395	<0.2[g]	1.0	33
Phenobarbital	240	380	1.8	0.10	30
Phenylalanine-E	270	385	--	0.4	28
Phenylephrine HCl-E	290	390	2.4	0.01	30
Phthalylsulfacetamide-E	290	415	0.6	0.001	31
Phthalylsulfathiazole-E	305	405	0.9	1.0	31
Procaine HCl-E	310	430	3.5	0.01	30
Propionaldehyde 4-nitrophenyl-hydrazone-EPA	395	525	0.50	0.06	26
Pyridine-E	310	440	1.4	0.0001	31
Pyrene-E	329	515	--	0.2	28
Quinidine sulfate-E	340	500	1.3	0.04	30
Quinine HCl-E	340	500	1.3	0.04	30

Compound					Ref.
Retene-E	265	510	--	0.001	23
Romnel[f]-E	300	475	<0.2g	0.006	21, 27
Rutonal-E	240	380	2.5	0.2	21, 27
Serotonin-E	w[d]	410	--	50	7
Salicylic acid-E	430	315	6.2	0.05	28
Sevin[f]-E	300	510	2.0	0.004	21, 27
Strychnine phosphate-E	290	440	1.2	0.050	29
Sodium sulfathiazole-E	315	410	1.4	1.0	31
Succinyl sulfathiazole-E	310	420	1.3	N.A.U.[e]	34
Sulfabenzamide-E	305	405	0.7	0.001	34
Sulfacetamide-E	280	410	1.3	0.0001	34
Sulfadiazine-E	275	410	0.7	0.001	34
Sulfaguanidine-E	305	405	0.7	0.01	34
Sulfamerazine-E	280	405	0.7	0.0001	34
Sulfamethazine-E	280	410	0.8	0.0001	34
Sulfanilamide-E	270	405	1.3	0.001	34
Sulfapyridine-E	310	440	1.4	0.0001	34
Sulfathiazole-E	310	420	0.9	1.0	34
Sulfenone-E	275	391	<0.2g	0.0005	34
Tedion[f]-E	295	410	<0.2g	0.002	21, 27
1,2,4,5-Tetramethylbenzene-EPA	275	392	4.5	1.8	36
Thebaine-E	315	500	1.0	1.0	29
2-Tolidine-E	310	510	2.2	0.02	28
Toxaphene[f]-E	240	390	1.9	0.20	21, 27
Triphenylene-E	291	461	15	0.0002	28
Trithion[f]-E	305	430	<0.2g	0.003	21, 27
2,4,5-Trichlorophenoxyacetic acid-E	300	480	<0.2g	0.005	21, 27
2,4,5-Trichlorophenol-E	305	485	<0.2g	0.03	21, 27

Compound-solvent[a]	λ_{ex} (nm)[b]	λ_{em} (nm)[b]	τ (sec)[c]	Limit of detection (μg/ml)	Refs.
Tromexan-E	295	460	0.6	0.01	33
Tronothane HCl-E	300	410	1.2	0.02	30
Tryptophan-E	295	440	1.5	0.002	28
Tyrosine-E	300	405	5.3	0.01	28
N-Acetyl-L-tyrosine ethyl ester-H	250	395	--	0.1	5
U.C. 10854[f]-E	270	385	3.9	1.9	21,27
Warfarin-E	305	460	0.8	0.01	33
Yohimbine HCl-E	290	410	7.4	0.01	29
Zechtran[f]-E	285	440	0.5	0.006	21,27

[a] Solvents: E = ethanol; EPA = 5:5:2 volume ratio of diethyl ether, isopentane, ethanol; H = 5:11:4 volume ratio of water, methanol, and ethanol; O = octane.

[b] Only the most intense wavelength peaks (uncorrected for instrumental response) are given.

[c] Decay times τ are given for the peak wavelengths.

[d] No monochromator or filter was used on the exciting radiation.

[e] Not analytically useful.

[f] These compounds are pesticides.

[g] Decay times τ shorter than 0.2 sec cannot be measured by using an ordinary time-base recorder.

Fig. 93 logarithmic decay curves are given for benzoic acid (3.00×10^{-4} M, $\tau = 2.4$ sec) benzaldehyde (6.5×10^{-5}, $\tau < 0.1$ sec) in ethanol, and for tryptophan (6.04×10^{-6} M, $\tau = 6.4$ sec) and tyrosine (5×10^{-6} M, $\tau = 1.4$ sec) dissolved in 5×10^{-3} M sodium methoxide in ethanol.

6. Sensitivity

A fairly complete listing of the minimum detectable concentrations of compounds that have been determined by phosphorescence techniques are listed in Table 35. The minimum detectable sample concentration is defined as the concentration that gives rise to a signal equal to the mean background signal under identical experimental conditions; that is, it can be determined by extrapolating the linear portion of the analytical curve to the level of the mean background signal. This definition is quite conservative because it eliminates any false sensitivity introduced by nonlinear analytical curves. The values in Table 35 were compiled from the literature and should only be used as guide-lines to the sensitivities of compounds in phosphorimetry.

A more statistically valid definition of the minimum detectable sample concentration is the concentration that results in a signal-to-noise ratio S/N of $t(\sqrt{2}/\sqrt{n})$.

7. Complementary Nature of Fluorescence and Phosphorescence

The techniques of phosphorescence and fluorescence are complementary rather than competitive. If a compound is strongly fluorescent, it will be weakly phosphorescent, and vice versa; that is, the larger Φ_F, the lower Φ_P. This was demonstrated earlier for the halofluoresceins. Unsubstituted fluorescein and its fluoro and chloro derivatives are fluorescent, and not phosphorescent. Bromofluorescein and iodofluorescein are phosphorescent, not fluorescent.

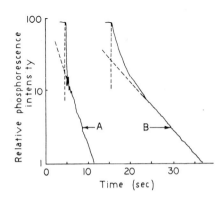

FIG. 93. Logarithmic decay curves for a mixture of (A) benzoic acid and benzaldehyde and (B) a mixture of tryptophan and tyrosine. Ethanol was the solvent for all solutions.

Sauerland and Zander [37] compared the phosphorimetric and fluorometric limits of detection of triphenylene and 1, 2:5, 6-dibenzanthracene; the former was most sensitively assayed by phosphorescence, the latter by fluorescence (Table 36).

TABLE 36[a]

Detection Limits for Aromatic Hydrocarbons

	Limit of detection (g/ml)	
Hydrocarbon	Phosphorescence	Fluorescence
Triphenylene	2×10^{-9}	1×10^{-7}
1, 2:5, 6-Dibenzanthracene	1×10^{-7}	2×10^{-8}

[a] Data from Sauerland and Zander [37].

8. Improved Instrumentation for Phosphorimetry

Zweidinger and Winefordner [38] have shown that, by using a rotating sample cell, a more stable power supply, and a better solvent-cleanup procedure, the detection limits for most organics can be lowered by more than a hundredfold and that precision and accuracy can be increased by more than tenfold. Furthermore, the time and effort involved in sampling and measuring can be considerably reduced. Precise, accurate, sensitive, selective, and rapid analysis could also be performed in solvents forming opaque or densely cracked glasses--no longer is it necessary that a clear solid be formed for good results (see Table 37 for typical results). This increases the applicability of phosphorimetry in biological analysis.

The rotating sample cell assembly currently being used is a modification of the one described by Hollifield and Winefordner [8]; the present system is shown in Fig. 94. The rotating assembly described in detail by Zweidinger and Winefordner [38] and modified by Lukasiewicz, et al. [18-20], consists primarily of a Varian A60-A high resolution NMR spinner assembly (Varian Associates, Palo Alto, Calif.) mounted on an AMINCO phosphoroscope sample compartment in place of the usual lid. The pressure cap of the spinner assembly is covered with black tape to ensure a completely light-tight sample compartment. The rotating sample cell is driven by a flow of nitrogen or air gas taken from a compressed gas tank. The sample cell must have a 6 mm

TABLE 37

Precision of Phosphorescence Measurements[a] for Clear Glasses

and Snows with the Varian Spinner Assembly[b]

N[c]	Nature of matrix	Stationary random orientation	Stationary aligned	Rotating
4	Clear[d]	8. 7	1. 3	0. 8
6	Clear[d]	6. 0	0. 5	0. 8
5	Cracked[d]	13. 7	3. 4	1. 4
11	Clear[d]	3. 1	0. 9	0. 3
10	Clear[d]	2. 9	2. 8	1. 0
5	Snow[e]	3. 6	1. 6	0. 9
10	Snow[e]	2. 8	0. 8	0. 6
10	Snow[e]	2. 7	2. 4	0. 7

[a] Phosphorescence measurements made on a $1. 6 \times 10^{-5}$ M sulfanilamide solutions, which gives s signal five orders of magnitude above the phototube dark current.

[b] Data from Zweidinger and Winefordner [38].

[c] Number of determinations.

[d] Ethanol solvent.

[e] Isooctane-ethanol, 4:' (v/v) mixture as solvent.

O. D. and be about 25 cm long. The sample rotation speed is easy to maintain constant at some speed between 450 and 1400 rpm; the actual speed is unimportant as long as it is maintained constant during a series of measurements, which can be assured by means of a normal two-stage regulator and a rotameter flow meter to monitor the gas flow rate. The normal quartz sample cells used by Zweidinger and Winefordner [38] were 5 mm O. D. and 3 mm I. D. , whereas the quartz capillary cells used by Lukasiewicz, et al. [18-20] were 5. 0 mm O. D. and 0. 90 mm I. D. The cells were made of synthetic, high-purity, optical-grade quartz (Quartz Scientific Co. , Eastlake, Ohio for the capillary tubing and Amersil, Inc. , Hillside, New Jersey for the normal tubing).

By use of elliptical source condensing system and a more stable xenon lamp power supply, it is possible to reduce drift and noise by about 10-fold as compared to the standard power supply used with a commercial phosphorometer. The elliptical source condensing system is commercially

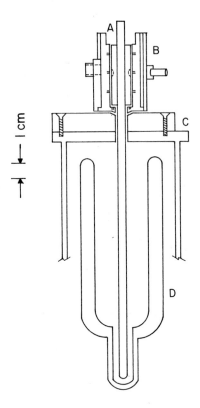

FIG. 94. Schematic diagram of rotating sample cell assembly: (A) quartz
sample cell; (B) Varian (909614-04) spinner assembly for NMR; (C) Aminco
light cover mount; (D) Aminco quartz dewar flask.

available (in this case, from American Instrument Co. , Inc. , Silver Spring,
Maryland), whereas, the xenon lamp power supply system was constructed from
a highly-regulated power supply (Harrison Lab Model 6268 A, Hewlett-Packard,
Palo Alto, Calif.) operating in the constant current mode at 7. 5A (150 W lamp)
and the starter circuit shown in ref [38].

Fisher and Winefordner [39] have described a pulsed source-gated
detector system for phosphorometry; such a system was shown to allow an
increase in selectivity of analysis due to the use of time resolution, which
will not be discussed here, and an increase in sensitivity of analysis due to the
increased signal-to-noise ratio. The latter increase is a result of an increased
signal and a decreased noise.

B. Analytical Determinations

1. Applications to Air Pollution

Numerous polycyclic aromatic hydrocarbons and heterocyclics are found in the atmosphere of industrial cities. Many of these compounds have been found to produce cancer in animals and humans. A number of thorough luminescent examinations of these compounds have been carried out by Sawicki and workers at the Air Pollution Control Center, Research Triangle Park, North Carolina.

Sawicki [40] showed that phosphorescence spectroscopy should be a valuable technique for trace analysis in a review in 1964. This technique was particularly valuable for the qualitative identification of aromatic and heterocyclic carbonyl compounds in atmospheric dust. Sawicki and Johnson [41] showed that polycyclic aromatics in air can be identified by phosphorescence on thin-layer-chromatography strips.

Thin-layer chromatography and phosphorimetry were used by Sawicki et al. [42] to characterize certain heterocyclic hydrocarbons. Sawicki and Pfaff [43, 44] demonstrated that phosphorimetric measurements can be made directly on glass-fiber chromatograms. Limits of detection (in ng) were as low when samples were collected on glass-fiber chromatograms at low temperatures as when samples were in EPA (0. 1 ml) rigid glasses or on cellulose thin-layer chromatograms. This was true for such molecules as benzo[e]pyrene, 4-hydroxyacetopheneone, anthraquinone, benzo[h]quinoline, triphenylene, and 2-nitrofluorene.

An extensive study of the room-temperature and low-temperature fluorescence and phosphorescence of compounds containing the 4-nitroaniline phosphor and analogous groups has been reported by Sawicki and Pfaff [26]. The limits of detection, lifetimes, and the excitation and emission spectra for 27 phosphorescent compounds were reported in EPA rigid-glass solution. The complementary nature of fluorescence and phosphorescence was emphasized. It was noted that aromatic nitro compounds usually yield low-intensity absorption spectra, and therefore luminescence was the technique of choice. Most 4-nitrophenylhydrazones of aliphatic aldehydes and ketones were nonfluorescent but strongly phosphorescent, whereas most 4-nitrophenylhydrazones of aromatic aldehydes and ketones were nonphosphorescent but fluorescent in solvents of low dielectric constant. It was explicitly stated that phosphorescence and room-temperature and low-temperature fluorescence were three powerful complementary analytical tools that achieved their greatest value when used together in microchemical analysis rather than separately.

2. Analysis of Petroleum Products

Manedov [45] used phosphorescence to identify a number of aromatic hydrocarbons in the wax distillates of petroleum. Khaluporskii [46] studied the phosphorescence of lubricating oils and showed that various key products could be identified. Sidorov and Rodomakina[47] demonstrated various applications of phosphorescence spectroscopy to the analysis of petroleum products.

Drushel and Sommers [48] have presented an extensive study of well over
100 aromatic compounds containing sulfur, nitrogen, or oxygen as assayed by
phosphorescence. The authors stated that the techniques used for the characterization of compounds containing sulfur, nitrogen, and oxygen has usually been
limited to three spectrometric methods (mass, infrared, and ultraviolet
absorption). Luminescence techniques, according to the authors, offered
significant advantages for the identification of unknown components; in particular
phosphorimetry offered the largest quantity of easily obtainable physical
information: excitation spectrum, emission spectrum, and easily determinable
lifetime. In addition, with sufficient monochromator resolution, the predominant vibrational spacing of the ground electronic state could be determined.
The authors demonstrated that phosphorimetry provided the requisite information
to identify the components from a gas-chromatography separation of a 430 to
$650^{\circ}F$ petroleum fraction. They also demonstrated that phosphorescence was
invaluable in the characterization of a nitrogen-compound concentrate from a
catalytic hydrogenation of quinoline. Compound type identification was easily
accomplished with the use of phosphorimetry.

Drushel and Sommers [49] have proposed a separation scheme for the
isolation and identification of nitrogen compounds from petroleum. The scheme
used gas chromatography, solid-liquid chromatography, extraction, infrared,
mass, ultraviolet, fluorescence, and phosphorescence spectrometry.
Phosphorimetry proved to be significantly more valuable in identification than
fluorometry. The use of phosphorescence lifetimes to confirm identification
was stressed by the authors.

3. Assay of Pesticides and Alkaloids

A method for the analysis of biphenyl, a widely used fungicide for citrus
fruits, was described by McCarthy and Winefordner [32] who worked with orange
juice or pulp. The cleanup procedure consisted of a prior ether extraction and
finally a thin-layer-chromatography separation of the biphenyl from orange
essential oils and waxes. Due to the extremely great sensitivity of biphenyl by
phosphorimetry (limit of detection is 4×10^{-3} µg/ml), an analysis could be
performed easily on a semimicroquantitative scale. The total time required for
the determination of the biphenyl content of four different samples with four
replicates on each sample was only 2 hours.

The phosphorescence characteristics of 52 pesticides (including several
known degradation products) were surveyed by Moye and Winefordner [27]. For
the 32 pesticides found to phosphoresce sufficiently to be analytically useful the
authors listed spectral characteristics, lifetimes, limits of detection, and the

range of the analytical curves over which approximate linearity was observed. Phosphorimetry was shown to be considerably more sensitive than other analytical methods (e. g. , gas chromatography, fluorometry, and absorption) for a number of the pesticides. The authors point out that the carbamates in particular showed great promise since colorimetric methods, which are normally used, have detection limits in the parts-per-million range, which in phosphorimetry is extended by a factor of 1000 to nanograms per milliliter (parts per billion).

Moye and Winefordner [35] in a subsequent paper successfully applied phosphorimetry to the determination of 4-nitrophenol (a major metabolic product of parathion) in urine again by combining thin-layer chromatography and phosphorimetry. The procedure required 40 min and at least 0. 01 μg of 4-nitrophenol in a 5-ml volume of urine. The average recovery of 4-nitrophenol from doped urine samples was 88%, with a relative standard deviation of 2. 5%.

The phosphorescence characteristics of pesticides and related compounds are listed in Table 38.

Hollifield and Winefornder [29] have reported the phosphorescence characteristics of a number of alkaloids in the isoquinoline, morphine, and indole groups. They examined spectral changes with solvent acidity of basicity; they also reported the decay times and limits of detection for a number of alkaloids. They suggested that phosphorimetry could provide both the phytochemist and the pharmaceutical chemist with a powerful tool for the qualitative and quantitative analysis of alkaloids; they noted that an increase in selectivity could be achieved by pH adjustment in order to shift the phosphorescence excitation and/or emission spectra in a desired direction to produce better excitation and emission resolution.

Winefordner and Moye [34] have developed a rapid quantitative method for the determination of the three alkaloids: nicotine, nornicotine, and anabasine in tobacco. Using the Aminco-Bowman spectrophotofluorometer with phosphoroscope attachment, detection limits of the three alkaloids were found to be approximately 0. 01 μg/ml.

4. Assay of Pharmaceuticals

Hollifield and Winefordner [31] have also presented a procedure for the analysis of several sulfonamides in serum. The phosphorescence charac-teristics of 15 sulfonamides were determined, and the phosphorescence excitation and emission spectra, decay times, and limits of detection were presented. Sulfadiazine, sulfamerazine, and sulfacetamide were added in vitro to serum samples. By using the single-step deproteinization-extraction

TABLE 38

Spectrophosphorimetric Determination of Pesticides
and Related Compounds[a,b]

Compound	Excitation wavelength (nm)	Bands[c] (nm)	Lifetime (sec)	Detection limit (g/ml)	Linear range of calibration curve (moles/liter)
DDT (p,p')	270	420	0.2	7×10^{-10}	1.9×10^{-9} - 7.4×10^{-4}
DDD (p,p')	265	415	0.2	1×10^{-9}	4.0×10^{-9} - 1.3×10^{-3}
DDE (p,p')	270	425	0.2	2×10^{-10}	1.0×10^{-9} - 8.8×10^{-4}
Kelthane	285	515	0.2	6×10^{-10}	1.5×10^{-9} - 7.1×10^{-4}
Methoxychlor	275	380, 395, 360	0.7	4×10^{-10}	1.3×10^{-9} - 9.6×10^{-5}
Chlorobenzilate	275	415, 425, 445, 400, 480	0.2	1×10^{-9}	3.0×10^{-9} - 1.2×10^{-3}
Toxaphene	240	390	1.9	2×10^{-8}	4.5×10^{-8} - 7.5×10^{-3}
Kepone	260	410	1.25	1×10^{-6}	2.0×10^{-6} - 9.2×10^{-3}
Sulfenone	275	390, 375	0.2	5×10^{-10}	2.0×10^{-9} - 9.1×10^{-5}
Tedion	295	410	0.2	2×10^{-10}	5.0×10^{-8} - 6.3×10^{-5}
Orthotran	260	395, 375	< 0.2	2×10^{-9}	8.0×10^{-9} - 7.5×10^{-4}
Parathion	360	515, 490	< 0.2	8×10^{-9}	3.0×10^{-8} - 9.0×10^{-4}
Ronnel	300	475	< 0.2	6×10^{-10}	2.0×10^{-9} - 1.0×10^{-3}
Co-Ral	335	510, 490	< 0.2	4×10^{-11}	1.0×10^{-10} - 8.3×10^{-6}
Diazinon	275	395, 375	5.0	3×10^{-9}	1.0×10^{-8} - 1.1×10^{-3}
Guthion	325	420, 400	0.6	6×10^{-8}	2.0×10^{-7} - 7.5×10^{-3}
Trithion	305	430	< 0.2	3×10^{-10}	8.0×10^{-10} - 8.5×10^{-5}
Aramite	285	400	3.3	3×10^{-10}	1.0×10^{-9} - 1.1×10^{-4}
Isolan	285	395	1.6	2×10^{-7}	1.0×10^{-6} - 1.3×10^{-2}

Sevin	300	510, 475, 485, 550	2.0	4×10^{-9}	$2.0 \times 10^{-8} - 1.0 \times 10^{-3}$
Zectran	285	440	0.45	5×10^{-9}	$2.5 \times 10^{-8} - 7.6 \times 10^{-4}$
Bayer 44646	290	460	0.60	1×10^{-8}	$6.0 \times 10^{-8} - 6.3 \times 10^{-4}$
Bayer 37344	275	435	<0.2	1×10^{-8}	$5.0 \times 10^{-8} - 7.4 \times 10^{-3}$
NIA 10242	285	400	1.6	7×10^{-10}	$3.0 \times 10^{-9} - 7.3 \times 10^{-4}$
U.C. 10854	270	385	2.9	2×10^{-9}	$1.0 \times 10^{-8} - 1.2 \times 10^{-3}$
Imidan	305	440, 420	0.75	6×10^{-10}	$2.0 \times 10^{-9} - 8.5 \times 10^{-5}$
2,4,5-Trichloro-phenoxyacetic acid	300	480	<0.2	5×10^{-10}	$2.0 \times 10^{-9} - 9.5 \times 10^{-4}$
2,4-Dichloro-phenoxyacetic acid	290	495	<0.2	4×10^{-9}	$2.0 \times 10^{-8} - 8.6 \times 10^{-4}$
p-Chlorophenol	290	505	<0.2	2×10^{-8}	$1.8 \times 10^{-7} - 1.1 \times 10^{-2}$
2,4,5-Trichloro-phenol	305	485	<0.2	3×10^{-9}	$1.5 \times 10^{-8} - 6.7 \times 10^{-4}$
p-Nitrophenol	355	520, 495	<0.2	2×10^{-11}	$1.7 \times 10^{-10} - 5.0 \times 10^{-4}$
1-Naphthol	320	475, 495, 520	1.15	2×10^{-10}	$1.7 \times 10^{-9} - 8.4 \times 10^{-4}$

[a] Data from Ref. 27.

[b] All measurements were made in absolute alcohol.

[c] The bands quoted first are suitable for use as key bands.

technique, recoveries of 92 to 105 % were obtained for drug concentrations of 10 to 0.1 μg %, with a relative standard deviation of ± 5 %.

Winefordner and Tin [30] developed experimental procedures that allow the phosphorimetric determination of procaine, phenobarbital, cocaine, and chlorpromazine in serum as well as cocaine and atropine in urine. After a single extraction, an EPA solution of the substance of interest was measured for phosphorescence. The time for a complete analysis was only 30 min. Tables 39 and 40 list the excitation and emission maxima of a number of drugs determined by Winefordner and Tin [30, 50] in urine and blood, as well as the fluorescence lifetimes and detection limits.

Winefordner and Latz [6] developed a simple single-extraction method for the analysis of aspirin by phosphorimetry. The complementary nature of fluoro-metry and phosphorimetry was particularly well demonstrated in this example since aspirin phosphoresced strongly but did not fluoresce, whereas salicylic acid (the metabolite of aspirin) fluoresced but did not phosphoresce. The pro-cedure involved a chloroform extraction from acidified serum, evaporation of the chloroform, and dissolving the residue in EPA. Phosphorimetric measurements were performed on the EPA solution. Recoveries were above 90 % after applying the blank correction. The time required for a complete analysis was only 10 min.

Hollifield and Winefordner [33] have emphasized the complementary nature of fluorometry and phosphorimetry in developing methods for the quantitative measurement of anticoagulants in whole blood. The anticoagulants studied were dicumarol, diphenadione, phenindione, ethyl biscoumacetate, and warfarin. The authors reached the following conclusions: both fluorometry and phosphorimetry are sensitive and selective techniques; fluorometry is some-what simpler than phosphorimetry, but it is not as selective. The anti-coagulants were extracted from whole blood with the ethanol single-step deproteinization extraction method. The phosphorescence of the sample was corrected for the blank. Fluorometry was more sensitive for warfarin, whereas for dicumarol, phosphorimetry was more sensitive. Fluorescence blanks were considerably greater than those found for phosphorescence. Recoveries of doped samples were excellent, and the relative standard deviation in all cases was less than 5 %.

Winefordner and Latz [6] investigated the possibility of trace analysis of pharmaceuticles in blood and urine. Latz [7] demonstrated that the background phosphorescence of an EPA solution of the residue resulting from the evaporation of a chloroform extract of acidified blood was probably due to the aromatic amino acids tyrosine, tryptophan, and phenylalanine. There was a negligible background contribution by all other normal constituents in blood in comparison with the aromatic amino acids.

TABLE 39

Spectrophosphorimetric Determination of Drugs in
Urine and Blood[a]

Compound	Medium	Excitation wavelength (nm)	Key band (nm)	Detection limit[b] (g/ml)	Relative standard deviation[c] (%)
Procaine	Blood	310	430	1×10^{-8}	2 - 5
Cocaine	Blood	240	400	1×10^{-8}	3 - 10
Cocaine	Urine	240	400	1×10^{-8}	2 - 10
Phenobarbital	Blood	240	380	1×10^{-7}	2 - 3
Atropine	Urine	240	380	1×10^{-7}	2 - 3
Chlorpromazine hydrochloride	Blood	320	490	- -	3 - 4

[a]Data from Ref. [30] and [50].

[b]The detection limits quoted refer to solutions in ethanol.

[c]To determine the relative standard deviations the test samples were each investigated five times. The standard deviation varies with the range of concentrations, so the figures above are limiting values.

Hollifield and Winefordner [31] have described a method for the analysis of sulfa drugs in blood. The addition of absolute ethanol (9:1, v/v) to whole blood results in deproteinization. After centrifugation, an aliquot of the supernatant is further diluted (1:100, v/v) with absolute ethanol to give a solution suitable for direct phosphorimetric measurement. This procedure avoids such separation steps as extraction by immiscible solvents and has the advantage of simplicity and speed. Recently these authors have extended this procedure to the analysis of anticoagulants in blood.

5. Assay of Biologically Active Molecules

Churchich [51] studied the phosphorescence of native muramidase. The phosphorescence spectrum (λ_{em} = 332 nm) is similar to that of tryptophan, but its quantum efficiency was much less than that of tryptophan. Muramidase possesses two discrete decay times, both shorter than that of tryptophan. In a study of the extent of involvement of tryptophan residues in the phosphorescence of muramidase, Churchich [52] found the lifetimes of natural and reoxidized muramidase to be 1 and 1.9 sec, respectively.

TABLE 40

Spectrophosphorimetric Determination of Various Drugs[a]

Compound[b]	Excitation wavelength (nm)	Bands[c] (nm)	Lifetime (sec)	Detection limit (g/ml)
Mebaral	240	380	2.2	1×10^{-8}
Rutonal	240	380	2.5	2×10^{-8}
Benzocaine	310	430, 420, 440	5.3	7×10^{-9}
p-Aminobenzoic acid	310	430, 420, 440	3.2	4×10^{-9}
Butacaine sulfate	310	430, 420, 440	5.7	5×10^{-8}
Cyclaine hydrochloride	240, 290	400, 410, 370	2.4	6×10^{-9}
Metycaine hydrochloride	240, 290	400, 410, 370	2.7	6×10^{-9}
Benzoic acid	240, 290	400, 410, 370	2.3	5×10^{-9}
Quinidine sulfate	340, 250	500, 470	1.3	5×10^{-8}
Quinine hydrochloride	340, 250	500, 470	1.3	4×10^{-8}
Lidacaine	265, 340	400	1.1	1.2×10^{-6}
Caffeine	285, 245	440	2.0	2×10^{-7}
Ephedrine	225, 410	390	3.6	2×10^{-7}
Phenylephrine hydrochloride	290, 240	390	2.4	1×10^{-8}
Tronothane hydrochloride	300, 240	410	1.2	2×10^{-8}
Cinchophen	350, 270	520, 490	0.8	2×10^{-8}
Physostigmine sulfate	315, 260	420	3.6	3×10^{-8}
Chlortetracycline	280	410	2.7	5×10^{-8}

[a] Data from Ref. 30.

[b] The compounds are arranged according to their structural similarities and the relationships between their spectral characteristics.

[c] The first band given is suitable for use as key band, and the phosphorescent lifetime and limits of detection have been determined for this.

Churchich [53] presented data on the phosphorescence of pyridoxal-5-phosphate and pyridoxamine-5-phosphate. It was found that Zn^{2+} quenched the phosphorescence.

The effect of solvents and the physical state on the phosphorescence emission of trypsin, ribonuclease, phenylalanine, tryptophan, and tyrosine were studied by Nag-Chaudhuri and Augenstein [54]. The spectra in several solvents were presented.

Bobrovich and Konev [55] reported the phosphorescence of amylase.

Vladimirov and Litvin [56] reported a detailed study of the relation of the phosphorescence of single amino acids to protein phosphorescence. The authors studied zein, human-serum albumin, and egg albumin, and prepared synthetic mixtures of tyrosine, tryptophan, and phenylalanine in the same relative abundance as in the proteins. The phosphorescence spectra of egg albumin and serum albumin contained what appeared to be tryptophan emission; similar results were obtained in the equivalent mixtures of amino acids. The phosphorescence spectrum of zein, however, was composed of tyrosine-like emission; similar results were reported for the equivalent amino acid mixture.

Results similar to those of Vladimirov and Litvin [56] were noted by Vladimirov and Burshtein [57], who studied gamma globulin and actinomycin. In addition, a number of other similar studies have been reported [58-62]. The phosphorescence behavior of proteins has been attributed to an intramolecular-energy-transfer mechanism [63]; energy absorbed by phenylalanine is transferred to tyrosine or tryptophan residues; energy absorbed by, or transferred to, tyrosine is transferred to tryptophan residues; finally, tryptophan emission is observed. For proteins where tryptophan emission is absent the transfer mechanism stops with tyrosine, and thus tyrosine emission is observed.

Grossweiner [64] studied the phosphorescence spectra of tyrosine, tryptophan, indole, phenol, and egg albumin. He concluded that the excited spectrum of egg albumin was the sum   of those from the tyrosine and tryptophan residues.

The phosphorescence of ribonuclease and insulin was investigated by Freed, Turnbull, and Salmre [65]. Both of these substances gave tyrosine emission, but no tryptophan emission. Freed and Salmre [2] reported a study of tryptophan and its indole-type metabolic products.

Douzou and Francq [58] compared the phosphorescence of human and horse serum albumin with the emission of phenylalanine, tyrosine, histidine, and tryptophan. Stauff and Wolf [61] studied the phosphorescence of albumin, lactoglobulin, alcohol dehydrogenase, xanthine oxidase, cytochrome c, peroxidase, and catalase. The phosphorescence was enhanced on addition of eosin.

Steele and Szent-Györgyi [66] give the phosphorescence characteristics of a number of purines and pyrimidines: adenine, adenosine, and adenosine mono-, di-, and triphosphate. Very strong phosphorescence was reported for the latter. The authors suggested that this observation might have quantitative analytical importance since the ATP content of muscle is high. Phosphorescence lifetimes for the above compounds were determined to be 0. 39, 0. 40, 0. 39, 0. 42, and 0. 39 sec, respectively. It was noted that the addition of glucose to DNA or RNA increased the observed phosphorescence intensity of these species.

Bersohn and Isenberg [67] performed a complete study of the phosphor-escence characteristics of DNA. Deoxycytidylic acid, uracil, cytosine, thymine, and polyuridylic acid displayed no phosphorescence.

Longworth [68] reported the phosphorescence and fluorescence quantum yields of purine, adenine, guanine, and caffeine as well as the phosphorescence of guanosine, guanylic acid, adenosine, and adenylic acid.

6. Calcified Tissues

The phosphorescence of teeth was reported in 1963 by Wisolzky [69], who also studied the effect of neotetrazolium chloride on tooth phosphorescence [70]. Quenching effects were found to be dependent on the structural region of the tooth being examined and on the presence of carious lesions. Hoerman and Manciewicz [15] have studied the phosphorescence characteristics of dentin, bone, and enamel. Decay times were reported to be 31 ± 2 sec at 93°K.

REFERENCES

1. G. N. Lewis and M. J. Kasha, J. Amer. Chem. Soc., 66, 2100 (1944).

2. S. Freed and W. Salmre, Science, 128, 1341 (1958).

3. R. J. Keirs, R. D. Britt, and W. E. Wentworth, Anal. Chem., 29, 202 (1957).

4. C. A. Parker and C. G. Hatchard, Analyst, 87, 664 (1962).

5. S. Freed and M. H. Vise, Anal. Biochem., 5, 338 (1963).

6. J. D. Winefordner and H. W. Latz, Anal. Chem., 35, 1517 (1963).

7. H. W. Latz, Ph.D. thesis, University of Florida, Gainesville, 1963.

8. H. C. Hollifield and J. D. Winefordner, Anal. Chem., 40, 1759 (1968).

9. G. N. Lewis and M. Calvin, J. Amer. Chem. Soc., 67, 1232 (1945).

10. C. A. Hitchinson and B. W. Mangun, J. Chem. Phys., 29, 952 (1958); ibid., 34, 908 (1961).

11. W. J. McCarthy and J. D. Winefordner, in Fluorescence. Theory, Instrumentation, and Practice, G. G. Guilbault, ed., Dekker, New York, 1967, Chapter 10.

12. P. A. St. John, W. J. McCarthy, and J. D. Winefordner, Anal. Chem., 38, 1828 (1966).

13. J. D. Winefordner, W. J. McCarthy, and P. A. St. John, J. Chem. Ed., 44, 80 (1967).

14. T. C. O'Haver and J. D. Winefordner, Anal. Chem., 38, 682 (1966).

15. K. C. Hoerman and S. A. Manciewicz, Arch. Oral Biol., 9, 517 (1964).

16. J. D. Winefordner, P. A. St. John, and W. J. McCarthy, in Fluorescence Assay in Biology and Medicine (S. Udenfriend, ed.), Vol. 2, Academic Press, New York, 1970.

17. W. J. Potts, J. Chem. Phys., 20, 809 (1952).

18. R. J. Lukasiewicz, J. J. Mousa, and J. D. Winefordner, Anal. Chem., in press, (1972).

19. R. J. Lukasiewicz, J. J. Mousa, and J. D. Winefordner, Anal. Chem., in press, (1972).

20. R. J. Lukasiewicz, J. J. Mousa, and J. D. Winefordner, Anal. Chem., in press, (1972).

21. H. A. Moye, Ph. D. thesis, University of Florida, Gainesville, 1965.

22. S. P. McGlynn, J. Daigre, and F. J. Smith, J. Chem. Phys., 39, 675 (1963).

23. L. V. S. Hood and J. D. Winefordner, Anal. Chem., 38, 1922 (1966).

24. R. J. Keirs, R. D. Britt, and W. E. Wentworth, Anal. Chem., 29, 202 (1957).

25. P. A. St. John and J. D. Winefordner, Anal. Chem., 39, 500 (1967).

26. E. Sawicki and J. Pfaff, Microchem. J., 12, 7 (1967).

27. H. A. Moye and J. D. Winefordner, J. Agr. Food Chem., 13, 516 (1965).

28. J. D. Winefordner et al., University of Florida, unpublished results.

29. H. C. Hollifield and J. D. Winefordner, Talanta, 12, 860 (1965).

30. J. D. Winefordner and M. Tin, Anal. Chim. Acta, 31, 239 (1964).

31. H. C. Hollifield and J. D. Winefordner, Anal. Chim. Acta, 36, 352 (1966).

32. W. J. McCarthy and J. D. Winefordner, J. Assoc. Off. Agr. Chemists, 48, 915 (1965).

33. H. C. Hollifield and J. D. Winefordner, Talanta, 14, 103 (1967).

34. J. D. Winefordner and H. A. Moye, Anal. Chim. Acta, 32, 278 (1965).

35. H. A. Moye and J. D. Winefordner, J. Agr. Food Chem., 13, 533 (1965).

36. S. P. McGlynn, B. T. Neely, and W. C. Neely, Anal. Chim. Acta, 28, 472 (1963).

37. H. D. Sauerland and M. Zander, Erdoel Kohle, 19, 502 (1966).

38. R. Zweidinger and J. D. Winefordner, Anal. Chem., 42, 639 (1970).

39. R. P. Fisher and J. D. Winefordner, Anal. Chem., in press (1972).

40. E. Sawicki, Chemist-Analyst, 53, 88 (1964).

41. E. Sawicki and H. Johnson, Microchem. J., 8, 85 (1964).

42. E. Sawicki, T. W. Stanley, J. D. Pfaff, and W. L. Elbert, Anal. Chim. Acta, 31, 359 (1964).

43. E. Sawicki and J. D. Pfaff, Anal. Chim. Acta, 32, 521 (1965).

44. J. D. Pfaff and E. Sawicki, Chemist-Analyst, 54, 30 (1965).

45. K. J. Manedov, Chem. Abstr., 53, 561g (1959).

46. M. K. Khaluporskii, Chem. Abstr., 57, 2494f and 8799 (1962).

47. N. K. Sidorov and G. M. Rodomakina, Chem. Abstr., 57, 12, 782e (1962).

48. H. V. Drushel and A. L. Sommers, Anal. Chem., 38, 10 (1966).

49. H. V. Drushel and A. L. Sommers, Anal. Chem., 38, 19 (1966).

50. J. D. Winefordner and M. Tin, Anal. Chim. Acta, 32, 64 (1965).

51. J. E. Churchich, Biochim. Biophys. Acta, 92, 194 (1964).

52. J. E. Churchich, Biochem. Biophys. Acta Previews, 6, No. 1 (1966).

53. J. E. Churchich, Biochim. Biophys. Acta, 79, 643 (1964).

54. J. Nag-Chaudhuri and L. Augenstein, Biopolymers Symp., No. 1, 441 (1964).

55. V. P. Bobrovich and S. V. Konev, Dokl. Akad. Nauk SSSR, 155, 197 (1964).

56. I. A. Vladimirov and F. F. Litvin, Biophysics, 5, 151 (1960).

57. I. A. Vladimirov and E. A. Burshtein, Biophysics, 5, 445 (1960).

58. P. Douzou and J. C. Francq, J. Chim. Phys., 59, 578 (1962).

59. G. C. Barenboim, Biofizika, 7, 227 (1962).

60. E. A. Chernitskii, S. V. Konev, and V. P. Bobrovich, Dokl. Akad. Nauk Belorussk. SSSR, 7, 628 (1963).

61. J. Stauff and H. Wolf, Z. Naturforsch, 19b, 87 (1964).

62. I. A. Vladimirov, S. L. Aksentsev, and V. I. Olensov, Biofizika, 10, 614 (1965).

63. J. Nag-Chaudhuri and L. Augenstein, Biopoylmers Symp., No. 1, 441 (1964)

64. L. I. Grossweiner, J. Chem. Phys., 24, 1255 (1956).

65. S. Freed, J. H. Turnbull, and W. Salmre, Nature, 181, 1731 (1958).

66. R. H. Steele and A. Szent-Gyorgyi, Proc. Natl. Acad. Sci. U.S., 43, 477 (1957).

67. R. Bersohn and I. Isenberg, Biochim. Biophys. Res. Commun., 13, 205 (1963).

68. J. W. Longworth, Biochem. J., 85, 104 (1962).

69. J. J. Wisolzky, J. Amer. Dent. Assoc., 64, 392 (1963).

70. J. J. Wisolzky, J. Dent Res., 43, 659 (1964).

Chapter 6

INORGANIC SUBSTANCES

I.   GENERAL CONSIDERATIONS

Inorganic ions have been determined in solution by the following three methods. 1. The luminescence of the ion is determined directly in solution after being placed in an appropriate solution containing an inorganic reagent (HCl, HBr, etc.).

2. The inorganic ion is combined with a nonfluorescent organic ligand to form a highly fluorescent metal chelate (over 40 different metals have been determined by this technique).

3. The ion is determined indirectly by measuring the amount of quenching of the fluorescence of a chelate or by causing the release of a ligand that can then react to form a fluorescent product, as in the determination of cyanide with the palladium complex of 8-hydroxyquinoline-5-sulfonic acid (8-HQ-5-S):

$$\text{Pd-8-HQ-5-S} \xrightarrow{\text{CN}^-} \text{8-HQ-5-S} + \text{Pd(CN)}_6^{4-} \xrightarrow{\text{Mg}^{2+}} \text{fluorescent chelate} \qquad (53)$$

nonfluorescent

The organic ligand should be an aromatic molecule, containing oxygen or

nitrogen, that is itself nonfluorescent.   The presence of nonbonding n electrons
makes it probable that the excited state will be n, π^*, which is nonfluorescent or
weakly fluorescent.   On complex formation with a metal ion, the n electrons are
utilized in bonding with the metal and thus become less accessible for excitation.
The π, π^* state then results on excitation, and strong fluorescence is observed.

The organic ligands that have most commonly been used to react with metal
ions to form fluorescent chelates are the 2,2'-dihydroxyazo dyes; 8-hydroxy-
quinoline and its derivatives; flavonols; salicylidene compounds and 2,2'-di-
hydroxymethines; salicylic acid; benzoin; Rhodamine B, 6G, and S; salicylalde-
hydes; ß-diketones; hydroxynaphthoic acid; and hydroxyanthraquinones.   Other
ligands have found specific applications, and new compounds are continuously
added to the list.

In the course of a study of azo compounds to determine the minimum
structural requirements for the combination of an azo compound with calcium
and magnesium, for example, Diehl et al [1] found that 2,2'-dihydroxyazoben-
zene (18) was unique in reacting with magnesium and not calcium.   At pH 10
this ligand reacts with magnesium to produce a stable, orange-fluorescent
chelate (19; λ_{ex} 470, λ_{em} 580 nm), but it forms no colored or fluorescent
complex with calcium.   The chelate is fluorescent in water solution at pH
values greater than 11, but the fluorescence intensity is increased in ethanol-
water solutions at pH values of 10 to 11.4.   At lower pH values protons compete
more effectively with the Mg^{2+} ion for the hydroxy oxygens, and consequently,
the chelate is not formed quantitatively.   At pH values greater than about 11.5
magnesium hydroxide may form in appreciable amounts.

(54)

(18)   (19)

Reference material on the analysis of inorganic ions by fluorescence can be
found in books by White [2], Guilbault [3], Hercules [4], Bozhevol'nov [5],
Udenfriend [6,7], and Konstantinova-Shlezinger [8].   Weissler and White [9]
have written a chapter on the analysis of inorganic and organic compounds in
Meites' book and have written review articles on the fluorometric analysis of
inorganic substances that appear every 2 years in Analytical Chemistry [10].

II.   DIRECT ANALYSIS OF INORGANIC SUBSTANCES

Many inorganic substances either fluoresce or phosphoresce directly in the solid state. Salts of rare-earth elements and uranyl salts fluoresce in solution. Other inorganic ions fluoresce in solutions after the addition of an inorganic reagent (HCl or HBr). We now consider each of these classes of analysis in turn.

A. Solid Crystalline Compounds

Although many crystalline compounds are fluorescent, those that are of analytical usefulness are few.

Two fundamentally different cases should be distinguished: the fluorescence of chemically pure substances and the fluorescence of multicomponent systems (crystalline substances with trace amounts of foreign ions that are the activators of the observed fluorescence).

In his review on the fluorescence of solids Randall [11] gave a survey of the inorganic compounds that may be regarded as fluorescent when chemically pure. In addition to the group of salts fluorescing in aqueous solutions, which has already been mentioned, manganese halides, as well as several tungstates and molybdates are luminescent. Some nonfluorescent salts become fluorescent at temperatures far below 0°C. Moreover at lower temperatures the fluorescence bands often become narrower, with the appearance of separate peaks.

A detailed investigation of the low-temperature fluorescence spectra of a number of "pure" salts (AgI, HgI_2, PbI_2, CdS) has been made [12], and the authors give a critical review of the prevailing conceptions on the nature of the fluorescence of "pure" substances. The specific differences from that of ordinary phosphor crystals were emphasized.

The fluorescence, in the solid state at low temperatures, of nonmetals that are gases under ordinary conditions has been known for a long time. Working at low temperatures has become a routine operation, and it may prove expedient to use low temperatures for the analysis of inorganic compounds, as is already being done with certain organic compounds.

According to Randall [11], the elements that are capable of fluorescence include phosphorus. The fluorescence of diamonds has been intensively studied.

The fluorescence of multicomponent systems (e.g., minerals) is successfully utilized for the determination of rare earths and uranium; this, however, is difficult to achieve for other elements, which lack the characteristic atomic structures of the rare earths. Their fluorescence spectra are mostly wide, amorphous bands, subject to considerable shifts, depending on the structure and composition of the crystalline substance as a whole. Thus, for example, manganese fluoresces orange in zinc sulfide, red in cadmium phosphate, and yellow-green in zinc silicate. Conversely the same main substance fluoresces

differently with different activators. Thus, for instance, zinc sulfide displays
an orange fluorescence when activated by manganese, a yellow-green fluorescence when activated by copper, and a blue fluorescence when activated by
silver.

Thus, as a general rule, the fluorescence of multicomponent crystalline
systems (i. e. , of crystalline inorganic compounds containing impurities) is not
sufficiently characteristic identification of the substance or the impurity; in
isolated cases this may be possible, and suitable reactions have been
developed [13,14].

B.   Rare Earths and Uranyl Elements

Uranium compounds are fluorescent in solution; the fluorescence of a
number of such compounds was studied by Vavilov and Levshin [15].

For fluorescence observations in solution it is necessary to convert
uranium compounds to uranyl compounds, since the salts of tetravalent
uranium and uranates do not fluoresce in solution. From an analytical point
of view the most important is the fluorescence of uranium in sodium fluoride
beads, which may be accompanied by the interfering fluorescence of niobium.
The latter does not occur in potassium fluoride beads, but the fluorescence
of uranium becomes weaker [16, 17]. Uranium determination in beads has
proved fully reliable and is now being widely used under both laboratory and
field conditions [18]. The method is not new: its exceptional sensitivity was
noted as early as 1935 [19]; nevertheless, it was further improved, and
modifications were developed for its application to the various kinds of
analysis [20]. The main objectives were to increase the sensitivity, to identify
the factors affecting the brightness of uranyl fluorescence in beads, and to
eliminate the necessity of purifying the uranium compounds isolated from the
sample and introduced into the bead.

The addition of 2% lithium fluoride to sodium fluoride instead of the
carbonate flux has been proposed [21]. A method for the determination of
uranium in natural waters, and a special apparatus for fusing the beads have
been described [22]. Since the bead method is used to determine trace amounts
of uranium (on the order of 1 μg of U in 1 g of rock), its content in the beads is
very low, and this necessitates measuring the brightness of weakly fluorescent
small areas (bead disks).

A method has been suggested for determining uranium from the fluorescence
of its solution in concentrated acid [23,24], To determine microamounts of
uranium in biological material the uranium should be first precipitated as an
uranium-protein complex [23].

Zaidel and Larionov have studied the fluorescence of rare-earth elements in solution [25]. Terbium, gadolinium, and cerium display the strongest fluorescence in solution. These workers regard fluorescence as the most sensitive method for the detection of the ions of these elements. The method is less sensitive with respect to europium and is not sensitive enough to be applied to dysprosium and samarium because of low sensitivity.

The fluorescence spectrum of gadolinium consists of a single, bright, narrow band at 311 nm.

The fluorescence spectra of solutions of cerium salts consist of a single broad band extending from 330 to 402 nm. The sensitivity of the fluorometric determination of cerium is very high. Zaidel and Larionov determined cerium concentrations in solutions on the order of 10 ppb.

The fluorescence spectra of europium and terbium consist of several characteristic bands. Solutions of europium and terbium salts display, respectively, red and yellow-green fluorescence. Europium salts are detectable at concentrations as low as 10 ppm. The sensitivity to terbium is much higher, so that concentrations on the order of 10 ppb to 1 ppm can be determined. In addition to the ordinary fluorescence terbium salts display an afterglow on the order of 0.001 sec.

Rare-earth elements were determined by means of a Beckman spectrophotometer with a photomultiplier; the sensitivity of the instrument makes it possible to study fluorescence spectra even at very low intensities [26]. As to the accuracy of their method, the authors point out that 88 repeated determinations of cerium and praseodymium in thorium, and of terbium in dysprosium, yttrium, and gadolinium, carried out at different concentrations, showed a mean deviation of $\pm$1.5 to 2%. As shown in Table 41, seven rare-earth elements could be determined with a high degree of accuracy [27,28].

Qualitative analysis can be conveniently performed by fluorescence observations of rare-earth elements in borax and phosphoric acid beads [29]. According to Haitinger [29], spark excitation makes it possible to observe, by means of a spectroscope eyepiece, three to six individual bands in the fluorescence spectra of many salts. For instance, europium has three bands: cherry red, orange, and yellow; samarium has six bands: dark red, cherry red, orange, yellow, green, and green-blue. A similar spectrum, though less characteristic, is produced by gadolinium salts. In borax beads all three elements samarium, europium, and gadolinium display a very bright fluorescence. The bands appear just as clearly in the fluorescence spectra of dysprosium and especially of terbium. Cerium borax beads display a bright-

TABLE 41

Direct Analysis of Rare-Earth Elements

Element	Minimum detectable quantity (g)
Praseodymium	2.5×10^{-10}
Neodymium	5×10^{-7}
Samarium	5×10^{-11}
Europium	2.5×10^{-10}
Gadolinium	10^{-6}
Terbium	2.5×10^{-9}
Dysprosium	2.5×10^{-9}
Erbium	5×10^{-9}
Thulium	5×10^{-9}

blue fluorescence, with a continuous spectrum, with maximum intensity at about 450 nm.

Haitinger [29] gives the sensitivity of the detection of the various rare-earth elements by this method (Table 42).

Zaidel and Malakhova [30] photographed the fluorescence spectra of gadolinium and samarium in borax beads and demonstrated that the visible fluorescence that had been ascribed by Haitinger [29] to gadolinium was due to the presence of an admixture of samarium.

For analytical purposes a detailed study was made of the fluorescence of samarium in calcium tungstate and sulfate [31]; a method was developed permitting the detection of 0.05 μg of samarium in 25 mg of calcium tungstate. The emitted radiation was measured by means of a monochromator with a photomultiplier. The fluorescence of samarium was found to become 50% more intense when the solid solvent contained 10% of lead tungstate. In calcium sulfate samarium fluoresced when excited by the near ultraviolet; the detectable minimum was 15 μg in 100 mg of sulfate, but it could be made much lower by preliminary electron irradiation. This rendered the fluorescence bright red, and its spectrum became one amorphous band (instead of nine bands).

TABLE 42

Fluorescence Analysis of Rare-Earth Elements in
Borax and Phosphoric Acid Beads [a]

Element	Detection limit (μg)	Minimum detectable concentration
Cerium	0. 4	1:10,000
Samarium	4. 5	1:1000
Europium	20. 0	1:500
Gadolinium	45. 0	1:100
Terbium	2. 0	1:5000
Dysprosium	4. 5	1:1000

[a] Data from Ref. [29].

A basically different method for the detection of rare-earth elements has been described [32]. Calcium oxide is introduced into the lower part of a colorless hydrogen flame; if a rare-earth element is present, there is a flareup of fluorescence. This indicates the presence of a definite rare-earth element. According to Neunhoeffer [32], the method is sensitive enough to detect 1 ng of yttrium, and the fluorescence itself is due to the excitation by the slow electrons in the hydrogen flame.

The concentration of rare-earth impurities in gadolinium oxide has been determined by Poluektov and Gava [33]. The fluorescence of samarium at 568 or 604 nm, europium at 617 nm, dysprosium at 573 nm, or terbium at 542 nm is measured and related to concentration.

Steele and Robert [34] determined small amounts of the uranyl ion by extracting uranyl nitrate into ethyl acetate in the presence of aluminum nitrate, which acts as a salting-out agent. An aliquot portion of the organic layer is pipetted into a pellet of sodium fluoride contained in a platinum dish and is fused after evaporation. The fluorescence of the fused cake is measured. As little as 30 ppb of uranium can be determined.

C. Use of Inorganic Reagents for the Analysis of Inorganic Ions

Belzi and Kushnirenko [35] assayed As(III) and As(V) in hydrochloric or hydrobromic acid as a frozen solution at -196°C. The sensitivity of assay of

As(III) and As(V) is 0.15 and 37 ppm in 7.6 M HCl, and 7.5 ppb and 3.7 ppm in 7.6 M HBr, respectively. In another paper these authors [36] described methods for the assay of bismuth, antimony, and selenium (Table 43). In HCl as little as 2.1 ppb of bismuth can be assayed or as little as 60 ppb of selenium and 1 ppb of antimony.

Solov'ev and Bozhevol'nov [37] assayed bismuth, lead, and antimony in concentrated HCl at -196°C. The reported wavelengths were Bi: λ_{ex} 312, λ_{em} 410 nm; Pb: λ_{ex} 260, λ_{em} 385 nm. The limits of detection were 10^{-5} % Pb^{2+}, 10^{-6}% Bi^{3+}, and 10^{-7}% Sb^{5+}.

Cerium(III) has been observed to fluoresce in inorganic acid solution. In dilute HCl, Ce(III) exhibits a characteristic fluorescence that has a λ_{ex} at 258 nm and an emission at 350 nm [38, 39]. As little as 1 ppm was determinable in the presence of lanthanum, yttrium, and europium. A strong quenching effect was observed by NO_3^-, Fe^{3+}, and Ce^{4+} [39]. Cukor and Weberling [40] determined cerium by reduction to Ce(III) with $Ti_2(SO_4)_3$ in $HClO_4$, followed by a measurement at 355 nm (λ_{ex} 260 nm). As little as 0.1 ppm was determinable in yttrium oxide.

Kirkbright, West, and Woodward [41] used the fluorescence of Ce^{3+} to indirectly assay for Fe^{2+}, As^{3+}, $C_2O_4^{2-}$, I^-, and Os^{8+} at smaller sensitivities of 5.6, 7.5, 8.8, 0.6, and 0.5 ppm, respectively. The method is based on the reduction of Ce^{4+} to Ce^{3+} by these compounds, and Ce^{3+} was measured at λ_{ex} 250 and λ_{em} 360 nm.

TABLE 43

Fluorescence Spectra of Bismuth, Antimony, and Selenium
in HCl and HBr[a]

| | Fluorescence spectra | | | |
| | λ_{ex}(nm) | | λ_{em} (nm) | |
Ion	In HCl	In HBr	In Hcl	In HBr
Bismuth	340	380	420	487
Antimony(III)	314	360	625	640
Selenium(IV)	305, 330	352	390	550

[a] Data from Ref. [36].

A fluorometric method for the iodide ion was described by Britton and Guyon [42]; it is based on the reaction of uranyl acetate with iodide. The fluorescence is measured at λ_{ex} 365 and λ_{em} 520 nm. From 2 to 20 µg of

I^- can be assayed in the presence of most other anions and cations. The most serious interference was from SCN^-.

Kirkbright and Saw [43] described a spectrofluorometric method for the assay of lead (0.1-0.6 ppm) in HCl (3:10) containing 0.8 M KCl. The fluorescence is measured within 15 min at λ_{ex} 270 and λ_{em} 480 nm. Interference was observed from Bi, Cr^{6+}, Cu^{2+}, Fe^{3+}, Mo^{6+} Tl^+, V^{5+}, ascorbic acid, and $S_2O_5^{2-}$ when present in fiftyfold excess of lead.

Kirkbright, Saw, and West [44] found that from 0.02 to 0.64 ppm of tellurium could be determined by means of the fluorescence at 586 nm (λ_{ex} 380 nm) of Te(IV) in a 9 M HCl "glass" at $-196°C$. Tellurium in lead was determined in the presence of many other ions; Fe^{3+}, Sn^{2+}, and I^- interfere in concentrations 50 times that of tellurium.

Shcherbov and Ivankova [45] found that 0.01 to 50 ppm of Tl(I) or Tl(III) can be assayed in 3 N HCl saturated with NaCl. Belzi and Kushnirenko [46] studied the fluorescence of Tl(I) and Pb(II) in glasslike frozen solutions of 8 N HCl, HBr, LiCl, and LiBr at $-196°C$. As little as 50 ppb of Tl and 10 ppb of Pb could be determined. The luminescence bands of Tl^+ and Pb^{2+} in HCl have maxima at 396 and 423 nm, with excitation bands at 242 and 272 nm, respectively; in HBr the maxima for Tl^+ and Pb^{2+} are 428 and 424 nm, respectively, with excitation bands at 223 and 262 nm for Tl^+, and 225 and 302 nm for Pb^{2+}.

Data on the assay of inorganic ions with inorganic reagents are presented in Table 44.

III. FLUORESCENT CHELATES AND QUENCHING REACTIONS

A. General

The formation of a highly fluorescent chelate by the combination of an ion with an organic ligand has proved to be one of the most sensitive and highly specific methods for the determination of many elements. Some of the elements for which analytical procedures have been described by this method are Al, As, Au, B, Be, Bi, Ca, Cd, Cu, Ga, Ge, Hf, Hg, Mg, Nb, Pd, Rh, Ru, S, Sb, Se, Sn, Si, Ta, Tb, Te, Th, Tl, Zn, Zr, W. In general, fluorescent chelates are formed primarily with diamagnetic ions for reasons discussed in Chapter 3, although methods have been proposed for some paramagnetic ions, such as Cu^+. Some anions, such as CN^- or F^-, can be determined by direct fluorophore formation or, alternatively, could be assayed by their quenching of the fluorescence of a chelate or the release of a ligand to form a fluorescent product.

TABLE 44

Assay of Inorganic Ions with Inorganic Reagents

Ion	Reagent	Sensitivity (ppm)	Refs.
As	Ce^{4+}	7.5	41
	HCl or HBr	0.15	35
Bi	HCl or HBr	0.002	36, 37
$C_2O_4^{2-}$	Ce^{4+}	8.8	41
Ce	HCl	1.0	38
	$HClO_4$	0.1	39
Fe^{2+}	Ce^{4+}	5.6	41
I^-	Ce^{4+}	0.6	41
Os	UO_2^{2+}	2.0	42
	Ce^{4+}	0.5	41
Pb	HCl	0.1	37, 43
	HCl, HBr, or LiCl	0.01	46
Sb	HCl or HBr	0.001	36
Se	HCl or HBr	0.06	36
Te	HCl	0.02	44
Tl	HCl or NaCl	0.01	45
	HCl, HBr, or LiCl	0.05	46

Some typical fluorometric methods for the determination of inorganic substances are listed in Table 45. Most of these methods are highly selective and sensitive, and represent the best analytical methods for these substances. We shall now go through these methods, element by element, alphabetically according to chemical symbol.

B. Assay of Ions

1. Silver (Ag)

El-Ghamry, Frei and Higgs [48] reported that the fluorescence of eosin (C.I. Acid Red 87, C.I. 45380) is quenched by Ag^+ in the presence of 1,10-phenanthroline with maximum effect in the pH range 3 to 8. As little as 4 ppb of Ag^+ can be assayed with a reproducibility of $\pm 2.7\%$. Most other cations were masked with EDTA. The λ_{ex} used was 300 nm, and λ_{em} was 545 nm. Only Pd^{2+} and CN^- interfere seriously.

Ryan and Pal [49] determined Ag^+ by oxidation to Ag^{3+} with $K_2S_2O_8$, followed by chelation with 8-hydroxyquinoline-5-sulfonic acid. The λ_{ex} was 375 nm, and λ_{em} was 485 nm. The fluorescence intensity is linear from 0.0125 to 5.0 ppm of Ag^+. Quenching is observed if more than 1 ppm of Cu^{2+}, Hg^{2+}, or Pd^{2+} is present; similar quantities of Zr^{4+} and Hf^{4+} cause a large increase in the fluorescence intensity.

Babko, Terletskaya, and Dubovenko [50] described a fluorometric method for Ag^+ based on its catalysis of the reaction of lucigenin with H_2O_2. From 0.08 to 1.6 ppm of Ag^+ could be assayed with a relative error of $\pm 7\%$. Interference occurs from Co, Pb, Cr, Cu, Ni, Mn, and Os.

Wheeler et al. [51] described the use of 1H-naphtho[2,3-d]triazole as a fluorometric reagent for Ag^+. The Ag^+ is assayed by its quenching of the fluorescence of this reagent at λ_{ex} of 362 and a λ_{em} of 406 nm. The method is subject to interferences.

TABLE 45

Fluorescence Methods for the Assay of Inorganic Ions

Ion	Reagent	Method[a]	Sensitivity(ppm)	Refs.
Ag	Butylrhodamine S	C	0.01	47
	Eosin + 1,10-phenanthroline	Q	0.004	48
	8-Hydroxyquinoline-5-sulfonic acid	C	0.013	49
	Lucigenin + H_2O_2	Cat	0.08	50
	1H-Naphtho[2,3-d] triazole	Q	0.1	51
	Resorufin	C	0.01	52
Al	Acid Alizarin Garnet R	C	0.007	53
	Coumarin derivatives	C	0.20	54
	Flazo Orange	C	0.001	55
	3-Hydroxy-2-naphthoic acid	C	0.01	56,57
	8-Hydroxyquinoline	C	0.10	58
	Lumogallion	C	0.014	59
	Mordant Blue 9	C	0.0005	60
	Morin	C	0.05	61,62
	Pontachrome BBR	C	0.02	63
	Pontachrome VSW	C	0.02	63
	Quercetin	C	0.05	64
	N-Salicylidene-2-amino-3-hydroxyfluorene	C	0.001	65-67
	Salicylidene-o-aminophenol	C	0.0003	68
As	Gutzeit test	Ch	1.0	69
	Uranyl nitrate	C	100	70-72

TABLE 45 -- Continued

Ion	Reagent	Method[a]	Sensitivity(ppm)	Refs.
Au	Butylrhodamine B	C	0.1	73
	Kojic acid	C	0.1	74
	Rhodamine B	C	0.02	75
B	Acetylsalicylic acid	C	0.01	76
	Alizarin Red S	C	1.0	77
	1-Amino-4-hydroxy-anthraquinone	C	1.0	78
	Benzoin	C	0.04	79-83
	Cochineal Red	C	1.0	84
	Dibenzoylmethane	C	0.0005	85
	Flavonol	C	1.0	86
	Morin	C	1.0	87
	Phenylfluorone	C	1.0	88
	Quercetin	C	1.0	89
	Quinalizarin	C	0.01	90
	Resacetophenone	C	1.0	91,92
	Rhodamine 6G + salicylic acid	C	0.001	93
	Thoron I	C	0.005	94
Ba	Curcumin (Turmeric Yellow)	I	20	70, 95
	Fluorexone	C	80.0	96,97
Be	1-Amino-4-hydroxy anthraquinone	C	0.2	98,99
	Benzoin	C	0.1	79
	1,4-Dihydroxyanthraquinone	C	0.2	99
	2-(2'-Hydroxyphenyl) benzothiazole	C	0.1	100
	8-Hydroxyquinaldine	C	0.001	101
	Morin	C	0.01	70-72, 99, 102-105
	Substituted 2-hydroxy-3-naphthoic acid	C	0.09	106
	Tetracycline + 5,5-diethyl-2-thiobarbituric acid	C	0.10	107
Br	Fluorescein	Ch	1.0	108
	Uranyl nitrate	Q	1.0	109
Ca	Calcein	C	0.2	110-114
	Curcumin	I	2.0	70,95
	Fluorexone	C	1.0	96,97
	8-Hydroxyquinoline	C	1.0	95,115
	8-Quinolylhydrazone	C	0.2	116

TABLE 45 -- Continued

Ion	Reagent	Method[a]	Sensitivity(ppm)	Refs.
Cd	2-(2'-Hydroxyphenyl)-benzoxazole	C	2	117
	8-Hydroxyquinoline	C	2	70, 118
	Morin	C	2	119
	p-Tosyl-8-aminoquinoline	C	0.02	120
Ce	Sulfonaphtholazoresorcinol	C	0.05	121
Cl$^-$	Uranyl nitrate	Q	1.0	109
CN$^-$	Chloramine T + nicotinamide	C	0.3	122
	Pd complex of 8-hydroxy-quinoline-5-sulfonic acid	Q	0.02	123
	Quinone	C	0.01	125, 126
	Pyridoxal	Cat	0.02	124
Co	Al-Pontachrome BBR	Q	0.001	127
Cr	Triazinylstilbexone	Q	0.004	128
Cu	Cochineal Red	C	2.0	70, 95
	2-(2'-Hydroxyphenyl)benzoxazole	Q	0.1	129
	1-(2-Hydroxypropyl)anabasine	Q	0.05	130
	Luminocupferron	C	0.1	131
	Rose Bengal Extra + 1,10-phenanthroline	C	0.1	132
	Salicylalazine	C	0.05	133
	Thiamine	C	0.1	134
	1,1,3-Tricyano-2-amino-1-propene	C	0.1	135
Cs	8-Hydroxyquinoline	C	0.1	136, 137
Eu	Benzoyltrifluoroacetone	C	0.003	138, 139
	Hexafluoroacetone-trioctyl phosphine oxide	C	0.0001	140
	Hexafluoro-2,4-pentanedione	C	0.001	141
	Tetradientate complex with 2-theonyltrifluoroacetone, collidine, and diphenyl-guanidine	C	1.0	142
	2-Theonyltrifluoroacetone	C	0.0001	143

TABLE 45 -- Continued

Ion	Reagent	Method[a]	Sensitivity(ppm)	Refs.
F⁻	Al-Acid Alizarin Garnet R complex	Q	0.001	53
	Al-morin complex	Q	0.2	144
	Mg-8-hydroxyquinoline complex	Q	0.01	145
	Ternary complex with Zr + Calcein Blue	C	0.01	146
	Zr-3-hydroxyflavone complex	Q	0.1	147
Fe	Cochineal Red	Q	1.0	70, 95
	α-naphthoflavone	Q	1.0	70, 95
	Rhodamine S	Q	1.0	148
	2, 2', 6', 2''-Terpyridyl	Q	0.01	149
Ga	5, 7-Dibromo-8-hydroxy-quinoline	C	0.1	150
	1-(2, 4-Dihydroxyphenylazo)-2-naphthol-4-sulfonic acid	C	0.01	151
	8-Hydroxyquinaldine	C	0.02	152, 153
	8-Hydroxyquinoline	C	0.05	154, 155
	Lumogallion	C	0.1	156-158
	Morin	C	0.1	119, 159, 160
	2-(2'-Pyridyl)benzimidazole	C	0.07	161
	Rhodamine B	C	0.01	162, 163
	Rhodamine 6G	C	0.1	164
	Salicylidene-o-aminophenol	C	0.1	165
	Solochrome Red ERS, Black AS, or 6BFA	C	0.01	166-168
	Sulfonaphtholazoresorcinol	C	0.001	169
	2, 2'-4'-Trihydroxy-5-chloro-1, 1'-azobenzene-3-sulfonic acid	C	0.001	170
Ge	Benzoin	C	2.0	171
	Resacetophenone	C	100	92
	Trihydroxyanthraquinone	C	2.0	172
Hf	Flavonol	C	0.1	173
	Quercetin	C	1.0	174
Hg	Rhodamine B	Q	0.1	175
I	Luminol	Q	1.0	176
	α-Naphthoflavone	Q	1.0	70-72
	Uranyl nitrate	Q	1.0	109

TABLE 45 -- Continued

Ion	Reagent	Method[a]	Sensitivity(ppm)	Refs.
In	8-Hydroxyquinaldine	C	0.2	177
	8-Hydroxyquinoline	C	0.04	178
	Morin	C	0.2	179
	2-(2'-Pyridyl)benzimidazole	C	0.1	161
	Pyronine Y	C	5.0	180
Ir	2,2':6',2''-Terpyridyl	C	2.0	181
K	8-Hydroxyquinoline	C	1.0	115,136, 182
	Zinc uranyl acetate	C	1.0	72
Li	Dibenzothiazolylmethane	C	0.5	183
	8-Hydroxyquinoline	C	0.1	184
	Quercetin	C	1.0	185
	Uranyl nitrate	C	1.0	71
Mg	Fluoran	C	0.01	186,187
	8-Hydroxyquinoline	C	0.01	188-192
	1-(8-hydroxyquinoline-7-azo)-2-naphthol-4-sulfonic acid	C	0.01	193
	1-(2-Hydroxy-3-sulfo-5-chlorophenylazo)-2'-hydroxynaphthalene	C	0.020	194
	Lumomagneson	C	0.004	195
	Bis(salicylideneamino)benzofuran	C	0.010	196
	Bissalicylideneethylenediamine	C	0.0002	197
Mn	8-Hydroxyquinoline-5-sulfonic acid	C	0.005	198
Mo	Carminic acid	C	0.9	199,200
	8-Hydroxyquinoline sodium-tetraphenylborate	C	0.2	201
	Primuline	C	20	202
NH_4^+	Hantzch reaction	C	0.01	203,204
	NADH	E	0.01	205
NO_3^-	2,3-Diaminonaphthalene	C	0.01	206
Na	8-Hydroxyquinoline	C	1.0	136,137
	Zinc uranyl acetate	Q	1.0	207
Ni	Al-1-(2-pyridylazo)-2-naphthol	Q	0.00003	208

TABLE 45 -- Continued

Ion	Reagent	Method[a]	Sensitivity(ppm)	Refs.
O_2	Acriflavine	Ox	0.01	209-212
	9,10-Dihydroacridine	Ox	0.01	213
	Epinephrine	Ox	2.0	214
	Fluorescein	Ox	0.01	215
O_3	9,10-Dihydroacridine	Ox	0.01	216,217
	2-Diphenylacetyl-1,3- indandione-1-hydrazone	C	0.02	218
H_2O_2	Diacetyl-2',7'-dichloro- fluorescein	Ox	0.001	219
	p-Hydroxyphenylacetic acid	Ox	0.001	220
	Scopoletin	Ox	0.001	221,222
PO_4^{3-}	Al-morin	Q	0.05	223
	Molybdophosphate- Rhodamine B	C	0.04	224
	NADPH	E	0.01	225
Pb	Morin	C	5.0	72
Rb	8-Hydroxyquinoline	C	5.0	226
Ru	5-Methyl-1,10-phenan- throline	C	1.0	227
S^{2-}	Fluorescein mercuriacetate	C	0.00005	228
	Pd complex with 8-hydroxy- quinoline-5-sulfonic acid	Q	0.2	123
SO_4^{2-}	Th-morin	Q	24	229
Sb	Luminol	C	0.05	230
	Rhodamine 6G	C	0.1	231
	2,4',7-Trihydroxyflavone	C	0.04	232
Sc	5,7-Dichloro-8-hydroxy- quinoline	C	0.1	233
	Morin + phenazone	C	0.01	234
	Salicylalsemicarbazide	C	1.0	235
Se	3,3'-Diaminobenzidine	C	0.02	236-238
	2,3-Diaminonaphthalene	C	0.02	239-244
Si	Ammonium molybdate	C	0.003	245

TABLE 45 -- Continued

Ion	Reagent	Method[a]	Sensitivity(ppm)	Refs.
Sm	Hexafluoroacetone-trioctyl-phosphine oxide	C	0.1	140
	1,10-Phenanthroline + 2-phenylcinchoninic acid ternary complex	C	0.5	246
	2-Theonyltrifluoroacetone	C	0.0001	143
	2-Theonyltrifluoroacetone ternary complex	C	10.0	142
Sn	Flavonol	C	0.1	247
	8-Hydroxyquinoline-5-sulfonic acid	C	0.005	248
	Morin	C	0.2	119
Sr	Fluorexone	C	80.0	96,97
Tb	Antipyrine + salicylate	C	0.1	249
	EDTA-sulfosalicylic acid	C	0.006	250
	Hexafluoroacetone-trioctyl-phosphine oxide	C	0.1	140
	4,4'-Methylenedi-[3-methyl-1-(2-pyridyl)pyrazol-5-ol]	C	0.025	251
	Phenyl salicylate	C	0.1	252
Th	1-Amino-4-hydroxyanthra-quinone	C	8	81
	Flavonol	C	0.01	253
	Morin	C	0.02	254
	Quercetin	C	0.02	255
Ti	Salicylic acid	C	1.0	72
Tl	Cochineal Red	Q	1.0	72
	Rhodamine B	C	0.1	256,257
	Uranyl sulfate	Q	1.0	72
U	Morin	Q	0.05	258
V	Resorcinol	C	2.5	259
W	Carminic acid	C	0.3	199,200
	3-Hydroxyflavone	C	1.0	260
	Rhodamine B	Q	1.0	261
Y	5,7-Dibromo-8-hydroxy-quinoline	C	0.1	262
	8-Hydroxyquinoline	C	0.02	263

TABLE 45 -- Continued

Ion	Reagent	Method[a]	Sensitivity(ppm)	Refs.
Zn	Benzoin	C	0.5	264
	8-Hydroxyquinoline	C	1.0	265,266
	Luminocupferron	C	0.2	267
	2,2'-Methylenedibenzo- thiazole	C	2.0	268
	Picolinaldehyde-2-quinolyl- hydrazone	C	0.026	269
	p-Tosyl-8-aminoquinoline	C	0.02	120
Zr	Flavonol	C	0.1	173
	Morin	C	0.02	270
	2,4',7-Trihydroxyflavone	C	0.05	271

[a] Key: C, chelate; Cat, catalytic; Ch, chemical; E, enzymatic; I, indicator; Ox, oxidation; Q, quenching.

Various workers have proposed to determine Ag^+ with an alkaline solution of resorufin, which combines with Ag^+ to form a precipitate that is detectable against a background of the fluorescent resorufin solution [52]. Similar reactions are given by Mg, Pb, and, in the presence of ammonia, Cu, Cd, Fe, Al, Cr, Ni, Zn, Sr, and Ba.

Perminova and Shcherbov [47] suggested the use of butylrhodamine S for the fluorometric determination of Ag in minerals. The fluorescence is measured at 580 nm.

2. Aluminum (Al)

More chelates have been described for Al than for any other inorganic ion. Some of the most important of these are listed in Table 45.

2,4,2'-Trihydroxyazobenzene-5'-sulfonic acid, sodium salt (20), commonly known as Acid Alizarin Garnet R (AAGR), gives a brilliant yellow fluorescence when complexed with Al (λ_{em} 580 nm) [53]. The chelate of Al with 2,2'-di-hydroxy-1-1'-azonaphthalene-4-sulfonic acid, sodium salt (21), also known as Pontachrome Blue Black Red (PBBR) or Superchrome Blue, fluoresces red (λ_{em} 640 nm) [63].

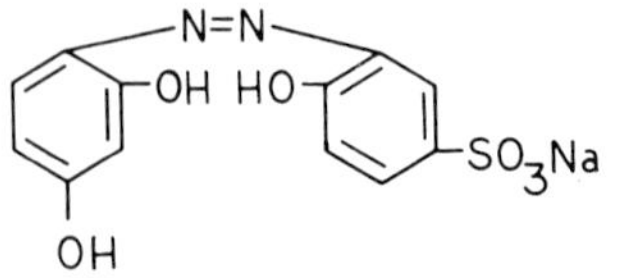

(20) 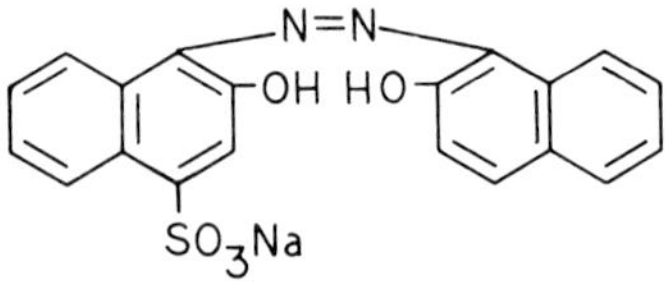   (21)

Aluminum with salicylidene-o-aminophenol (SOAP, 22) produces a greenish fluorescence [68], and the chelate of Al with morin (2',3,4',5,7-pentahydroxy-flavone, 23) has a greenish-blue fluorescence.   Morin and PBBR have been used as fluorescence reagents for many years.   Salicylidene-o-aminophenol was reported in about 1954 as a reagent for Al and has been more thoroughly investigated by Dagnall, Smith, and West [68], who claim a sensitivity of 0.27 ppb.   The effect of substituents on the fluorescence of Al chelates of SOAP are shown in Table 46.   The most fluorescent derivative of SOAP is the phenyl-substituted compound.

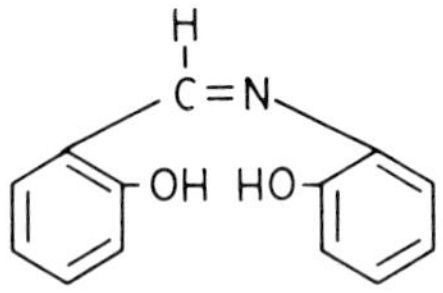

(22) 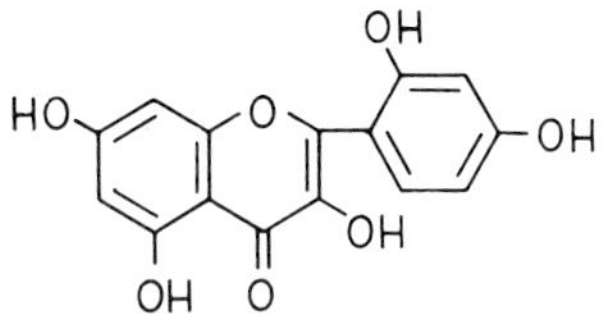   (23)

Of all these fluorometric reagents for the determination of Al in the parts-per-million range, purified AAGR is best for use with photoelectric apparatus because of its availability, sensitivity, and stability, and because the emission is in the range of the normal phototube and is near a high response of the emission grating often used.   For visual observation PBBR is also satisfactory.   The Al-AAGR mixture reaches a maximum fluorescence in a few minutes, but the PBBR complex requires 30 min.

A comparison of the responses of five reagents for Al is shown in Fig. 95.   Pontachrome Violet SW (PVSW) is a commercial dye with the chemical name 2,2'-dihydroxy-1,1'-naphthaleneazobenzene-5-sodium sulfonate (24). This reagent gives an orange-red fluorescence with Al.

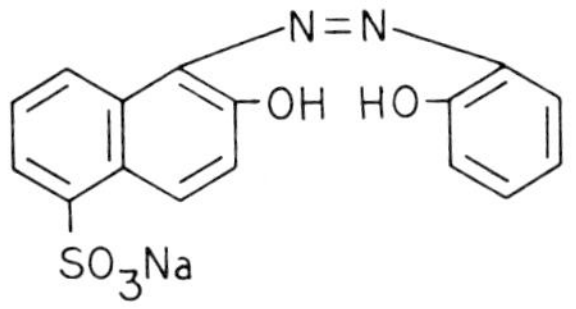

(24)

A new reagent for Al is N-salicylidene-2-amino-3-hydroxyfluorene (NSAHF, 25), which forms a very brilliant yellow chelate with Al [65-67]. This is the first time that the fluorene nucleus has been used in a chelate.

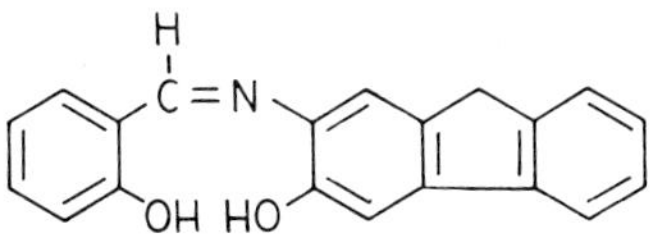

(25)

TABLE 46

Substituted Salicylidene-o-Aminophenols: Effect of Various Substituent Groups on the Fluorescence Intensity F of the Aluminum Chelates[a]

R' = H		R = H				
R	F	R'	F	R'	R	F
C_6H_5	380	H	251	C_6H_5	Cl	53
CH_3	281	Cl	156	Cl	Br	142
$(CH_3)_2CCH_2CH_3$	276	CH_3	142	Cl	Cl	142
H	251	C_6H_5	108	CH_3	CH_3	138
Cl	222			C_6H_5	C_6H_5	138
Br	177					
CO_2CH_3	167					
OCH_3	124					
I	120					
OH	69					
NO_2	1					

[a] Reprinted from Ref. [18] by courtesy of the American Chemical Society.

The preparation and characteristics of compound (25) have been the subject of a recent report [70]. This is one of the most sensitive reagents for Al and belongs to a type of molecule that might have further applications. The compound is not very stable in alcohol solution and must therefore be made fresh each day. Dale, Jones, and Radley [67] in England use this reagent with acetic acid in dimethylformamide (DMF) solution and find it much more stable and more sensitive in this medium. As little as 2 ng of Al in 1 ml of 10% alcohol solution may be measured, and in DMF the limit is 0.8 ppb. If NSAHF were included in Fig. 95, it would show about five times the intensity

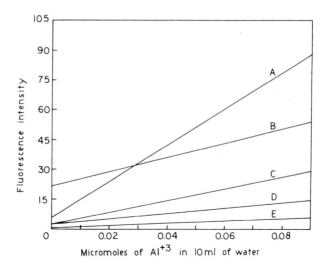

FIG. 95.   Relative fluorescence intensity of five fluorescence reagents for Al: (A) Acid Alizarin Garnet R (20), (B) 3-hydroxy-2-naphthoic acid (27), (C) Pontachrome Violet SW (24), (D) morin (23), (E) Pontachrome Blue Black Red (21).

of AAGR.   The emission curve of Al-NSAHF extends from 475 to 600 nm and is well within the sensitive range of a 1P28 electron multiplier tube coupled with the emission grating of the Aminco-Bowman spectrofluorometer.

Another new reagent for Al is Flazo Orange, 2-hydroxy-5-chlorophenylazo-2-hydroxynaphthalene (26), which gives a red fluorescence with Al in an alcohol solution [55].   Within the last 2 years results with compounds similar to Flazo Orange but containing an SO_3Na group have been described by other authors for several elements and are discussed elsewhere in this section.

The Al chelate of 3-hydroxy-2-naphthoic acid (27) gives a blue fluorescence. This compound was introduced by Cherkesov and Zhegalkena [56] as a reagent for Be and Al and has been studied in detail by other workers [57].   It is able to detect 0.2 ng of Al and 0.2 ng of Be.

(26)   (27)

Compound (27) illustrates another type of group that deserves further investigation.   The hydroxy acids have been given little attention in studies of fluorescent metal chelates.

The linkage of an azo dye to 8-hydroxyquinoline has proved to be a productive new grouping. This compound (28) has been shown to be extremely sensitive reagent for both Al at pH 6 to 6.5 and Mg at pH 10 to 11.5. The commercial dye known as Mordant Blue 9, C.I. 14855 [6-(5-chloro-2-hydroxy-3-sulfophenylazo)-5-hydroxy-1-naphthalenesulfonic acid] is reportedly able to detect 0.5 ppb of Al [60].

(28)

Since morin (23) is not readily available, Davydov and Devekki [64] suggested that it be replaced by its isomer quercetin (3,5,7,3',4'-pentahydroxy-flavone), which is easily extracted by 85% alcohol from onion peel. Quercetin, like morin develops an intense green fluorescence with Al. According to Bozhevol'nov, Pryanishnikov, and Pal'chits [272], morin may be replaced by datiscin or by an extract of the root bark of plants of the genus Datisca. Datiscin (3,5,7,2'-tetrahydroxyflavone) retains the complex-forming hydroxy group in the position 2', and it is closer in its chemical properties to morin than is quercetin.

At present 8-hydroxyquinoline is widely used in determinations of Al. The reaction has been known for a long time [58]. A method has been described in detail for the determination of Al in steel by this reagent [273]. The authors made a careful check of all possible sources of error. They found that the presence of an excess of the reagent and the time that has elapsed from the moment of addition of the reagent to the measurement (up to 24 h) have no appreciable effect on the fluorescence. The fluorescence is quenched by Ti, V, and Fe impurities, but the error is still less than in other methods if the content of these elements is of the same order as that of Al. The fluorescence of the solution is still measurable at the fluorescing-complex concentrations of 0.2 µg per 50 ml. Therefore the separation of impurities can be carried out in extremely small volumes of the solution, which simplifies the analysis considerably.

The method is suitable for Al determinations in solutions; it has also been used for Al determinations in phosphate rocks [274] and is apparently of considerable interest in geochemical investigations. A method has been developed for the determination of Al in Mg metal by 8-hydroxyquinoline [275].

Nishikawa et al. [59] proposed the use of lumogallion [5-chloro-3-(2,4-di-hydroxyphenylazo)-2-hydroxybenzenesulfonic acid] as a reagent for the assay of Al and Ga. In weakly acidic solution (pH 5) lumogallion forms an Al complex that has a green fluorescence. From 0.004 to 0.08 ppm of Al is determinable. The reaction is complete within 20 min at $80^\circ C$, and the complex is extracted into amyl alcohol.

Aguila [54] found that the dyes obtained by coupling the diazonium salt of 4-amino-3-hydroxynaphthalene-1-sulfonic acid with a number of fluorescent coumarin derivatives can be used as fluorescent indicators for Al. The compound obtained with 4-methylumbelliferone gives a pink fluorescence (λ_{ex} 540, λ_{em} 590 nm) with 0.2 to 10 ppm Al.

Babko, Volkova, and Getman [276] compared fluorescein, 8-hydroxy-quinoline, SOAP, morin, quercetin, lumogallion, and Mordant Black as reagents for Al, using the product of the molar absorptivity and the quantum yield as criteria of sensitivity. The preferred reagents were found to be morin and SOAP.

Almost all metal chelates have an optimum pH for maximum fluorescence. In the case of the azo-dye chelates this is pH 4.8, and, since they contain the sulfonic acid group, they are soluble in water. Ligands that do not contain a sulfonic acid group are best used in 30% ethanol, DMF, or formamide.

A precaution should be noted on the meaning of sensitivity values given in the literature for Al on two points, the reagents and the method. All common reagents, including distilled water, contain Al in appreciable quantities. In dealing with nanogram quantities the Al in reagents must be removed. Distilled water run through Al pipes or allowed to stand in glass containers may contain as much as 32 ppb of Al. Ammonia, acetic acid, sodium acetate, etc., all contain many parts per billion of Al. The second point of precaution is to consider how the author obtained his sensitivity value. For example, if the reagent is added to 1 ml of Al solution of 10 ng/ml and then diluted to 10 ml, a sensitivity of 1 ng/ml might be reported. However, the chances are that actually 1 ng/ml would not produce a measurable fluorescence.

3. Arsenic (As)

As pointed out by Haitinger [69], the sensitivity of the Gutzeit test for As (paper strips impregnated with corrosive sublimate) is improved considerably by observation of the fluorescence of the forming arsenious acid. The detectable minimum of arsenious anhydride can be reduced to 1 μg.

Arsenic can be determined according to Goto [70-72] with uranyl nitrate. The precipitate that separates out in the presence of arsenic has a yellow-green fluorescence. The sensitivity is low, 1:10,000.

4.   Gold (Au)

Kojic acid, 5-hydroxy-2-hydroxymethyl-1,4-pyrone (29), produced an intense green fluorescence with Au at a pH of 5.6 to 6.8.   The fluorescence intensity is linear from 0.01 to 1.0 ppm Au [74].

Marienko and May [75] described the use of Rhodamine B (C. I. Basic Violet 10) as a reagent for the assay of Au.   The $AuCl_4^-$ in 0.4 M HCl forms with Rhodamine B^+ (30) a complex that has a λ_{ex} of 550 and a λ_{em} of 575 nm. As little as 0.02 ppm of Au can be determined with a deviation of ±0.003 ppm. The method is reported to be 25 times more sensitive than atomic absorption.

Podberevskaya and Sushkova [73] described butylrhodamine B for the assay of Au.   The fluorescence and limit of detection were about the same as for Rhodamine B.

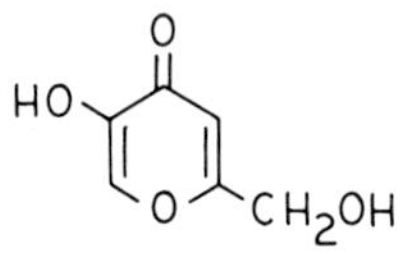

(29)

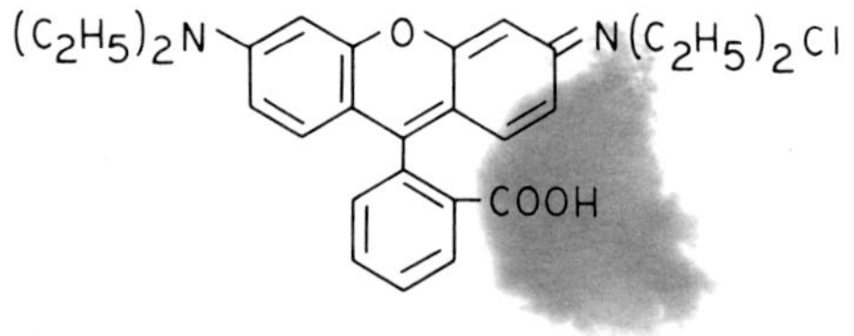

(30)

5.   Boron(B)

The hydroxyanthraquinones quinizarin (31), 1-amino-4-hydroxyanthra-quinone (32), and quinalizarin (33) form chelates with B in concentrated H_2SO_4.   The highly hydroxylated anthraquinones form color complexes that are not fluorescent.   The compounds with one or two OH groups adjacent to the carbonyl group form highly fluorescent complexes with B.

Holme [90] reported that a solution of quinalizarin in 91 to 96% H_2SO_4 could be used for the fluorometric determination of B in concentrations down to 0.01 ppm.   The intensity is measured at 595 nm, with excitation at 365 nm.

The determination of B with Alizarin Red S (Na salt of 1,2-dihydroxy-anthraquinone-3-sulfonic acid) is well known [77].   This reagent is used as 0.002% solution in concentrated H_2SO_4.   Equal volumes of the reagent and the test solutions are used.   In the presence of B the yellow fluorescence of

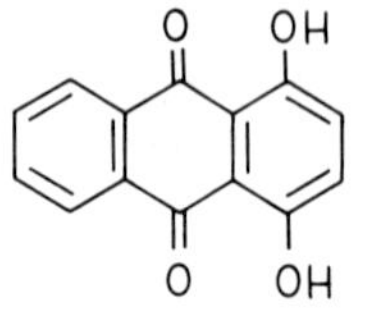

(31)

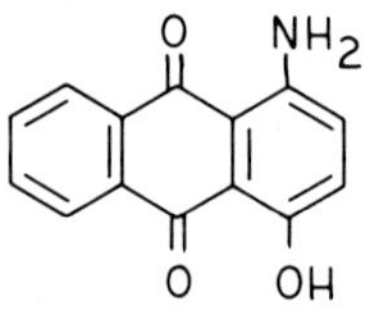

(32)

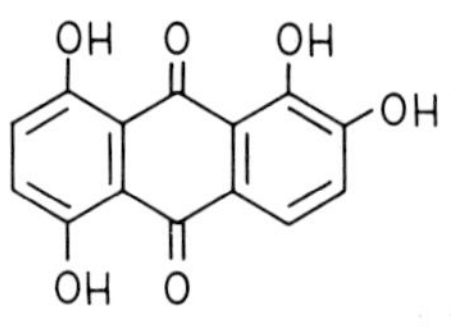

(33)

the solution turns pink-red.   The sensitivity is 5 μg in 5 ml.   Boron can be
determined in the presence of a thousandfold excess of most cations and
anions; only iodides, chlorates, Sb, and Fe interfere.   The interference of
iodides is eliminated by the addition of Ag_2SO_4 powder, that of chlorates by
the introduction of a 30% formaldehyde solution, that of Sb by chlorine water,
and that of Fe by reduction with $SnCl_2$ powder [77].   A chelate ester with a
six-membered ring (34) is formed in the reaction.

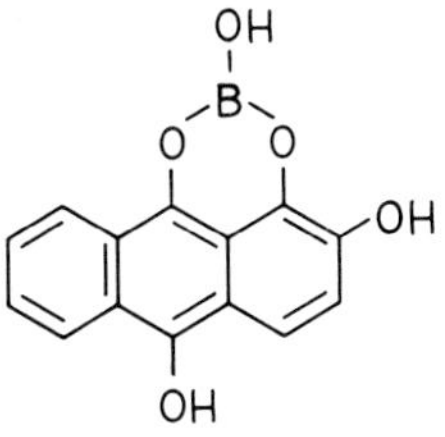

(34)

Other hydroxyanthraquinones can also be used as reagents, provided their
OH groups are in position 1, 4, 5, or 8.   Similar chelate compounds with B
are also formed by aminoanthraquinones, aminohydroxyanthraquinones (e. g.,
1-amino-4-hydroxyanthraquinone [78] and by the 2-hydroxycarbonyl compounds
(e. g., resacetophenone [91, 92].

Another widely used fluorometric reagent for the detection and
quantitative determination of B is benzoin (benzoylphenylcarbinol, 35).   As
distinguished from Zn, B is determined by benzoin in a 85% alcohol-alkaline
solution [79-83].   In the presence of B a blue fluorescence flares up,
reaching its maximum intensity in 4 to 5 min.

Benzoin (35) is a highly specific reagent for B and will detect 10 ppb.
The boron-benzoin complex has a λ_{ex} at 365 and a λ_{em} at 480 nm.   Since
benzoin in alkaline solution is oxidized in air, nitrogen gas is bubbled through
the solution, and the reaction mixture is covered with nitrogen.   The intensity
of the complex is more intense in formamide than it is in ethanol.   In using
any reagent for B it must be remembered that some plastics, Pyrex glass,

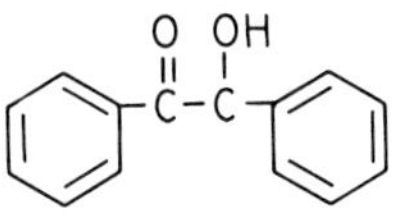

(35)

and other glasses contain this element [79-83].

An interesting method for the determination of B by phenylfluorone has been suggested by Shcherbov and Korzheva [88]. The method makes it possible to carry out the reaction in aqueous medium. The authors discovered an intense green fluorescence of phenylfluorone in the presence of B in an alkaline medium. They showed that the optimum pH for the reaction is 9.5. On standing, the fluorescence intensity in solutions containing the reagent changes; in the blank solution it decreases, but in the solution containing B it becomes higher; after a 24-h period 1 μg of B can be detected in 1 ml. This fluorescence persists for 2 or 3 days. Many cations interfere with the reaction reaction by quenching the fluorescence. Of the anions, phosphates weaken the fluorescence, whereas acetates, nitrates, sulfates, chlorides, and iodides have no effect.

Marcantonatos, Gamba, and Monnier [85] found that boric acid forms a highly sensitive luminescent complex with dibenzoylmethane in concentrated H_2SO_4. The fluorescence is measured at 410 nm, with excitation at 385 nm. The limit of detection is 0.5 ppb of B.

Boron was determined by Babko and Vasilevskaya [93] as a ternary complex with salicylic acid and Rhodamine 6G. The complex is extracted into benzene, and the fluorescence of the solution is measured at 440 nm, with excitation at 366 nm. The method is sensitive to 1 ng of B.

The fluorometric determination of B in high-purity $SiCl_4$ with Thoron I has been described by Rigin and Melnichenko [94]. The fluorescence intensity at 586 nm is measured; from 0.005 to 5 μg of B is determinable.

Acetylsalicylic acid has been proposed as a fluorometric reagent for B [78]. The intensity of the fluorescent complex reaches a maximum in concentrated H_2SO_4-acetic acid (2:3). As little as 10 ppb of B can be determined.

Finally, Monnier and Marcantonatos [277] have described the direct fluorometric assay of B in steel with 4'-chloro-2-hydroxy-4-methoxybenzophenone. The complex is excited at 365 nm and has a λ_{em} at 490 nm.

6. Barium (Ba)

A complexometric determination of Ca, Sr, and Ba has been reported [96, 97] both separately and in the presence of Mg in a strongly alkaline solution with the fluorescent complexometric indicator bis[N, N-di-(carboxymethyl)-aminomethyl]fluorescein, which was named "fluorexone" by the authors. This compound may also be used for the fluorometric determination of traces of these metals. In this case Ca can be quantitatively determined when its amount exceeds 80 μg in 1 ml; the same applies to the corresponding

amounts of Sr or Ba.   Titration of Ca, Sr, or Ba can be carried out in the
presence of Cu, Zn, Cd, Co or Ni, after a solution of KCN has been added to
the titrated solution; an addition of triethanolamine is required in the presence
of Fe, Al, or Mn.

Barium can be detected by precipitation as $BaSiF_6$ in the presence of
turmeric, which is a fluorescent adsorption indicator [70, 95].   The indicator
is adsorbed on the precipitate and emits a brown fluorescence.   The
sensitivity is low, 1:20,000.   A similar reaction is also given by Mg [95].
Calcium and strontium produce a yellow-green fluorescence with the same
reagent in an alkaline alcohol-aqueous medium at dilutions of 1:500,000 and
1:20,000 respectively [70, 95].

7. Beryllium (Be)

The fluorometric reactions of Be have been well studied and are used in
practical work.   Various reagents have been tested, including anthraquinone
derivatives.   Thus, for instance, in the presence of Be an intense red
fluorescence is displayed by 1-amino-4-hydroxyanthraquinone (32) in
alkaline solution [98, 99].   The reaction makes it possible to determine Be
in dilutions of 0.2 ppm.   The optimum concentration of alkali is 0.25 N; above
0.3 N, the intensity of fluorescence decreases.   The reaction is very specific.
Under the conditions of the reaction a similar effect is produced only in the
presence of high concentrations of Li, 3.5 mg in 5 ml.   Sodium chloride
quenches the fluorescence of the complex of Be with 1-amino-4-hydroxy-
anthraquinone when its content is 1 g in 5 ml.   Colored cations, for example,
Cr, interfere with the reaction only at high concentrations.   Anions of HCl,
HNO_3, H_2SO_4, and H_3BO_3 do not interfere.   Beryllium is precipitated in
alkaline media by phosphates, arsenates, molybdates, tungstates, and
uranates, but this can be prevented by the addition of tartrate.   Chromates
interfere with the reaction by oxidizing the reagents.   The fluorescence
emission of the Be complex with 1,4-dihydroxyanthraquinone is at 630 to
650 nm for the complex with 1-amino-4-hydroxyanthraquinone [99].   As little
as 0.2 ppm Be is determinable.

A more sensitive, though less specific, reagent for Be is 8-hydroxy-
quinaldine [101].   As little as 1 ppb of Be is detectable with this reagent.

A large number of publications [70-72, 99, 102, 103] deal with the
determination of Be by morin (23).   In an alkaline solution morin gives a
bright green fluorescence with Be.   According to Goto [70-72], Be can be
determined at 0.006 ppm.   This reaction has been applied [103] to the
quantitative determinations of Be in biological materials; a detailed study was
made of the possible sources of error, and a method was developed for the

determination of 10^{-8} % Be in urine. A disadvantage of the method is that it
involves a complicated procedure for the removal of impurities that affect the
fluorescence of the final solution. The method could be simplified somewhat
[104]. Morin when used as a reagent must satisfy certain standards of purity.
It can be used to determine traces of Be present as dust; a photomultiplier is
used for measuring the fluorescence. According to Welford and Harley [105],
0.005 µg of Be can be detected. The main disadvantage of morin is its low
specificity. On the other hand, if the interfering impurities are masked, the
sensitivity of morin to Be also becomes considerably reduced.

The determination of Be by 2-(2'-hydroxyphenyl)benzothiazole developed
by Holzbecher [100], is both more sensitive and more specific than any other
reaction. The peak fluorescence of the reaction product is observed at pH
6.0. Neither Zn nor Al gives the fluorescence, which is another advantage.
Beryllium can be determined at 0.1 ppm. The fluorescence is weakened by
the presence of Fe, Cr, Sn, Al, Bi, Sb, Zr, and Ti, but these elements can
be masked by Rochelle salt. Copper also interferes.

Another reagent for Be is benzoin (35), with which as little as 0.1 ppm is
determinable [79].

Naito, Nagano, and Yosui [107] found that tetracycline and 5,5-diethyl-2-
thiobarbituric acid together give a blue-violet fluorescence with Be at pH 9.0
(λ_{ex} 406, λ_{em} 506 nm). As little as 0.1 ppm of Be is detectable. The molar
ratio of the complex is 1:1:1. Interferences are observed from Fe, Mg, Ca,
Ba, and Al at greater than twofold amounts relative to Be.

Budesinsky and West [106] described 1-(dicarboxymethylaminomethyl)-2-
hydroxy-3-naphthoic acid as a sensitive and selective fluorometric reagent for
Be, and also for La and Lu. The Be complex shows a maximum fluorescence
at pH 6.8, with excitation and emission wavelengths at 360 and 450 nm.
Linear calibration plots are obtained in the range of 0.09 to 1.8 µg Be.

8. Bromine (Br)

Traces of Br can be detected by the quenching of the fluorescence of
fluorescein (λ_{ex} 440 nm; λ_{em} 470 nm) in glacial acetic acid [108]. Bromine
substitutes for two hydrogen atoms in the fluorescein molecule, forming eosine,
which has a characteristic yellow fluorescence. No interference from Cl was
observed; as little as 2 ng/ml is determinable in aerosols and particulates.

Bromine (Br⁻) can be assayed by its quenching effect on the fluorescence
of uranyl nitrate [109].

9. Calcium (Ca)

Calcium can be determined complexometrically with fluorexone [96, 97] or with curcumin [70, 95], as described for Ba. The sensitivity for Ca is about 2 ppm with curcumin (yellow-green fluorescence) and about 1 ppm with fluorexone.

Calcium can also be determined with 8-hydroxyquinoline, which reacts with Ca in an ammoniacal solution to give a bright fluorescence [95, 115]. A similar reaction occurs with Mg, but not with Sr nor with Ba.

Calcein has been proposed as an excellent reagent for the assay of Ca in biological material such as blood [110-114]; as little as 0.2 μg/ml is determinable. Several wavelengths have been used, the most common being 405- or 435-nm excitation and 485- or 515-nm emission.

Bozhevol'nov et al. [116] have proposed the use of 8-quinolylhydrazone as a reagent for Ca in the pH range 11 to 13. As little as 0.2 ppm is determinable with an SS-5 primary filter and an OS-11 secondary filter.

10. Cadmium (Cd)

8-Hydroxyquinoline has been reported as a fluorometric reagent for Cd. A light-yellow fluorescence is observed; as little as 2 ppm is detectable [70, 118]. It is reported that with 2-(2-hydroxyphenyl)benzoaxole it is possible to detect 2 ppm of Cd in NH_3 solution (λ_{ex} 365, λ_{em} 430 nm) [117].

Morin has also been described as a nonselective reagent for the assay of 2 ppm of Cd [119]. Cadmium can also be determined with p-tosyl-8-amino-quinoline (36) in concentrations of about 20 ppb [120].

(36)

11. Cerium (Ce)

The fluorometric determination of Ce(III) with sulfonaphtholazoresorcinol . [4-(2,4-dihydroxyphenylazo)-3-hydroxynaphthalene-1-sulfonic acid] was described by Huu, Volkova, and Getman [121]. A red-fluorescent chelate is formed at pH 4 to 5 and permits the assay of 50 ppb of Ce(III). The λ_{ex} is 500 and the λ_{em} is 620 nm. Thorium interferes and must be separated.

12.   Chloride (Cl⁻)

Chloride (Cl⁻) can be detected by its quenching effect on the fluorescence
of uranyl nitrate [109] similar to that described for the bromide ion.

13.   Cyanide (CN⁻)

A fluorescent derivative is formed by conversion of cyanide to cyanogen
chloride with subsequent reaction with nicotinamide [122]. The fluorescence
is excited with an Hg lamp at 365 nm and the blue fluorescence formed is
measured. From 0.30 to 5 ppm CN⁻ is determinable.

Cyanide can also be measured by its demasking of the nonfluorescent Pd
complex of 8-hydroxyquinoline-5-sulfonic acid, followed by addition of Mg^{2+}
to form the highly fluorescent Mg complex with the released reagent. The
resulting green fluorescence is measured. As little as 0.02 ppm CN⁻ is
determinable. This reaction is also given by SCN^-, S^{2-}, and thiols, which
interfere [123].

Guilbault and Kramer [125] noted that CN⁻ reacts with quinone monoxime
benzenesulfonate ester to yield a highly fluorescent product (excitation
maximum at 440 nm and fluorescence at 500 nm). As little as 0.5 μg of CN⁻
could be detected. They tested over 30 anions and found that only CN⁻ yielded
a fluorescent product. Sulfide, thiocyanate, and ferrocyanide, which interfere
in most tests for cyanide, do not interfere in this test. In a subsequent report
Guilbault and Kramer [126] found that fluorescent products are formed on the
condensation of CN⁻ with many p-benzoquinone derivatives, all of them having
essentially the same fluorescence characteristics. Fluorescence is developed
by adding CN⁻ in 0.1 ml of phosphate buffer (pH 7.0) to 3 ml of a 3.4×10^{-4} M
solution of the p-benzoquinone derivative in dimethyl sulfoxide. Fluorescence
develops rapidly at room temperature. Of all the p-benzoquinone derivatives
tested, p-benzoquinone itself appears to be the most satisfactory for CN⁻
detection. The reaction is most rapid, fluorescence is proportional to CN⁻
concentration over a wide range, and as little as 0.2 μg of CN⁻ can be detected
in the final reaction volume. Guilbault and Kramer [126] presented good
evidence that the reaction involves the addition of 2 equivalents of CN⁻ to the
quinone:

$$O=\!\!\!\!\begin{array}{c}\\ \bigcirc \\ \end{array}\!\!\!\!=O \; + \; 4CN^- \longrightarrow \quad \underset{NC}{\overset{NC}{\text{HO}}}\!\!\!\!\begin{array}{c} \\ \bigcirc \\ \end{array}\!\!\!\!\underset{CN}{\overset{CN}{\text{OH}}} \qquad (55)$$

Takanashi and Tamura [124] described a fluorometric method for CN⁻
based on its catalysis of the oxidation of pyridoxal to the lactone of 4-pyridoxic

acid. The fluorescence intensity is measured at 432 nm, with excitation at
365 nm. As little as 0.02 ppm CN^- is determinable. The method is about 10
times more sensitive than the cyanogen bromide-benzidine procedure.

14. Cobalt (Co)

Traces of Co^{2+} are determined by the decrease in fluorescence intensity
of the 1:2 Al-Pontachrome BBR complex. The decrease is attributed to the
formation of a complex between Co^{2+} and Pontachrome BBR. A sensitivity of
1 ppb is attained. Better sensitivity is obtained by extracting the Al complex
with amyl alcohol; CrO_4^{2-} and Al^{3+} interfere [127].

15. Chromium (Cr)

A new luminescent reagent for Cr(III), triazinylstilbexone, was proposed by
Temkina et al. [128]. The reagent reacts with Cr to give a complex at pH 2.5
to 3.5, resulting in a quenching of the complex on fluorescence at a λ_{em} of
450 nm. The sensitivity is 4 ppb.

A chelometric titration of Cr with EDTA using Calcein Blue and a fluorometric end point was described by Iritani et al. [129]. The end point is
detected at 745 nm.

16. Copper (Cu)

In the presence of Cu in alkaline solution, Cochineal Red changes its
fluorescence from red to white-blue [70, 95]. The sensitivity is 2 ppm.
Interference can occur from Pd, Pt, and Au.

A very specific method for the determination of 0.1 ppm of Cu is based on
titration with a 0.01% solution of 2-(2-hydroxyphenyl)benzoxazole in acetone [129].
At pH > 6 the reagent possesses a green fluorescence, which decreases in the
presence of Cu due to formation of a nonfluorescent chelate (37). After the
reagent is added in excess the fluorescence increases again.

(37)

The reaction between H_2O_2 and 1-(2-hydroxypropyl)anabasine produces a
green fluorescence (λ_{em} 525 nm) when exposed to ultraviolet light. In the
presence of Cu^{2+} the fluorescence maximum is hypsochromically shifted and

the intensity is decreased. From 0.05 to 5.0 ppm of Cu is determinable. Interference is observed from UO_2^{2+} and Pd^{2+} [130].

A luminescence method for determining Cu in skin was described by Konstantinov et al. [131]. Luminocupferron forms a fluorescent dimer with Cu in alkali, permitting the assay of 0.1 ppb.

Bailey, Dagnall, and West [132] dexcribed a direct fluorometric method for Cu based on its reduction to Cu^+ with hydroxylammonium chloride followed by formation of a ternary complex of Cu^+ with Rose Bengal Extra (C.I. Acid Red 94) and 1,10-phenanthroline (λ_{ex} 560, λ_{em} 570 nm). From 0.1 to 0.6 µg of Cu is determinable with a precision of about 6%.

Salicylalazine has been proposed by Bozhevol'nov [133] as a luminescent reagent for Cu. The sensitivity of the reaction is 0.05 ppm. A bright blue fluorescence appears at pH 12 in the presence of Cu; the intensity of the fluorescence is proportional to the Cu content in the range 0.05 to 1.0 ppm. Many ions do not interfere: Li, Na, K, Rb, Sr, Ca, Ba, Al, Hg, Pb, Sb, Cr, Se, Co, Tl, Pd, Cd, Ru, As, Ag. Only Fe, Mn, Ni, and Mg interfere.

Thiamine as a new fluorometric reagent for Cu was described by Yamane et al. [134]. The reaction is good for as little as 0.1 ppb and up to 30 ppm. An excitation of 290 nm and an emission at 650 nm are used.

A method for determining Cu in animal tissue and human urine was described by Ritchie and Harris [135]. The method is based on a yellow fluorophor that is formed with 1,1,3-tricyano-2-amino-1-propene. From 0.1 to 0.5 ppm is determinable.

17. Cesium (Cs)

A sensitive fluorometric method for Sc and other Group I elements is based on its reaction with 8-hydroxyquinoline to give a bright green fluorescence. The sensitivity is 0.1 ppm [136,137].

18. Europium (Eu)

Both Eu(II) and Sm(III) form chelates with the ß-diketone 2-theonyltrifluoro-acetone in DMF solution (linear range 0.1-100 ppb) [143]. The excitation and emission peaks for Eu are 390 and 615 nm, those of Sm are 370 and 545 nm. No chelate is formed between Tb(III) and 2-theonyltrifluoroacetone.

A method for the assay of Eu and Sm based on the fluorescence of their tetradentate complexes with 2-theonyltrifluoroacetone, collidine, and diphenyl-guanidine was described by Melenteva et al. [142]. The fluorescence is measured at 590 to 620 nm. Sensitivity ranges are 1-5 ppm Eu_2O_3 and 10 to 50 ppm Sm_2O_3.

The fluorometric determination of Eu with benzoyltrifluoroacetone was described by Shigematsu et al [138]. Europium is determined as the trioctylphosphine oxide adduct of the benzoyltrifluoroacetone chelate of Eu (λ_{em} 612, λ_{ex} 365 nm). The detection limit is 3 ppb of Eu. Only Sm and Fe^{3+} interfere [139].

Fisher and Winefordner [140] optimized the experimental conditions for the spectrofluorometric assay of Eu, Sm, and Tb as their hexafluoroacetone-trioctylphosphine oxide complexes. The wavelengths of excitation and emission of each complex together with the range of concentrations are presented in Table 47.

TABLE 47

Analytical Determination of Sm, Eu, and Tb[a]

Ion	λ_{ex} (nm)	λ_{em} (nm)	Range (M)
Samarium	350	565	10^{-4} to 5×10^{-7}
Europium	360	615	10^{-4} to 10^{-9}
Terbium	350	550	10^{-4} to 10^{-7}

[a] Reprinted from Ref. [140] by courtesy of the American Chemical Society.

Williams and Guyon [141] determined Eu and Tb based on the chelates with hexafluoro-2,3-pentanedione. The Eu calibration curve is linear from 1 to 50 ppb.

19. Fluoride (F^-)

Powell and Saylor [53] have shown that the Al-Acid Alizarin Garnet R complex is an excellent quantitative reagent for F^-. The decrease in fluorescence of the complex provides a quantitative measure of the amount of F^- present. As little as 1 ppb is detectable; Be, Co, Cr, Cu, Fe, Ni, Th, Zr, and PO_4^{3-} interfere.

Alternatively, F^- has been studied by its quenching of the fluorescence of the Al-morin complex at (λ_{ex} 365, λ_{em} 500 nm). As little as 0.2 ppm is detectable [144]. Guyon, Jones, and Britton [147] assayed F^- by its quenching of the Zr-3-hydroxyflavone chelate at 460 nm. Only Al^{3+}, MoO_4^{2-}, citrate, tartrate, and oxalate interfere. From 0.1 to 10 ppm is determinable.

Air-pollution control devices for measuring HF in the air are commercially available from such companies as Stanford Research [145]. Air is drawn past a moving 35-mm tape freshly impregnated with Mg-8-hydroxyquinoline. Variations in the fluorescence are recorded continuously for the assay of F^- in the parts-per-billion region.

Har and West [146] have described a direct-assay procedure for F^- based on the ternary complex formed with Zr and Calcein Blue. The complex

(λ_{ex} 350, λ_{em} 410 nm) was found to have a Zr-Calcein Blue-F ratio of 1:1:1.
The increased fluorescence of the complex was attributed to conformational
stabilization of the excimer of the fluorophor in the ternary complex. As little
as 10 ppb of F^- can be detected, and of all the common anions, only phosphate
interferes seriously.

20. Iron (Fe)

Cochineal [70, 95], α-naphthoflavone [70, 95], resorufin [149], and
Rhodamine S [148] have been described as reagents for Fe in the parts-per-
million region. Generally it can be concluded that spectrophotometric methods
are much more sensitive than fluorescence methods for Fe(II) or Fe(III) because
of the intense paramagnetism of Fe. Fink, Pivnichny, and Ohnesorge [149]
described a very sensitive method for Fe at 10 to 500 ppb based on its quenching
of the luminescence of 2, 2':6', 2''-terpyridyl (λ_{ex} 313, λ_{em} 350 nm). Interferences are observed from Co, Cu, and Ni.

21. Gallium (Ga)

Most reagents for Al also are useful for Ga assay at lower pH, although less
sensitively. Morin [119, 159, 160], 8-hydroxyquinoline [154, 155], 8-hydroxy-
quinaldine [152, 153], and 5, 7-dibromo-8-hydroxyquinoline [150] have been
proposed for Ga in the parts-per-billion range.

The best reagent for the analysis of Ga is Rhodamine B (30, 38a) or 6G
(reported as 6Zh in the Soviet literature) [162-164]. These dyes form
fluorescent chelates with Ga(III), Au(III), Tl(III), and Al(III), which are extracted
by benzene from HCl solution. The optimum HCl concentration for Ga is 4 M
[162, 163]. The ratio of Ga to Rhodamine B is 1 mole of $GaCl_4^-$ to 1 mole of
RhB^+. The ring with the carboxyl is unnecessary since Acridine Red (39a)
and Thiopyronine (39b) also form complexes. The $GaCl_4^-$ probably combines
with the $-NR_2Cl$ or $-NR_2$ group since fluorescein (38b), which is similar to

(38)   (39)

a: R = N(C_2H_5)_2; R' = N(C_2H_5)_2Cl   a: X = O; R = NH(CH_3);
                                                   R' = NHCH_3Cl

b: R = OH; R' = O   b: X = S; R = N(C_2H_5)_2;
                                                   R' = N(C_2H_5)_2Cl

Rhodamine B (38a) but lacks these groups, does not form a complex with Ga in
HCl.

Various azo dyes containing the 2, 2'-dihydroxyazo group as the functional
analytical group and similar azomethine compounds have been proposed as
fluorometric reagents for Al and Ga. The azo dyes include a large class of the
so-called Solochrome dyes, the best known of which is 2, 2'-dihydroxy-(1-azo-1')
-4'-naphthalenesulfonic acid (Zn salt), or Acid Chrome Blue-Black [151]. The
same dye is known as Pontachrome BBR. The reaction is carried out in an
acetic acid medium. In the presence of Ga a bright red fluorescence appears on
heating (0.01 ppm is detectable).

Ladenbauer, Korkis, and Hecht [167] and Oshima [168] suggested
Solochrome Red ERS (4-sulfonic acid-ß-naphthol-α-azo-1-phenyl-3-methyl-5-
hydroxypyrazole) and Solochrome Black AS (5-sulfonic acid-2-hydroxyphenyl-
azo-ß-naphthol) as fluorometric reagents for Ga. With the latter reagent the
sensitivity is 0.05 μg in 5 ml of solution. The optimum pH is 4.7.

The determination of Ga by sulfonaphtholazoresorcinol is even more
sensitive (0.001 ppm) [169]. It is carried out in aqueous alcohol at pH 3.0,
with monochloroacetic acid as buffer. The difference between Solochrome
Black and sulfonaphtholazoresorcinol is that the complex formed by Solochrome
Black with Ga fluoresces only after extraction from aqueous solution with an
alcohol that is sparingly soluble in water, such as butyl, amyl, or hexyl alcohol.

At present an even more sensitive method has been developed for the
quantitative determination of Ga. Bozhevol'nov, Lukin, and Gradinarskaya [170]
studied the effect of substituents on the fluorescent properties of chelate
compounds of Ga with dihydroxyazo compounds, and they found 2, 2', 4'-trihyd-
roxy-5-chloro-1, 1'-azobenzene-3-sulfonic acid to be a more sensitive reagent
for Ga (when used in an aqueous medium) than sulfonaphtholazoresorcinol;
moreover its Ga complex is extracted with isoamyl alcohol and the fluorescence
is more intense after the extraction. The fluorescence intensity of the Ga
complex of this reagent is practically constant in the pH range 1.7 to 3.5. The
fluorescence of the extracted complex is increased by a factor of 3.5 if equal
volumes of isoamyl alcohol and the aqueous sample solution are used. The
fluorescence intensity of solutions of this reagent is proportional to the Ga
concentration, both in aqueous solutions and in isoamyl alcohol, provided the Ga
concentration does not exceed 100 ppb. In aqueous solution the sensitivity of
the reaction is 1 ppb. If isoamyl alcohol is used for the extraction of the
samples, and if the volume ratio of isoamyl alcohol to aqueous solution is 1:10,
Ga can be detected in amounts of 0.1 ppb. A detailed study of the effect of
various cations and anions on the fluorescence intensity of the Ga complex
showed that the fluorescence is quenched by Sn, Zr, and Pr when their amount is

100 times that of Ga, and by Cu, Fe, V, and Mo when their amount is 10 times that of Ga. Other cations do not quench even when present in amounts a thousand times larger than that of Ga. Aluminum is capable of forming a fluorescent complex, but its fluorescence is less intense. When the Ga-Al ratio is 1:1, the presence of Al can be ignored, and the measurements can be carried out at pH 1.7 to 3.5. If a tenfold excess of Al is present, the pH of the solution should be 1.7 to 2.7, and in the case of a hundredfold excess of Al the working pH range is narrower, between 1.7 and 2.2. The method of additions makes it possible to carry out the determination also in the presence of quenching impurities.

Bark and Rixon [161] described the spectrofluorometric assay of Ga with 2-(2'-pyridyl)benzimidazole. From 70 to 700 ppb is determinable (λ_{ex} 347, λ_{em} 413 nm). The fluorescence is quenched by Co, Ce, Cu, Fe, Hg, Ni, Pt, Pd, Ag, S^{2-}, and MoO_4^{2-}.

Babko, Volkova, and Getman [278] compared several fluorescent reagents for Ga and found Rhodamine B, lumogallion, and 3-hydroxy-4-(2,4-dihydroxy-phenylazo)naphthalene-1-sulfonic acid to be the best reagents.

22. Germanium (Ge)

Benzoin is a highly specific reagent for B, but under proper conditions it can also be made a reagent for Ge. The fluorescence is yellow-green; as little as 2 ppm is detectable [171]. Interfering ions include As, B, Be, CrO_4^{2-}, NO_2^-, and SiO_3^{2-}. Resacetophenone [92] reacts with Ge to give a bright yellow-green fluorescence in solutions of concentrated acids. The reaction is selective but only sensitive to 100 ppm. Boric acid produces a blue fluorescence.

Germanium also reacts with trihydroxyanthraquinone to give a red fluorescence sensitive to 2 ppm of Ge [172]. Aluminum, boron, and thallium interfere.

23. Hafnium (Hf)

Flavanol reacts with Hf to give a fluorophore with λ_{ex} 365 and λ_{em} 460 nm [173]. As little as 0.1 ppm is detectable, but Zr, Al, F^-, Fe, and PO_4^{3-} interfere.

A more specific reagent is quercetin [174], which reacts with Hf in 9 M $HClO_4$ to give a green fluorophore (λ_{em} 505, λ_{ex} 340 nm). As little as 1 ppm is determinable in the presence of Zn.

24. Mercury (Hg)

A fluorometric method for Hg(II) is based on its quenching of the fluorescence of Rhodamine B (λ_{ex} 486, λ_{em} 586 nm) [175]. Concentrations of 0.1 to

10 ppm are determinable. Interfering compounds include cysteine, albumin, Tl, Pd, Pt, Bi, Cd, Fe, and Sb.

25. Iodine (I)

Since iodinated aromatic compounds are nonfluorescent due to a facilitated transition to the triplet state, several methods have been proposed for the assay of I by its quenching of the fluorescence of such highly fluorescent compounds as luminol [176], α-naphthoflavone [70-72], and uranyl nitrate [109]. These methods are sensitive only in the parts-per-million region.

26. Indium (In)

Indium has been assayed by complexation with 8-hydroxyquinaldine [177], 8-hydroxyquinoline [178], and morin [179]. The sensitivities are 0.2, 0.04, and 0.2 ppm, respectively.

Bark and Rixon [161] described the assay of In with 2-(2'-pyridyl)benzimid-azole. From 110 to 1000 ppb is assayable with a deviation of 2.3%. Bordea [180] described the use of Pyronine Y [3,6-bis(dimethylamino)xanthylium chloride] for the assay of 10 to 100 ppm In. The fluorescence was at a λ_{em} of 585 nm.

Volkova et al. [279] studied the complexes of In with morin, quercetin, SOAP, luminogallion, Rhodamine B, and Rhodamine 6G.

27. Iridium (Ir)

A luminescence method for Ir was described by Fink and Ohnesorge [181]. From 2 to 20 ppm is determined by measuring the emission intensity of the Ir complex with 2,2':6',2''-terpyridyl (λ_{ex} 365, λ_{em} 520 nm). Interferences are observed from Fe, Rh, Ru, and Os.

28. Potassium (K)

Fluorometric methods for K based on chelates with 8-hydroxyquinoline [115,136,182] and zinc uranyl acetate [72] have been proposed. Sensitivity is only in the parts-per-million region.

29. Lithium (Li)

Pitts and Ryan [183] assayed Li with dibenzothiazolylmethane. From 0.5 to 20 ppm of Li is assayable.

8-Hydroxyquinoline reacts with Li to form an intensely fluorescent chelate (λ_{ex} 365, λ_{em} 540 nm). No other alkali metals react in the basic region; Mg and Zn interfere, but the interference of the former can be suppressed by the addition of NaF [184]. As little as 0.2 ppm is determinable.

Other fluorometric methods are based on the complexation of Li with quercetin [185] and uranyl acetate [71].

30. Magnesium (Mg)

Alcoholic solutions of Mg^{2+}-8-hydroxyquinoline exhibit a fluorescence (λ_{ex} 420, λ_{em} 530 nm) at pH 6.5 [188]. Subsequently Schachter [189] suggested the use of the more-water-soluble 8-hydroxyquinoline-5-sulfonic acid for assay of Mg^{2+}. Small aliquots of serum are assayable without the need of deproteinization. Hill [190] adapted the 8-hydroxyquinoline procedure to an automated assay in serum and urine. He used an autoanalyzer system in conjunction with a Model 111 Turner fluorometer. Analyses were performed on serum dialyzates; potassium oxalate was added to prevent Ca^{2+} interference. Hill investigated the specificity of the automated method and showed that the zinc and phosphate normally found in serum and urine do not interfere. The method was shown to be highly reproducible and to yield serum values that compare almost exactly with those obtained by flame photometry. Klein and Oklander [191] utilized 8-hydroxyquinoline-5-sulfonic acid in an automated fluorometric determination of Mg in serum.

The 8-hydroxyquinoline method for Mg is now considered to be a standard clinical procedure. A modification of the Schachter procedure has appeared in Standard Methods of Clinical Chemistry [192]. More recently Pruden et al. [280] compared serum values obtained by fluorometry with those obtained by photometry, atomic absorption, and flame emission. They reported that fluorometry and photometry gave values that were slightly higher than those obtained by flame emission or atomic absorption. It is quite likely that the higher values were the result of Ca interference and that the use of potassium oxalate, as in the automated procedure of Hill [190], would have yielded correct values. An interesting observation made by Pruden et al. [280] is that repetitive freezing and thawing of serum increases the fluorescence obtained in the 8-hydroxyquinoline assay. This results from release of interfering substances and not from increments in Mg.

Wallach and Steck [186] have reported that the fluorescein complexone can be used as a specific reagent for Mg if the pH is kept at 12 and ethylene glycol bis(ß-aminoethyl ether)-N,N'-tetraacetate is added to bind the Ca.

Hill [187] used fluoran (also referred to as "fluorescein complexone") to develop an automated fluorometric method for the determination of serum Ca. The fluorescein complexone that he used was found to be highly stable in alkaline solution when stored in polyethylene bottles in the laboratory. This is in contrast to the calcein used by Kepner and Hercules. Hill [187] used comparable equipment to that used in the assay for Mg [190]. He evaluated the specificity of the method and showed that it was highly reproducible and specific when applied to human serum. Since several urines were found to contain interferences, the method is not yet applicable for urine assay.

Many of the ligands used for the analysis of Al in acid can be used for Mg in alkaline solution. For example, 2, 2'-dihydroxyazobenzene is a good reagent for Mg [281]. The compound 1-(8-hydroxyquinoline-7-azo)-2-naphthol-4-sulfonic acid (40), formed by the linkage of 2, 2'-dihydroxyazobenzene to 8-hydroxyquinoline is a good reagent for Mg at pH 10 to 11.5 [193]. Lumomagneson (41) formed by the addition of an hydroxyazo compound to barbituric acid, can detect 4 ppb of Mg [194]. Compound (42) 1-(2-hydroxy-3-sulfo-5-chlorophenylazo)-2'-hydroxynaphthalene, will detect 20 ng/ml of Mg [194]. Gusev [195] assayed Mg^{2+} in blood serum and urine with lumom-agneson (λ_{ex} 365, λ_{em} 610 nm).

Bissalicylideneethylenediamine [α, α'-(ethylenedimetrilo)di-o-cresol] (43) forms a 1:1 chelate with Mg^{2+} in DMF solution and can detect 0.17 ppb of Mg [197]. This is one of the most sensitive reagents for Mg. The fluorescence spectra of this chelate are shown in Fig. 96.

Dagnall, Smith, and West [196] described a fluorometric method for the assay of Mg with 2, 3-bis(salicylideamino)benzofuran (λ_{ex} 525, λ_{em} 555 nm); as little as 0.01 ppm is determinable.

31. Manganese (Mn)

Pal and Ryan [198] described a fluorometric method for Mn based on its complexation of 8-hydroxyquinoline-5-sulfonic acid in concentrations as low as 5 ppb. The λ_{ex} was 375 and λ_{em} was 485 to 490 nm. The only serious interference under the reaction conditions used came from Ce.

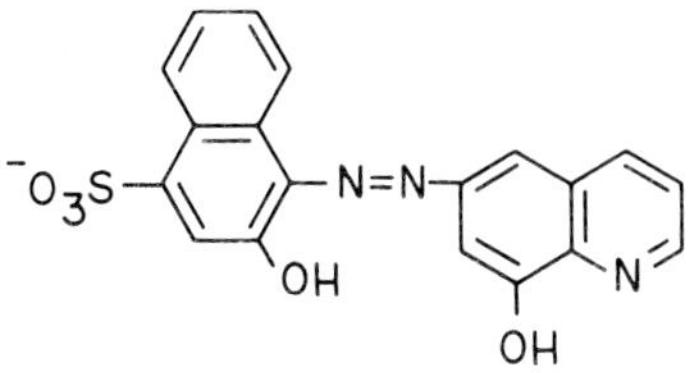

(40)

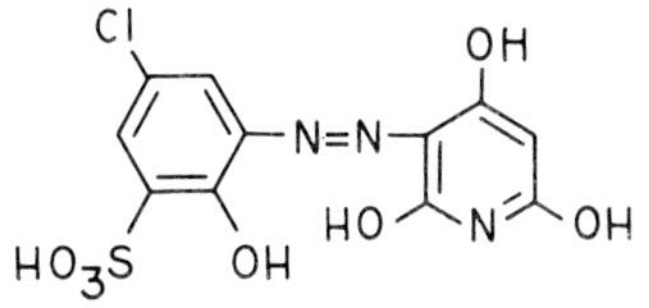

(41)

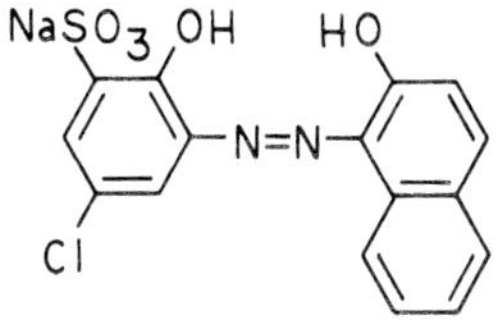

(42)

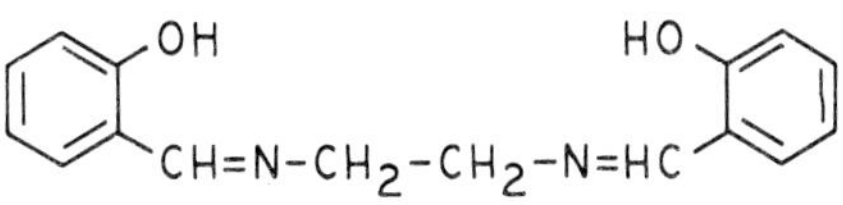

(43)

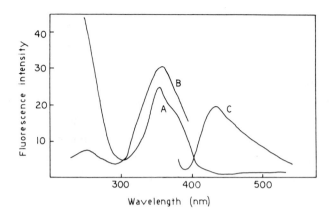

FIG. 96. Absorption (A), fluorescence excitation (B), and fluorescence emission (C) spectra of the Mg-bissalicylideneethylenediamine chelate in dimethylformamide.

32. Molybdenum (Mo)

Kirkbright, West and Woodward [199, 200] have described carminic acid (<u>44</u>) as a reagent for assay of Mo in steel. As little as 0. 9 ppm is determinable.

(44)

A fluorometric method for Mo with 8-hydroxyquinoline and sodium tetraphenylborate was described by Titkov [201]. The wavelengths used were λ_{ex} 395 and λ_{em} 506 nm. From 0. 2 to 1 ppm is determinable.

A titrimetric luminescence method for Mo was described by Andrushko et al. [202]. The MoO_4^{2-} ion is titrated with Pb^{2+} in the presence of the indicator primuline. As little as 20 ppm is determinable.

33. Nitrogen Compounds (N)

a. Ammonia. A fluorometric method for NH_4^+ is based on the reaction of a formaldehyde plus diketone with NH_4^+ to give a lutidine derivative (the Hantzsh reaction)(λ_{ex} 405, λ_{em} 510 nm). From 0. 01 to 0. 25 ppm N is determinable [203, 204].

Rubin and Knott [205] introduced an enzymatic procedure to measure NH_4^+ fluorometrically. This was achieved by utilizing the enzyme glutamic dehydrogenase, α-ketoglutarate, and reduced diphosphopyridine nucleotide,

NADH. By suitable choice of conditions the reaction was made dependent on the NH_4^+ concentration and was followed by the disappearance of NADH fluorescence. The extreme sensitivity of pyridine nucleotide fluorescence permits assay of submicrogram quantities of NH_4^+. The rapidity of assay eliminates errors caused by NH_4^+ generation from biological material, which occurs during lengthy procedures of isolation of digestion.

b. Nitrate. Two methods have been described for the NO_3^- ion. One is based on the quenching of the fluorescence of fluorescein due to formation of nitro-fluorescein, which is nonfluorescent. The sensitivity is 0.10 ppm, and the fluorescence is measured at λ_{ex} 495, λ_{em} 570 nm.

Sawicki [206] described a fluorometric method for NO_3^- based on its reduction to nitrite with hydrazine sulfate and the subsequent determination of the nitrite formed with 2,3-diaminonaphthalene. Solvent effects on the fluorescence intensity and optimum reaction conditions were discussed. The wavelengths used are λ_{ex} 364 and λ_{em} 412 nm. As little as 0.01 ppm to 1.13 ppm is determinable.

34. Sodium (Na)

Fluorometric methods for Na based on its chelation with 8-hydroxy-quinoline [136, 137] or quenching of fluorescence of zinc uranyl acetate [207] have been described. The sensitivity is only 1 ppm.

35. Nickel (Ni)

Schenk, Dilloway, and Coulter [208] described a fluorescent method for Ni(II) based on its quenching of the fluorescence of Al-1-(2-pyridylazo)-2-naphthol in the range 10^{-9} to 10^{-7} M. The method was described as far more sensitive than atomic absorption methods for Ni and has few inter-ferences.

36. Oxygen-Containing Compounds

a. Oxygen (O_2). A sensitive reagent for O_2 is the alkaline solution of fluorescein leucobase (fluorescin) [215], which is obtained by reducing an alkaline solution of fluorescein with Zn dust in an atmosphere of nitrogen or other inert gas. When the gaseous mixture to be analyzed is bubbled through this reagent, fluorescin is oxidized to fluorescein in the presence of O_2 and the solution begins to fluoresce. The intensity of the fluorescence can be used to estimate the amount of the oxidized leucobase and therefore of the amount of O_2 contained in the volume of the gas that has passed through the reagent. Quantitative measurements are possible when the O_2 content is on the order of a few parts per million.

A method has been described for the detection of traces of O_2 dissolved in water [214]. A solution of epinephrine in 25 % alkali does not exhibit its typical yellow-green fluorescence unless it contains dissolved O_2. Dilution of the epinephrine solution with water containing traces of dissolved O_2 causes the appearance of a typical yellow-green fluorescence; for equal concentrations of epinephrine the intensity of fluorescence is proportional to the amount of dissolved O_2. The detectable minimum is below 1 ppm. The method may be used for quantitative determinations.

Parker and Barnes [82] have shown that the borate-benzoin complex in alcohol is quite sensitive to O_2.

Konstantinova-Shlezinger [213] utilized 9,10-dihydroacridine to measure O_2 fluorometrically. In the presence of O_2 this compound is oxidized to acridine. The extremely high intensity of the dihydroacridine fluorescence and the rapidity of its reaction with O_2 make it possible to determine as little as 1 μg of O_2 in a 5-ml volume of gas, with a precision of $\pm 15\%$. The method has actually been applied. It is much more sensitive to ozone.

A much more sensitive method for detecting O_2 utilizing the phosphorescence of adsorbed acriflavine, was first suggested by Kautsky and Hirsch [209] and later studied by Franck and Pringsheim [210] and by Kautsky and Muller [211]. It was found that acriflavine, adsorbed on silica gel (or other solids), has a greater yield of both fluorescence and phosphorescence than it does in solution. The green phosphorescence of acriflavine is extremely sensitive to O_2 and is half-quenched by it at 5×10^{-5} mm Hg pressure. Less than 10^{-11} mole of O_2 can be detected. The method of Tolmach [212] involves continuous sweeping with pure nitrogen (or other inert gas) of the sample in which O_2 is produced. The carrier gas transports the O_2 to a tube containing acriflavine adsorbed on silica gel. Variations in intensity of phosphorescence with varying amounts of O_2 are measured to provide a calibration curve. The sensitivity of this phosphorescence method is such that 10^{-9} atm of O_2 can be detected in an atmosphere of carrier gas. The changes in O_2 tension can be recorded continuously. The sensitivity is so great that samples of leaf juice or chloroplast suspensions in a volume contained in a 2-mm-diameter loop are sufficient for studying O_2 production during metabolism.

b. Ozone (O_3). Ozone, like hydrogen peroxide, can be detected by its selective oxidizing properties. Many years ago Konstantinova-Shlezinger [216] suggested that the nonfluorescent compound 9,10-dihydroacridine is fairly selectively oxidized by O_3 to yield the highly fluorescent compound acridine. The latter is excited maximally at 355 nm and fluoresces maximally at 482 nm when dissolved in acetic acid-ethanol mixtures. Later

Konstantinova-Shlezinger [213] used 9,10-dihydroacridine to determine O_3 in air (1µg of O_3 in 100 ml of air to within ±5%). By this fluorometric procedure she found that the O_3 content of the air in Moscow was 0.02 µg/l at a height of 100 meters.

Watanabe and Nakadoi [217] have used this reaction to measure O_3 in the parts-per-hundred-million range. They showed that at room temperature 1 equivalent of O_3 rapidly converts 1 equivalent of 9,10-dihydroacridine to acridine. The fluorescence of the latter can therefore be used for rapid and continuous monitoring of atmospheric O_3. The fluorometric procedure is more sensitive than the colorimetric method and although neither is entirely specific for O_3, the fluorometric procedure is the more specific of the two. Oxygen, sulfur dioxide, and hydrogen peroxide do not interfere with the acridine procedure. Nitrogen dioxide interferes with ozone assay by both procedures and is difficult to remove. Independent assay for nitrogen dioxide can be used to correct for its interference in the fluorometric assay.

A fluorometric method for atmospheric O_3 was described by Amos [218]. Ozone is collected in a solution of 1,2-di-(4-pyridyl)ethylene in chloroform giving pyridine-4-aldehyde, which is estimated fluorometrically with 2-diphenylacetyl-1,3-indandione-1-hydrazone. The method is extremely sensitive and can detect 0.02 ppm of O_3.

c. Hydrogen Peroxide (H_2O_2). Keston and Brandt [219] used diacetyl-2',7'-dichlorofluorescin (45) as a nonfluorescent substrate for H_2O_2. The substrate is oxidized to diacetyldichlorofluorescein (46), which is highly fluorescent.

(45)   (46)

The assay can be carried out by following the rate of appearance of fluorescence with a recording instrument or by stopping the reaction at a given time and measuring the resulting fluorescence. The excitation maximum of the oxidized diacetyldichlorofluorescein is at 503 nm, and the fluorescence maximum is at 525 nm. The method has a high degree of specificity. However, at extremely low levels of H_2O_2, oxidizing substances in biological materials may interfere. Certain substituted phenols at high concentration can bring about oxidation of the reagent. If this occurs or if there is a need to increase sensitivity, then separation of H_2O_2 by procedures such as distillation may be introduced.

Guilbault and Hackley [282] published a report on the use of homovanillic acid as a fluorescent reagent for assay of peroxidase enzymes. Homovanillic acid is a metabolite of 3,4-dihydroxyphenethylamine and is found in tissues and urine. Procedures for its assay based on oxidation to a fluorophor with ferricyanide were previously described [283].

Guilbault and Hackley [282] substituted peroxidase and H_2O_2 for the ferricyanide and developed assays for peroxidases as well as for H_2O_2. Guilbault and Hackley [220] subsequently found that p-hydroxyphenylacetic acid can be used instead of homovanillic acid and offers many advantages over the latter.

Andreae [221] and Perschke and Broda [222] used peroxidase and 6-methyl-7-hydroxycoumarin (scopoletin) to assay H_2O_2. Peroxidation resulted in a loss of fluorescence. As little as 5×10^{-12} mole/ml of H_2O_2 was detectable.

37. Phosphate (PO_4^{3-})

Land and Edmonds [223] determined phosphate by its quenching effect on the fluorescence of the Al-morin chelate at λ_{em} 510 nm. From 0.05 to 1 ppm is determinable. Other chelates were found to be less satisfactory than the Al-morin.

From 0.04 to 0.6 ppm of P were determined by Kirkbright and West [224], who used the ion-association reaction complex of molybdophosphate with Rhodamine B. The fluorescence intensity of the complex is measured at λ_{ex} 350 and λ_{em} 575 nm. Large amounts of silicate do not interfere, and As and V are tolerable at twenty-fivefold and fiftyfold excesses.

Lowry, Passonneau, and Schultz [225] described an enzymatic method for phosphate, sensitive to 2×10^{-11} mole:

$$\text{glycogen} + PO_4^{3-} \xrightarrow{\text{phosphorylase}} \text{glucose-6-phosphate} \tag{57}$$

$$\text{glucose-1-phosphate} \longrightarrow \text{glucose-6-phosphate} \tag{58}$$

$$\text{glucose-6-phosphate} + NADP \xrightarrow{E} NADPH + \text{6-phosphogluconolactone} \tag{59}$$

The fluorescence of NADPH is measured and equated to the PO_4^{3-} concentration.

38. Lead (Pb)

Very sensitive methods for Pb are not known. At 5 ppm Pb can be determined by its yellow-green fluorescence with morin [72].

39. Rubidium (Rb)

A fluorometric method for Rb is based on its chelation with 8-hydroxy-quinoline. The method is not particularly sensitive [226].

40. Ruthenium (Ru)

A sensitive method for Ru is based on its reduction to Ru(III) followed by its chelation with 5-methyl-1,10-phenanthroline (λ_{ex} 465, λ_{em} 577 nm). The sensitivity is 1 ppm; interfering ions include Ag, Ce, Cr_2O_7, Fe, MnO_4^-, and Pd [227].

41. Sulfur (S)

a. Sulfide (S^{2-}). Sulfide has been determined fluorometrically by its quenching of the fluorescence of the Pd complex of 8-hydroxyquinoline-5-sulfonic acid [123] and of fluorescein mercuriacetate [284]. The former method is sensitive to only 0.2 ppm. The latter is sensitive to 1 to 10 ppb of S^{2-}. Interference occurs from Co, Ni, Fe, and I^-. A method for the direct assay of H_2S in the atmosphere at concentrations below the parts-per-billion range was described by Axelrod et al. [228]. Sulfide is collected in NaOH solution and fluorescein mercuriacetate is added. The fluorescence is recorded at λ_{ex} 499, λ_{em} 519 nm. As little as 0.05 ppb is detectable.

b. Sulfate (SO_4^{2-}). A fluorometric method for SO_4^{2-} was described by Morgen et al. [229]. The method is based on the quenching by SO_4^{2-} of the fluorescence of the Th-morin complex. As little as 24 ppm of SO_4^{2-} is determinable within $\pm 1.3\%$.

42. Antimony (Sb)

A fluorometric method for Sb is based on the chemiluminescence produced in the reaction of luminol with Sb(V) at pH 11 to 12 [230]. Only Sb(V) present as $SbCl_6^-$ is active in this reaction. The best oxidant for the conversion of Sb(III) to Sb(V) is $NaNO_2$. As little as 0.05 ppm is detectable; small amounts of Fe and Cu quench the chemiluminescence.

Ivankova and Shcherbov [231] studied the fluorescence reaction of Sb with Rhoadamine S and 6G. The greatest sensitivity for the fluorometric determination of Sb(III) (0.1 ppm) is obtained with Rhodamine 6G.

A fluorometric procedure for the Sb determination with 2, 4', 7 -trihydroxy-
flavone has been developed and is much more sensitive than other common
methods. The fluorescence of the Sb(III) complex is measured in perchloric
acid solution, using phosphate as a masking agent. The method has a detection
limit of 0. 04 μg and a precision of about 2%. After decomposition of the sample
by pyrosulfate fusion, the Sb is extracted as the triiodide into methylisobutyl
ketone from sulfuric acid solution [232].

43. Scandium (Sc)

Nishikawa, Hiraki, and Shigematsu [233] found that solutions of 8 -hydroxy-
quinoline and the 5, 7 -dichloro, 5, 7 -dibromo, 5, 7 -diiodo, and 2 -methyl
derivatives form fluorescent chelates with Sc, with λ_{em} at 510, 526, 520, 515,
and 505 nm, respectively. For Sc assays 5, 7 -dichloro -8 -hydroxyquinoline
gave the best results. As little as 0. 1 ppm is determinable.

In the presence of anions of strong acids Sc forms an extractable complex
with morin and phenazone. As little as 0. 01 ppm is determinable at λ_{ex} 425 nm.
Interference occurs from Al, Ga, In, Ti, Th, Zr, and Lu [234].

The fluorometric determination of Sc with salicylalsemicarbazide was described by Shcherbov and Nikolaeva [235]. The sensitivity is only about 1 ppm.

44. Selenium (Se)

Methods for the assay of Se in biological material, based on the fluorescence
of the complex with 3, 3' -diaminobenzidine, have recently been published by
Cousins [236] and Watkinson [237]. Both workers found that excitation must be
carried out above 420 nm to minimize blank fluorescence. The fluorescence
maximum was reported as being approximately 580 nm. A fluorescence
spectrometer was not used for these measurements. Recoveries were from
85 to 100%. Various grasses, fruits, and bulbs were found to contain from
0. 01 to 0. 08 μg of Se per gram. Soil content (New Zealand) was reported as
1. 4 to 1. 5 μg/g.

Parker and Harvey [238] investigated the product formed by the condensation of Se with 3, 3' -diaminobenzidine. The product 3', 4' -diaminophenyl-
piazselenol (47), contains an additional ring (piazselenol), which, through
additional conjugation with the benzene ring, endows the molecule with its
altered and increased fluorescence.

(47)

Parker and Harvey [238] found that diaminobenzidine has many drawbacks as a reagent for Se assay. First the fluorescence efficiency of the product is low (ϕ = 0.055), and its molar absorptivity is also quite low. As a result, fluorescence sensitivity is very low. Furthermore 3,3'-diaminobenzidine contains two pairs of orthodiamine groups, only one of which reacts with Se. The resulting piazselenol is still highly basic and can only be extracted into organic solvents from basic solutions. This brings up problems of stability and precipitation of metal ions that may be present in the sample.

In a subsequent study Parker and Harvey [239] investigated the piazselenols formed by the condensation of Se with several other diamines: o-phenylene-diamine, which yields 3,4-benzo-1,2,5-selenodiazol (48); 1,2-diaminonaphtha-lene, which yields 3,4-benzopiazselenol (49); 2,3-diaminonaphthalene, which gives 4,5-benzopiazselenol (50). The absence of residual amine groups in derivatives (48), (49), and (50) permits their extraction directly from the acid reaction mixture into organic solvents.

(48)

(49)

(50)

The fluorescence characteristics of these four derivatives are presented in Table 48. It is apparent that derivative (50) formed on condensation with 2,3-diaminonaphthalene, is the most highly fluorescent and that its fluorescence emission is not as far into the red as that of the diaminobenzidine product (47).

This is very important because photomultiplier detectors in most commercial spectrometers are least sensitive in the red end of the spectrum. Parker and Harvey [239] used decalin as a solvent because of its low volatility and the sharp spectra obtained with it. They found that the fluorescence of 4,5-benzopiaz-selenol (50) was substantially diminished by the dissolved oxygen in the organic solvent. However, even with this oxygen quenching the fluorescence of (50) was far more intense than that of the other derivatives.

Allaway and Cary [240] used 2,3-diaminonaphthalene for the assay of Se in agricultural products (milk, alfalfa, etc.) and compared it with diaminobenzi-dine. They corroborated the findings of Parker and Harvey concerning the

TABLE 48

Fluorescence Characteristics of Various Piazselenol Derivatives [a]

Derivative	Solvent	Excitation maximum (nm)	Fluorescence maximum (nm)	Relative fluorescence intensity[b]
(47)[c]	Toluene	436	600	5
(48)[d]	Cyclohexane	313	--	Extremely low
(49)[e]	Cyclohexane	366	400	0.2
(50)[f]	Cyclohexane	366	520 and 580	123
(50)[f]	Decalin	366	520 and 580	100

[a] After Parker and Harvey [239].

[b] Fluorescence intensity of 4,5-benzopiazselenol (50) in decalin arbitrarily taken as 100. Excitation and fluorescence spectra were obtained on a corrected instrument.

[c] 3',4'-Diaminophenylpiazselenol.

[d] 3,4-Benzo-1,2,5-selenodiazol.

[e] 3,4-Benzopiazselenol.

[f] 4,5-Benzopiazselenol.

advantages of diaminonaphthalene and showed that the method gave values in good agreement with neutron-activation assay. Allaway and Cary pointed out that the major difficulties were in the combustion steps, which would, of course, influence any method. Their studies and those of Lott et al. [241] have helped establish the specificity of the 2,3-diaminonaphthalene assay for Se.

2,3-Diaminonaphthalene was also used as a reagent for Se by Clarke [242]. Wilkie and Young [243] obtained a more stable compound by using a 10% solution of concentrated HCl to prepare 2,3-diaminonaphthalene. Olson [244] reported a sensitivity of 0.02 ppm for the assay of Se in plants.

45. Silicon (Si)

Kasiura [245] described a fluorometric amplification method for the determination of Si, based on the complex formed with ammonium molybdate in NH_3. The fluorescence is measured at 605 nm, with excitation at 475 nm. As little as 3 ppb of Si is determinable.

46. Samarium (Sm)

Samarium can be determined by its complexation with 2-theonyltrifluoro-
acetone [143], by a 2-theonyltrifluoroacetone ternary complex [142], and by a
complex with hexafluoroacetone-trioctylphosphine oxide [140], as described for
europium (see also Table 47).

A method for the assay of traces of Sm in the form of a ternary complex
with 1,10-phenanthroline and 2-phenylcinchoninic acid was described [246].
The sensitivity is about 0.5 ppm.

47. Tin (Sn)

Tin in concentrations as low as 0.2 ppm can be determined by its
complexation with flavanol [247]. A blue-white fluorescence is observed. Both
Hf and Zr give a similar test.

Pal and Ryan [248] assayed Sn by the fluorescence of the 8-hydroxyquino-
line-5-sulfonic acid complex at λ_{ex} 360 and λ_{em} 515 nm. As little as 5 ppb to
0.25 ppm is assayable. Quenching of the fluorescence occurs in the presence
of Fe, Cu, Hg, Ni, F^-, EDTA, citrate, oxalate, and tartrate. Traces of Al,
Zn, Hf, and Zr enhance the fluorescence.

Tin can be assayed by its reaction with morin to give a bright blue-green
fluorescence [119]. It is possible to determine as little as 0.2 ppm.

48. Strontium (Sr)

Fluorexone has been proposed as a chelometric reagent for Sr [96, 97].
As little as 80 ppm are determinable. Interference is observed from other
alkaline earth elements.

49. Terbium (Tb)

Terbium can be assayed by a ternary complex formed with hexafluoro-
acetone and trioctylphosphine oxide [140], as described for europium (see also
Table 47). As little as 0.1 ppm is detectable.

Terbium has been assayed by the fluorescence of the complex with
antipyrine and salicylate at 530 nm. As little as 0.1 ppm is determinable [249].

A specific spectrofluorometric method for Tb as a ternary complex with
EDTA and sulfosalicylic acid has been described [250]. From 0.0064 to 3.2
ppm is determinable. The λ_{ex} is 320 and the λ_{em} is 410 and 545 nm. No
interference was noted from fiftyfold excesses of 33 metal ions and 14 anions.

A new reagent for the assay of Tb is based on the reaction with
4,4'-methylenedi-[3-methyl-1-(2-pyridyl)pyrazol-5-ol]. As little as 0.025
ppm is determinable. The λ_{em} is 550 nm for Tb and 575 nm for Dy [251].

Kononenko et al. [252] studied the fluorescent product of the reaction of Tb with phenyl salicylate. The λ_{em} is 535 to 570 nm, and as little as 0.1 ppm is determinable.

50. Thorium (Th)

Thorium reacts with 1-amino-4-hydroxyanthraquinone to give a chelate at pH 2.3 that has a λ_{ex} of 550 to 580 and a λ_{em} of 660 nm. As little as 8 ppm is determinable; Fe, Ga, Pr, and Zr interfere [81].

Thorium reacts with flavanol [253] to give a fluorescent chelate (λ_{ex} 390, λ_{em} 460 nm). As little as 0.01 ppm is determinable.

Thorium can be assayed with morin [254] in 0.01 M HCl in 50% ethanol. The λ_{ex} is 420 and the λ_{em} is 520 nm, with as little as 0.02 ppm determinable. The elements Al, Ca, Fe, La, and Zr interfere. Babko et al. [255] described the assay of Th with quercetin (λ_{ex} 425, λ_{em} 541 nm). As little as 0.02 ppm is determinable.

51. Titanium (Ti)

Titanium reacts with salicylic acid to give a complex sensitive to about 1 ppm [72]. The method has not been extensively used.

52. Thallium (Tl)

Goto [72] described a method for the assay of 1 ppm of Tl based on its quenching of the fluorescence of Cochineal Red and of uranyl sulfate.

Rhodamine B was used as a fluorometric reagent for the detection of Tl by Feigl, Gentil, and Goldstein [256]. This reagent combines with trivalent Tl to form an orange-red fluorescent compound in benzene solution. The detection of 0.1 ppm of Tl in the presence of 500 ppm Au, Hg, and Sb is possible by this method.

It has been shown [257] that the Tl-Rhodamine B complex contains two atoms of Tl to one molecule of Rhodamine B. Thallium combines with the Rhodamine B molecule in the form of a negatively charged complex anion.

53. Uranium (U)

Uranium can be determined by its quenching of the fluorescence of morin. As little as 0.05 ppm is determined [258]. Interfering substances are removed by ether extraction.

54. Vanadium (V)

A fluorometric method for V is based on a complex formed with resorcinol in 10 M H_2SO_4. The fluorescence is red; as little as 2.5 ppm is determinable. Interference by Ce(IV) is found [259].

55. Tungsten (W)

A fluorometric method for W with carminic acid was described for concentrations as low as 0. 3 ppm [199, 200].

A fluorometric procedure for W based on the reaction with 3-hydroxy-flavone was described by Bottei and Trusk [260]. As little as 1 ppm can be determined.

A method based on the quenching of Rhodamine B fluorescence (λ_{ex} 365, λ_{em} 570-640 nm) was described [261]. As little as 1 ppm is determinable.

56. Yttrium (Y)

A fluorometric method for Y in La salts with 5, 7-dibromo-8-hydroxy-quinoline was described by Kirillov et al. [262]. As little as 0. 1 ppm can be assayed.

Yttrium forms with 8-hydroxyquinoline a complex that is extracted with $CHCl_3$ from pH 9. 5. As little as 0. 02 ppm is assayable. Interference from Ce and La is observed [263].

57. Zinc (Zn)

Zinc reacts with benzoin to give a green fluorescence; as little as 0. 5 ppm is determinable [264].

Another reagent for Zn is 8-hydroxyquinoline, which displays a light-yellow fluorescence (λ_{ex} 420 nm); as little as 1 ppm can be detected [265, 266].

Zinc can be determined with p-tosyl-8-aminoquinoline (36) in concentrations of about 20 ppb [120]. A more sensitive reagent is 2, 2'-methylenedi-benzothiazole (51) which will detect 2 ppb [268]. Picolinaldehyde-2-quinolyl-hydrazone (52) will detect 26 ppb [269].

(51)   (52)

A method for Zn using lumocupferron (a-benzamido-4-dimethylaminocin-namic acid) was described [267]. The intensity of the bright-green fluores-cence is measured. As little as 0. 2 ppm is determinable.

58. Zirconium (Zr)

The blue-white fluorescence (λ_{ex} 390, λ_{em} 465 nm) exhibited by flavanol in the presence of Zr in H_2SO_4 solution is the basis of a method for determining as little as 0. 1 ppm. Hafnium is the only cation that interferes; F^- and PO_4^{3-} also interfere [173].

Zirconium reacts with morin to give a yellow-green fluorescence. As little as 0.02 ppm is determinable. Interferences are noted from F^-, Fe, Ge, Hf, Nb, PO_4^{3-}, Sc, and Ta [270]. A quantitative method has been described for Zr determination with morin in 2 N HCl in the presence of Al, Be, Ba, Sb, Sn, Th, and U; the method is based on a comparison of fluorescence intensities before and after the addition of complexone III [270].

The reaction for Zr [173] has been studied in detail and seems to be of practical importance. The proposed reagent is 3-hydroxyflavone. The fluorescence of the compound formed with Zr is extremely bright; the reagent itself also fluoresces, but its green fluorescence is eliminated by means of a filter. The reaction is highly specific; of the 53 tested cations and anions, only Th, Al, and Hf as well as Zr fluoresce with 3-hydroxyflavone in an acid medium. The hydroxyflavone-Zr compound is stable in 0.2 N H_2SO_4; neither Al nor Th fluoresces under these conditions. The fluorescence spectra due to Zr and Hf overlap, and the overall content of both elements is determined from the measured brightness of fluorescence. The method simplifies considerably the determination of Zr in ores and is especially useful when the ZrO_2 content in the ore is below 0.25%.

A fluorometric method for Zr with 2,4',7-trihydroxyflavone in H_2SO_4 was described by Filer [271]. The method has a detection limit of 0.05 ppm and a precision of 1% or 5 μg. There are few interferences.

REFERENCES

1. H. Diehl, R. O. Olsen, G. Speilholtz, and R. Jensen, Anal. Chem., 35, 1144 (1963).

2. C. E. White, Fluorometric Analysis. A Practical Approach, Dekker, New York, 1970, Chapters 4-6.

3. G. G. Guilbault (ed.), Fluorescence. Theory, Instrumentation, and Practice, Dekker, New York, 1967, Chapter 7.

4. D. M. Hercules (ed.), Fluorescence and Phosphorescence Analysis, Interscience, New York, 1966, Chapter 4.

5. E. A. Bozhevol'nov, Fluorometric Analysis of Inorganic Materials (in Russian), Khimiza, Moscow, 1960.

6. S. Udenfriend, Fluorescence Assay in Biology and Medicine, Vol. 1, Academic Press, New York, 1966, Chapter 12.

7. S. Udenfriend, op. cit., Vol. 2, 1970, Chapter 16.

8. M. A. Konstantinova-Shlezinger, Fluorometric Analysis (N. Kaner, transl.), Davey, New York, 1965, Chapter 12.

9. A. Weissler and C. E. White, in Handbook of Analytical Chemistry (L. Meites, ed.), McGraw-Hill, New York, 1963, Chapter 6, pp. 176-196.

10. C. E. White, biennial reviews on fluorometric analysis in Analytical Chemistry, 1948-1962 and, with A. Weissler, 1964-1968.

11. J. T. Randall, Trans. Faraday Soc., 35, 1 (1939).

12. V. Arkhangelskaya and P. Feofilov, Opt. Spektrosk., 2, 107 (1957).

13. V. Levshin, E. Arapova, and E. Baranova, Tr. Kom. Anal. Khim. SSSR, 12, 393 (1960).

14. K. Stolyarov and N. Grigorev, Zh. Anal. Khim., 14, 71 (1959).

15. S. Vavilov and V. Levshin, Z. Physik, 48, 397 (1928).

16. M. Haitinger, Die Fluoreszenzanalyse in der Mikrochemie, Wien-Leipzig, 1937.

17. I. Papsih and L. Hoag, Proc. Natl. Acad. Sci. U.S., 13, 726 (1927).

18. C. E. White, Anal. Chem., 24, 1952 (1965).

19. F. Hernegger and B. Karlik, Sitzber. Akad. Wiss. Wein, Part IIa, 144, 217 (1935).

20. G. Price, R. Ferritti, and S. Schwartz, Anal. Chem., 25, 322 (1953).

21. F. Centappi, A. Ross, and M. Sesa, Anal. Chem., 28, 1651 (1956).

22. L. Thatcher and F. Barker, Anal. Chem., 29, 1575 (1957).

23. C. E. White, Anal. Chem., 26, 129 (1954).

24. C. E. White, Anal. Chem., 24, 87 (1952); 30, 729 (1958).

25. A. N. Zaidel and Y. Larionov, Dokl. Akad. Nauk. SSSR, 16, 443 (1937).

26. F. Huki, R. Heidel, and R. Huke, Anal. Chem., 24, 606 (1952).

27. V. Fassel, R. Heidel, and R. Huke, Anal. Chem., 24, 606 (1952).

28. V. Fassel and R. Heidel, Anal. Chem., 26, 1134 (1954).

29. M. Haitinger, Die Fluoreszenzanalyze in der Mikrochemie, Wien-Leipzig, 1937.

30. A. Zaidel and G. P. Malakhova, Dokl. Akad. Nauk. SSSR, 85, 591 (1952).

31. C. G. Peattie and L. B. Rogers, Anal. Chem., 25, 518 (1953).

32. O. Neunhoeffer, Z. Anal. Chem., 132, 91 (1951).

33. N. S. Poluektov and S. A. Gava, Zavod. Lab., 35, 1458 (1969).

34. T. W. Steele and R. V. Robert, Nucl. Sci. Abstr., 22, 2754 (1968).

35. M. U. Belzi and I. Kushnirenko, Referat. Zh. Khim. 19GD, 1969, Abstr. No. 16G92.

36. M. U. Belzi and I. Kushnirenko, Referat. Zh. Khim. 19GD, 1969, Abstr. No. 6G67.

37. A. Solov'ev and E. A. Bozhevol'nov, Chem. Abstr., 66, 82105n (1967).

38. M. Furukawa, S. Sasaki, R. Nakashima, and S. Shibata, Nagoya Kogyo Gijutsu Shikensho Hokoku, 17, 251 (1968).

39. N. S. Poluektov, A. Kirillov, M. A. Tishichenko, and Y. Zelyukova, Zh. Anal. Khim., 22, 707 (1967).

40. P. Cukor and R. P. Weberling, Anal. Chim. Acta, 41, 404 (1968).

41. G. F. Kirkbright, T. S. West and C. Woodward, Anal. Chim. Acta, 36, 208 (1966).

42. D. A. Britton and J. C. Guyon, Microchem. J., 14, 1 (1969).

43. G. F. Kirkbright and C. G. Saw, Talanta, 15, 570 (1968).

44. G. F. Kirkbright, C. G. Saw, and T. S. West, Analyst, 94, 457 (1969).

45. D. P. Shcherbov and A. I. Ivankova, Prom. Khim. Reakt. Osob. Chist. Veshchestv., 8, 191 (1967).

46. M. U. Belzi and I. Kushnirenko, Referat. Zh. Khim. 19GD, 1969 (6), Abstr. No. 6G66.

47. D. N. Perminova and D. P. Shcherbov, Prom. Khim. Reakt. Osob. Chist. Veshchestv., 8, 181 (1967).

48. M. T. El-Ghamry, R. W. Frei, and G. W. Higgs, Anal. Chim. Acta, 47, 41 (1969).

49. D. E. Ryan and B. K. Pal, Anal. Chim. Acta, 44, 385 (1969).

50. A. K. Babko, A. V. Terletskaya, and L. I. Dubovenko, Zh. Anal. Khim., 23, 932 (1968).

51. G. L. Wheeler, J. Andrejack, J. H. Wiersma, and P. F. Lott, Anal. Chim. Acta, 46, 239 (1969).

52. H. Eichler, Z. Anal. Chem., 96, 22 (1934).

53. W. Powell and J. Saylor, Anal. Chem., 25, 960 (1953).

54. J. F. Aguila, Talanta, 14, 1195 (1967).

55. C. E. White, in Fluorescence, Theory, Instrumentation, and Practice (G. G. Guilbault, ed.), Dekker, New York, 1967, p. 281.

56. I. Cherkesov and V. Zhegalkena, Dokl. Akad. Nauk. SSSR, 118, 309 (1958).

57. G. F. Kirkbright, T. S. West, and C. Woodward, Anal. Chem., 37, 137 (1965).

58. C. Gentry and L. Scherrington, Analyst, 71, 432 (1946).

59. Y. Nishikawa, K. Hiraki, K. Morishige, and T. Shigematsu, Japan Analyst, 16, 692 (1967).

60. J. de Albinati, Anales Assoc. Quim. Argentina, 53, 61 (1965); Anal. Abstr., 5432 (1966).

61. C. E. White and C. S. Lowe, Ind. Eng. Chem. , Anal. Ed. , 12, 229 (1940).

62. F. Will, Anal. Chem. , 33, 1360 (1961).

63. A. Weissler and C. E. White, Anal. Chem. , 18, 530 (1946).

64. A. Davydov and A. Devekki, Zavod. Lab. , 10, 134 (1941).

65. C. E. White, H. McFarlane, J. Fogt, and R. Fuchs, Anal. Chem. , 39, 367 (1967).

66. Z. Holzbecher, Chem. Listy, 47, 680, 1023 (1953).

67. A. Dale, P. Jones, and J. Radley, Inst. Rept, U. S. Dept. of Army Contract DA91-591-3309, 1965.

68. R. M. Dagnall, R. Smith, and T. S. West, Talanta, 13, 609 (1966).

69. M. Haitinger, Die Fluoreszenzanalyse in der Mikrochemie, Wien-Leipzig, 1937.

70. H. Goto, Sci. Rept. Tohoku Imp. Univ. , Ser. 1, 29, 204 (1940).

71. H. Goto, Sci. Rept. Tohoku Imp. Univ. Ser. 1, 29, 287 (1940).

72. H. Goto, Chem. Zb. , 1, 1068 (1941).

73. N. K. Podberevskaya and V. Sushkova, Zavod. Lab. , 36, 1048 (1970).

74. A. Murata and T. Ujaihara, Buneski Kagaku, 10, 497 (1961).

75. J. Marienko and I. May, Anal. Chem. , 40, 1137 (1968).

76. V. Podchainova, L. Skonyakova, and B. Dvinyaninov, Referat. Zh. Khim. 19GD, 1968, Abstr. No. 16G13.

77. L. Szebellady and S. Tamay, Z. Anal. Chem. , 107, 26 (1936).

78. J. Radley, Analyst 69, 47 (1944); Anal. Chem. , 21, 1345 (1949).

79. C. E. White, J. Chem. Educ. , 28, 369 (1951).

80. C. E. White, A. Weissler and D. Busker, Anal. Chem. , 19, 802 (1947).

81. C. E. White and D. E. Hoffman, Anal. Chem. , 29, 1105 (1957).

82. C. A. Parker and W. J. Barnes, Analyst, 82, 606 (1957).

83. C. A. Parker and W. J. Barnes, Analyst, 85, 828 (1960).

84. L. Szebelledy and F. Gaal, Z. Anal. Chem. , 98, 255 (1934).

85. M. Marcantonatos, G. Gamba, and D. Monnier, Helv. Chim. Acta, 52, 538 (1969).

86. K. Tanbock, Naturwiss. , 30, 439 (1942).

87. C. E. White, Fluorescence Analysis. A Practical Approach, Dekker, New York, 1970, p. 61.

88. D. P. Shcherbov and R. Korzheva, Tezisy dokladov soveshchaniyapo lyuminestsenta, 1958, p. 65.

89. L. Kommenda, Chem. Listy, 47, 531 (1953).

90. A. Holme, Acta Chem. Scand., 21, 1679 (1967).

91. K. Neelakantam and L. Row, Proc. Ind. Acad. Sci., 16a, 349 (1942).

92. N. Raju and G. Rao, Nature, 174, 400 (1954).

93. A. K. Babko and A. Vasilevskaya, Ukr. Khim. Zh., 33, 314 (1967).

94. V. Rigin and N. Melnichenko, Zavod. Lab., 33, 3 (1967).

95. P. Dancknortt and J. Eisenbrand, Lumineszenzanalyze in filtrtierten ultravioletten Licht, Leipzig, 1956.

96. Y. Korbl and F. Vydra, Chem. Listy, 51, 1457 (1957).

97. Y. Wilkins, Talanta, 4, 80 (1960).

98. C. E. White and C. S. Lowe, Ind. Eng. Chem., Anal. Ed., 13, 809 (1941).

99. M. H. Fletcher, C. E. White, and M. S. Sheftel, Ind. Eng. Chem., Anal. Ed., 18, 179 (1946).

100. Z. Holzbecher, Coll. Czech. Chem. Commun., 20, 193 (1955).

101. K. Motojinia, Bull. Chem. Soc. Japan, 29, 75 (1956).

102. C. W. Sill and C. P. Willis, Anal. Chem., 31, 598 (1959).

103. F. W. Klempeter and A. Martin, Anal. Chem., 22, 828 (1950).

104. H. Laitinen and P. Kivalo, Anal. Chem., 24, 1467 (1952).

105. G. Welford and J. Harley, J. Amer. Ind. Hyg. Assoc. Quart., 13, 332 (1952).

106. B. Budesinsky and T. S. West, Anal. Chim. Acta, 42, 455 (1968).

107. T. Naito, H. Nagano, and T. Yosui, Japan Analyst, 18, 1068 (1969).

108. H. Axelrod, J. Bonelli, and J. Lodge, Env. Sci. Technol., 5, 420 (1971).

109. V. Volman, Bull. Soc. Chim., 53, 385 (1933) (Ref. 8).

110. A. B. Borle and F. Briggs, Anal. Chem., 40, 339 (1968).

111. M. Lewin, M. Wills, and D. Baron, J. Clin. Pathol., 22, 222 (1969).

112. B. Fingerhut, A. Poock, and H. Miller, Clin. Chem., 15, 870 (1969).

113. H. G. Classen, P. Marquardt, and M. Spath, Arzneimittel Forsch., 18, 211 (1968).

114. T. Uemura, Sci. Rept. Tohoku Imp. Univ., Ser. 4, 34, 31 (1968).

115. C. Miller and R. Magee, J. Chem. Soc., 3183 (1951).

116. E. A. Bozhevol'nov, L. Federova, I. Krasavin, and V. Dziomko, Zh. Anal. Khim., 24, 531 (1969).

117. N. Louis and A. Reber, Anal. Chem., 26, 936 (1954).

118. J. Eisenbrand, Pharma. Ztg., 75, 1003 (1930).

119. V. Patrovsky, Chem. Listy, 47, 676 (1953).

120. E. A. Bozhevol'nov, Chem. Abstr., 65, 7989 (1966).

121. C. Ti Huu, A. I. Volkova, and T. Getman, Zh. Anal. Khim., 24, 688 (1969).

122. J. S. Hanker, R. M. Gamson, and H. Klapper, Anal. Chem., 29, 879 (1957).

123. J. S. Hanker, A. Gelberg, and B. Whitten, Anal. Chem., 30, 93 (1958).

124. S. Takanashi and Z. Tamura, Chem. Pharm. Bull. Tokyo, 18, 1633 (1970).

125. G. G. Guilbault and D. N. Kramer, Anal. Chem., 37, 918 (1965).

126. G. G. Guilbault and D. N. Kramer, Anal. Chem., 37, 1395 (1965).

127. J. de Albinati, Anales Assoc. Quim. Argentina, 55, 61 (1967).

128. V. Temkina, E. A. Bozhevol'nov, and N. Dyatlova, Zh. Anal. Khim., 22, 1830 (1967).

129. N. Iritani, T. Miyahara, and I. Takahashi, Japan Analyst, 17, 1075 (1968).

130. L. Zeltser, Z. Maksimyeheva, and S. Talipov, Referat. Zh. Khim. 19GD, 1970, Abstr. No. 1G85.

131. A. V. Konstantinov, L. M. Korobochkim, and G. V. Anastasina, Tr. Novoi Appl. Metod, 5, 167 (1967).

132. B. W. Bailey, R. N. Dagnall, and T. S. West, Talanta, 13, 1661 (1966).

133. E. A. Bozhevol'nov, Tr. VN11 Khim. Reakt., No. 24, Goskhimydat, 1960.

134. Y. Yamane, M. Miyazaki, and M. Ohtawa, Japan Analyst, 18, 750 (1969).

135. K. Ritchie and J. Harris, Anal. Chem., 41, 163 (1969).

136. F. Pollard, J. McOmie, and J. Elberh, J. Chem. Soc., 466 (1951).

137. F. Pollard, J. McOmie, and J. Elberh, J. Chem. Soc., 470 (1951).

138. T. Shigematsu, M. Matsui, and T. Suimida, Bull. Inst. Chem. Res. Kyoto Univ., 46, 249 (1968).

139. T. Shigematsu, M. Matsui, and R. Wake, Anal. Chim. Acta, 46, 101 (1969).

140. R. P. Fisher and J. D. Winefordner, Anal. Chem., 43, 454 (1971).

141. D. E. Williams and J. C. Guyon, Anal. Chem., 43, 139 (1971).

142. E. Melenteva, N. Poluektov, and L. Kononenko, Zh. Anal. Khim., 22, 187 (1967).

143. R. Belcher, R. Perry, and W. I. Stephen, Analyst, 94, 26 (1969).

144. H. Willard and C. Horton, Anal. Chem., 24, 862 (1952).

145. S. W. Chaikin, Res. Ind., 5, No. 3 (1953).

146. T. L. Har and T. S. West, Anal. Chem., 43, 136 (1971).

147. J. C. Guyon, B. E. Jones, and D. A. Britton, Mikrochim. Acta, 1180 (1968).

148. Tableaux des reactifs pour l'analyse minerale, report of the International Commission on New Reactions and Analytical Reagents, Paris, 1948.

149. D. Fink, J. Pivnichny, and W. Ohnesorge, Anal. Chem., 41, 833 (1969).

150. G. Beck, Mikrochim. Acta, 47 (1939).

151. V. Nazarenko and S. Vinkovetskaya, Zh. Anal. Khim., 13, 327 (1958).

152. M. Ichihashi, T. Shigematsu, and T. Nishikawa, J. Chem. Soc. Japan, 78, 1139 (1957).

153. T. Nishikawa, J. Chem. Soc. Japan, 79, 236 (1958).

154. E. B. Sandell, Anal. Chem., 19, 63 (1947).

155. J. Collat and L. Rogers, Anal. Chem., 27, 961 (1955).

156. E. A. Bozhevol'nov, A. Lukin, and M. Gradinarskaya, USSR Pat. 116,838 (1958).

157. A. Lukin and E. A. Bozhevol'nov, Zh. Akh., No. 1 (1960).

158. E. A. Bozhevol'nov, A. Lukin, V. Yanishevskaya, and E. Kholod, USSR Pat. 119,287 (1958).

159. L. Bradaks, F. Feigl, and F. Hecht, Mikrochim. Acta, 269 (1951).

160. E. Herzfeld, Z. Anal. Chem., 115, 131 (1938).

161. L. Bark and L. Rixon, Anal. Chim. Acta, 45, 425 (1969).

162. H. Onishi, Anal. Chem., 27, 832 (1955).

163. H. Orighi and E. B. Sandell, Anal. Chim. Acta, 13, 159 (1955).

164. D. P. Shcherbov, I. Solovyan, and A. Drobachenko, Tezisy dokladov 6-go soveshchaniya lyuminestentu, Leningrad, 1958.

165. V. Patrovsky, Chem. Listy, 48, 537 (1954).

166. J. Radley, Analyst, 68, 369 (1943).

167. I. Ladenbauer, J. Korkis, and F. Hecht, Mikrochim. Acta, 1076 (1955).

168. G. Oshima, Japan Analyst, 7, 549 (1958).

169. A. Lukin and E. A. Bozhevol'nov, J. Anal. Chem. USSR (English Transl.), 15, 45 (1960).

170. E. A. Bozhevol'nov, A. Lukin, and M. Gradinarskaya, Anal. Abstr., 7, 3164 (1960).

171. N. Raju and G. Rao, Nature, 175, 167 (1955).

172. Tr. po Khim. i. Khim. Tekhnol., 1, 134 (1958).

173. W. C. Alford, L. Shapiro, and C. E. White, Anal. Chem., 23, 1149 (1951).

174. A. Brookes and A. Townshend, Chem. Commun., 24, 1660 (1968).

175. G. Oshima and K, Nagasawa, Chem. Pharm. Bull. Tokyo, 18, 687 (1970).

176. A. Ponomarenko, N. Markar'yan, and A. Komlev, Dokl. Akad. Nauk SSSR, 86, 115 (1952).

177. N. Shinagawa, H. Imai, and H. Sunabala, J. Chem. Soc. Japan, 77, 1479 (1956).

178. R. Bock and K. Hochstein, Z. Anal. Chim., 138, 337 (1953).

179. V. Patrovsky, Chem. Listy, 47, 1338 (1953); G. Beck, Mikrochim. Acta, 287 (1937).

180. A. Bordea, Bull. Inst. Politech. Iasi, 13, 209 (1967).

181. D. Fink and W. Ohnesorge, Anal. Chem., 41, 39 (1969).

182. H. Block, Paper Chromatography, 1955.

183. A. Pitts and D. Ryan, Anal. Chim. Acta, 37, 460 (1967).

184. C. E. White, M. H. Fletcher, and J. Parks, Anal. Chem., 23, 478 (1951).

185. J. Michal, Chem. Listy, 50, 77 (1956).

186. D. Wallach and T. Steck, Anal. Chem., 35, 1035 (1963).

187. J. B. Hill, Clin. Chem., 11, 122 (1965).

188. D. Schachter, J. Lab. Clin. Med., 54, 763 (1959).

189.   D. Schachter, J. Lab. Clin. Med., 58, 495 (1961).

190.   J. B. Hill, Ann. N. Y. Acad. Sci., 102, 1 (1962).

191.   B. Klein and M. Oklander, Clin. Chem., 13, 26 (1967).

192.   R. E. Thiers, in Standard Methods of Clinical Chemistry (S. Meites, ed.), Vol. 5, Academic Press, New York, 1965, p. 131.

193.   A. Badrinas, Talanta, 10, 704 (1963).

194.   E. A. Bozhevol'nov, Oesterr. Chem. Ztg., 66, 74 (1965); Chem. Abstr., 65, 7989 (1966).

195.   G. Gusev, Lab. Delv, No. 3, 157 (1968).

196.   R. M. Dagnall, R. Smith, and T. S. West, Analyst, 92, 20 (1967).

197.   C. E. White and F. Cuttitta, Anal. Chem., 31, 2083 (1959).

198.   B. K. Pal and D. Ryan, Anal. Chim. Acta, 47, 35 (1969).

199.   G. F. Kirkbright, T. S. West, and C. Woodward, Talanta, 13, 1637 (1966).

200.   G. F. Kirkbright, T. S. West, and C. Woodward, Talanta, 13, 1645 (1966).

201.   Y. Titkov, Ukr. Khim. Zh., 36, 613 (1970).

202.   G. Andrushko, Z. Maksimycheva, and S. Talipov, Referat. Zh. Chim., 1969, Abstr. No. 18G96.

203.   S. Belman, Anal. Chim. Acta, 29, 120 (1965).

204.   V. Sardesai and H. Provido, Mikrochem. J., 14, 550 (1969).

205.   M. Rubin and L. Knott, Clin. Chim. Acta, 18, 409 (1967).

206.   C. Sawicki, Anal. Letters, 4, 761 (1971).

207.   F. Feigl, Spot Tests in Inorganic Analysis. (tr. by R. E. Oesper), 5th ed., Elsevier, Amsterdam, 1958.

208.   G. Schenk, K. Dilloway, and J. Coulter, Anal. Chem., 41, 510 (1969).

209.   H. Kautsky and A. Hirsch, Z. Anorg. Allgem. Chem., 222, 126 (1935).

210.   J. Franck and P. Pringsheim, J. Chem. Phys., 11, 21 (1943).

211.   H. Kautsky and G. Müller, Z. Naturforsch., 2a, 167 (1947).

212.   L. J. Tolmach, Arch. Biochem. Biophys., 33, 120 (1951).

213.   M. A. Konstantinova-Shlesinger, Tr. Fig. Inst. Akad. Nauk SSSR Fiz. Inst. imi P. N. Lebedevo, 2, 7 (1942).

214.   M. A. Konstantinova-Shlesinger, and V. Krasnova, Zavod. Lab., 6, 567 (1945).

215.   M. A. Konstantinova-Shlesinger, Zh. Fiz. Khim., 9, 6 (1938).

216.   M. A. Konstantinova-Shlesinger, Chem. Abstr., 30, 2521 (1936).

217.   H. Watanabe and T. Nakadoi, J. Air Pollut. Control Assoc., 16, 614
       (1966).

218.   D. Amos, Anal. Chem., 42, 842 (1970).

219.   A. S. Keston and R. Brandt, Anal. Biochem., 11, 1 (1965).

220.   G. G. Guilbault and E. Hackley, Anal. Chem., 40, 1256 (1968).

221.   W. A. Andreae, Nature, 175, 859 (1955).

222.   H. Perschke and E. Broda, Nature, 190, 257 (1961).

223   D. B. Land and S. Edmonds, Mikrochim. Acta, 1013, (1966).

224.   G. F. Kirkbright and T. S. West, Anal. Chem., 43, 1434 (1971).

225.   O. H. Lowry, J. V. Passonneau, and S. Schultz, Anal. Biochem., 19,
       300 (1967).

226.   M. Haitinger, Die Fluoreszenzanalyse in der Mikrochemie, Wien,
       Leipzig, 1937.

227.   H. Veening and W. Brandt, Anal. Chem., 32, 1426 (1960).

228.   H. Axelrod, J. Cary, J. Bonelli, and J. Lodge, Anal. Chem., 41, 1856
       (1969).

229.   E. Morgen, N. Vlasov, and V. Tyutin, Referat. Zh. Khim. 19GD,
       1969, Abstr. No. 23G162.

230.   O. Komlev and V. Zinchuk, Referat. Zh. Khim. 19GD, 1968, Abstr.
       No. 4G66.

231.   A. I. Ivankova and D. P. Shcherbov, Referat. Zh. Khim., 1968, Abstr.
       No. 18G104.

232.   T. D. Filer, Anal. Chem., 43, 725 (1971).

233.   Y. Nishikawa, K. Hiraki, and T. Shigematsu, J. Chem. Soc. Japan,
       90, 483 (1969).

234.   V. Nazarenko and V. Antonovich, Zh. Anal. Khim., 24, 358 (1969).

235.   D. P. Shcherbov and V. Nikolaeva, Prom. Khim. Reakt. Osob. Chist.
       Veshchestv., 186 (1967).

236.   E. B. Cousins, Austral. J. Expt. Biol. Med. Sci., 38, 11 (1960).

237.   J. H. Watkinson, Anal. Chem., 32, 981 (1960).

238.   C. A. Parker and L. G. Harvey, Analyst, 86, 54 (1961).

239.   C. A. Parker and L. G. Harvey, Analyst, 87, 558 (1962).

240.   W. H. Allaway and E. E. Cary, Anal. Chem., 36, 1359 (1964).

241.   P. F. Lott, P. Cukor, G. Moriber, and J. Solga, Anal. Chem., 35,
       1159 (1963).

242.   W. E. Clarke, Analyst, 95, 65 (1970).

243.   J. B. Wilkie and M. Young, J. Agr. Food Chem., 18, 946 (1970).

244.   O. Olson, J. Assoc. Off. Anal. Chem., 52, 627 (1969).

245.   K. Kasiura, Chemia Anal., 14, 1325 (1969).

246.   L. I. Kononenko, E. Melenteva, R. Vitkin, and N. Polvektov,
       Prom. Khim. Reakt. Osob. Chist. Veshchestv., 223 (1967).

247.   C. F. Coyle and C. E. White, Anal. Chem., 29, 1486 (1957).

248.   B. K. Pal and D. Ryan, Anal. Chim. Acta, 48, 227 (1969).

249.   M. Tishchenko, L. I. Kononenko, N. Poluektov, Prom. Khim. Reakt.
       Osob. Chist. Veshchestv., 231 (1967).

250.   R. M. Dagnall, R. Smith, and T. S. West, Analyst, 92, 358 (1967).

251.   E. Butter, I. Kolowos, and H. Holzapfel, Talanta, 15, 901 (1968).

252.   L. I. Kononenko, S. Mishchenko, and N. Poluektov, Zh. Anal. Khim.,
       21, 1392 (1966).

253.   R. S. Bottei and A. D'Alessio, Anal. Chim. Acta, 37, 405 (1967).

254.   R. Milkey and M. Fletcher, J. Amer. Chem. Soc., 79, 5425 (1957).

255.   A. Babko, C. Hzeu, A. Volkova, and T. Getman, Ukr. Khim. Zh.,
       35, 292 (1969).

256.   F. Feigl, V. Gentil, and D. Goldstein, Anal. Chim. Acta, 9, 393
       (1953).

257.   M. A. Konstantinova-Shlesinger, Fluorometric Analysis (N. Kaner,
       transl.), Davey, New York, 1965, p. 159.

258.   E. Tomic and F. Hecht, Mikrochim. Acta, 896 (1955).

259.   V. Rao and G. Rao, Z. Anal. Chim., 161, 406 (1958).

260.   R. S. Bottei and A. Trusk, Anal. Chim. Acta, 41, 374 (1968).

261.   A. Murata and F. Yamaguchi, J. Chem. Soc. Japan, 77, 1259 (1956).

262.   A. Kirillov, R. Lauer, and N. Polvektov, Zh. Anal. Khim., 22, 1333
       (1967).

263.   M. Ichihashi, T. Shigematsu, and T. Nishikawa, J. Chem. Soc.
       Japan, 77, 1474 (1956).

264.   C. E. White and M. Neustadt, Ind. Eng. Chem., Anal. Ed., 15, 599
       (1943).

265.   L. Merritt, Ind. Eng. Chem., Anal. Ed., 16, 758 (1944).

266.   G. Smith, R. Jenkins, and J. Gough, J. Histochem. Cytochem., 17, 749 (1969).

267.   A. Konstantinov, L. M. Korobochkin, and G. V. Anastasina, Referat. Zh. Biol. Khim., 1968, Abstr. No. 8F89.

268.   R. R. Trenholm and D. E. Ryan, Anal. Chim. Acta, 32, 317 (1965).

269.   E. R. Jensen and R. T. Pfaum, Anal. Chem., 38, 1268 (1966).

270.   R. Geiger and E. Sandell, Anal. Chim. Acta, 16, 346 (1957).

271.   T. Filer, Anal. Chem., 43, 469 (1971).

272.   E. A. Bozhevol'nov, A. Pryanishnikov, and B. Palchits, USSR Pat. 114,463 (1958).

273.   E. Goon, J. Petley, W. McMullen, and S. Wiberley, Anal. Chem., 25, 608 (1953).

274.   F. Grimaldi and G. Levine, U. S. Geological Survey Trace Elements Inv. Rest., 60 (1950).

275.   A. Fioletova, Zh. Anal. Khim., 14, 739 (1959).

276.   A. K. Babko, A. Volkova, and T. Getman, Zh. Anal. Khim., 22, 1004 (1967).

277.   D. Monnier and M. Marcantonatos, Anal. Chim. Acta, 36, 360 (1966).

278.   A. K. Babko, A. Volkova, and T. Getman, Ukr. Khim. Zh., 35, 69 (1969).

279.   A. Volkova, T. Getman, and T. Kukibaev, Ukr. Khim. Zh., 35, 844 (1969).

280.   E. L. Pruden, R. Meier, and D. Plant, Clin. Chem., 12, 613 (1966).

281.   R. Olsen and H. Diehl, Anal. Chem., 35, 1142 (1963).

282.   G. G. Guilbault and E. Hackley, Anal. Chem., 40, 190 (1968).

283.   D. F. Sharman, Brit. J. Pharmacol., 20, 204 (1963).

284.   A. Grunert, K. Ballschmiter, and G. Tolg, Talanta, 15, 451 (1968).

Chapter 7

ASSAY OF ORGANIC COMPOUNDS

I. INTRODUCTION

There are many thousands of organic compounds that can be assayed by fluorescence at levels below 1 ppm (1 μg/ml). Some of the more fluorescent molecules, such as fluorescein, umbelliferone, and quinine, can be assayed at concentrations of 1 part per trillion (10^{-6} μg/ml).

In order to be fluorescent a molecule must be highly absorbing (high molar absorbtivity); hence only highly conjugated nonaromatic, aromatic, and hetero-cyclic compounds are fluorescent (see Chapter 3 for discussion of structural effects).

Excellent reference material on the assay of organics has been published by Udenfriend [1, 2] in his two books in which he discusses fluorescent methods for the assay of compounds of interest in biology and medicine; Phillips and Elevitch [3], who discuss fluorometric techniques in clinical pathology; White and Weissler [4, 5], who give reference material on the analysis of organic compounds; White and Weissler [6] in their biannual reviews on analysis; Berlman [7], who presents the fluorescence spectra of about 100 aromatic compounds; Pringsheim [8], who presents information on the luminescence characteristics of various classes of organic compounds; Passwater [9], who lists bibliographical data on fluorescence analysis; Konstantinova-Shlesinger [10], who discusses applications of fluorescence in the assay of organics; and White and Argauer [11], who present data on the assay of several organic compounds. De Ment [12] lists over 2800 compounds that possess native fluorescence either in the ultraviolet or the visible region and can be assayed.

II. AROMATIC HYDROCARBONS

The fluorescence spectra of some benzene derivatives in alcohol and hexane, and in water are presented in Tables 49 and 50, respectively. Benzene itself is only weakly fluorescent in the ultraviolet. However, when benzene is substituted with electrophilic groups, such as amino or hydroxyl, the fluorescence is increased. A measurement of the fluorescence wavelengths allows one to differentiate qualitatively the type of compound present.

The absorption and fluorescence bands of the hydrocarbons in which a series of benzene rings are fused in a straight chain exhibit increasing wave-lengths and increased fluorescence with increase in the number of rings and of the conjugated double bonds (Table 51). The fluorescence yield increases from benzene to naphthalene to anthracene and then drops rapidly for the aromatic hydrocarbons containing more than three phenyl rings in a straight chain. A blue shift in the wavelength of fluorescence is observed when the aromatic system is branched; for example, pentacene is red, benzochrysene is blue-green, and coronene is blue-violet (Table 52). Thus the fluorescence of the compounds consisting of five rings changes from red to violet with shortening of the longest chain occurring in the molecules, as shown in Table 52. Even the fluorescence of the seven-ring compound coronene, which has no more than three rings in any straight chain, is blue-violet and very similar to that of anthracene. By the addition of still more benzene rings to coronene, the fluorescence color is again shifted toward the red, but without much regularity. The fluorescence of the nine-ring compounds dibenzocoronene and violanthrene

TABLE 49

Fluorescence Spectra of Benzene Derivatives in Liquid Solution[a]

Compound	Solvent[b]	Band limits (nm)	Maxima (nm)
Benzene	A	255-300	260, 263, 264, 275, 283, 291
Toluene	A	261-300	262, 264, 265, 274, 280, 289
o-Xylene	A	260-320	260, 268, 271, 280, 290, 304, 313
m-Xylene	A	267-282	268, 271, 280
p-Xylene	A	265-290	268, 274, 280, 286
Mesitylene	A	265-300	270, 271, 275, 279, 286, 297
Durene	H	280-340	Continuous
Phenol	A	287-350	Continuous
o-Cresol	A	287-385	Continuous
m-Cresol	A	286-385	Continuous
p-Cresol	A	292-385	Continuous
o-Hydroxybenzoic acid	A	376-480	Continuous
m-Hydroxybenzoic acid	A	328-444	Continuous
p-Hydroxybenzoic acid	A	323-408	Continuous
Aniline	A	300-410	Continuous, weak maxima at 305, 336
o-Anisidine	A	313-429	Continuous
p-Anisidine	A	339-423	Continuous
o-Tolunitrile	A	287-376	Continuous
p-Tolunitrile	A	280-351	Continuous
Biphenyl	H	294-365	294, 305, 314, 319, 327, 340, 355
Diphenylmethane	H	272-320	275, 279, 285, 293, 301
Bibenzyl	H	270-320	275, 279, 284, 291, 304
Dibenzylethylene	H	270-320	278, 283, 296, 306, 314
Diphenyl ether	H	284-368	Continuous
Diphenylamine	H	326-415	Continuous

[a] Data from Ref. [10].

[b] Key: A, alcohol; H, hexane.

TABLE 50

Fluorescence of Some Benzene Derivatives (1 ppm) in Water[a]

$$R^1 - \bigcirc - R^2$$

R^1	R^2	λ_{ex} (nm)	λ_{em} (nm)	Quantum efficiency[b] (%)
H	OH	270	330	3.2
H	OCH_3	270	303	3.4
H	NH_2	280	350	2.5
H	$N(CH_3)_2$	286	365	9.7
H	F	257	289	0.7
H	Cl	N.F.[c]		
H	Br	N.F.		
H	I	N.F.		
H	NO_2	N.F.		
H	$NHCOCH_3$	N.F.		
H	COOH	N.F.		
NH_2	F	289	362	12.3
NH_2	SO_3H	254	352	5.0
NH_2	OCH_3	297	375	4.2
NH_2	CH_3	289	357	2.8
NH_2	Cl	290	362	1.7
NH_2	$NHCOCH_3$	290	352	0.018
NH_2	NO_2	N.F.		
OH	CH_3	278	313	8.8
OH	OCH_3	289	328	5.9
OH	F	276	315	3.0
OH	Cl	280	317	0.89
OH	Br	N.F.		
OH	NO_2	N.F.		
Quinine				55.0

[a] Data from Ref. [13].

[b] Related to quinine sulfate in 0.1 N H_2SO_4.

[c] N.F. = nonfluorescent.

TABLE 51

Fluorescence Spectra of Aromatic Hydrocarbons in Solution[a]

Hydrocarbon	Fluorescence wavelength (nm)
Benzene	250-300
Naphthalene	300-365
Anthracene	372-460
Naphthacene	460-580 (468, 498, 533, 574)
Pentacene	Red
Rubrene	545-623 (maxima 560, 590)
Phenanthrene	348-407 (348, 366, 385, 407)
Chrysene	360-400
Pyrene	370-400
Perylene	Blue (440, 470)
Fluorene	302-370 (maxima 302, 325)
Cholanthrene	Blue-violet (400-500)
Decacyclene	477-600 (476, 510, 552, 595)
Fluorocyclene	410-540 (415, 440, 466, 504, 435)

[a] Data from Ref. [10].

is yellow, and the fluorescence of dinaphthocoronene is green, although the
latter contains five benzene rings fused in a straight row, as in pentacene, with
three more rings added on each side.

Changes in concentration have also been reported to affect the fluorescence
spectrum. As the concentration of pyrene in xylene increases, the fluorescence
changes from blue to green, but the absorption spectrum remains unchanged
[14].

Fluorene is a three-ring compound: the most probable structure
determining its ground state corresponds to the form shown in compound (53)
so that, as in biphenyl, no double bond links the two benzene rings. If this is
correct, the absorption and fluorescence spectra would be due mainly to the
two disconnected benzene rings. As a matter of fact, the fluorescence of
fluorene does not differ greatly from that of biphenyl.

Cholanthrene (54) can be derived from 1,2-benzanthracene by addition of a
five cornered ring.

When dissolved in hexane, it emits a broad, continuous fluorescence band. The
solid compound is distinguished by the same peculiarity as benzopyrene: it
exists in a blue- and a green-fluorescing modification. The same holds for
methylcholanthrene.

(53)   (54)

TABLE 52

Fluorescence of Polycyclic Aromatic Hydrocarbons
Dissolved in Hexane[a]

Compound	Formula	Fluorescence color
Pentacene		Red
1,2-Benztetracene		Yellow-green
8,9-Benzochrysene		Blue-green
1,2,5,6-Dibenz-anthracene		Blue (3900, 4150, 4500)
1,2-Benzopyrene		Blue-violet (3947, 4033, 4089, 4134-4160, 4286-4319, 4354)
Perylene		Blue
Coronene		Blue-violet

[a] Reprinted from Ref. [10] by courtesy of D. Davey and Co.

Benzopyrene, cholanthrene, and methylcholanthrene are known to be highly carcinogenic; this was assumed at first to be related to their ability to fluoresce. There are, however, many other hydrocarbons with similar fluorescence yield and spectra that have not shown carcinogenic activity. The existence of two modifications with different fluorescence spectra, on the other hand, has been observed only with the three carcinogenic compounds.

Decacyclene (55) and fluorocyclene have much more complicated structures, consisting of a central ring surrounded by a number of naphthalene groups.

(55)

The absorption spectra of these compounds are divided into two parts: a band near 300 nm, practically coinciding with the naphthalene band, and a long-wavelength band that in the case of decacyclene stretches from 340 to 450 nm. The visible fluorescence is excited only by light absorption in the second band, whereas absorption in the naphthalene band is ineffective.

Schenk and Wirz [15] described a fluorometric method for the assay of mixtures of aromatic hydrocarbons. The method is based on the removal of the lowest electronic transition of anthracene by a Diels-Alder reaction of this molecule with maleic anhydride to form a nonfluorescent compound. Following this such aromatic compounds as pyrene, diphenylstilbene, and perylene could be assayed.

III. HETEROCYCLIC COMPOUNDS

In the old chromophor-fluorogen theory heterocyclic rings, such as pyridine, pyrrole, diazine, and pyrone, were classified as chromophors, and since they were found to be not fluorescent by themselves, it was assumed that a "fluorogen" had to be attached to them in order to provide the ability to fluoresce. Thus in the acridine molecule the central pyridine ring would be the chromophor and the two benzene rings would act as fluorogens. The analogy between the structures of anthracene and acridine is obvious, however, and since a similar analogy prevails between their absorption and emission spectra, it cannot be doubted that these originate from the same mechanism: the resonance between the various structural formulas of the three-ring system. The spectra belong to the whole molecule, not to an isolated chromophor group. The fluorescence bands of acridine in neutral solutions are located in the same spectral region as those of anthracene, and, though the spectrum of acridine is more diffuse at room temperature, it is resolved at $-180°C$ into a sequence of four separated bands with nearly the same spacing that occurs in the anthracene spectrum.

Xanthone is another compound consisting of three fused rings, but without any double bonds in the central pyrone ring: its fluorescence is exclusively ultraviolet. The same is true for carbazole (56) and dibenzofuran; the last two compounds are quite analogous to fluorene (53), which has been treated in the preceding section, but with the additional influence of the imido group and the oxygen, respectively, on the fluorescence of the benzene rings. Similar considerations should also apply to methylacridone (57). Instead, its absorption and fluorescence bands in neutral and acidified solutions coincide almost exactly with those of acridine. To provide an anthracenelike resonance structure for methylacridone, one must assume that one of its most probable forms is that

(56)   (57)

of a zwitterion (58) with a negative charge on the oxygen and a positive charge on the nitrogen.

(58)

Table 53 lists the fluorescence of various pure and substituted heterocyclic ring compounds. Although nothing seems to be known concerning the fluorescence of coumarin itself, most of its hydroxy derivatives, beginning with umbelliferone, are fluorescent in liquid solutions. 6,7-Dihydroxycoumarin is the substance that (as esculin) produces the strong blue fluorescence of chestnut

TABLE 53

Fluorescence of Heterocyclic Compounds in Liquid Solution

Compound	Fluorescence (nm)
Quinoline	385-490
Umbelliferone	Blue
Acridine	425-454[a]
Methylacridone	425-454[a] (maxima 433, 445)
Xanthone	Ultraviolet
Carbazole	340-420 (maxima 347, 359, 376)
Diphenyl oxide	310-370 (maxima 316, 328, 345)
Alloxazine	Violet
Lumazine	Green[b]
Methylaminocitraconic methylimide	Yellow
Quinine sulfate	410-500[c] (maxima 437)

[a] In acidified solution, blue-green.

[b] In acidified solution, blue.

[c] In acidified solutions, whitish blue (400-675 nm; maxima 555, 466).

extracts, one of the earliest examples of fluorescent plant extracts. Many coumarin derivatives corresponding to the general formula (59) have been investigated; the fluorescence is violet or blue in all cases. Compounds in which R' is a phenyl ring, however, are not fluorescent [16, 17].

Sherman and Robins [18] studied the fluorescence characteristics of a number of substituted 7-hydroxycoumarins; some of their data are summarized in Table 54. The 3-carbethoxy and 3-cyano derivatives were found to be the most fluorescent.

Quinoline and numerous quinoline derivatives are fluorescent. 8-Hydroxy-quinoline, which is frequently used for the determination of aluminum and magnesium, is not fluorescent in the solid state, but it exhibits a fairly strong

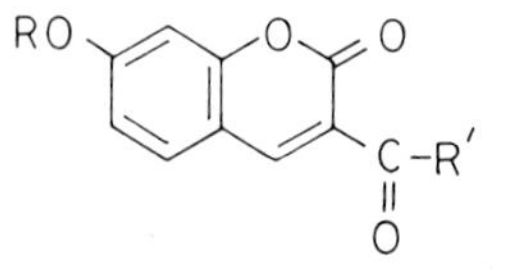

(59)

$R = H$ or CH_3; $R' = H$, CH_3, or C_2H_5

TABLE 54

Fluorescence of Substituted 7-Hydroxycoumarins[a]

Substituent	Excitation maximum (nm)	Fluorescence maximum (nm)	Fluorescence intensity[b]	Absorption maximum (nm)	Molar absorptivity
3-Benzoyl-	415	468	6.4×10^{-4}	412	43,000
4-Phenyl-	365	515	1.2×10^{-2}	372	17,700
Unsubstituted	376	454	0.96	365	18,500
4-Methyl-	367	449	1.0	359	17,000
3-Carboxy-	396	450	1.9	385	36,700
3-Carboxamido-	398	445	2.5	400	39,300
3-Phenyl-	420	456	2.7	412	37,700
3-Acetyl-	419	458	3.1	413	43,500
3-Carbethoxy-	398	445	3.6	402	38,700
3-Cyano-	408	450	3.6	407	41,600

[a]Reprinted from **Ref. 18** by courtesy of the American Chemical Society.
[b]Relative to 4-methyl-7-hydroxycoumarin.

green fluorescence in alcoholic solution. This luminescence is completely quenched by the addition of a few drops of a weak acid, but the compounds that are formed with aluminum and magnesium and are precipitated from an acidified aqueous solution as small crystals emit a very brilliant green fluorescence under ultraviolet excitation. The bright fluorescence of tetrahydroquinoline is blue-violet in neutral and yellowish-green in alkaline solution [19].

Nikolic and Sablic [20] studied the fluorescence characteristics of some quinoline derivatives. Nine derivatives with a MeO or an EtO group in the 6-position and COOH in the 4-position were studied. Fluorescence was observed only in acid solution, and a linear relation between pH and the intensity of fluorescence was observed in a definite concentration range.

Balemans and Van den Veerdonk [21] studied the fluorescence behavior of indole and 15 of its 3- or 5-substituted derivatives. All of the compounds showed fluorescence in the 380- to 460-nm region in strong base-formaldehyde solution. The indoles unsubstituted in the 5-position showed weak fluorescence at 360 nm; the 5-methoxy compounds showed strong fluorescence at 545 nm.

The fluorescence of all indole compounds is markedly diminished above pH 11.5, the residual fluorescence having a maximum at 420 nm [22]. According to Konev [23], the quenching results from a transfer of a proton from the imino nitrogen of the excited indole molecule to the hydroxyl ion of the solvent. In the presence of formaldehyde there is little decrease in the fluorescence even up to pH 13. Formaldehyde appears to combine with the ring nitrogen of indole to form a product that does not dissociate in the excited state [24]. Alkylation of the indole nitrogen leads to compounds that show little fluorescence quenching in alkali.

Guilbault and Kramer [25] have shown that 3-hydroxyindole is strongly fluorescent in the red (λ_{ex} 495, λ_{em} 570 nm), whereas N-methylindoxyl [26] is fluorescent in the green (λ_{ex} 430, λ_{em} 501 nm).

Maickel and Miller [27] described a fluorometric method for indole derivatives based on the reaction of these compounds with phthalaldehyde to give a fluorescence. Only indoles substituted with -OH or -OCH$_3$ substituents gave a fluorescence. 5-Hydroxytryptamine reacted with phthalaldehyde to give a strongly fluorescent compound, allowing the detection of as little as 5 ng of the amine.

Alloxazine and its alkyl derivatives are interesting because they are the compounds from which vitamin B$_2$ is derived; the fluorescence of this compound plays an important part in modern vitamin analysis. They contain one benzene ring in addition to a pyrazine and a pyrimidine ring in their three-ring system. In the azoflavines the benzene ring is replaced by a pyridine ring without impairing the fluorescence. Another substance consisting of two heterocyclic rings only (a pyrazine and a pyrimidine ring) is lumazine (60). Lumazine itself

(60)

and many of its alkylated derivatives exhibit a brilliant fluorescence that
varies from blue to green with increasing pH of the solution.

Quinine and its salts are of special interest because they have been
investigated very exhaustively and provide a great variety of phenomena, many
of which have not yet been explained satisfactorily. Although the fluorescence
of quinine salts is violet in neutral and whitish blue in acidified aqueous solutions (λ_{ex} 250 and 350; λ_{em} 450 nm), an intense green fluorescence is obtained
if the bisulfate, hydrochloride, or valerate of quinine is heated to the melting
point and subsequently dissolved at high concentration in water. Although the
salts are dehydrated by the heating process, it must be assumed that still
another change occurs in the constitution of the compounds, since the mere loss
of water of crystallization would not influence their behavior in aqueous
solutions. If the solutions are acidified, their fluorescence recovers the
characteristic whitish blue color, and dilution to a very low concentration has
the same effect. On the other hand, addition of alkali does not affect the green
fluorescence, and thus it becomes probable that a negative ion or a neutral
molecule is the carrier of this luminescence. The excitation spectrum of the
green fluorescence is also shifted in the direction of longer wavelengths. No
explanation is available for the phenomena observed when the heat-treated salts
are dissolved in organic solvents; in chloroform the fluorescence of the pre-
heated hydrochloride and the valerate is also green, whereas the fluorescence
of the bisulfate is blue under these conditions, and the fluorescence of valerate
dissolved in benzene is violet [28].

Perkampus, Knop, and Knop [29] studied the fluorescence and phosphorescence of 1,10-phenenthrolines in heptane at 20°C.

Perry et al. [30] reported the fluorescence and phosphorescence characteristics of several alkyl carbazoles. These data are summarized in Table 55.
The study indicated that luminescence is useful for assay of these compounds.

Derivatives of benzofuran are interesting because the addition of nonfused
benzene rings to a compound seems to enhance its fluorescence and shifts the
emission spectrum toward longer wavelengths. Benzofuran itself is not
fluorescent in the visible region, but when two of its hydrogen atoms in the 1-

TABLE 55

Fluorescence and Phosphorescence Characteristics of Carbazole and Alkyl Carbazoles[a]

Species	Excitation maximum (nm)[b]		Emission maximum (nm)[b]		Range of linearity (decade)[c]		Limit of detection (μg/ml)[d]		Phosphorescence lifetime[e,h] (sec)
	Fluor[f]	Phos[g]	Fluor[f]	Phos[g]	Fluor	Phos	Fluor	Phos	
Carbazole:									
In ethanol	340	341	360	436	4	4	0.0003	0.001	7.8
In cyclohexane	340	297	360	435	4	4	0.0005	0.001	7.2
n-Methylcarbazole:									
In ethanol	346	336	360	437	4	4	0.0008	0.001	8.4
In cyclohexane	346	298	360	431	4	4	0.0005	0.001	7.5
n-Ethylcarbazole:									
In ethanol	339	340	369	437	4	4	0.001	0.001	7.8
In cyclohexane	340	298	364	433	4	4	0.0008	0.001	8.1
2-Methylcarbazole:									
In ethanol	346	333	357	442	4	4	0.001	0.001	8.1
In cyclohexane	346	332	356	443	4	4	0.001	0.001	7.5

[a]Reprinted from Ref. 30 by courtesy of the American Chemical Society.

[b]Maxima are uncorrected for instrumental characteristics. Accuracy of wavelength setting is ±2nm.

[c]The range of linearity could be considerably greater because the highest measured concentration was still on the linear portion.

[d]Limits of detection were taken as those concentrations producing a signal twice the background.

[e]Shortest measurable lifetime is 0.5 sec.

[f]Measurements taken at 298°K.

[g]Measurements taken at 77°K.

[h]Lifetimes are taken as time for phosphorescence to decay to 1/e of the original signal. All decays were exponential and precise to ±0.2 sec.

and 3-positions are substituted by benzene rings, the compounds exhibit a strong greenish fluorescence in the crystalline state as well as in benzene solution. The fluorescence is still further displaced toward the red end of the spectrum if the substituents are biphenyl groups instead of benzene rings forming so-called dixenylbenzofurans. On the other hand, the introduction of methyl groups in the 5- and 6-positions has practically no effect on the fluorescence spectra. The behavior of the dihydrobenzofurans is similar, the emission bands being displaced somewhat toward the violet and showing some structure [31].

TABLE 56

Fluorescence Band Peaks of Benzofurans[a]

Substituent	Fluorescence band peak (nm)	
	Benzofuran	Dihydrobenzofuran
1,3-Diphenyl	486	384, 405
1,3-Diphenyl-5,6-dimethyl	486	384, 408, 459
1,3-Dixenyl	525	429, 484, 496
1,3-Dixenyl-5,6-dimethyl	625	429, 491, 502

[a] Reprinted from Ref. [31] by courtesy of the American Chemical Society.

IV. ORGANIC DYES

Of more than 2000 dyes of the Color Index, less than 200 are listed as fluorescent in liquid solutions. A superficial survey suffices to show that the intensity and hue of color are in no way related to the ability to fluoresce. On the other hand, it is obvious that the most strongly fluorescent dyes all belong to a few restricted classes (Table 57) of which those derived from the xanthene group (fluorescein, rhodamine), from the acridine group (euchrysine), and from the phenazine group (safranine, Magdala Red) are the most important, whereas there is not a single instance of fluorescence in liquid solution to be found among the nearly 700 azo dyes. The latter consist of a number of phenyl rings linked to each other by -N=N- groups; in the former two benzene rings are fused to a central heterocyclic ring (namely, a pyridine, a pyrone, or a pyrazine ring).

Even more instructive is a comparison between dyes of the acridine or xanthene class with diphenylmethane or triphenylmethane dyes. Thus the

TABLE 57

Absorption and Fluorescence Bands of Dyes

in Aqueous or Alcoholic Solution[a]

Compound[b]	First absorption band[b] (nm)	Fluorescence Band[b] (nm)	Color
Fluoran	U. V.	290-460 (320)	Violet, strong
Fluorescein	440-520 (494)	510-590 (515)	Yellow-green, very strong
Eosin (tetrabromofluorescein)	450-560 (517)	520-600 (540)	Yellow, strong
Erythrosin (tetraiodofluorescein)	460-556 (516)	518-588 (537)	Yellow, weak
Rose Bengal (tetraiodotetrachlorofluorescein)	(544)	550-670 (600)	Orange, very weak
Rhodamine B Extra	480-600 (550)	550-700 (605)	Red, strong
Rhodamine 6G	480-590 (526)	536-602 (555)	Yellow, strong
Acridine Red	455-600	560-680	Orange, medium
Pyronine B	540-590	560-650	Orange, medium
Acridine	300-450	400-480	Blue-violet, medium
Acridine Yellow	UV, 520	475-640	Green
Euchrysine	UV, 540	505-670 (585)	Greenish-yellow medium
Rheonine A	UV, 510	470-650	Green, weak
Acriflavine	UV, 500	485-660	Yellowish-green strong
Magdala Red	400-600 (524)	550-700 (600)	Red, strong
Safranine	(539)		Yellow-red
Thionine	480-630 (580)		Orange, medium
Methylene Blue	550-700 (658)	650-700	Red, medium

[a]Data from Ref. [10].

[b]Approximate limits of bands; peaks of bands in parentheses.

formulas for rhodamine and Malachite Green differ mainly by the absence of
the central oxygen atom in the latter and by the ensuing fact that the two benzene
rings are no longer connected through a closed ring but through an open carbon
bridge.

A similar example is provided by fluorescein and phenolphthalein.
Fluorescein itself is very slightly fluorescent in alcoholic solution. Addition
of alkali to the solution produces the alkali salt, which dissolves easily in
water and exhibits the well-known yellow-green fluorescence characteristic
of the negative fluorescein ion (uranin). Each of these ions (61) and (62) has
two resonance structures of identical energy with the electric charge on one or
the other of the two symmetrical oxygen atoms, and both have about the same
yellow color in transmitted light. If in an alcoholic solution of uranin the
solvent is replaced stepwise by ether, the color and the fluorescence disappear
completely because the dye goes over to the lactoid modification. The molecule
corresponding to formula (63) has no resonance structure of equal energy apart

(61)   (62)   (63)

from the various Kekule forms of the benzene rings.

Phenolphthalein differs from fluorescein in the same way as Malachite
Green differs from rhodamine. It is colorless in neutral and deep red in
alkaline solutions, but is equally nonfluorescent in all liquid solutions.
However, when the alkaline solution is solidified by the addition of gelatin, it
exhibits a vivid-orange fluorescence.

Examples of diphenylmethane and triphenylmethane dyes that are fluorescent only in rigid media are rosamine, auramine, Methyl Violet, and fuchsin.
The closed-ring structure protects the excited molecules of certain dyes against
internal conversion, and a similar protection is provided by the fixation of the
excited molecules on a rigid support. Apparently intramolecular vibrations that
favor the occurrence of internal conversion are inhibited. When adsorbed on
textile fibers many of the "nonfluorescent" azo dyes are also fluorescent. The
anthracene dyes, which have been mentioned as being fluorescent in xylene and
tetralin solutions, consist exclusively of fused benzene rings.

The fluorescent properties of a series of Naphthol AS dyes (64) were studied
by **Vaughn**, Guilbault, and Hackney [32]. Some of the many dyes studied

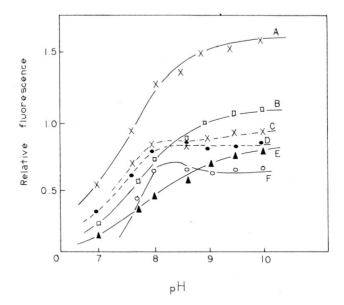

$$(64)$$

include Naphthol AS (3-hydroxy-2-naphthanilide), Naphthol AS-BI (6-bromo-3-hydroxy-2-naphthyl-o-anisidide), Naphthol AS-D (3-hydroxy-2-naphthyl-o-toluidide), Naphthol AS-GR (3-hydroxy-2-anthryl-o-toluidide), Naphthol AS-LC (4'-chloro-3-hydroxy-2',5'-dimethoxy-2-naphthanilide), Naphthol AS-MX (2',4'-dimethyl-3-hydroxy-2-naphthanilide), and Naphthol AS-TR (4'-chloro-3-hydroxy-2-naphthyl-o-toluidide). The effect of pH on the fluorescence of the Naphthol AS derivatives (3.2×10^{-6} M) is shown in Fig. 97. At pH $<$ 6 these compounds are nonfluorescent and precipitate from solution. At a pH of about 10 a limiting fluorescence is reached. Naphthol AS-D and AS-TR were found to have the strongest fluorescence. Naphthol AS-GR had no fluorescence in aqueous solution, but its phosphate ester had an intense green fluorescence. The fluorescent properties of the Naphthol AS derivatives are shown in Table 58.

FIG. 97. Effect of pH on fluorescence of Naphthol AS derivatives (3.2×10^{-6} M): (A) Naphthol AS-D, (B) Naphthol AS-TR, (C) Naphthol AS-LC, (D) Naphthol AS-MX, (E) Naphthol AS-BI, (F) Naphthol AS.

TABLE 58

Fluorescence Properties of Naphthol AS Derivatives at pH 9[a]

Naphthol	Wavelength (nm)		Fluorescence coefficient[b]	Residual fluorescence of phosphate ester[c]
	λ_{ex}	λ_{em}		
AS	388	516	2.2×10^5	0.133
AS-BI	405	515	2.5×10^5	0.11
AS-D	388	515	6.6×10^5	0.210
AS-GR	388[d]	488[d]	--	
AS-GR Phosphate	388	488	1.3×10^6	26.1[c]
AS-LC	388	522	2.8×10^5	0.09
AS-MX	388	512	2.6×10^5	0.076
AS-TR	388	512	3.1×10^5	0.014

[a]Reprinted from Ref. [32] by courtesy of the American Chemical Society.

[b]Fluorescence of hydrolyzed substrate divided by the concentration of original substrate in M. Value for quinine sulfate in 0.1 N H_2SO_4 at λ_{ex} 350 nm, λ_{em} 450 nm was 2.5×10^6.

[c]3.5×10^{-5} M.

[d]In Methyl Cellosolve.

[e]Fluorescence of 3.2×10^{-4} M Naphthol AS phosphates at pH 9.0.

V. ACIDS, ALCOHOLS, ALDEHYDES, AND KETONES

A. Acids

Table 59 lists organic acids that are determinable by fluorescence methods and gives the wavelengths, sensitivity, and necessary reagents.

One of the most common methods for the assay of polycarboxylic acids involves their condensation with resorcinol to yield fluorescent products. For example, malic acid condenses to yield umbelliferone-4-carboxylic acid (65).

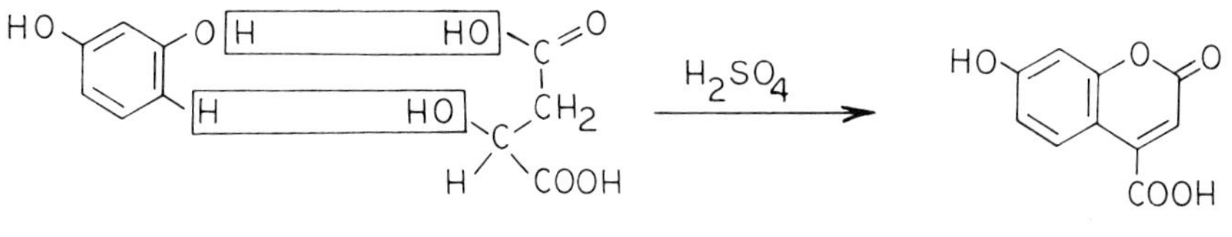

$$\xrightarrow{H_2SO_4}$$

(61)

(65)

The various carboxylic acids have different colors and intensities of fluorescence, as shown in Table 60.

Hummel [56] used orcinol (3,5-dihydroxytoluene) instead of resorcinol to

TABLE 59

Fluorometric Methods for the Assay of Organic Acids

Acid	Reagent, pH, or method	λ_{ex} (nm)	λ_{em} (nm)	Sensitivity (ppm)	Ref.
Acetic	Resorcinol-HCl	330	440	0.1	33
Adenylic	pH 1	285	395	0.07	34
Anthranilic	pH 7	300	405	0.001	34
Ascorbic	Dihydroxynaphthalene-sulfonic acid	360	465	0.01	35
Citric	Thionyl chloride, NH_3	365	440	10	36
	Acetic anhydride-pyridine	475	525	10	37
3,4-Dihydroxyphenyl-acetic	pH 7	280	330	0.04	34
Gallic	$CuSO_4$	UV	Green	10	38
Gibberellic	H_2SO_4	405	465	5	39
Guanidoacetic	Ninhydrin	305, 390	495	0.5	40
Homogentisic	pH 7	290	340	0.02	34
Homovanillic	pH 7	270	315	0.2	34
	$KMnO_4$	315	420	0.01	41
3-Hydroxyanthranilic	pH 7	320	415	0.001	34
5-Hydroxyanthranilic	pH 7	340	430	0.001	42
m-Hydroxybenzoic	pH 12	314	430	1	43
p-Hydroxycinnamic	pH 7	350	440	0.02	34
p-Hydroxymandelic	pH 7	300	380	0.006	34
p-Hydroxyphenyl-acetic	pH 7	380	310	0.03	34
p-Hydroxyphenyl-pyruvic	pH 7	290	340	0.1	34
Indoleacetic	pH 7	285	360	0.001	44
Kynurenic	pH 7	325	405	0.003	34
	pH 11	325	440	0.005	34
Malic	2-Naphthol	357	441	0.1	45
	Resorcinol	365	485	0.01	46
Naphthaleneacetic	pH 11	282	327	0.02	44
Oxalic	Quenching	360	460	0.1	47
Pyruvic	NAD	360	485	0.1	48
Salicylic	pH 11	310	435	0.01	49
Sebacic	Resorcinol	365	430	1	50

TABLE 59--Continued

Acid	Reagent, pH, or method	λ_{ex} (nm)	λ_{em} (nm)	Sens-itivity (ppm)	Ref.
Succinic	Resorcinol	470	Green	0.01	51
Terephthalic	Conversion to amine	364	438	0.1	52
Thioglycolic	Naphthoquinone	365	Blue-white	0.3	53
Thymidylic	Br_2, o-amino-benzaldehyde	365	420	0.02	54
Uric	pH 11	325	370	0.7	34
Xanthurenic	pH 11	350	460	0.005	34

TABLE 60

Fluorescence of Resorcinol Polycarboxylic Acid Derivatives[a]

Acid	Fluorescence color	Relative fluorescence
Fumaric	Blue-violet	24
α-Ketoglutaric	Blue-green	35
Oxalacetic	Blue-green	12
Succinic	Yellow-green	20
Malic	Blue-violet	22
Isocitric	Light blue	58
Citric	Sky blue	89

[a]Reprinted from Ref. [55].

determine malic acid specifically in the presence of other tricarboxylic acid intermediates. The condensation product, 7-hydroxy-5-methyl coumarin, fluoresces with an intense blue color in strong acid.

Lowry et al. [57] and Maitra and Estabrook [58] determined pyruvic acid by the use of lactic dehydrogenase (LDH) and NADH:

$$\text{pyruvate + NADH} \xrightarrow{\text{LDH}} \text{NAD + lactic acid} \qquad (62)$$
$$\text{(fluorescent)}$$

Naruse et al. [59] used a similar procedure to assay citric acid:

$$\text{citrate} \xrightarrow{\text{enzymes}} \text{di-isocitrate} \qquad (63)$$

$$\text{d-isocitrate + NADP} \xrightarrow{\text{enzyme}} \text{α-ketoglutarate} + CO_2 + \text{NADPH} \qquad (64)$$
$$\text{(Fluorescent)}$$

Reaction (63) is catalyzed by aconitase, reaction (64) by isocitrate dehydro-genase; NADPH is measured fluorometrically.

Spiker and Towne [60] developed a fluorescence method for the assay of α-keto acids based on condensation with o-phenylenediamine. The resulting quinoxaline derivatives (66) are stable and highly fluorescent.

$$(65)$$

$$(\underline{66})$$

From 0.05 to 2.0 μg of keto acids (Table 61) is determinable.

TABLE 61

Fluorescence Characteristics of the Quinoxaline Derivatives
of α-Keto Acids[a]

α-Keto acid	λ_{ex} (nm)	λ_{em} (nm)	Relative intensity
Glyoxylic	338	518	16
α-Ketobutyric	365	495	69
α-Ketocaproic	365	498	66
α-Ketoglutaric	360	503	90
α-Ketoisocaproic	365	500	83
α-Ketoisovaleric	367	500	56
Oxalacetic	368	498	23
Phenylpyruvic	368	503	25
Pyruvic	360	490	100

[a]Reprinted from Ref. [60] by courtesy of the American Chemical Society.

The reader is referred to Chapter 8 for other enzymatic methods for the assay of organic acids.

B. Alcohols

Methanol has been determined by an oxidation to formaldehyde [61], followed by an assay of the latter by condensation with acetylacetone and ammonia, as described by Belman [62]. The product, 3,5-diacetyl-1,4-dihydro-2,6-lutidine (67) has a λ_{ex} of 405 and a λ_{em} of 505 nm. As little as 0.05 ppm of methanol or formaldehyde is assayable.

$$CH_3-OH \longrightarrow H-\overset{\underset{\displaystyle O}{\|}}{C}-H \qquad (66)$$

$$(67)$$

$$(\underline{67})$$

The same reaction has been used for the assay of mannitol and sorbitol, 1,2-diols, and α-amino primary alcohols, which are oxidized to formaldehyde by $NaIO_4$.

Acetol, an intermediate in the metabolism of acetone, 1,2-propanediol, lactic acid, and pyruvic acid, can be assayed by its condensation with o-amino-benzaldehyde to yield the fluorophor 3-hydroxyquinaldine (68) [63].

$$(68)$$

$$(\underline{68})$$

The excitation source was the 365-nm mercury line; the λ_{em} was 400 to 440 nm.

A fluorometric method for glycerol in plasma was described by Ko and Royer [64]. A coupled enzyme system is used:

$$ATP + glycerol \xrightarrow{\quad glycerokinase \quad} ADP + \alpha\text{-glycerophosphate} \xrightarrow{\quad NAD \quad}$$

$$DHAP + \underset{(fluorescent)}{NADH} \qquad (69)$$

The fluorescent NADH is measured and is proportional to the glycerol content. Hydrazine is added to precipitate the dihydroxyacetone (DHAP) formed and shift the equilibrium toward completion.

Triglycerides are determined by saponification to glycerol after separation from phospholipids. The glycerol produced is then assayed in one of two ways. One method involves a condensation with o-aminophenol in the presence of sulfuric acid and an oxidizing agent to yield 8-hydroxyquinoline [65]. Addition of Mg^{2+} results in an intense fluorescence, allowing the determination of as little as 20 μg of triglyceride. Kessler and Lederer [66] described a fluoro-metric method for glycerol, based on its oxidation by periodate to formaldehyde, followed by the condensation of formaldehyde with acetylacetone and ammonia

to yield the highly fluorescent product 3,5-diacetyl-1,4-dyhydro-2,6-lutidine
(67). Triglycerides can be assayed in microliter quantities of serum.

C. Aldehydes and Ketones

An analytical method for acetaldehyde, described by Velluz et al. [67], is
based on the reaction with 3,5-diaminobenzoic acid to yield the fluorescent
derivative 2-methyl-5-carboxy-7-amino quinoline. Salvador and Albers [68]
used the same reaction to assay for the α-aminobutyric acid metabolite succinic
semialdehyde.

Aldehydes and ketones undergo condensation with aromatic hydrazides to
yield fluorescent hydrazines. For example, Camber [69] introduced salicyloyl
hydrazide as a reagent for carbonyl compounds. Chen [42] found that most of
the hydrazones had the same λ_{ex} (350 nm) and the same λ_{em} (425 nm) as the
original hydrazide, but that the hydrazide could be separated from the hydrazone
by solvent extraction.

Brandt and Cheronis [70] used the reagent 2-diphenylacetyl-1,3-indandione-
1-monohydrazone (69) for the assay of carbonyl compounds:

(69)   (70)

The fluorescence characteristics of compounds (69) and (70) are similar,
although (70) fluoresces more intensely. The λ_{ex} for both are at 295 and 400
to 415 nm, with λ_{em} at 525 nm. The azines are generally separated from the
hydrazone by paper chromatography since the latter does not migrate in most
solvents.

Surveys of reagents for the fluorometric assay of formaldehyde were
presented by Sawicki et al. [71,72]. Nanogram quantities of formaldehyde can
be detected by condensation with 2-nitro-1,3-indandione. Kamel and Wizinger
[73] found that highly fluorescent products were obtained on condensation of
formaldehyde with 6-amino-1-naphthol-3-sulfonic acid.

The Hantzsh reaction for assay of formaldehyde by its condensation with
acetylacetone and ammonia was described in Eq. (67) for assay of methanol
[62].

Sawicki and Carnes [74] compared dimedone, acetylacetone, and 7-amino-

4-hydroxynaphthalene-2-sulfonic acid (J-acid) as reagents for formaldehyde. The acetylacetone procedure is highly selective (0.6 μg detectable); the J-acid and dimedone procedures are highly sensitive (0.06 and 0.04 μg, respectively), the former being the more selective of the two.

Abbott and Hercules [75] described a fluorometric method for the assay of aromatic carbonyl compounds. Analysis is based on the triplet-singlet energy transfer from aromatic carbonyl donors to a hydrocarbon acceptor. In the assay of benzophenone using perylene as acceptor as little as 10^{-5} M benzophenone could be assayed.

The acroleins are a major group of irritants produced during combustion. They are highly reactive and can be made to condense with many reagents to yield fluorescent derivatives.

A fluorometric procedure for acrylaldehyde (acrolein) and related compounds was described by Alarion [76]. The method is based on a condensation with m-aminophenol to yield a fluorophor (λ_{ex} 400, λ_{em} 510 nm) that has been identified as 7-hydroxyquinoline. As little as 50 picomoles per ml is determinable.

Sawicki et al. [77] have reported several fluorometric methods for the assay of ß-hydroxyacrolein, which is a tautomer of malonaldehyde. This compound reacts with many amines to give intensely fluorescent derivatives. The products formed with ethyl p-aminobenzoate and 4,4'-sulfonyldianiline yield the most intense and stable fluorescence. With ethyl p-aminobenzoate, λ_{ex} is 500, λ_{em} is 550 nm; with 4,4'-sulfonyldianiline, λ_{ex} is 475, λ_{em} is 545 nm. As little as 0.3 ng of malonaldehyde is detectable. Acrolein gave no fluorescence.

Bredereck et al. [78] found that most aromatic aldehydes are nonfluorescent in hydrocarbon solution but do fluoresce on addition of polar solvents. Pyrene-3-aldehyde, naphthalene-2-aldehyde, anthracene-9-aldehyde, and acenaphthalene-3-aldehyde all fluoresce in hydroxylic solvents, but not in hydrocarbon solvents.

A study of 26 aromatic aldehydes in methanol indicated [79] that the aldehyde group exhibits a strong quenching effect even in aromatic systems substituted with fluorescence promoting groups such as -OH and $-NH_2$. Fluorescence is observed when acid is added to the methanol because the aldehyde group is converted to the methyl acetal, thus eliminating the quenching effect.

VI. AMINES AND AMINO ACIDS

A. Amines

1. General Reactions

Several fluorophor reactions have been described for the assay of amines.

One such reagent, 7-chloro-4-nitro-benzoxadiazole (71), which is nonfluorescent, reacts with amines and amino acids in the presence of potassium acetate
to yield the highly fluorescent compound (72) [80].   Methylamine, dimethyl-
amine, benzylamine, and ammonia all give highly fluorescent derivatives.

Primary aliphatic and alicyclic amines react with fluorescein chloride and
zinc chloride to give orange-fluorescent rhodamine dyes λ_{ex} 546, λ_{em} 605 nm)
with a detection limit of 10 to 30 μg of amine [81].   As little as 10 ng is
determinable.

Dombrowski and Pratt [82] described a sensitive fluorometric method for
the determination of primary aromatic amines.   The procedure requires
diazotization of the amino group, followed by coupling with 2,6-diaminopyridine

(71)   (72)

(71)

(DAP).   The resulting azo dye is reacted with cupric sulfate to produce an
intensely fluorescent derivative (λ_{ex} 360, λ_{em} 420 nm).   As little as 2 to 6
ng/ml of most aromatic primary amines can be determined.

Primary aliphatic amines can be assayed by reaction with 9-isothiocyanato-
acridine to yield the corresponding thiourea derivatives.   These derivatives are
highly fluorescent in base, allowing the assay of nanogram concentrations of
amines [83].

Pesez and Bartos [84] described a fluorometric method for primary and
secondary amines based on their replacement of the sulfonic acid group in
1,2-naphthoquinone-4-sulfonic acid; the resulting product, when treated when
KBH_4, yields a fluorescent material.

2.   Catecholamines

Figure 98 shows the various pathways for the biosynthesis and metabolism
of the catecholamines.   Fluorometric procedures for the assay of each of the
catecholamines have been described.   Two general procedures are the trihydroxyindole and the ethylenediamine reactions.

In the trihydroxyindole method epinephrine (73a) and norepinephrine (73b) are
oxidized to fluorescent products (74) [85-91].

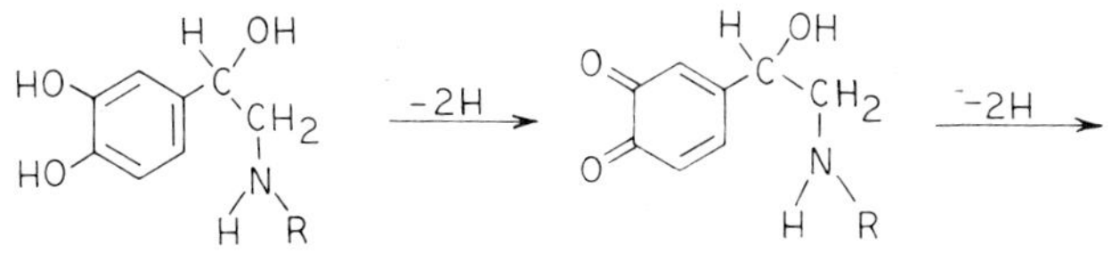

(72)

(73)

a :   R = CH₃

b :   R = H

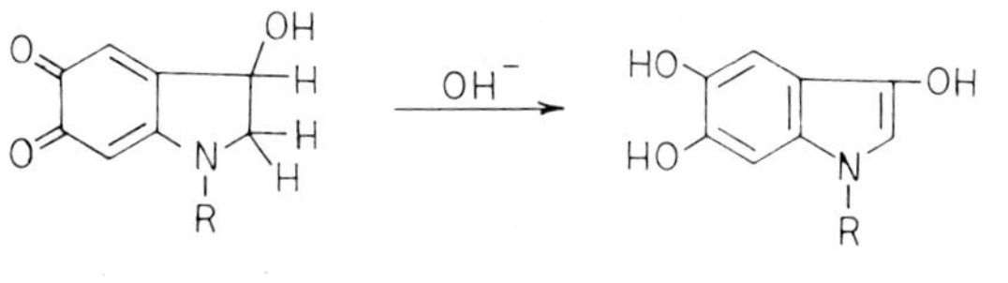

(74)

Crout [90] devised a procedure for the simultaneous assay of epinephrine and norepinephrine in urine and tissues. The procedure is based on the fact that both compounds are oxidized to the aminochrome stage at pH 6.5, whereas only epinephrine undergoes significant oxidation at pH 3.5. Epinephrine yields a fluorophor with λ_{ex} 410, λ_{em} 520 nm; norepinephrine only with λ_{ex} 395, λ_{em} 505 nm.

Merrills [92, 93] described a semiautomatic method for the determination of these two metabolites in mixtures. The compounds are isolated by a chromatographic procedure, and the final eluates are transferred to a Technicon AutoAnalyzer in combination with a flow-cell Locart fluorometer.

The second reaction that has been widely used in the assay of catecholamines is oxidation with adenochrome (75) followed by condensation with ethylene-diamine [94, 95]:

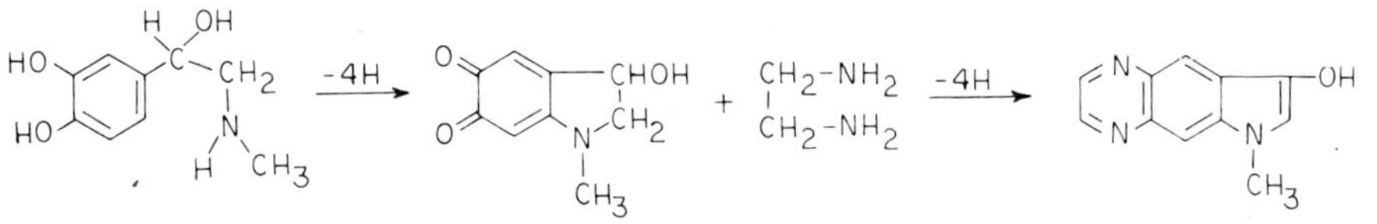

(73a)   (75)   (76)

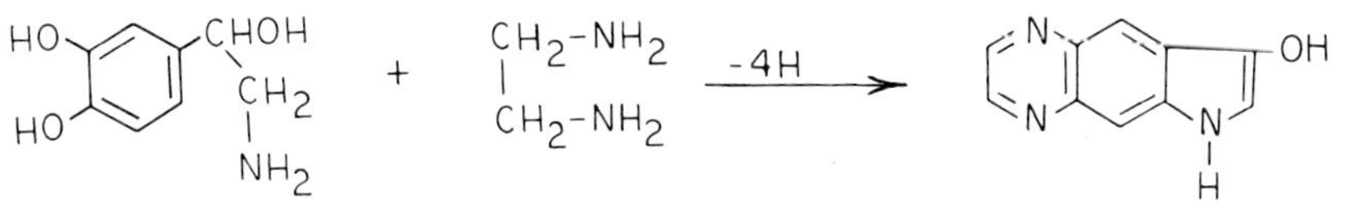

(73)

(73b)   (77)

(74)

The fluorescence bands of compound (76) are reported to be λ_{ex} 420, λ_{em} 525 nm; those of compound (77) are λ_{ex} 420, λ_{em} 485 to 495 nm [96-98].

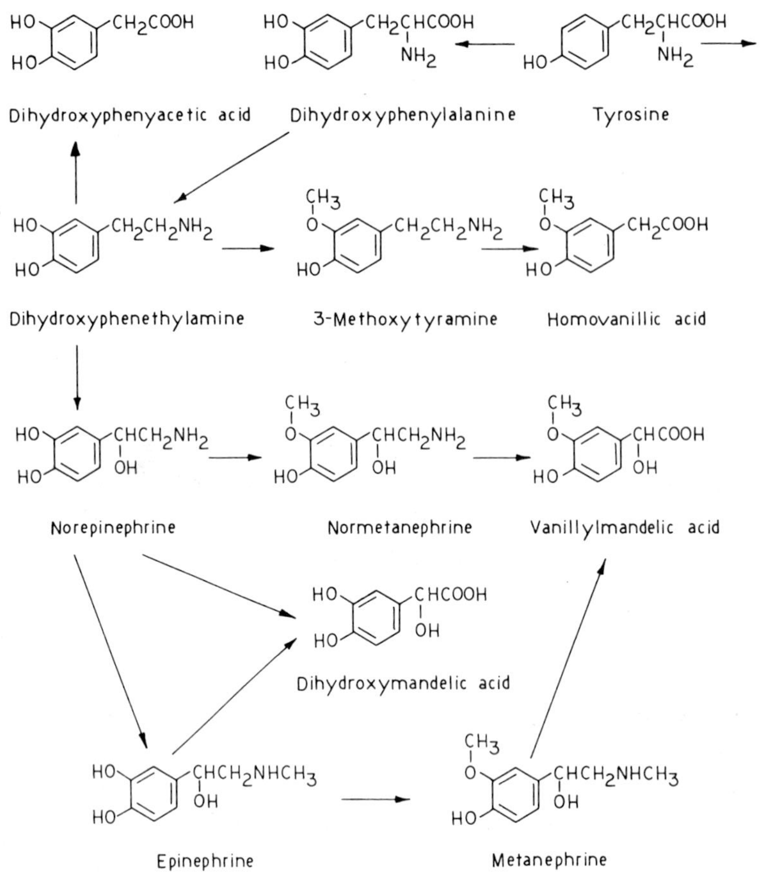

Dihydroxyphenyacetic acid   Dihydroxyphenylalanine   Tyrosine

Dihydroxyphenethylamine   3-Methoxytyramine   Homovanillic acid

Norepinephrine   Normetanephrine   Vanillylmandelic acid

Dihydroxymandelic acid

Epinephrine   Metanephrine

FIG. 98. Pathways for the biosynthesis and metabolism of catecholamines.

Another interesting reaction [99, 100] for the assay of the catecholamines involves a condensation with formaldehyde to yield the fluorescent 6, 7-dihydroxy-3, 4-dihydroisoquinolines (78: λ_{ex} 410, λ_{em} 470 to 480 nm).

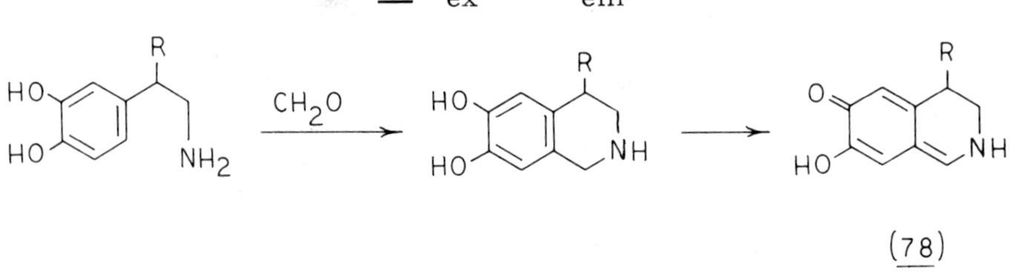

$$\text{(75)}$$

$$(\underline{78})$$

Dopa (3, 4-dihydroxyphenylalanine) fluoresces itself, albeit weakly (λ_{ex} 285, λ_{em} 325 nm). It can be assayed by the trihydroxyindole reaction (fluorophor λ_{ex} 365, λ_{em} 495 nm) [101]. Dopa is usually separated from other catecholamines by passage over a Dowex 50 (Na$^+$) column.

Dopamine (3, 4-dihydroxyphenethylamine) can be converted to the highly fluorescing dihydroxyindole (79: λ_{ex} 345, λ_{em} 410 nm) by oxidation with iodine, followed by rearrangement [102]. Dopa undergoes a similar reaction to give a

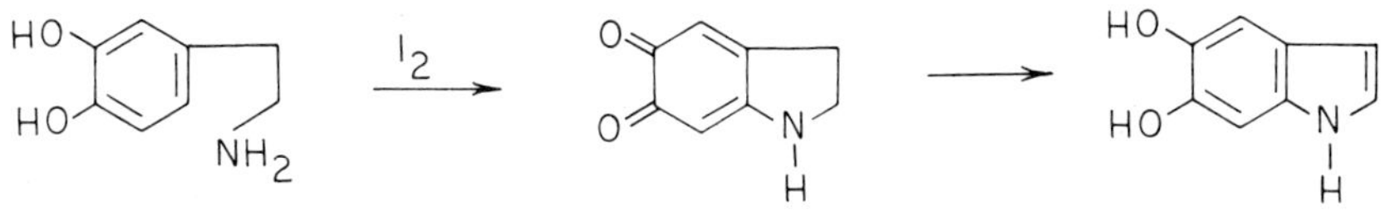

(76)

(79)

product with identical fluorescence characteristics; it can be separated, however, by ion exchange.

Eichorn, Rutenberg, and Kott [103] described a fluorometric method for dopamine, dopa, and dopac (dihydroxyphenylacetic acid) in mixtures. The compounds are separated by paper electrophoresis at low voltage and are then condensed with ethylenediamine to form fluorophores.

Oberman et al. [104] described a method for dopamine in urine based on a reaction with dansyl chloride to give a Dansyl-dopamine (λ_{ex} 335, λ_{em} 485 nm).

Metanephrine can be assayed by its native fluorescence (λ_{ex} 285, λ_{em} 335 nm).

Smith and Weil-Malherbe [105] described a fluorometric method for metanephrine and normetanephrine in urine, based on their oxidation to fluorescent indoles at two different pH values.

3. Histamine

A sensitive method [106] for histamine involves condensation with o-phthalaldehyde to a highly fluorescent product (80: λ_{ex} 360, λ_{em} 450 nm).

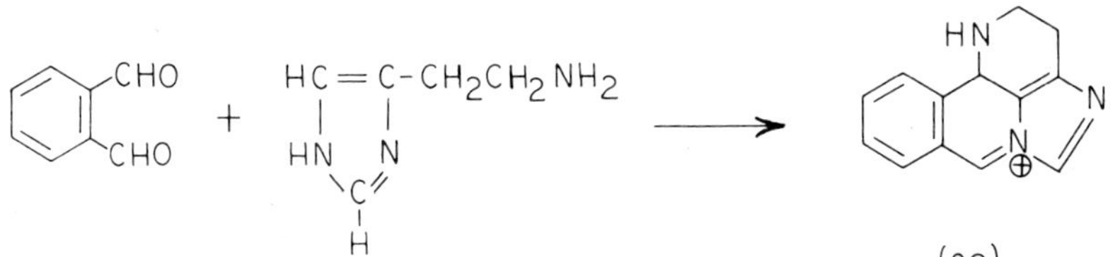

(80)

The range of concentrations determinable is 0.005 to 0.5 µg/ml.

Six colorimetric and two fluorometric procedures for histamine were compared by Sawicki et al [107]. For studies of histamine released in tissues by airborne allergens the only method with adequate sensitivity and selectivity is a fluorometric method using phthalaldehyde. The detection limit was 7 ng.

Alkon et al. [108] measured various imidazoles (e. g., methylhistamine,

histamine, histidine, 3-methylhistidine, 1-methylhistidine, and imidazoleacetic acids) by reaction with N-bromosuccinimide and o-phenylenediamine.

4. Tryptamine and 3-Hydroxytryptamine

Tryptamine can be assayed by its native fluorescence (λ_{ex} 285, λ_{em} 360 nm).

Hess and Udenfriend [109] described a procedure for tryptamine based on its condensation with formaldehyde to tetrahydronorharman (81) and dehydrogenation to the highly fluorescent norharman (82: λ_{ex} 310 and 365, λ_{em} 440 nm).

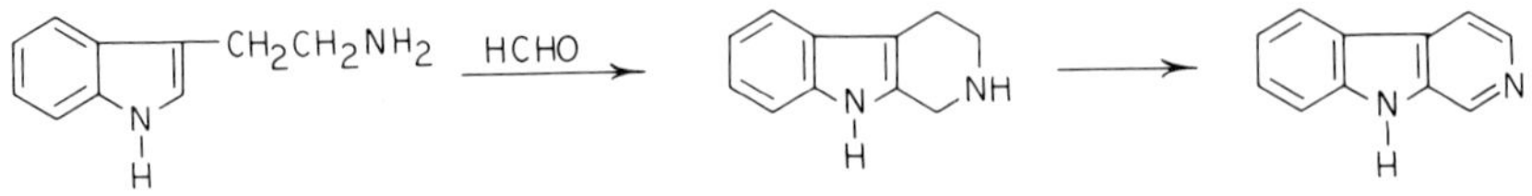

(78)

(81)   (82)

Jonsson and Sandler [110] described the fluorescence of a number of indolethylamines condensed with formaldehyde. Methods were described for tryptamine, 5-hydroxytryptamine, 6-hydroxytryptamine, and 5,6-dihydroxy-tryptamine.

Jason and Stevens [111] have reported the condensation of ninhydrin with tryptamines to yield highly fluorescent ß-carbolines. Quay [112] described a method for 5-hydroxytryptamine (serotonin) by means of its product with ninhydrin. 5-Hydroxytryptamine possesses a fluorescence in 3 M HCl (λ_{ex} 330, λ_{em} 550 nm) that can be used for its assay [112].

Vanable [113] found that 5-hydroxytryptamine would condense with ninhydrin in solution to give a strong fluorescence. The resulting fluorophor is excited at 380 nm and fluoresces at 500 nm. The major advantage of this method is increased sensitivity.

Maickel and Miller [27] reported a highly sensitive procedure for the assay of 5-hydroxytryptamine by measuring the fluorescence of the product formed by condensation with o-phthalaldehyde.

5. Tyramine

Tyramine is assayed by the fluorescent derivative (83: λ_{ex} 465, λ_{em} 565 nm) it forms with α-nitroso-ß-naphthol [114]. Spector et al. [115] combined solvent extraction and condensation with nitrosonaphthol to develop a method for tyramine in tissues:

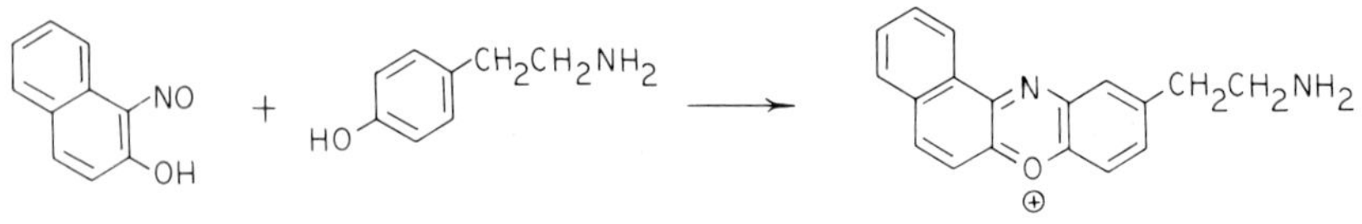

(79)

(83)

6. Other Amines

A fluorometric method for ethylenediamine in milk, described by Pasarela and Waldrin [116], is based on its condensation with adrenochrome (75) to yield a fluorophor (λ_{ex} 460, λ_{em} 510 nm).

McCleskey [117] developed a fluorometric method for urea based on its condensation with diacetylmonoxime. The λ_{ex} is 380 and the λ_{em} is 415 nm. Roch-Ramel [118] has developed an ultramicro method for determining as little as 2×10^{-11} mole of urea. The urea is converted to ammonia with urease. In the presence of glutamate dehydrogenase NADH is converted to NAD. The fluorescence of the latter is measured after treatment with strong base.

Guanidine can be assayed by its reaction with ninhydrin to form a fluorophore (λ_{ex} 305, 390 and λ_{em} 495 nm). As little as 0.5 µg/ml is assayable [40].

B. Amino Acids

Table 62 lists some of the important amino acids and the wavelengths used in their fluorometric determination.

TABLE 62

Amino Acids and Their Fluorescence Characteristics

Amino Acid	Method	λ_{ex}(nm)	λ_{em}(nm)	Sensitivity (µg/ml)	Ref.
Arginine	Ninhydrin	305, 390	495	0.3	40
Dopa	Trihydroxyindole	365	495	0.01	101
Histidine	o-Phthalaldehyde	340	480	0.01	106
Phenylalanine	Direct, H_2O	260	282	0.1	119
Tryptophan	Direct, pH 11	287	348	0.003	120
Tyrosine	Direct, pH 7	280	310	0.01	24
	a-Nitroso-ß-naphthol	460	570	0.005	121

Of the amino acids, phenylalanine (84) [119], tyrosine (85) [24], and tryptophan (86) [120] are the only ones of sufficient fluorescence intensity to be measured directly in solution (relative fluorescence 100 for tryptophan, 0.5 for phenylalanine, and 9 for tyrosine).

(84)   (85)   (86)

The fluorescence of these amino acids demonstrates the effect of structure on luminescence. Phenylalanine (84), with only a benzene ring and a -CH$_2$- side chain, is weakly fluorescent. Adding a hydroxy group, as in tyrosine (85), causes a twentyfold increase in the fluorescence. If an indole ring is added, as in tryptophan (86), the relative fluorescence increases to 200 times that of phenylalanine.

The measurement of phenylalanine in blood has become the accepted diagnostic method for phenylketonuria (PKU). Diagnosis must be made soon after birth, and the required volumes of sample are extremely small. The normal phenylalanine content of blood is 149 μg/ml. Values for PKU patients can be as high as 600 μg/ml.

Teale and Weber [119] reported the spectral characteristics of phenylalanine in water. Vladimirov [122] reported that the fluorescence of phenylalanine is fifty to sixty times higher in the solid state than in solution and suggested the possible use of this observation for analysis. Lowe [123] reported that ninhydrin reacts with phenylalanine to give a very fluorescent product and proposed this as an analytical method for the amino acid.

The most sensitive method for phenylalanine is that of McCaman and Robins [124, 125], which is based on the reaction of phenylalanine and ninhydrin [123]; peptides are added to enhance the fluorescence. By the proper choice of peptide (Table 63) and pH the reaction can be made highly specific for phenylalanine. Hill et al. [126] automated this reaction using Auto Analyzer modules manufactured by Technicon. At pH 5.8 the reaction is essentially specific for phenylalanine.

Tryptophan is usually assayed by its native fluorescence at pH 11 (λ_{ex} 287,

TABLE 63

Effect of Peptide on the Fluorescence Intensity in the
Phenylalanine-Ninhydrin Reaction

Peptide	Relative fluorescence
No peptide	0.5
Glycyl-L-tryptophan	28.0
L-Alanyl-glycine	30.0
L-Alanyl-L-alanine	70.0
Glycyl-DL-phenylalanine	75.0
Glycyl-L-alanine	95.0
L-Leucyl-L-alanine	100.0

λ_{em} 348 nm). The fluorescence is 10 times greater than that of tyrosine, and the fluorescence maximum is almost 40 nm higher [120]. Thus it is possible to assay tryptophan in the presence of a large excess of tyrosine.

Another method for tryptophan involves its condensation with formaldehyde, followed by oxidation to norharman (82) [109]. The method is highly specific and it is so sensitive that it can be applied to the assay of tryptophan in fingertip blood [127].

Brand and Shaltiel [128] reported that N-bromosuccinimide reacts with tryptophan to give a fluorophor (λ_{ex} 400, λ_{em} 530 nm).

The native fluorescence of tyrosine (λ_{ex} 280, λ_{em} 310 nm)[24] is not sufficiently intense nor specific to be used for the direct assay of tyrosine in crude tissue extracts. However, such an assay can be made in hydrolyzates of purified proteins. Duggan and Udenfriend [120] demonstrated that the fluorescence of tyrosine could be resolved from that of tryptophan. Moreover in acid solution the tryptophan is largely destroyed, so that tyrosine can be measured.

A better method for tyrosine, developed by Waalkes and Udenfriend [121] is based on its condensation with α-nitroso-ß-naphthol to yield a highly fluorescent product (λ_{ex} 460, λ_{em} 570 nm). Modifications of this procedure for the assay of tyrosine in blood [125] and serum [129], as well as micro-adaptations of the procedure [125], have been proposed.

Roth [130] described a fluorescence reaction that is general for all amino acids except cysteine, proline, and hydroxyproline. The method is based on the reaction with o-phthalaldehyde in the presence of the reducing agent 2-mercaptoethanol to yield a fluorescent product (λ_{ex} 340, λ_{em} 455 nm).

Conn and Davis [40] have reported fluorescence methods for arginine, creatine, and other guanidinium compounds -- methods that are more specific and sensitive than any other available ones. They are based on the reaction with ninhydrin in strongly alkaline solution to yield fluorophors (Table 62). Conn [131] has used this method for a sensitive assay of these amino acids in blood and urine.

Sawicki and Carnes [132] described a fluorometric method for the assay of glycine and amino acids; it is based on condensation with acetylacetone and formaldehyde to yield 1-substituted derivatives of 3,5-dimethyl-1,4-dihydro-2, 6-lutidine. As little as 50 ng of glycine, other amino acids, and primary amines is determinable.

Histidine is separated from other amino acids on a Dowex 50 column and then is determined by condensation with o-phthalaldehyde. The product is fluorescent (λ_{ex} 340, λ_{em} 480 nm), although the fluorescence is not as intense as that produced by histamine [106].

Guilbault and Hieserman [133] have described a fluorometric method for the assay of amino acids. The method is based on the oxidation of the non-fluorescent homovanillic acid to the highly fluorescent dimer by the peroxide formed in the oxidation of amino acids by amino acid oxidase:

$$\text{amino acid} \xrightarrow{\text{oxidase}} H_2O_2 \xrightarrow[\text{peroxidase}]{\text{homovanillic acid}} \text{fluorescence} \qquad (80)$$

As little as 0.1 µg of amino acids is determinable.

VII. CARBOHYDRATES

A. General Methods

A rapid spectrofluorometric method for determining nanogram quantities of carbohydrates was described by Rogers, Chambers, and Clarke [134]. The method is based on the hydrolysis of carbohydrates to monosaccharides, followed by a dehydration to 2-furaldehydes, which then condense with resorcinol to yield xanthenone derivatives. The fluorescence produces has a λ_{ex} of 488 and a λ_{em} of 508 nm. The range of concentrations determinable is 4 to 15 ng per 100 ml.

Guilbault et al. [135] have described fluorometric methods for the assay of such carbohydrates as glucose, fructose, maltose, cellobiose, and lactose in mixtures. The methods used are described in detail in Chapter 8.

B. Glucose

Guilbault, Brignac, and Zimmer [136] proposed a sensitive fluorometric method for the determination of glucose in biological samples like blood and urine. The peroxide formed on enzyme action is monitored with homovanillic acid, which is oxidized to a highly fluorescent product:

$$\text{glucose} \xrightarrow{\text{oxidase}} H_2O_2 \qquad (81)$$

$$H_2O_2 + \text{homovanillic acid} \xrightarrow{\text{peroxidase}} \text{fluorescence} \qquad (82)$$

The rate of production of fluorescence is proportional to the concentration of glucose in the concentration range 0.01 to 10 µg/ml. Phillips and Elevitch [137] used this procedure to assay for glucose in plasma. As little as 1 µl of sample is needed.

Guilbault and co-workers have devised a fluorometric reaction to monitor the peroxide formed in the glucose-glucose oxidase enzyme reaction [138]. Phthalic anhydride (87) is used to react with peroxide to form a peroxyphthalate (88); this peroxy compound is a strong oxidant and oxidizes indole (89), which is nonfluorescent, to the highly fluorescent indigo white (90). The rate of fluorescence production is a measure of the glucose present.

$$H_2O_2 + \text{(87)} \longrightarrow \text{(88)} \tag{83}$$

(87)   (88)

$$\text{(89)} \longrightarrow \text{(90)} \tag{84}$$

(89)   (90)

Momose and Ohkura [139, 140] proposed to assay glucose by its condensation with 5-hydroxy-1-tetralaone (91) to yield the fluorophor benzonaphthalene-
dione (92). Hexoses or polysaccharides containing hexoses react.

$$\text{(91)} + \text{glucose} \xrightarrow{H_2SO_4} \longrightarrow \text{(92)} \tag{85}$$

(91)   (92)

Momose and Ohkura [140] used the 365-nm mercury line for excitation and a
λ_{em} of 535 nm. Less than 1 µg/ml can be determined.

Bourne [141] found that a λ_{ex} of 470 and a λ_{em} of 550 nm should be used
for best results. She scaled the method down to use one-tenth the amount of
blood (2 µl) used by Momose and Ohkura [139, 140]

Weber et al. [142] applied this procedure to the assay of hexoses in
hydrolyzates of oligosaccharide units. The procedure was automated to
permit continuous monitoring of effluents from Dowex-1 borate columns. As
with the ninhydrin reaction with amino acids, each hexose gave a different
response with the reagent. The relative fluorescent intensities were as
follows: trehalose 1.00, galactose 0.60, and mannose 0.41.

C. 2-Deoxysugars

2-Deoxy-D-glucose (94), an analog of glucose, can be assayed by condensation with 3,5-diaminobenzoic acid (93). The fluorophor is presumed to be 2-(1-glycerol)-5-carboxy-7-aminoquinoline (95: λ_{ex} 410, λ_{em} 495 nm) [143]. The method is highly precise and specific. Glucose and related hexoses do not react. Only 2-deoxy-D-glucose does, yielding a fluorophor of 10% the intensity of that produced from 2-β-D-glucose. Aldehydes of the type R-CH$_2$-CHO do react, and hence compounds like acetaldehyde do interfere.

$$(86)$$

(93)   (94)   (95)

D. Pentoses

Towne and Spikner [144] studied the reaction of hexoses and pentoses with o-phenylenediamine in 50% H_2SO_4 and found that almost all of these compounds give highly fluorescent derivatives (λ_{ex} 360, λ_{em} 460 nm). The pentoses yielded the more intense derivatives. As little as 0.2 to 2 μg/ml could be assayed.

Anthrone condenses with pentoses in 70% H_2SO_4 to yield fluorescent products (λ_{em} 510 nm) [145]. The procedure can be used to distinguish between pentoses and hexoses, but the sensitivity is not good.

E. Ketoses

Ketoses condense with dimedone (5,5-dimethyl-1,3-cyclohexanedione) in 85% phosphoric acid to give highly fluorescent products [146]. The reaction is specific for keto sugars. Vurek and Pegram [147] have developed a micro-adaptation of the dimedone method for ketoses permitting the assay of nanogram quantities of inulin. The λ_{ex} is 366 nm, and the fluorescence is at 400 nm.

A highly specific procedure [148] for ketohexoses, such as fructose and fructose phosphates, is based on a condensation with resorcinol to form a fluorophor (λ_{ex} 430, λ_{em} 478 nm).

VIII. COENZYMES, NUCLEIC ACIDS, AND PYRIMIDINES

A. General Remarks

Many purines, pyrimidines, and coenzymes possess a native fluorescence in solution that can be used for their assay (Table 64). The fluorescence spectra of adenosine, AMP, ADP, and ATP are shown in Fig. 99, those of the pyridine nucleotides in Fig. 100, and those of riboflavin, FMN, and FAD as a function of pH in Fig. 101. Alternative methods for the assay of coenzymes involve the use of enzyme systems; some of these methods are presented in Table 65. In the following sections we briefly cover some of those analytical methods.

TABLE 64

Fluorescence Characteristics of Bases and Derivatives[a]

Compound	Medium	Absorption maximum (nm)	Excitation maximum (nm)	Fluorescence maximum (nm)
Adenine	pH 1	263	265[b]	380
Adenosine	$5 \text{ N } H_2SO_4$	257	272	390
Adenylic acid	$5 \text{ N } H_2SO_4$	257	272	390
ADP	$5 \text{ N } H_2SO_4$	257	272	390
ATP	$5 \text{ N } H_2SO_4$	257	272	390
FAD	pH 6.6	--	365	520
FMN	$HOAc\text{-}CHCl_3$	--	365	520
Guanine	pH 1	272 (sh)[c]	275[b]	360
	pH 11	273	275[b]	350
Guanosine	pH 1	277 (sh)[c]	285	390
Guanylic acid	pH 1	277 (sh)[c]	285	390
1-Methylguanine	pH 1	274 (sh)[c]	290	370
2-Methylamino-6-hydroxypurine	pH 11	278	280[b]	355
2-Dimethylamino-6-hydroxypurine	pH 11	281	290[b]	368
NADH	pH 8-11	--	340	460
NAD	NaOH (7 M)	--	340	460
NADPH	pH 8-11	--	340	460
NADP	NaOH (7 M)	--	340	460
Purine	pH 13	271	285	370
2,6-Diaminopurine	pH 7	280	300	360
Isoguanine	pH 11	285	300	360
8-Azaguanine	pH 13	280	290	360
Thymine	pH 11	291	290[b]	380

[a]Some data from Ref. [149].

[b]Corrected values.

[c]Shoulder or point of inflection.

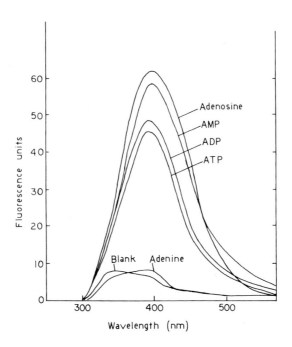

FIG. 99. Fluorescence spectra of adenine and its nucleoside and nucleotides in 7 N H_2SO_4. Excitation was at 265 nm, and the concentration was 5 μg/ml in each instance. Reprinted from Ref. [150] by courtesy of P. Greengard.

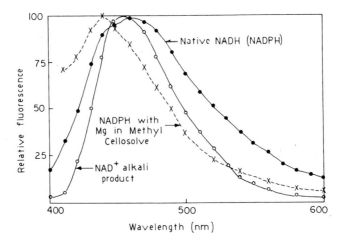

FIG. 100. Fluorescence spectra of pyridine nucleotides. Values are plotted as percentage of the maximum fluorescence. On an absolute scale NADH fluorescence would have to be reduced by a factor of about 8. Reprinted from Ref. [175] by courtesy of H. Borresen.

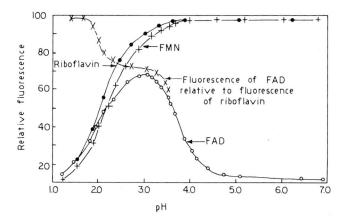

FIG. 101. Influence of pH on the fluorescence of riboflavin, flavin mononucleotide (FMN), and flavin-adenine dinucleotide (FAD). Reprinted from
Ref. [34] by courtesy of D. Duggan, R. Bowman, B. Brodie, and S. Udenfriend.

TABLE 65

Enzyme Systems Used in the Assay of Coenzymes

Coenzyme	Enzyme system	Refs.
ADP	Pyruvate kinase	150-152
AMP	Myokinase	150-152
ATP	Luciferase	153-158
	Hexokinase	150, 159, 160
	Phosphoglycerate kinase	150, 160
Creatine phosphate	Creatine phosphokinase, hexokinase	161, 162
	Creatine phosphokinase, luciferase	163, 164
FAD	D-Amino acid oxidase	165, 166
FMN	Lactic oxidase	167
	Cytochrome c reductase	168
NAD	Alcohol dehydrogenase	169, 170
	Glutamate dehydrogenase	171
NADH	Lactate dehydrogenase	172
	Diaphorase	170
NADP	Glucose-6-phosphate dehydrogenase	172
	Glutamate dehydrogenase	171
	Isocitric dehydrogenase	173
NADPH	Glutamate dehydrogenase	172
	Glutathione reductase	173

B.   Adenine and Guanine Nucleosides and Nucleotides

The fluorescence-spectral characteristics of a number of purine derivatives have been reported in aqueous solution [34].   No fluorescence was
observed in pure water, but a significant fluorescence was observed in strong
acid (Table 64).   Over 100 biologically important compounds were studied.
Udenfriend and Saltzman [149] repeated many of these measurements and
reported the quantum yields of fluorescence of many of these purines.   For
adenine and its derivatives quantum yields of about 0.2% were obtained; for
guanine the value was about 1.5 to 2.0%.

The fluorescence of the nucleosides and nucleotides of adenine shows a
second peak as the acidity is increased.   In 7 N H_2SO_4 adenine shows no
fluorescence above the blank, but adenosine and the nucleotides do.

Börresen [175] published an excellent study on the native fluorescence of
purines and pyrimidines.   The instrument was calibrated and the spectra were
corrected.   Enzymatic methods for ATP are presented in Chapter 8.

C.   Flavine Coenzymes

1.   Native-Fluorescence Methods

The coenzyme derivatives of riboflavin, FMN, and FAD are generally
determined by their native fluorescence, by conversion to lumiflavin, or by
using enzyme systems.

The fluorescence of the flavines was first investigated by Bessey et al.
[176] and Weber [177], who reported that the fluorescence of FAD is much
weaker than that of free riboflavin due to the quenching effect of the adenine
part of the molecule.   The influence of pH on the fluorescence of riboflavin,
FMN, and FAD is shown in Fig. 99.   The spectral characteristics are shown
in Table 64.

A method for the assay of riboflavin, FMN, and FAD in tissues was
reported by Bessey et al. [176] and Burch [178].   It is based on the differences
in the fluorescence of the three compounds as a function of pH and the difference
in the partition coefficients of the three compounds between benzyl alcohol and
aqueous solutions at pH 6.8.

2.   Enzymatic Methods

Flavine mononucleotide is a cofactor for the enzyme lactic oxidase from
pneumococci:

$$\text{lactate} + O_2 + \text{FMN} \longrightarrow \text{acetate} + CO_2 + H_2O + \text{FMNH} \qquad (87)$$

The FMN content of a sample is determined by the activation of the enzyme [167]. The reaction is usually monitored by following the oxygen uptake in a Warburg manometer. A plot of rate of oxygen uptake versus FMN concentration is linear up to about 10^{-7} M FMN.

Alternatively FMN can be assayed by using NADP and cytochrome c reductase [168]. The change in NADPH concentration measured spectrophotometrically at 340 nm, is used to monitor the reaction.

Flavine adenine dinucleotide is the coenzyme of D-amino acid oxidase from pig kidney and can be determined specifically by its activation of this enzyme:

$$\text{D-alanine} + \text{FAD} \xrightarrow[\text{oxidase}]{\text{D-amino acid}} \text{pyruvic acid} + \text{FADH} + \text{NH}_3 \qquad (88)$$

$$\text{FADH} + \text{O}_2 \longrightarrow \text{FAD} + \text{H}_2\text{O}_2 \qquad (89)$$

Warburg and Christian [165] and Straub [166] were the first to describe this method for FAD. The oxygen uptake, measured manometrically, is proportional to the FAD in concentrations of up to 25 μg/ml. The activation of D-amino acid oxidase by FAD must be compared with standard solutions since the Michaelis constant for the enzyme-FAD complex varies with temperature and with the preparation of enzyme.

D. Nicotinamide Nucleotides

1. Methods Based on Native Fluorescence

The pyridine nucleotides exist in oxidized and reduced forms (Fig. 102).

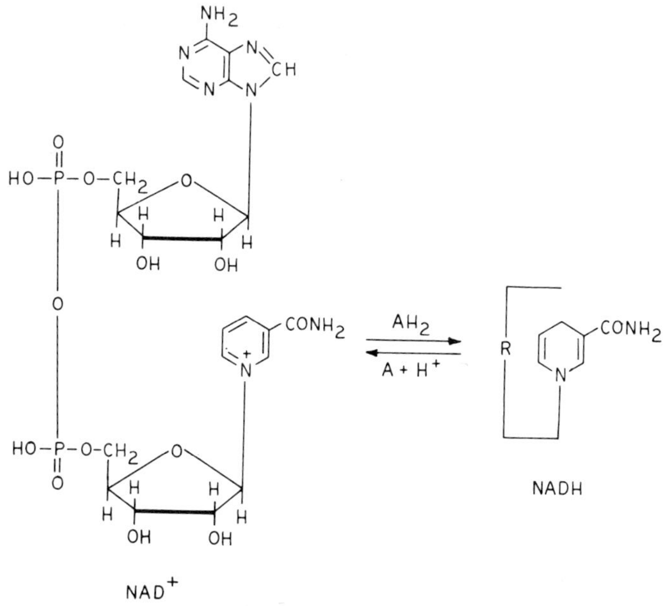

FIG. 102. Oxidized and reduced forms of nicotinamide adenine dinucleotide.

Reduced NAD (NADH) and NADP (NADPH) have maximum absorbance at 340 nm and a high fluorescence (λ_{ex} 340, λ_{em} 460 nm) [174]. Lowry et al. [174] have described procedures for using the native fluorescence of NADH and NADPH for their assay in tissues and have discussed the effect of solvents, pH; and trace metals on the fluorescence.

The fluorescence spectra of NADH and NADPH are shown in Fig. 98.

Kaplan et al. [179] found that the oxidized pyridine nucleotides (NAD and NADP) are converted to highly fluorescent products when treated with alkali. Lowry et al. [174] reported the spectra of the NAD^+ and $NADP^+$ alkali product (Fig. 98), which are almost identical with those of NADH or NADPH.

Huff and Perlzweig [180] have shown that NAD and NADP condense with acetone in alkali solution to form highly fluorescent products. Methods were developed for these coenzymes in blood and urine.

2. Enzymatic Methods

a. Nicotinamide Adenine Dinucleotide (NAD) and Its Reduced Form (NADH).

Nicotinamide adenine dinucleotide can be determined by quantitative reduction to NADH by ethanol and alcohol dehydrogenase:

$$\text{ethanol} + \text{NAD} \xrightarrow[\text{dehydrogenase}]{\text{alcohol}} \text{NADH} + \text{acetaldehyde} + H^+ \quad (90)$$

The equilibrium constant of this reaction favors the oxidation of NADH to NAD, but the reduction of NAD can be effected by using a pH of 9 to 10, a high ethanol concentration, and by trapping the acetaldehyde formed with hydrazine or semi-carbazide [169]. A pyrophosphate buffer is generally used because pyrophosphate binds heavy-metal ions, which may inhibit alcohol dehydrogenase.

The reaction can be monitored by measuring the NADH fluorometrically at a λ_{ex} of 340 nm and a λ_{em} of 460 nm. Similarly, the fluorometric resazurin indicator reaction of Guilbault and Kramer [170] can be used:

$$\text{NADH} + \text{resazurin} \xrightarrow[\text{sulfate}]{\text{phenazine methyl}} \text{resorufin} \quad (91)$$

(nonfluorescent)   (fluorescent)

The rate of production of the highly fluorescent resorufin is proportional to the NAD concentration.

Reduced NAD (NADH) can be assayed by any NAD-specific dehydrogenase reaction in which NADH is quantitatively oxidized. The reverse of reaction (90) can be used for NADH assay:

$$\text{acetaldehyde} + H^+ + \text{NADH} \xrightarrow[\text{dehydrogenase}]{\text{alcohol}} \text{NAD} + \text{ethanol} \quad (92)$$

Alternatively one can use the lactate dehydrogenase system:

$$\text{NADH} + \text{H}^+ + \text{pyruvate} \xrightarrow{\text{lactate dehydrogenase}} \text{NAD} + \text{lactate} \qquad (93)$$

The equilibrium constants for both these reactions favor the oxidation of NADH even with only a small excess of substrate. The reaction can be monitored fluorometrically by noting the decrease in the fluorescence of NADH [172].

Muscle lactate dehydrogenase reacts with NADPH as well as with NADH, but this interference can be eliminated by working at a pH of about 7.8. At this pH the rate of oxidation of NADH is about 2000 times faster than that of NADPH. Alcohol dehydrogenase preparations may contain small amounts of NADP-specific alcohol dehydrogenase, resulting in the oxidation of NADPH at high enzyme concentrations.

Guilbault and Kramer [170] described a fluorometric method for the analysis of 2×10^{-7} M to 2×10^{-5} M concentrations of NADH. The method is based on the production of the highly fluorescent resorufin (Eq. 91)).

b. Nicotinamide Adenine Dinucleotide Phosphate (NADP) and Its Reduced Form (NADPH). The assay of NADP can be carried out by its reduction to NADPH by glucose-6-phosphate dehydrogenase (G-6-PDH):

$$\text{G-6-P} + \text{NADP} \xrightarrow{\text{G-6-PDH}} \text{6-phosphogluconate} + \text{NADPH} + \text{H}^+ \qquad (94)$$

The equilibrium constant for this reaction favors the quantitative reduction of NADP, and the reaction can be monitored fluorometrically by the production of NADPH. The reaction is completely specific for NADP [172].

Similarly NADP can be assayed by using the isocitric dehydrogenase system or any dehydrogenase system requiring NADP and not NAD. The equilibrium for this reaction also favors the formation of NADPH, and the enzyme is specific for NADP [173].

Reduced NADP (NADPH) can be assayed with the glutamate dehydrogenase system:

$$\alpha\text{-oxoglutarate} + \text{NH}_4^+ + \text{NADPH} \xrightarrow{\text{glutamate dehydrogenase}} \text{glutamate} + \text{NADP} \qquad (95)$$

The equilibrium favors the formation of NADP, and the reaction can be monitored fluorometrically by the decrease in the fluorescence of NADPH [172]. Glutamate dehydrogenase is not specific for NADPH: it also reacts with NADH. The interference from NADH can be removed by oxidation with lactate dehyrogenase and pyruvate (Eq. (93)).

The assay of NADPH can also be carried out with glutathione reductase, which is specific for NADPH:

$$\text{glutathione} + \text{NADPH} + \text{H}^+ \xrightarrow{\text{reductase}} 2\text{-glutathione} + \text{NADPH} \qquad (96)$$

The equilibrium is in favor of NADP formation, but the activity of glutathione reductase is low compared with glutamate dehydrogenase [173].

Rhodes and Wooltorton [181] described a method for the assay of NAD, NADH, NADP, and NADPH by measuring the increase in fluorescence due to the reduction of resazurin to its highly fluorescent form resorufin in a recycling system involving a specific dehydrogenase, its substrate, 5-methyl-phenazinium methyl sulfate, and resazurin.

E. Creatine Phosphate

1. Hexokinase Procedure

Creatine phosphokinase (CPK) catalyzes the transfer of phosphate from creatine phosphate to ADP:

$$\text{creatine phosphate} + \text{ADP} \underset{\text{Mg}^{2+}}{\overset{\text{CPK}}{\rightleftarrows}} \text{creatine} + \text{ATP} \qquad (97)$$

The reaction can be monitored by coupling the hexokinase and glucose-6-phosphate dehydrogenase systems [161, 162].

The ATP produced phosphorylates glucose in the presence of hexokinase. The glucose-6-phosphate produced is oxidized catalytically by the glucose-6-phosphate dehydrogenase in the presence of NADP. One mole of NADPH is liberated for each mole of creatine phosphate, and the increase in the fluorescence due to NADPH produced is measured. The overall reaction is

$$\text{creatine phosphate} + \text{glucose} + \text{NADP} \rightleftarrows \text{creatine} + 6\text{-phospho-}$$
$$\text{gluconate} + \text{NADPH} + \text{H}^+ \qquad (98)$$

The creatine phosphokinase procedure is specific for creatine phosphate. Inosine phosphates are practically inactive. In the reverse reaction ADP cannot replace ATP, and compounds related to creatine (creatinine or arginine) are not substrates [182].

2. Assay with Luciferase

The ATP formed from ADP in the creatine phosphokinase reaction (Eq. (97)) can be assayed by the firefly reaction described in Chapter 8. The production of the chemiluminescence is a measure of the creatine phosphate present [163, 164]. The peak luminescence produced after 30 sec is linearly proportional to creatine phosphate in the 10- to 100-μg concentration region.

F. Pyridoxal and Derivatives

Coursin and Brown [183] have made use of the native fluorescence of pyridoxine, pyridoxal, and pyridoxamine to assay for these substances. All three fluoresce maximally in 0.1 M phosphate buffer, pH 6.75. The excitation and emission wavelengths for pyridoxal are 330 and 385 nm, those for pyridoxamine are 315 and 400 nm.

Bonavita [184] has shown that pyridoxal-5-phosphate and pyridoxal react with cyanide in alkaline solution to give fluorescent derivatives. The cyano-hydrin of pyridoxal-5-phosphate exhibits maximal fluorescence at pH 3.8 (λ_{ex} 315, λ_{em} 420 nm). The pyridoxal derivative exhibits maximum fluorescence at pH 9 to 10 (λ_{ex} 358, λ_{em} 430 nm). By proper control of pH and spectral wavelengths each compound can be determined in the presence of the other.

Yamada et al. [185, 186] showed that the cyanide method could be adapted to the very sensitive assay of cyanide in blood and tissue. The pH dependence was thoroughly investigated, as were other assay parameters.

G. Nucleic Acids

Udenfriend and Saltzman [149] showed that nucleic acids possess no fluorescence in neutral or alkaline solutions. Fluorescence was observed in 0.1 and 5 N H_2SO_4. This fluorescence is similar to that of the nucleosides and nucleotides of adenine and guanine. Instrumentation and techniques for measuring the luminescence of nucleic acids were reviewed by Longworth [187]. Special emphasis was paid to analysis at 77°K.

A fluorometric method for RNA and DNA was proposed by LePacq and Paoletti [188]. The method is based on the observation that the dye ethidium bromide (3, 8-diamino-5-ethyl-6-phenylphenanthridinium bromide) increases the fluorescence quantum efficiency of nucleic acids on binding. A corresponding bathochromic shift in the excitation spectra is also observed. Fluorometric measurements are made at a λ_{ex} of 546 and a λ_{em} of 590 nm. Mixtures of RNA and DNA can be analyzed by measuring the total nucleic acids; RNase is added and the DNA remaining is measured. The method is specific for nucleic acids and has a high sensitivity (0.01 μg/ml of DNA).

Van Dyke and Szostkiewicz [189] adapted the procedure of LePacq and Paoletti [188] for use with the AutoAnalyzer; a flow diagram of the system was given.

A unique component of DNA is the sugar deoxyribose. Kissane and Robins
[190] developed an extremely sensitive fluorometric method for DNA based on
the formation of a fluorescent quinaldine by the reaction of deoxyribose with
3,5-diaminobenzoic acid (λ_{ex} 420, λ_{em} 520 nm). Deoxyribose is liberated
from the purine residues by heating with a strong acidic solution of 2,5-diamino-
benzoic acid.

Another method for the assay of DNA involves a measurement of thymine
in DNA [190]. Thymine is converted to acetol (96) which is condensed with
o-aminobenzaldehyde to form the fluorophor 3-hydroxyquinaldine (97).

$$(99)$$

(96)   (97)

IX. DRUGS AND MEDICINAL AGENTS

Fluorescence has proved to be a powerful tool for the analysis of drugs and
medicinal agents because of its great sensitivity and high specificity. Pharm-
acology requires analytical methods that can distinguish between the various
drugs and their metabolic products. Since many drugs are administered at
very low doses, sometimes as little as 100 μg a day, the analytical method
must be highly sensitive, sensitive enough to pick up the small amounts
excreted from the body. Fluorescence has this capability.

Fluorometric methods for many of the common drugs and medicinal agents
are listed in Table 66.

Detailed procedures for the assay of all these drugs are not presented
here, but the reader is referred to the original references for details or to
Udenfriend's Fluorescence Assay in Biology and Medicine for details [230].

X. PESTICIDES

Some of the fluorometric methods for the assay of pesticides are presented
in Table 67.

MacDougall has published detailed reviews [231,232] on the use of
fluorescence in the assay of pesticide residues in animal and plant tissue.
Discussed are instrumentation, structure-fluorescence relationships, and
chemical mechanisms of some of the procedures. Also covered are the
problems involved in isolating pesticides from tissues free of interfering
substances.

TABLE 66

Fluorometric Methods for the Assay of Drugs and Medicinal Agents

Substance	Conditions	λ_{ex} (nm)	λ_{em} (nm)	Sensitivity (ppm)	Ref.
Actinomycin D	H_2O_2-OH^-	370	420	0.10	191
5-Alkyl-2-thiohydantoins	2,6-Dichloroquinone chlorimide	365	520	0.5	193
N-Allylnormorphine	pH 1	285	355	0.1	192
Aminopterin	pH 7	280, 370	460	0.02	192
p-Aminosalicylic acid	pH 11	300	405	0.004	192
Amobarbital	pH 13	265	410	0.1	192
Ampicillin	Hydrolysis	346	422	0.05	194
Antimycin	pH 7-9	350	420	0.1	195
Aspirin	HOAc-$CHCl_3$	280	335	0.01	196
Atabrine	Caffein-0.05 M H_2SO_4	365	540	0.01	197
Atropine	Eosin Y	365	556	1.0	198
Berberine	DMF	380	510	0.20	199
Bromolysergic acid diethylamide	pH 1	315	460	0.10	192
Chlorpromazine	pH 11	350	480	0.01	192
Clortetracycline	pH 11	355	445	0.02	192
Cinchonidine	pH 1	315	420	0.01	192
Cinchonine	pH 1	320	420	0.01	192
Codeine	pH 1	245, 285	350	0.1	200
Desmethylimipramine	pH 14	295	415	0.1	201, 202
Diethylstilbestrol	Ethanol	360	435	0.2	203
Digitalis	HCl-glycerol	350	465	0.1	204
Diphenhydramine	Tinopal GS	365	450	0.20	205
Dipyridamole	$CHCl_3$	438	540	0.1	206
Emetine	pH 1	290	320	0.05	207
Epinephrine	Ferricyanide	365	Yellow-green	0.002	208
Eserine	pH 1-7	265, 315	350	0.04	209
Estrogens	pH 13	490	546	0.1	210
Harmine	pH 1	300, 365	400	0.002	192
Hydroxyamphetamine	pH 1	275	300	0.05	192

TABLE 66 (Continued)

Substance	Conditions	λ_{ex} (nm)	λ_{ex} (nm)	Sensitivity (ppm)	Ref.
3-Hydroxy-N-methyl-morphinan	pH 1	275	320	0.05	192
Isoniazid	Cyanogen bromide	300	405	0.10	211
	H_2O_2	320	415	0.05	212
Lysergic acid diethyl-amide	pH 7	325	365	0.002	192
Menadione	Ethanol	335	480	0.07	192
Meperidine	Formaldehyde-H_2SO_4	275	425, 440	0.10	213
Mephenesin (Tolserol)	pH 1	280	315	0.05	192
Metaraminol	Aldehyde condensation	370	500	0.002	214
Methotrexate	pH 7	280, 375	460	0.02	192
Morphine	Ferricyanide	250	440	0.1	215
	pH 1	285	350	0.1	200
Neocinchophen	pH 7	275, 345	455	0.004	192
Neosynephrine	pH 1	270	305	0.01	192
Norepinephrine	Ferricyanide	365	Yellow-green	0.003	208
Oxychloroquin	pH 11	335	380	0.08	192
Oxytetracycline	pH 11	390	520	0.05	192
Pamaquine	pH 13	300, 370	530	0.06	192
Penicillin	2-Methoxy-6-chloro-9-(ß-aminoethyl)-aminoacridine	365	540	0.05	216, 217
Pentobarbital	pH 13	265	440	0.10	192
Phenobarbital	pH 13	265	440	0.5	192
Phenothiazines	H_2O_2	--	--	0.25	218
Piperoxan	pH 7	290	325	0.05	192
Podophyllotoxin	pH 11	280	325	0.005	192
Procaine	pH 11	275	345	0.01	192
Protripyline	H_2SO_4	310	386	3.0	219
	HCl, HClO$_4$, H$_3$PO$_4$	295	360	0.003	220
Pyribenzamine	CNBr	370	460	1.0	192, 221
Pyrithyldione	$NH_2OH \cdot HCl$	365	460	0.2	222
Quinacrine	pH 11	285, 420	500	0.02	192

TABLE 66--Continued

Substance	Conditions	λ_{ex} (nm)	λ_{em} (nm)	Sensitivity (ppm)	Ref.
Quinapyramine	Eosin	365	550	0.01	223
Quinidine	Acid	360	460	0.05	224
Quinine	pH 1	250, 350	450	0.002	192
Rescinnamine	pH 1	310	400	0.008	192
Reserpine	pH 1	300	375	0.008	192
Sarcolysine	pH 7	260	365	--	225
Streptomycin	pH 13	366	445	0.1	226
Sulfanilamide	pH 3-10	275	350	0.1	227
Tetracycline	HCl-NaOH-AlCl$_3$	475	550	0.10	228
	pH 11	390	515	0.02	192
Thiamylal	pH 13	310	530	0.05	192
Thiopental sodium	pH 13	315	530	0.1	192
Trimethoprim	KMnO$_4$	275	350	0.04	229
Yohimbine	pH 1	270	360	0.01	192
Zoxazolamine	pH 11	280	320	0.1	192

TABLE 67

Assay of Pesticides

Substance	Conditions	λ_{ex} (nm)	λ_{em} (nm)	Ref.
Amprolium	Ferricyanide	400	455	233, 234
Bayer 22,408	Oxidation	372	480	235
Benomyl	Hydrolysis	285	335	236
Buquinolate	H$_2$O	265	375	237
Carbaryl	0.25 N NaOH	285	340	238
CoRal	Hydrolysis	330	410	239
Guthion	Hydrolysis	340	400	240
Malvin	HCl, NaNO$_2$	366	490	241
Maretin	Hydrolysis	372	480	242
Potasan	Methanol	320	385	44
Terephthalic acid	Conversion to amine	364	438	52
Zinophos	Hydrolysis	315	375	243

The excitation and emission spectra, decay times, analytical curves and limits of detection for 32 pesticides were reported by Moye and Winefordner [244]. Ragab [245] reported the thin-layer-chromatography separation and detection after exposure to Br_2 vapor, $FeCl_3$, and 2-(2'-hydroxyphenyl)benzoxazole of 25 organothiophosphorus pesticides and 5 sulfur-containing degradation products.

Bowman and Beroza [246] determined five carbamate pesticides in milk in methanol-tetramethylammonium hydroxide solution. The range of detectability varied from 0.3 ppm for the most sensitive carbamate to 3.0 ppm for the least sensisitve. Carbaryl was detected in bees, pollen, honey, beeswax, and sugar solution by measuring the fluorescence in 0.25 N NaOH [238].

Hornstein [44] has studied the fluorescence characteristics of a number of compounds and showed that only a few possess a native fluorescence. Compounds that did not fluoresce at a concentration of 1 ppm include DDT, methoxychlor, Diazinon, rotenone, Aramite, toxaphene, aldrin, dieldrin, chlordane, and heptachlor.

Other pesticides that are not fluorescent themselves can be converted to fluorophors. Bayer 22,408 can be oxidized with hydrogen peroxide to a highly fluorescent derivative [235].

CoRal (98) can be converted to coumaric acid (99) and coumarilic acid (100) by hydrolysis in hot alkali [239].

(98)   (99)   (100)

(101)   (102)

Guthion (100) is hydrolyzed in hot alkali to anthranilic acid (102) [240].

Benomyl is tracted from animal tissues, fruits, etc., with ethyl acetate and is then converted to 2-aminobenzimidazole by alkaline hydrolysis. The latter is determined at λ_{ex} 285, λ_{em} 335 nm. As little as 0.1 ppm is determinable [236].

Terephthalic acid is assayed by nitration, followed by a reduction to the fluorescent amino derivative. The lower limit of detection is 0. 1 ppm [52].

Zinophos is a good nematocide. It is converted to the highly fluorescent 2-pyrazinol on hydrolysis in alkali. A method for its assay in plant produce at concentrations of 5 μg/100 g of material has been developed [243].

XI. PORPHYRINS

A. General Remarks

The porphyrins, which are among the most highly fluorescent compounds in nature, are derivatives of porphine (four-pyrazole rings joined into a ring system by four methane bridges). The porphyrins are excited in the visible and emit fluorescence in the red and infrared. This fluorescence can be detected with standard instruments equipped with 1P21 and 1P28 photomultiplier tubes, but the use of a red-sensitive detector (RCA 7102 tube) increases the sensitivity markedly.

The naturally occurring porphyrins include hemoglobin, myoglobin, cytochromes, chlorophyll, and other pigments. Chlorophyll will be discussed separately in another chapter.

Dhere [247] has carried out a thorough investigation on the fluorescence of porphyrins. He described and classified the fluorescence spectra of the porphyrins in a variety of solvents. The same spectra are observed in alcohol, dioxane, ammonia, and pyridine.

In acid solution only three bands appear. In pyridine protoporphyrin has a λ_{em} at 634 nm, whereas hematoporphyrin has a λ_{em} at 625 nm. Copropor-phyrin, uroporphyrin, mesaporphyrin, and etioporphyrin have almost the same fluorescence characteristics as hematoporphyrin.

A survey of the luminescence characteristics of porphyrins was written by Schwartz et al. [248], who discussed factors that affect the fluorescence and quenching agents. Sharp emission bands were observed in organic solvents; as little as 10^{-7} g was determinable.

Solov'ev et al. [249,250] related the spectral characteristics of porphyrins to molecular structure. The porphyrins were found to have two emission bands with mirror-image symmetry to the two absorption bands.

Runge [251] developed a microfluorospectrophotometer that was able to detect as little as 10^{-10} g of porphyrin in tissue sections. Specific porphyrins in tissues and plasma were identified.

Martinez and Mills [252] developed a rapid procedure for the assay of total porphyrins in urine based on a combination of anion exchange and spectrophoto-fluorometry. After the porphyrins are eluted from the resin with 3 M HCl,

their fluorescence at 646 nm is measured. A good precision, 2.9% coefficient
of variation, is obtained.

Doss [253] described a procedure for the separation of porphyrins as the
methyl esters on thin-layer chromatograms. The fluorescence intensity of the
porphyrin zones is measured continuously with a thin-layer-chromatography
scanner. The detection limit was less than 1 ng.

B. Bilirubin

Roth [254] developed a bilirubin-assay procedure based on the observation
that this compound develops a fluorescence on standing in H_3PO_4. The
fluorescence intensity is increased more than 60 times by addition of serum
albumin. The fluorescence intensity is measured at λ_{ex} 435, λ_{em} 500 nm.
Biliverdin yielded 60% the fluorescence of bilirubin; cholesterol and hemoglobin
gave no fluorescence.

C. Biliverdin

Biliverdin is a tetrapyrrole pigment derived from the metabolic degradation of hemoglobin. Rodriguez-Garay and Argerich [255] developed a procedure
for biliverdin in serum, bile, and urine based on the red fluorescence formed
by treating biliverdin with iodine in the presence of Zn^{2+}.

D. Coproporphyrin

Coproporphyrin is the predominant porphyrin in urine and feces. Reviews
on the fluorescence characteristics of coproporphyrin and procedures for its
assay in urine and feces have been prepared by Wranne [256] and Schwartz
et al. [248].

Schwartz et al. [257] and Neve and Aldrich [258] described procedures for
the assay of coproporphyrin I and III in mixtures. The compounds are
separated from other porphyrins, and their total fluorescence is measured in
0.1 M phosphate buffer, pH 6.0, at λ_{ex} 405, λ_{em} 618 nm. The mixture is then
stored in the dark for 12 to 24 h; the coproporphyrins coprecipitate and settle
out of solution. The residual fluorescence depends on the composition of the
mixture.

The chromatographic separation of coproporphyrin I and III was reported
by Eriksen [259]. This method was used by Sweeney and Eales [260] for the
assay of these two compounds.

E. Heme

Metalloporphyrin complexes of iron and copper are nonfluorescent. The
fluorescent assay of heme is based on the removal of iron and measurement of
the resulting porphyrin.

A highly sensitive method for the assay of heme proteins, primarily
hemoglobin, was described by Morrison [261]. The tissue samples are
solubilized in NaOH, oxalic acid is added, and the porphyrin fluorescence is
measured. As little as 0. 01 μg of heme protein could be assayed.

F. Protoporphyrin

Protoporphyrin assay is of diagnostic value in lead poisoning, iron
deficiency, and other conditions. Schwartz and Wikoff [262] and Wranne [256]
have described sensitive procedures for the assay of protoporphyrin
(λ_{ex} 415, λ_{em} 605 nm). A detailed study of the fluorescence characteristics of
protoporphyrin was presented by Schwartz et al. [248].

G. Urobilinogen and Urobilin

Urobilinogen is excreted in urine in two isomeric forms. It is nonfluorescent, but is oxidized to the fluorescent urobilin in air. The urobilin formed is
then measured fluorometrically. Adler [263] determined urobilinogen by oxidation with an alcoholic solution of iodine in the presence of zinc acetate. The
strongly green-fluorescent urobilin-zinc complex is then determined.

A procedure for the quantitative assay of urobilinogen and urobilin in urine
was devised by Kirkpatrick [264] and modified by Schmidt and Scholtis [265].
Persulfate is added at alkaline pH to oxidize urobilinogen. A filter instrument is
used with the 365-nm mercury line for excitation, with an emission at 530 nm.

H. Uroporphyrin

Schwartz [266] described a procedure for uroporphyrin based on the
conversion of porphobilinogen to uroporphyrin with iodine and the acetate ion.
The red-fluorescent porphyrin is extracted, and the fluorescence is measured
directly.

Talman [267] described a procedure for the assay of coproporphyrin and
uroporphyrin in urine. Hsia and Inouye [268] suggested a λ_{ex} of 365 and a
λ_{em} of 626 nm for the assay of uroporphyrin.

XII. STEROIDS

Steroids can be divided into four important classes on physiological and

structural considerations: cholesterol, estrogens, adrenal steroids and andro-
gens, and bile acids. Only the estrogens possess an aromatic ring and phenolic
substitution, which results in ultraviolet absorption and native fluorescence. In
concentrated sulfuric acid or in sulfuric-chloroform-acetic anhydride mixtures,
however, steroids are converted to highly colored and fluorescent products.
The former method is called the Salkowski reaction, the latter the Lieberman-
Burchard reaction. Fluorescence in acid media occurs with almost all steroids,
although the intensity varies with the specific type of acid and many other
factors.

   A review on the fluorescence of steroids was prepared by Braunsberg and
James [269], who stated that the best conditions for developing the fluorescence
of a given steroid must be arrived at empirically. Figure 103 shows the
structures of some of the more common steroids.

   Fluorescence is used routinely for the assay of cholesterol, the estrogens,

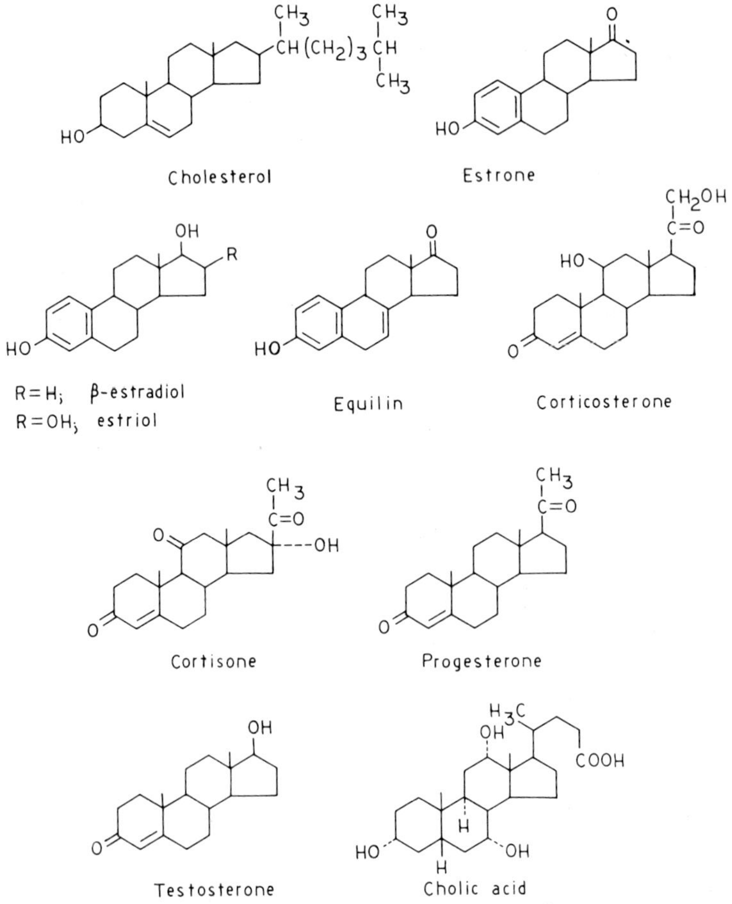

FIG. 103.   Common steroids.

testosterone, corticosteroids, bile acids, and steroid drugs. Some fluorometric methods for the assay of steroids are presented in Table 68. Almost all
of these methods are based on the fluorescent derivatives formed in strong
acid or alkali. Since many of the steroids are excreted into the urine in

TABLE 68

Fluorometric Methods for the Assay of Steroids

Substance	Reagent	λ_{ex} (nm)	λ_{em} (nm)	Sensitivity (ppm)	Ref.
Aldosterone	Conc. H_2SO_4	365	Blue	0.1	270
Bile acids	H_3PO_4	365	430	1.0	272
	H_2SO_4	436	500	0.1	273
Cholesterol	Trichloro-methane-H_2SO_4	546	600	0.1	274
Corticosterone	H_2SO_4	436	525	0.01	276
Cortisone	t-BuOK	380	580	0.01	271
11-Dehydrocortico-sterone	t-BuOK	380	580	0.01	271
Equilenin	EtOH	290, 340	370	0.001	34
Equilin	EtOH	290	345, 420	0.1	34
Estradiol-17ß	$POCl_3$-H_2SO_4	440	490	0.001	277
Estriol	$POCl_3$-H_2SO_4	440	480	0.001	277
Estrone	$POCl_3$-H_2SO_4	440	480	0.001	277
Hydrocortisone	H_2SO_4	436	530	0.10	276
	t-BuOK	380	580	0.01	271
	Periodate	475	530	0.01	278
	H_2SO_4	468	524	0.1	279
17-Hydroxycortico-sterone	H_2SO_4-EtOH	420	570	0.05	280
17-Hydroxy-11-deoxy-corticosterone	t-BuOK	380	580	0.01	271
Methyl testosterone	H_3PO_4	365	430	0.10	281
Prednisolone	H_2SO_4-EtOH	420	570	0.50	280
Progesterone	t-BuOK	380	580	0.01	271
Testosterone	t-BuOK	380	580	0.01	271
	Enzyme	450	540	0.001	282
Tetrahydrocortisone	H_2SO_4	436	525	0.2	276

conjugated form, it is frequently necessary to hydrolyze the conjugates with glucuronidase or acid prior to extraction and assay.

Fazekas and Webb [283] found that a fluorescent complex is formed between steroids and nucleotide coenzymes. For example, when hydrocortisone is treated with NAD, an intense greenish-yellow fluorescence appears within 5 min. The fluorescent complex also can be formed with all Δ^4-3-oxosteroids.

Excellent reviews of fluorescent methods for steroids have been prepared by Udenfriend in his Fluorescence Assay in Biology and Medicine [284].

XIII. VITAMINS

A. General Remarks

The vitamins were one of the first groups of biologically active compounds for which fluorometric methods were described. Many of these compounds either have aromatic structures containing substituent groups that possess native fluorescence or can be converted to fluorophors by simple procedures. The structures of some of the vitamins are shown in Fig. 104.

B. Vitamin A

Vitamin A is a carotenoid substance found in plant and animal tissues. One international unit of vitamin A is equivalent to 0.6 µg of ß-carotene, 0.344 µg of vitamin A acetate, or 0.3 µg of vitamin A alcohol. Assays for vitamin A hence must distinguish among these various forms.

Passannante and Avioli [285] found that vitamin A and carotene yield highly fluorescent derivatives when heated in 20% sulfuric acid in ethanol. Excitation maximum is at 425 nm, and the fluorescence maximum is at 475 nm.

A simple method for the assay of vitamin A in blood was suggested by Kahan [286] who measured the native fluorescence at λ_{ex} 345 and λ_{em} 490 nm. Hansen and Warwick [287] described a similar procedure for vitamin A in blood. A linear relationship to vitamin A acetate exists over the range 25 to 1600 µg/ml (λ_{ex} 340, λ_{em} 480 nm).

Garry et al. [288] described a fluorometric vitamin A assay in which a silicic acid column is used to separate the various vitamin A components.

Hansen and Warwick [289] suggested a fluorometric micromethod for assay of vitamins A and E in blood. The vitamins are extracted with hexane. Vitamin A is assayed by fluorescence with 340-nm excitation and 480-nm emission; vitamin E acetate is reduced by lithium aluminum hydride to the alcohol before assay with 295-nm excitation and 340-nm emission.

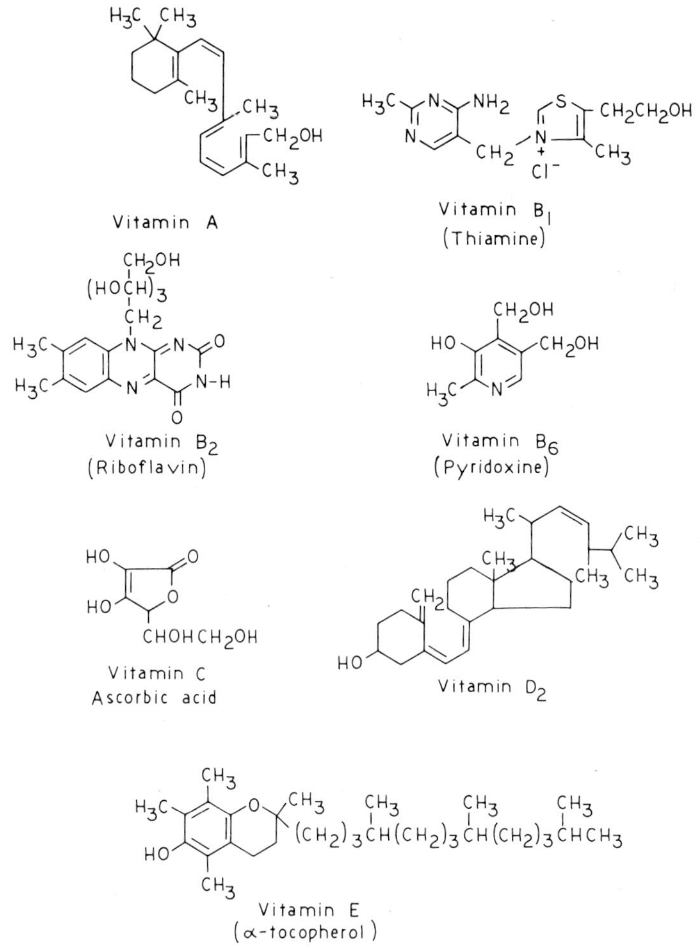

FIG. 104. Structures of vitamins.

C. Vitamin B

1. Vitamin B$_1$ (Thiamine)

Thiamine itself is nonfluorescent and is assayed by oxidation to the fluorophor thiochrome (103). This reaction method is used for the assay of B$_1$ in blood, urine, tissues, vitamin preparations, etc. [290-292]. The λ_{ex} is 365 to 370 nm, and λ_{em} is 445 nm.

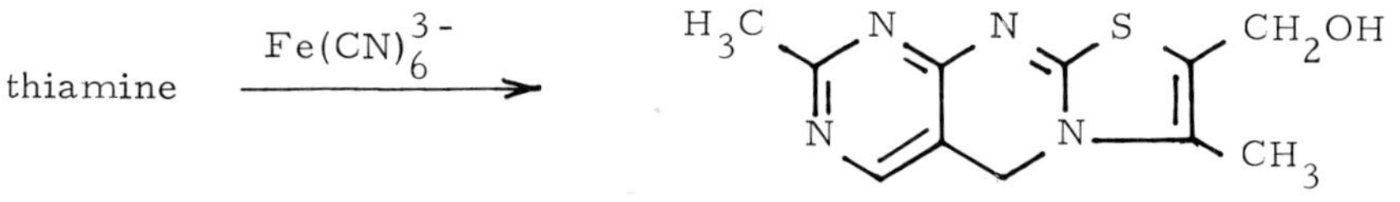

(102)

(103)

Fujita et al. [293] showed that pyrithiamine undergoes an oxidation similar to thiamine. The oxidation product, called pyrichrome, has a λ_{ex} of 410 and a λ_{em} of 480 nm. A method for the assay of thiamine and pyrithiamine in mixtures was described.

2. Vitamin B_2 (Riboflavin)

A plot of the fluorescence of riboflavin as a function of pH was shown in Fig. 99. The fluorescence of riboflavin and its nucleotide forms can be differentiated by their relative intensities, by changes in pH, and by the effects of quenching agents.

Riboflavin is assayed by its native fluorescence at pH 7 (λ_{ex} 370 and 440 λ_{em} 565 nm) at a sensitivity of 0.0012 µg/ml [294-296].

Alternatively riboflavin can be converted to the highly fluorescent derivative lumiflavin by irradiation with UV light in 0.5 M NaOH for 30 min (λ_{ex} 440, λ_{em} 560 nm) [297]. The method is suitable for the assay of total flavins in tissues [178].

3. Vitamin B_6 (Pyridoxine)

Pyridoxine, pyridoxal, pyridoxic acid, and pyridoxamine can be assayed by oxidation with $KMnO_4$ to the fluorescent pyridoxic lactone (λ_{ex} 350, λ_{em} 450 nm) [298]. Bridges et al. [299] have made a thorough study of the fluorescence of the hydroxypyridine analogs including the B_6 derivatives.

The native fluorescence of pyridoxine (λ_{ex} 340, λ_{em} 400 nm), pyridoxal (λ_{ex} 330, λ_{em} 385 nm), and pyridoxamine (λ_{ex} 335 nm, λ_{em} 400 nm) can be used for their assay in blood [183].

Pyridoxal reacts with cyanide in alkaline solution to form the fluorescent cyanohydrin [184]. Pyridoxamine and pyridoxine can be converted to pyridoxal before assay [300].

Detailed procedures for the assay of vitamin B_6 and its metabolites can be found in Udenfriend's Fluorescence Assay in Biology and Medicine [301].

4. Vitamin B_{12} (Cyanocobalamin)

Vitamin B_{12} has a native fluorescence at pH 7 (λ_{ex} 275, λ_{em} 305 nm). The sensitivity is 0.003 µg/ml [34].

Benzimidazole and 5,6-dimethylbenzimidazole are found in the combined form in vitamin B_{12}. These compounds can be released by acid hydrolysis and extracted into organic solvents prior to fluormetric assay [302,303]. The concentration of these substances is then proportional to the B_{12} content.

D. Vitamin C (Ascorbic Acid)

Archibald [304] described a condensation of dehydroascorbic acid with o-phenylenediamine to yield the blue fluorophor (104):

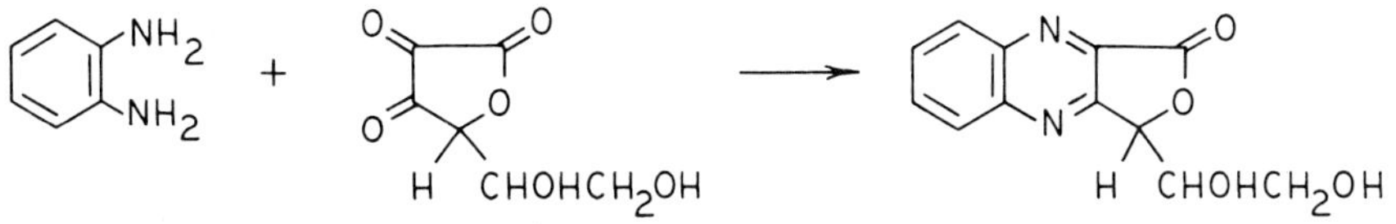

$$\text{(o-phenylenediamine)} \quad + \quad \text{(dehydroascorbic acid)} \quad \longrightarrow \quad \text{(104)} \qquad (103)$$

(104)

Ascorbic acid is first converted to dehydroascorbic acid with 2,6-dichloroindophenol. From 2 to 25 μg can be determined on paper chromatograms [305].

Deutsch and Weeks [306] and Polansky et al. [307] have described this procedure for the assay of ascorbic acid in solution. The wavelength of maximum excitation is 350 nm, the emission maximum is at 430 nm. The authors report that as little as 0.15 μg/ml is detectable.

E. Vitamin D

Jones et al. [308] observed that both vitamins D_2 and D_3 yield highly fluorescent products when treated with trichloroacetic acid and other acids. Both vitamins yield fluorophors with λ_{ex} of 390 and λ_{em} of 480 nm.

Chen et al. [309] investigated the fluorescence of vitamin D analogs in strong acids. The fluorescence characteristics they observed for some of these D vitamins are reported in Table 69.

TABLE 69

Fluorescence of D Vitamins and Derivatives[a]

Compound	λ_{ex} (nm)	λ_{em} (nm)	Fluorescence Intensity[b]
Vitamin D_2	390	470	42
Vitamin D_3	390	470	42
Cholesterol	350	415	16
Ergosterol	475	510	25

[a] Reprinted from Ref. [309] by courtesy of Academic Press.

[b] At optimum wavelengths.

Chen [310] introduced a number of dyes to detect vitamin D analogs on paper chromatograms. Acridine Yellow, Auramine O, and safranin gave useful differentiation.

F. Vitamin E (Tocopherol)

The three major substances with vitamin E activity that have been isolated from natural sources are α-ß- and γ-tocopherol, which are methyl derivatives of the parent substance tocol (see Fig. 102).

Tocopherol can be determined in blood and tissues by measuring its native fluorescence in hexane (λ_{ex} 295, λ_{em} 340 nm) [311, 312]. Only the free form of tocopherol fluoresces; the esters must be reduced to the free alcohol for total tocopherol assays ($LiAlH_4$). As little as 0.6 μg of tocopherol per milliliter of serum is assayable.

Alternatively tocopherol can be condensed with o-phenylenediamine after oxidation to yield a fluorescent phenazine derivative [313]. With α-tocopherol a single product is formed, with excitation at 270 and 370 nm and green emission. With ß- and γ-tocopherol a mixture of products distinguishable from α-tocopherol is formed.

G. Vitamin K

Duggan et al [34] studied the fluorescence of the K vitamins in a variety of solvents and found that none of them fluoresce. Beyer and Kennison [314] found that vitamin K is photodecomposed by ultraviolet light. A method for menadione (105a), similar in structure to K_1 (105b) and K_2 (105c), in blood and urine is based on its condensation with o-phenylenediamine to yield a phenazine derivative in a manner similar to that for tocopherol [315].

Menadione is also fluorescent in ethanol (λ_{ex} 335, λ_{em} 480 nm).

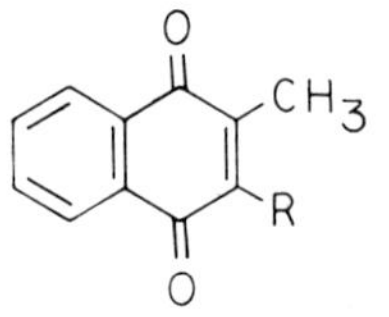

(105)

a: R = H

b: R = phytyl

c: R = difarnesyl

REFERENCES

1. S. Udenfriend, Fluorescence Assay in Biology and Medicine, Vol. 1, Academic Press, New York, 1966.

2. S. Udenfriend, op. cit. , Vol. 2, 1970.

3. R. E. Phillips and F. R. Elevitch, in Progress in Clinical Pathology (M. Stefanni, ed.), Grune and Stratton, New York, 1966, Chapter 4.

4. C. E. White and A. Weissler, in Handbook of Analytical Chemistry (L. Meites, ed.), McGraw-Hill, New York, 1963, Chapter 6.

5. C. E. White and A. Weissler, in Standard Methods of Chemical Analysis, Vol. IIIA (F. J. Welcher, ed.), Van Nostrand, Princeton, New Jersey, 1966, Chapter 5.

6. C. E. White and A. Weissler, Anal. Chem. , 36, 116R (1964); 38, 115R (1966); 40, 114R (1968); 42, 57R (1970).

7. I. B. Berlman, Handbook of Fluorescence Spectra of Aromatic Molecules, Academic Press, New York, 1965.

8. P. Pringsheim, Fluorescence and Phosphorescence, Interscience, New York, 1949.

9. R. A. Passwater, Guide to Fluorescence Literature, Plenum Press, New York, 1967.

10. M. A. Konstantinova-Shlesinger, Fluorometric Analysis (N. Kaner, transl.), Davey, New York, 1965.

11. C. E. White and R. Argauer, Fluorescence Analysis. A Practical Approach, Dekker, New York, 1970.

12. J. De Ment, Fluorochemistry, Chemical Publishing, Brooklyn, 1945.

13. J. W. Bridges and R. T. Williams, Nature, 196, 4849 (1962).

14. T. Forster and K. Kasper, Z. Elektrochem. , 59, 976 (1955).

15. G. Schenk and D. Wirz, Anal. Chem., 42, 1754 (1970).

16. W. Czapska-Narkiewicz, Bull. Intern. Cracow, A1935, 445.

17. S. Rangaswami, T. Seshandi, and V. Venkatesvarlu, Proc. Ind. Acad. Sci., A13, 316 (1941).

18. W. Sherman and E. Robins, Anal. Chem. , 40, 803 (1968).

19. U. Kubli and E. Schmidt, Helv. Chim. Acta, 28, 213 (1945).

20. K. Nikolic and L. Sablic, Arh. Farm. (Belgrade), 17, 13 (1967).

21. M. Balemans and F. van den Veerdonk, Experientia, 23, 906 (1967).

22. S. Konev and E. Chernitskii, Biofizika, 9, 520 (1964).

23. S. Konev, Fluorescence and Phosphorescence of Proteins and Nucleic Acids, Plenum Press, New York, 1967.

24. A. White, Biochem. J. , 71, 217 (1959).

25. G. G. Guilbault and D. N. Kramer, Anal. Chem. , 37, 120 (1965).

26. G. G. Guilbault, M. H. Sadar, R. Glazer, and C. Skou, Anal. Letters, 1, 365 (1968).

27. R. Maickel and F. Miller, Anal. Chem. , 38, 1937 (1966).

28. G. Schmidt, Z. fur Physik, 8, 160 (1922).

29. H. Perkampus, A. Knop, and J. Knop, Z. Naturforsch. , A23, 804 (1968).

30. A. Perry, P. Tidwell, J. Cetorelli, and J. Winefordner, Anal. Chem. , 43, 781 (1971).

31. R. Adams and M. Gold, J. Amer. Chem. Soc. , 62, 2038 (1940).

32. A. Vaughn, G. G. Guilbault, and D. Hackney, Anal. Chem., 43, 721 (1971).

33. G. Leonhardi and I. Glasenapp, Z. Physiol. Chem. , 286, 145 (1951).

34. D. Duggan, R. Bowman, B. Brodie, and S. Udenfriend, Arch. Biochem. Biophys. , 68, 1 (1957).

35. B. Hubmann-Ballabey, D. Mannier, and M. Roth, Mitt Geb Lebensmittel-unter Hyg. , 59, 482 (1968).

36. E. Leininger and S. Katz, Anal. Chem. , 21, 810 (1949).

37. M. Pellet, C. Seigner, and H. Cohen, Path. Biol. (Paris), 17, 909 (1969).

38. G. Kisilevich, Anal. Abstr. , 7, 3337 (1960).

39. F. Kavanagh and N. Kuzel, J. Agr. Food Chem. , 6, 459 (1958).

40. R. Conn and R. Davis, Nature, 183, 1053 (1959).

41. T. J. Mellinger, Amer. J. Clin. Pathol. , 49, 200 (1968).

42. P. Chen, Anal. Chem. , 31, 296 (1959).

43. G. Thommes and E. Leininger, Anal. Chem. , 30, 1361 (1958).

44. I. Hornstein, J. Agr. Food Chem. , 6, 32 (1958).

45. G. D. Christian and J. Moody, Anal. Chim. Acta, 41, 269 (1968).

46. M. Strassman, L. Ceci, and A. Tucci, Anal. Biochem. , 23, 484 (1968).

47. D. A. Britton and J. C. Guyon, Anal. Chim. Acta, 44, 397 (1969).

48. J. Fonichon, Y. Mimaire, and C. Studievic, Path. Biol. (Paris), 15, 40 (1967).

49. H. Weissbach, T. Waalker, and S. Udenfriend, J. Biol. Chem. , 230, 865 (1958).

50. Z. Gregorowicz and P. Gorka, Z. Anal. Chem. , 230, 431 (1967).

51. C. Barr, Plant Physiol. , 23, 443 (1948).

52. P. A. Giang, M. S. Schechter, and L. Weissbecker, J. Agr. Food Chem. , 15, 95 (1967).

53. H. Freytag, Z. Anal. Chem. , 139, 263 (1953).

54. D. W. Roberts and M. Friedkin, J. Biol. Chem. , 233, 483 (1958).

55. C. E. Frohman and J. Orten, J. Biol. Chem. , 205, 717 (1953).

56. J. P. Hummel, J. Biol. Chem. , 180, 1225 (1949).

57. O. Lowry, J. Passonneau, F. Hasselberger, and D. Schulz, J. Biol. Chem. , 239, 18 (1964).

58. P. K. Maitra and R. Estabrook, Anal. Biochem. , 7, 472 (1964).

59. H. Naruse, S. Cheng, and H. Waelsch, Exptl. Brain Res. , 1, 40 (1966).

60. J. Spiker and J. Towne, Anal. Chem. , 34, 1468 (1962).

61. V. Sandesai and H. Provido, J. Lab. Clin. Med. , 64, 977 (1964).

62. S. Belman, Anal. Chim. Acta, 29, 120 (1963).

63. O. Bandisch and H. Deuel, J. Amer. Chem. Soc., 44, 1586 (1922).

64. H. Ko and M. Royer, Anal. Biochem., 26, 18 (1968).

65. D. Mendelsohn and A. Antonis, J. Lipid Res., 2, 45 (1961).

66. G. Kessler and H. Lederer, Clin. Chem., 11, 809 (1965).

67. L. Velluz, M. Pesez, and M. Herbain, Bull Soc. Chim. France, 15, 681 (1948).

68. R. Salvador and R. Albers, J. Biol. Chem., 234, 922 (1959).

69. B. Camber, Nature, 174, 1107 (1954).

70. R. Brandt and N. Cheronis, Microchem. J., 5, 110 (1961).

71. E. Sawicki, in International Symposium on Microchemical Techniques (N. Cheronis, ed.), Wiley, New York, 1962, p. 59.

72. E. Sawicki, T. Stanley, and I. Pfaff, Chemist-Analyst, 51, 9 (1962).

73. M. Kamel and R. Wizinger, Helv. Chim. Acta, 43, 594 (1960).

74. E. Sawicki and R. Carnes, Mikrochim. Acta, 1968, 148.

75. S. Abbott and D. Hercules, Anal. Chem., 42, 171 (1970).

76. R. Alarion, Anal. Chem., 40, 1704 (1968).

77. E. Sawicki, T. Stanley, and H. Johnson, Anal. Chem., 35, 199 (1963).

78. K. Bredereck, T. Förster, and H. Oesterlin, in Luminescence of Organic and Inorganic Materials (H. Kallmann and G. Spruch, eds.), Wiley, New York, 1962, p. 161.

79. E. Crowell and C. Varsel, Anal. Chem., 35, 189 (1963).

80. P. Ghosh and M. Whitehouse, Biochem. J., 108, 155 (1968).

81. J. Eisenbrand and H. Hauprich, Arch. Pharm. Berl, 303, 201 (1970).

82. L. Dombrowski and E. Pratt, Anal. Chem., 43, 1042 (1971).

83. J. Sinsheimer, D. Hong, J. Stewart, M. Fink, and J. Burckhalter, J. Pharm. Sci., 60, 141 (1971).

84. M. Pesez and J. Bartos, Ann. Pharm. France, 27, 161 (1969).

85. H. Price and M. Price, J. Lab. Clin. Med., 50, 769 (1957).

86. G. Cohen and M. Goldenberg, J. Neurochem., 2, 58 (1957).

87. A. Vendsolv, Acta Physiol. Scand., 49, Suppl. 173 (1960).

88. P. Shore and J. Olin, J. Pharmacol. Exptl. Therap., 122, 295 (1958).

89. C. DuToit, Wright Air Development Center Tech. Rept. 59-175, 1959.

90. E. Crout, in Standard Methods of Clinical Chemistry (D. Seligson, ed.), Vol. 3, Academic Press, New York, p. 62.

91. U. Euler and F. Lishajko, Acta Physiol. Scand., 45, 122 (1959).

92. R. J. Merrills, Nature, 193, 988 (1962).

93. R. J. Merrills, Anal. Biochem., 6, 272 (1963).

94. S. Natelson, J. Lugovoy, and J. Pincus, Arch. Biochem., 23, 157 (1949).

95. H. Weil-Malherbe and A. Bone, Biochem. J., 51, 311 (1952).

96. H. Weil-Malherbe, Biochim. Biophys. Acta, 40, 351 (1960).

97. G. Mangan and J. Mason, Science, 126, 562 (1957).

98. G. Nadeau and L. Joly, Nature, 182, 180 (1958).

99. C. Bell and A. Somerville, Biochem. J., 98, 1c (1966).

100. H. Corrodi and N. Hillarp, Helv. Chim. Acta, 46, 2425 (1963).

101. A. Bertler, A. Carlsson, and E. Rosengren, Acta Physiol. Scand., 44, 273 (1958).

102. H. Drujans, T. Sourkes, S. Layne, and G. Murphy, Can. J. Biochem. Physiol., 37, 1154 (1959).

103. F. Eichhorn, A. Rutenberg, and E. Kott, Clin. Chem., 17, 296 (1971).

104. Z. Oberman, R. Chayen, and M. Herzberg, Clin. Chem. Acta, 29, 391 (1970).

105. E. Smith and H. Weil-Malherbe, Federation Proc., 20, 182 (1961).

106. P. Shore, A. Burkhalter, and V. Cohn, J. Pharmacol. Exptl. Therap., 127, 182 (1959).

107. E. Sawicki, C. Sawicki, C. Golden, and T. Kober, Microchem. J., 15, 25 (1970).

108. D. Alkon, A. Goldberg, J. Green, P. Levi, and K. Liao, Anal. Biochem., 40, 192 (1971).

109. S. Hess and S. Udenfriend, J. Pharmacol. Exptl. Therap., 127, 175 (1959).

110. G. Jonsson and M. Sandler, Histochem., 17, 207 (1969).

111. J. Jason and B. Stevens, Nature, 172, 772 (1953).

112. W. Quay, J. Pharm. Sci., 57, 1568 (1968).

113. J. Vanable, Anal. Biochem., 6, 393 (1963).

114. J. Oates, in Methods of Medical Research (J. Quastel, ed.), Vol. IX, Year Book, Chicago, 1961, p. 169.

115. S. Spector, K. Melmur, W. Lovenberg, and A. Sjoerdsma, J. Pharmacol. Exptl. Therap., 140, 229 (1963).

116. N. Pasarela and A. Waldrin, J. Agr. Food Chem., 15, 221 (1967).

117. J. McCleskey, Anal. Chem., 36, 1646 (1964).

118. F. Roch-Ramel, Anal. Biochem., 21, 372 (1967).

119. F. Teale and G. Weber, Biochem. J., 65, 476 (1956).

120. D. Duggan and S. Udenfriend, J. Biol. Chem., 178, 53 (1949).

121. T. Waalkes and S. Udenfriend, J. Lab. Clin. Med., 50, 733 (1957).

122. Y. Vladimirov, Dokl. Akad. Nauk SSSR, 116, 780 (1957).

123. I. Lowe, E. Robins, and G. Eyerman, J. Neurochem., 3, 8 (1958).

124. M. McCaman and E. Robins, J. Lab. Clin. Med., 59, 885 (1962).

125. P. Wong, M. O'Flynn, and T. Inouye, Clin. Chem., 10, 1098 (1964).

126. J. Hill, G. Summer, M. Pender, and N. Roszel, Clin. Chem., 11, 541 (1965).

127. D. Duggan and S. Udenfriend, J. Biol. Chem., 223, 313 (1956).

128. L. Brand and S. Shaltiel, Biochim. Biophys. Acta, 88, 338 (1964).

129. J. Ambrose, P. Sullivan, A. Ingerson, and R. Brown, Clin. Chem., 15, 611 (1969).

130. M. Roth, Anal. Chem., 43, 880 (1971).

131. R. B. Conn, Clin. Chem., 6, 537 (1960).

132. E. Sawicki and R. Carnes, Anal. Chim. Acta, 41, 178 (1968).

133. G. G. Guilbault and J. Hieserman, Anal. Biochem., 24, 135 (1968).

134. C. Rogers, C. Chambers, and N. Clarke, Anal. Chem., 38, 1851 (1966).

135. G. G. Guilbault, M. H. Sadar, and K. Peres, Anal. Biochem., 31, 191 (1969).

136. G. G. Guilbault, P. Brignac, and M. Zimmer, Anal. Chem., 40, 190 (1968).

137. R. E. Phillips and F. R. Elevitch, Amer. J. Clin. Pathol., in press.

138. G. G. Guilbault and G. Lubrano, Anal. Chim. Acta, 43, 253 (1968).

139. T. Momose and Y. Ohkura, Talanta, 3, 155 (1959).

140. T. Momose and Y. Ohkura, Chem. Pharm. Bull. (Tokyo), 7, 31 (1959).

141. B. Bourne, Clin. Chem., 10, 1121 (1964).

142. P. Weber, I. Bornstein, and R. Winzler, Anal. Biochem., 14, 100 (1966).

143. M. Belcher, Anal. Biochem., 2, 30 (1961).

144. J. Towne and J. Spikner, Anal. Chem., 35, 211 (1963).

145. R. Sawamura and T. Koyama, Chem. Pharm. Bull. (Tokyo), 12, 706 (1965).

146. S. Adachi, Anal. Biochem., 9, 224 (1964).

147. G. Vurek and S. Pegram, Anal. Biochem., 16, 409 (1966).

148. G. Morrison, Anal. Biochem., 12, 150 (1965).

149. S. Udenfriend and P. Saltzman, Anal. Biochem., 3, 49 (1962).

150. P. Greengard, Nature, 178, 632 (1956).

151. II. Holmsen, I. Holmsen, and A. Bernhardsen, Anal. Biochem., 17, 456 (1966).

152. F. Kubowitz and P. Ott, Biochem. Z., 314, 94 (1943).

153. R. Wahl and L. Kozloff, J. Biol. Chem., 237, 1953 (1962).

154. G. Lyman and J. de Vincenzo, Anal. Biochem., 21, 435 (1967).

155. S. Yokoyama and Y. Nose, Seikagaku, 39, 46 (1967).

156. E. Beutler and M. Baluda, Blood, 23, 688 (1964).

157. L. Aledort, R. Weed, and S. Troup, Anal. Biochem., 17, 268 (1966).

158. E. Beutler and C. Mathai, Blood, 30, 311 (1967).

159. A. Kornberg, J. Biol. Chem., 182, 779 (1950).

160. P. Greengard, Photoelec. Spectry. Group Bull., 11, 292 (1958).

161. A. Kibrick and A. Milhorat, Clin. Chim. Acta, 14, 201 (1966).

162. L. Noda, S. Kuby, and H. Lardy, in Methods of Enzymology, Vol. 2 (S. Colowick and N. Kaplan, eds.), Academic Press, New York, 1955, p. 605.

163. B. Streher and J. Totter, in Methods of Biochemical Analysis, Vol. 1, Interscience, New York, 1954, p. 341.

164. B. Streher and J. Totter, Anal. Biochem. Biophys., 22, 420 (1949).

165. O. Warburg and W. Christian, Biochem. Z., 298, 150 (1938).

166. F. Straub, Biochem. J., 33, 787 (1939).

167. S. Udaka, J. Koukol, and B. Vennesland, J. Bacteriol., 78, 714 (1959).

168. E. Haas, B. Horecker, and T. Hogness, J. Biol. Chem., 136, 747 (1940).

169. E. Racker, J. Biol. Chem., 184, 313 (1950).

170. G. G. Guilbault and D. N. Kramer, Anal. Chem., 36, 2497 (1964).

171. O. H. Lowry, J. Passonneau, D. Schulz, and M. Rock, J. Biol. Chem., 236, 2746 (1961).

172. J. Cooper, P. Srere, M. Tabachnik, and E. Racker, Arch. Biochem. Biophys., 74, 306 (1958).

173. M. Klingeberg and W. Slenczka, Biochem. Z., 331, 486 (1959).

174. O. H. Lowry, N. R. Roberts, and J. I. Kapphahn, J. Biol. Chem., 224, 1047 (1957).

175. H. Börresen, Acta Chem. Scand., 17, 921 (1963).

176. O. Bessey, O. Lowry, and R. Love, J. Biol. Chem., 180, 755 (1949).

177. G. Weber, Biochem. J., 47, 114 (1950).

178. H. Burch, in Methods of Enzymology, Vol. III, (S. Colowick and N. Kaplan, eds.), Academic Press, New York, 1957, p. 960.

179. N. Kaplan, S. Colowick, and C. Barnes, J. Biol. Chem., 191, 461 (1951).

180. J. Huff and W. Perlzweig, J. Biol. Chem., 167, 157 (1947).

181. M. Rhodes and L. Wooltorton, Phytochemistry, 7, 337 (1968).

182. M. Tanzer and C. Gilvarg, J. Biol. Chem., 234, 3201 (1959).

183. D. Coursin and V. Brown, Proc. Soc. Exptl. Biol. Med., 98, 315 (1958).

184. V. Bonavita, Arch. Biochem. Biophys., 88, 366 (1960).

185. M. Yamada, A. Saito, and Z. Tamura, Chem. Pharm. Bull. (Tokyo), 14, 482 (1966).

186. M. Yamada, A. Saito, and Z. Tamura, Chem. Pharm. Bull. (Tokyo), 14, 488 (1966).

187. J. Longworth, Photochem. Photobiol., 8, 589 (1968).

188. J. LePacq and C. Paoletti, Anal. Biochem., 17, 100 (1966).

189. K. Van Dyke and C. Szostkiewicz, Anal. Biochem., 23, 109 (1968).

190. J. Kissane and E. Robins, J. Biol. Chem., 233, 184 (1958).

191. J. Finkel and K. Knapp, Anal. Biochem., 25, 465 (1968).

192. S. Udenfriend, D. Duggan, B. Vasta, and B. Brodie, J. Pharmacol. Exptl. Therap., 120, 26 (1957).

193. M. E. Auerbach and E. Angell, J. Pharm. Pharmacol., 10, 776 (1958).

194. W. Jusko, J. Pharm. Sci., 60, 728 (1971).

195. S. Sehgal, K. Singh, and C. Vezina, Anal. Biochem., 12, 191 (1965).

196. C. Miles and G. Schenk, Anal. Chem., 42, 656 (1970).

197. M. Auerbach and H. Eckert, J. Biol. Chem., 154, 597 (1944).

198. S. Ogawa, M. Morita, K. Nishiura, and K. Fujisawa, J. Pharm. Soc. Japan, 85, 650 (1965).

199. G. Carlle, V. Leclec-Chevalier, and J. Mackle, Clin. J. Pharm. Sci., 5, 55 (1970).

200. R. Bowman, P. Caulfield, and S. Udenfriend, Science, 122, 32 (1955).

201. J. Gillette, J. Dingell, F. Sulser, R. Kuntzman, and B. Brodie, Experientia, 17, 377 (1961).

202. J. Dingell, F. Sulser, and J. Gillette, J. Pharmacol. Exptl. Therap., 143, 14 (1964).

203. J. Goodyear and N. Jenkinson, Anal. Chem., 32, 1203 (1960).

204. Acta Pharmacol. Toxicol., 8, 101 (1952).

205. A. Glazko, W. Dill, and R. Fransway, Federation Proc., 21, 269 (1962).

206. P. Labadie, Pathol. Biol. Semaine Hop., 12, 24 (1964).

207. B. Davis, M. Dodds, and E. Tomich, J. Pharm. Pharmacol., 14, 249 (1962).

208. S. Roston, Anal. Chem., 30, 1363 (1958).

209. R. Williams, in Spectrophotofluorimetric Techniques in Biology, NATO Advanced Studies Institute, Milan, 1964, p. 223.

210. H. Strickler and P. Stanchak, Clin. Chem., 15, 137 (1969).

211. J. Peters, Amer. Res. Respir. Diseases, 81, 485 (1960).

212. M. Hedrick, J. Rippon, L. Decker, and V. Bernsohn, Anal. Biochem., 4, 85 (1962).

213. L. Dal, M. Cortivo, and S. Weinberg, Anal. Chem., 42, 941 (1970).

214. P. Shore and H. Alpers, Life Sci., 3, 551 (1964).

215. H. Kupferberg, A. Burkhalter, and E. Way, J. Pharmacol. Exptl. Therap., 145, 247 (1964).

216. J. Scudi, J. Biol. Chem., 164, 183 (1946).

217. J. Scudi, J. Biol. Chem., 164, 195 (1946).

218. S. Tompsett, Acta Pharmacol. Toxicol., 26, 298 (1968).

219. G. Carlle, C. Sauriol, J. Mackle, and J. Panisset, Can. J. Pharm. Sci., 5, 72 (1970).

220. B. A. Persson, Acta Pharm., 7, 337 (1970).

221. E. Pearlman, J. Pharmacol. Exptl. Therap., 95, 465 (1949).

222. E. DeRitter, F. Jahns, and S. Rubin, J. Amer. Pharm. Assoc., Sci. Ed., 38, 319 (1949).

223. A. Spinks, Biochem. J., 47, 299 (1950).

224. A. Edgar and M. Sokolov, J. Lab. Clin. Med., 36, 478 (1950).

225. M. Chirigos and J. Mead, Anal. Biochem., 7, 259 (1964).

226. F. Faure and P. Blanquet, Clin. Chem. Acta, 9, 292 (1964).

227. R. Williams, in Spectrophotofluorimetric Techniques in Biology, NATO Advanced Studies Institute, Milan, 1964, p. 247.

228. R. Kelly, L. Peets, and K. Hoyt, Anal. Biochem., 28, 222 (1969).

229. D. Schwartz, B. Koechlin, and R. Weinfeld, Chemotherapy, 14, 22 (1969).

230. S. Udenfriend, Fluorescence Assay in Biology and Medicine, Academic Press, New York, Vol. 1, 1962, Chapter 13; Vol. 2, 1969, Chapter 17.

231. D. MacDougall, Residue Rev., 1, 24 (1962).

232. D. MacDougall, Residue Rev., 5, 119 (1964).

233. J. Kanora and C. Szalkowski, J. Assoc. Off. Agr. Chem., 47, 209 (1964).

234. C. Szalkowski, J. Assoc. Off. Agr. Chem., 48, 285 (1965).

235. P. Giang, J. Agr. Food Chem., 9, 42 (1961).

236. H. Pease and J. Gardiner, J. Agr. Food Chem., 17, 267 (1969).

237. H. Borfitz, J. Para, J. Stickles, G. Gunther, and B. Southworth, J. Assoc. Off. Agr. Chem., 50, 264 (1967).

238. R. Argauer, H. Shimanuki, and C. Alvarez, Abstracts 158th National ACS Meeting, New York, Sept. 1969.

239. C. Anderson, J. Adams, and D. MacDougall, J. Agr. Food Chem. , 7, 256 (1959).

240. J. Adams and C. Anderson, J. Agr. Food Chem. , 14, 53 (1966).

241. H. Bieber, Dt. Lebensmitt. Rdsch. , 63, 44 (1967).

242. R. Anderson, C. Anderson, and M. Yagelowich, J. Agr. Food Chem. , 14, 43 (1966).

243. U. Kugernagi and L. Terriere, J. Assoc. Off. Agr. Chem. , 11, 293 (1963).

244. H. Moye and J. Windfordner, J. Agr. Food Chem. , 13, 516 (1965).

245. M. Ragab, J. Assoc. Off. Agr. Chem. , 50, 1089 (1967).

246. M. Bowman and M. Beroza, Residue Rev. , 17, 23 (1966).

247. C. Dhéré, La fluorescence en biochimie, Presses Universitaires, Paris, 1937.

248. S. Schwartz, M. Berg, I. Bossenmauer, and H. Dinsmore, Methods Biochem. Anal. , 8, 221 (1960).

249. K. N. Solov'ev, Opt. Spectry. (English transl.), 10, 389 (1961).

250. K. Solov'ev, S. Shkirman, and T. Kachura, Izv. Akad. Nauk SSSR, Ser. Fiz. , 27, 767 (1963).

251. W. Runge, Science, 15, 1499 (1966).

252. C. Martinez and G. Mills, Clin. Chem. , 17, 199 (1971).

253. M. Doss, Z. Anal. Chem. , 252, 104 (1970).

254. M. Roth, Clin. Chim. Acta, 17, 487 (1967).

255. R. Rodriguez-Garay and T. Argerich, J. Lab. Clin. Med. , 62, 141 (1963).

256. L. Wranne, Acta Paediat. Suppl. , 124, 1 (1960).

257. J. Schwartz, V. Hawkinson, J. Cohen, and C. Watson, J. Biol. Chem. , 168, 133 (1947).

258. R. Nevé and R. Aldrich, Pediatrics, 15, 553 (1955).

259. L. Eriksen, J. Clin. Lab. Invest. , 10, 319 (1958).

260. G. Sweeney and L. Eales, Scand. J. Clin. Lab. Invest. , 16, 250 (1964).

261. G. Morrison, Anal. Chem. , 37, 1124 (1965).

262. S. Schwartz and H. Wikoff, J. Biol. Chem. , 194, 563 (1952).

263. A. Adler, Dent. Arch. Klin. Med. , 138, 309 (1922).

264. H. Kirkpatrick, Lancet, i, 71 (1953).

265. N. Schmidt and R. Scholtis, Clin. Chim. Acta, 10, 574 (1964).

266. S. Schwartz, Veterans Administration Tech. Bull. TB 10-94, Washington, D. C., 1953.

267. E. L. Talman, in Standard Methods of Clinical Chemistry, (D. Seligson, ed.), Vol. 2, Academic Press, New York, 1958, p. 137.

268. D. Hsia and T. Inouye, Inborn Errors in Metabolism, Part 2, Year Book, Chicago, 1966.

269. H. Braunsberg and V. James, Anal. Biochem., 1, 452 (1960).

270. H. Kalant, Biochem. J., 69, 93 (1958).

271. D. Abelson and P. Bondy, Arch. Biochem. Biophys., 57, 208 (1955).

272. M. Pesez, Ann. Pharm. France, 11, 670 (1953).

273. S. Levin, J. Irvin, and C. Johnston, Anal. Chem., 33, 856 (1961).

274. R. Albers and O. Lowry, Anal. Chem., 27, 1829 (1955).

275. E. Solow and L. Freeman, Clin. Chem., 16, 472 (1970).

276. J. Goldzieher and P. Besch, Anal. Chem., 30, 962 (1958).

277. J. Touchstone, J. Greene, and W. Kukovetz, Anal. Chem., 31, 1693 (1959).

278. B. Clark and R. Rubin, Anal. Biochem., 29, 31 (1969).

279. J. Jansen, E. Hvidberg, and J. Schou, Scand. J. Clin. Lab. Invest., 20, 49 (1967).

280. J. McLaughlin, T. Kaniecki, and I. Gray, Anal. Chem., 30, 1517 (1958).

281. Tokyo Jikeikai Ika Daigaku Zasshi, 72, 505 (1957).

282. M. Finkelstein, E. Forchielli, and R. Dorfman, J. Clin. Endocrinol., 21, 98 (1961).

283. A. Fazekas and J. Webb, Can. J. Biochem., 45, 1479 (1967).

284. S. Udenfriend, Fluorescence Assay in Biology and Medicine, Academic Press, New York, Vol. 1, 1962, Chapter 10; Vol. 2, 1969, Chapter 14.

285. A. Passannante and L. Avioli, Anal. Biochem., 15, 287 (1966).

286. J. Kahan, Scand. J. Clin. Lab. Invest., 18, 679 (1966).

287. L. Hansen and W. Warwick, Amer. J. Clin. Pathol., 50, 525 (1968).

288. P. Garry, J. Pollack, and G. Owen, Clin. Chem., 16, 766 (1970).

289. L. Hansen and W. Warwick, Amer. J. Clin. Pathol., 51, 538 (1969).

290. U. S. Pharmacopeia, 16th rev., Mack, Easton, Pa., 1960, p. 909.

291. Official Methods of Analysis, 9th ed., Assoc. Off. Agr. Chemists,
      Washington, D. C., 1960, p. 655.

292. H. Burch, in Methods of Enzymology (S. Colowick and N. Kaplan, eds.),
      Vol. III, Academic Press, New York, 1957, p. 946.

293. A. Fujita, Y. Nose, K. Ueda, and H. Eiichi, J. Biol. Chem., 196, 297
      (1952).

294. K. Giri and S. Balakrishnan, Anal. Chem., 27, 1178 (1955).

295. W. Ohnesorge and L. Rogers, Anal. Chem., 28, 1017 (1956).

296. Official Methods of Analysis, 10th ed., Assoc. Off. Agr. Chemists,
      Washington, D. C., 1965, p. 762.

297. K. Yagi, Bull. Soc. Chem. France, 1543 (1957).

298. S. Reddy, M. Reynolds, and J. Price, J. Biol. Chem., 233, 691 (1958).

299. J. W. Bridges, D. Davis, and R. T. Williams, Biochem. J., 98, 451
      (1966).

300. E. Toepfer, M. Polansky, and E. Hewston, Anal. Biochem., 2, 463
      (1961).

301. S. Udenfriend, Fluorescence Assay in Biology and Medicine, Academic
      Press, New York, Vol. 1, 1962, Chapter 7; Vol. 2, 1969, Chapter 8.

302. H. Barker, R. Smyth, H. Weissbach, J. Tooley, J. Ladd, and B.
      Volcani, J. Biol. Chem., 235, 480 (1960).

303. H. Weissbach, J. Tooley, and H. Barker, Proc. Natl. Acad. Sci. U.S.,
      45, 521 (1959).

304. R. Archibald, J. Biol. Chem., 158, 347 (1945).

305. U. Imhoff, Ernaehrungswiss., 5, 135 (1964).

306. M. Deutsch and C. Weeks, J. Assoc. Off. Agr. Chem., 48, 1248 (1965).

307. M. Polansky, R. Camarra, and E. Toepfer, J. Assoc. Off. Agr. Chem.,
      47, 827 (1964).

308. S. Jones, J. Wilkie, W. Morris, and L. Friedman, Abstracts 138th
      Meeting ACS, New York, 1960, p. 600.

309. P. Chen, A. Terepka, and K. Lane, Anal. Biochem., 8, 34 (1964).

310. P. Chen, Anal. Chem., 37, 301 (1965).

311. D. Duggan, Arch. Biochem. Biophys., 84, 116 (1959).

312. L. Hansen and W. Warwick, Amer. J. Clin. Pathol., 46, 133 (1966).

313. M. Kofler, Helv. Chim. Acta, 28, 26 (1945); 30, 1053 (1947).

314. R. Beyer and B. Kennison, Arch. Biochem. Biophys., 84, 63 (1959).

315. M. Kofler, Helv. Chim. Acta, 28, 702 (1945).

Chapter 8

FLUORESCENCE IN ENZYMOLOGY

I.   INTRODUCTION

A.   General Considerations

Enzymes are biological catalysts that enable the many complex chemical reactions, on which depends the very existence of life as we know it, to take place at ordinary temperatures.   One of the outstanding properties of enzymes is specificity.   An enzyme is capable of catalyzing a particular reaction of a particular substrate even though other isomers of that substrate or similar substrates may be present.

An example of the specificity of enzymes with respect to a particular substrate is found in luciferase, which catalyzes the oxidation of luciferin (106) to oxyluciferin [1].   A rather complete study of many compounds similar in structure to luciferin showed that the catalytic oxidation resulting in the production of the green luminescence occurs only with luciferin.   Substitution of an amino group for a hydroxyl group or addition of another hydroxyl group to the luciferin molecule alters the enzymatic action, and the green luminescence is not produced.

Another example of the specificity of enzymes is glucose oxidase, which

$$\text{(106)} + O_2 \xrightarrow[\text{Mg}^{2+},\ \text{ATP}]{\text{luciferase}} \text{oxyluciferin} + ADP \quad (104)$$

catalyzes the oxidation of ß-D-glucose to gluconic acid.   A rather complete study of about 60 oxidizable sugars and their derivatives showed that only 2-deoxy-D-glucose is oxidized at a rate comparable with that of ß-D-glucose. The anomer α-D-glucose is oxidized less than 1% as rapidly as the ß-anomer [2].   Urease, which catalyzes the hydrolysis of urea, is even more specific.

Enzymes exhibit specificity with respect to a particular reaction.   If one attempted to determine glucose by uncatalyzed oxidation, for example, by heating a solution of glucose and an oxidizing agent like ceric perchlorate, other side reactions would occur uncontrollably to yield products in addition to gluconic acid.   With glucose oxidase, however, catalysis is so effective at room temperature and a pH near 7 that the rates of the other thermodynamically possible reactions are negligible.

This specificity of enzymes and their ability to catalyze reactions of substrates at low concentrations are of great use in chemical analysis. Enzyme-catalyzed reactions have been used for a long time for the determination of substrates, activators, inhibitors, and also of enzymes themselves. Until recently, however, the disadvantages associated with the use of enzymes have seriously limited their usefulness. Frequently cited objections to the use of enzymes for analytical purposes have been their unavailability and instability, and the poor precision and laboriousness of the analyses. Although these objections were valid earlier, numerous enzymes are now available in purified form, with high specific activity, at reasonable prices. The instability of enzyme is, of course, always a potential hazard; yet, if this instability is recognized and reasonable precautions are taken, the difficulty may be minimized. Again, the poor precision, slowness, and labor that have made enzyme-catalyzed reactions unappealing as a means of analysis may be more a consequence of the methods and techniques than the fault of the enzymes. With the advent of new techniques, fluorometric and electrochemical, many of the previous difficulties have been resolved. Moreover, the automation of enzymatic reactions has increased the speed, ease, and reproducibility of assays utilizing enzymes.

Another problem in the use of enzymes in analytical chemistry lies in the cost of using large amounts of these materials, especially in routine analysis. This problem has been solved to some extent by the development of an immobilized (insolubilized) enzyme technique allowing continuous use of the enzyme for several days.

The following is a Michaelis-Menten equation for the enzyme kinetics:

$$E + S \xrightleftharpoons[k_{-1}]{k_1} ES \xrightarrow{k_2} E + P \qquad (105)$$

In this mechanism the substrate S combines with enzyme E to form an intermediate complex ES, which subsequently breaks down into products P and liberates the enzyme. The equilibrium constant for the formation of the complex, K_m, the Michaelis constant, is defined as

$$K_m = \frac{k_2 + k_{-1}}{k_1}. \qquad (106)$$

The rate of reaction V_0 is then some function of the enzyme and substrate (see Eq. (8)) as well as of activator and inhibitor concentration, if the latter two are present. At a fixed enzyme concentration

$$V_0 = \frac{V_{max}[S]_0}{K_m + [S]_0}. \qquad (107)$$

The initial rate increases with substrate until a nonlimiting excess of substrate is reached, after which additional substrate causes no increase in rate.

The concentration of material participating in an enzymatic reaction can be calculated in one of two ways: (a) by measuring the total change that occurs by chemical, physical, or enzymatic analysis of the product or unreacted starting material; or (b) from the rate of the enzyme reaction. In the first method large amounts of enzyme and small amounts of substrate are used to ensure a relatively complete reaction. The reaction is allowed to reach equilibrium, and the amount of substrate S in the sample can be calculated from the amount of P formed (S $\xrightarrow{\text{E}}$ P). The product P is chemically and physically distinguishable from S; for example,

$$\text{ethanol} + \text{NAD} \xrightarrow{\text{alcohol dehydrogenase}} \text{acetaldehyde} + \text{NADH} \qquad (108)$$

The molar absorptivity of NADH is $6.22 \times 10^6 \text{ cm}^2/\text{mole}$ at 340 nm.

Alternatively, a coupled reaction can be used to indicate how much substrate has been decomposed:

Enzyme reaction:

$$\text{glucose} + H_2O + O_2 \xrightarrow{\text{glucose oxidase}} \text{gluconic acid} + H_2O_2 \qquad (109)$$

Indicator reaction:

$$H_2O_2 + \text{leuco dye} \xrightarrow{\text{peroxidase}} H_2O + \text{dye} \qquad (110)$$

The intensity of the dye produced is a measure of the concentration of glucose present.

In the second method, the kinetic method, the initial rate of reaction V_0 is measured in one of many conventional ways, by following either the production of product or the disappearance of the substrate. The rate is a function of the concentrations of the substrate [S], enzyme [E], inhibitor [I], and activator [A]. For example, the concentration of glucose can be determined by measuring the initial rate of production of dye in reactions (109) and (110).

Because it is more reliable, the total-change method is generally favored over the rate method. However, the former technique can only be used for substrate analysis and not for determinations of [E], [A], and [I] because the effects of these are catalytic in nature, so that they affect only the rate and not the equilibrium. Furthermore the rate method is faster because the rate can be measured inititially without having to wait for the reaction to go to completion. The conditions that affect the rate (pH, temperature, ionic strength) must be carefully controlled in the kinetic method for maximum sensitivity. The temperature coefficient of the enzyme reaction rate is roughly 10% per $^\circ$C [3], and a $10^{\circ}C$ rise in temperature causes a 100% increase in the reaction rate. Hence constant temperature is essential in the assay of enzyme activity. Recent work by Guilbault et al. [4, 5] and Pardue et al. [6] has indicated that, with reasonable care, precision and accuracies of better than 1% can be obtained. Furthermore some of the difficulties encountered because of side reactions are eliminated in rate methods, and greater sensitivities can be obtained in many cases. With the automated equipment now available for performing rate methods, such techniques will probably be the ones of choice in the future.

B. Books and Reviews

Guilbault [7] has authored a book, Enzymatic Methods of Analysis, that includes sections on the immobilization of enzymes and the automation of enzymatic reactions, in addition to a discussion of the uses of enzymes in the analysis of substrates, activators, and inhibitors.

The use of fluorescence in the assay of enzymes and substrates is discussed in books by Purdy [8], Guilbault [9], and Udenfriend [10]. Ruyssen and Vandenriesche [11] have published a book on enzymes in clinical chemistry, and Phillips and Elevitch [12] have written a chapter on fluorometric techniques in clinical pathology in Steffani's Progress in Clinical Pathology. Both of these contain valuable procedures for enzyme analysis. Bergmeyer's Methods of Enzymatic Analysis [13] contains practical details of a vast range of enzymic methods. Blaedel and Hicks [14] have written a chapter on the analytical applications of enzyme-catalyzed reactions in Reilley's book.

Reviews on enzymatic analysis have been written by Guilbault [15], Bergmeyer [16], Devlin [17], and Roth [18]. Oldham [19] has reviewed radio-chemical methods for enzyme assay.

C. Sources of Reagents

A comprehensive list of the sources of all commercially available enzymes was compiled by Guilbault [7]. There are over 200 different enzymes available today, from several companies, including Boehringer (Mannheim,Germany), Sigma (St. Louis, Mo.), Calbiochem (Los Angeles), Mann Laboratories (New York), Worthington Biochemical (Freehold, N.J.), Nutritional Biochemical (Cleveland), General Biochemical (Chagrin Falls, Ohio), Gallard-Schlesinger (Long Island, N.Y.), Merck (Darmstadt, Germany), and Miles Labs (Elkhart, Ind.). Chromogenic substrates are offered by many of these; fluorogenic substrates are available from only a few companies, such as Isolab, Inc. (Akron, Ohio), which offers a full line of these compounds.

II. DETERMINATION OF ENZYMES

A. Introduction

Since the enzyme is a catalyst, theoretically one molecule of this material would eventually produce a sufficient change in the substrate to be measured. Hence high sensitivities can be realized in enzyme analysis. Because the concentration of enzyme is so small, it always limits the rate of reaction, and the rate can be taken as a measure of the enzyme concentration. In Eq. (109) the oxidation of glucose by oxygen to give peroxide and gluconic acid is catalyzed by glucose oxidase. The rate of peroxide production is measured by a second coupled reaction, the oxidation of a leuco dye, such as o-dianisidine, to yield a highly colored dye. When glucose, leuco dye, and oxygen are not rate limiting, the overall rate of reaction, as indicated by the rate of dye production, is proportional to the glucose oxidase activity.

B. Fluorescence Methods

Because of limitations in molar absorptivities, measurements of gas volumes, or of changes in pH, most methods previously described for measuring components in enzyme reactions are limited to reactions of reagents present at concentrations greater than 10^{-6} M. Because fluorometric methods are generally several orders of magnitude more sensitive than chromogenic ones, a large increase in the sensitivity of measurement should result. Thus much lower concentrations of reactants would be needed, and one could devise methods for substances at 10^{-9} M concentrations and lower. Moreover fluorometric methods are quite useful in biochemical work, in the localization of enzymes and related substrates (activators), within organs, and even within individual cells.

Because of their sensitivity and specificity, fluorescent methods have found increasing use in enzymology. For example, NADH and NADPH, the reduced forms of nicotinamide adenine dinucleotide (NAD) and nicotinamide adenine dinucleotide phosphate (NADP), are highly fluorescent. Thus all NAD- and NADP-dependent reactions involved in enzymatic analysis can be measured fluorometrically, with an increase of two to three orders of magnitude in sensitivity over colorimetric techniques [20].

In fluorometric assay methods for enzymes generally no fluorescence is initially observed. On addition of enzyme the fluorescence increases (Fig. 105). The rate of change in the fluorescence with time, $\Delta F/min$, is proportional to the concentration of enzyme.

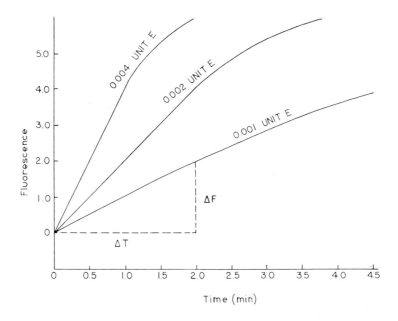

FIG. 105.   Rate of increase in fluorescence with time on addition of enzyme.

1.   Hydrolytic Enzymes

Fluorescence methods have also been used extensively for the determination of hydrolytic enzymes, based on the enzyme-catalyzed hydrolysis of a nonfluorescent ester to a highly fluorescent alcohol or amine.

a.   Cholinesterase.   Guilbault and Kramer have described four new fluorogenic substrates that can be used for a rapid and specific determination of cholinesterases: resorufin butyrate [21], indoxyl acetate [22], and 1- and 2-naphthyl acetate [23].   All of these are nonfluorescent, but they are hydrolyzed by cholinesterase to highly fluorescent compounds.

The substrates resorufin acetate and butyrate are hydrolyzed by cholinesterase, acid phosphatase, and chymotrypsin to the highly fluorescent resorufin (λ_{ex} 540, λ_{em} 580 nm).   The rate of resorufin production with time is proportional to the concentration of cholinesterase from 0.0001 to 0.123 unit/ml, with a deviation of only 1.0%.   By the choice of substrates, some specificity is possible.   For example, resorufin acetate is not hydrolyzed by lipase nor by acetylcholinesterase (from bovine erythrocytes or electric eel), but resorufin butyrate is hydrolyzed by lipase.   Similar concentrations of cholinesterase can be determined with a slightly lower deviation (0.9%) by using indoxyl acetate as substrate.   Various authors [24] have reported the

colorimetric assay of cholinesterase, based on the formation of indigo blue
from indoxyl acetate. Kramer and Gelman [25] and Gehauf and Goldenson [26]
have reported that indoxyl is a highly fluorescent compound and is easily
air-oxidized to indigo blue. Guilbault and Kramer [22] found that indoxyl
acetate could be used as a fluorogenic substrate for cholinesterase, by a proper
control of experimental conditions. Indoxyl acetate is hydrolyzed first to
indoxyl, which is fluorescent. Air then rapidly effects the oxidation of indoxyl
to indigo white, which is twice as fluorescent as indoxyl, and then to the
nonfluorescent indigo blue. An oxygen scavenger, ascorbic acid, prevents the
oxidation of indoxyl, so that one can obtain a stable fluorescent product.
However, in the absence of ascorbic acid below pH 7 indigo white forms and is
not oxidized to indigo blue. These results are supported by the research of
Gehauf and Goldenson [26]. Acylase, lipase, and acid phosphatase also catalyze
the hydrolysis of indoxyl acetate and are determinable by this procedure.

Chymotrypsin does not effect the hydrolysis and hence can be differentially
determined in the presence of cholinesterase by using two substrates, such as
indoxyl acetate and resorufin butyrate.

In attempting to develop an automatic procedure for monitoring anticholinesterase compounds in the air the author encountered many difficulties in using
resorufin butyrate and indoxyl acetate as substrates. These substrates, though
stable in air under ordinary laboratory conditions, were rapidly decomposed in
the presence of the large volumes of air sampled in a continuous device.

Naphthyl esters (107) have been used as colorimetric substrates for years
[27-29]. Since Hercules and Rogers [30] have shown that 1-naphthol (108a) and
2-naphthol (108b) are fluorescent, it seemed logical that the naphthyl esters
could serve as fluorogenic substrates for esterases.

$$(107) \qquad\qquad (108) \qquad\qquad (111)$$

a: $R^1 = OCOCH_3, OCOC_3H_7;$ a: $R^1 = OH ; R^2 = H$
   $R^2 = H$

b: $R^1 = H ; R^2 = OCOCH_3,$ b: $R^1 = H ; R^2 = OH$
   $OCOC_3H_7$

In order to determine which substrate would be best for the assay of
cholinesterase (ChE) the various 1- and 2-naphthyl esters were compared with
respect to stability, rate of enzymatic hydrolysis, and fluorescence of the final
product. The latter was determined by recording the fluorescence coefficient
(fluorescence reading on the Aminco-Bowman spectrofluorometer divided by the
concentration in moles per liter). Average values were obtained by using
several concentrations of pure material.

The rate of increase in fluorescence was found to be proportional to the
cholinesterase concentration in the range 0.0003 to 0.12 unit/ml, with a deviation of 1%, using 2-naphthyl acetate as substrate. Acylase, lipase, and
phosphatase also catalyze the hydrolysis of 2-naphthyl acetate and can be
determined in concentrations of about 0.0001 to 0.10 unit/ml with a deviation
of about 1.2%.

All values of fluorescence coefficients were referred to a reference value
of 1.4×10^6 using quinine sulfate in 0.1 M sulfuric acid (λ_{ex} 350, λ_{em} 450 nm).
The fluorescence coefficients obtained for 1- and 2-naphthol were 1.0×10^3 and
1.2×10^4, respectively, at a pH of 7.40. This can be compared with values of
1.6×10^7 and 1.0×10^5 for resorufin and indigo white, respectively. Overall,
2-naphthyl acetate was found to be the best substrate (of the naphthyl esters) for
cholinesterase.

Guilbault et al. [31] prepared several esters as substrates for cholinesterase: the acetate, propionate, and butyrate esters of N-methylindoxyl,
umbelliferone, and 4-methylumbelliferone. Comparison of these substrates
with other fluorogenic esters (indoxyl acetate, indoxyl butyrate, resorufin
acetate, 2-carbonaphthoxycholine, and 2-naphthyl acetate) indicated that N-
methylindoxyl acetate and butyrate were the best substrates for true and pseudo-
cholinesterase, respectively. Analysis of as little as 5×10^{-5} unit/ml of
cholinesterase can be performed by a direct initial-reaction-rate method in
2 to 3 min with an accuracy and precision of about 1.5%. All three N-methylindoxyl esters are very stable in solution, with a very low rate of spontaneous
hydrolysis and a high rate of enzymatic hydrolysis. All have good K_m values,
and the N-methylindoxyl formed is not easily air-oxidized to indigo derivatives.

A substrate for the fluorescence assay of acetylcholinesterase was reported
by Prince [32]. The compound, 1-methyl-7-acetoxyquinolinium iodide, is
hydrolyzed by acetylcholinesterase from a variety of sources to the highly
fluorescent 1-methyl-7-hydroxyquinolinium iodide (λ_{ex} 406, λ_{em} 505 nm at
pH 7.0). The latter can be detected at a concentration of 3×10^{-9} M making it
possible to assay quantities of the enzyme as low as 3×10^{-4} unit; 1 unit
catalyzes the hydrolysis of 1 micromole of acetylcholine per minute. There are
two problems with this substrate. There is a spontaneous hydrolysis leading to

fluorescence. This becomes almost negligible with a suitable choice of
substrate concentration and pH. The second problem pertains to specificity.
Since other hydrolytic enzymes attack this substrate, it cannot be used for the
specific assay of acetylcholinesterase.

   b. Cellulase. Guilbault and Heyn [33] tested several fluorogenic substrates
for cellulase: fluorescein dibutyrate, 1- and 2-naphthyl acetate, indoxyl
acetate, and resorufin butyrate. The latter is cleaved by cellulase to the highly
fluorescent resorufin (λ_{ex} 540, λ_{em} 560 nm). By the use of this substrate,
from 0.00010 to 0.060 unit of cellulase can be determined in only 2 min with an
accuracy and precision of 1.5%. This compares most favorably with other
methods, which take 1 to 5 h and have a precision and accuracy of only 5 to 10%.

   c. ß-D-Galactosidase. Rotman [34] reported an extremely sensitive
procedure for the fluorometric assay of ß-D-galactosidase activity. The
substrate was reported to be 6-hydroxyfluoran-ß-D-galactopyranoside, which on
enzymatic hydrolysis supposedly yielded the highly fluorescent 6-hydroxyfluoran.
Further investigation by Rotman et al. [35] indicated that the substrate used in
the earlier studies [34] was really fluorescein-di-(ß-D-galactopyranoside) (109),
and the product liberated by galactosidase action was fluorescein:

(109)

   The use of fluorescein-di-(ß-D-galactopyranoside) as a substrate for
galactosidase offers many advantages when the assay is carried out under the
usual laboratory conditions. Furthermore Rotman [34] was able to demonstrate
the unique applications that are made possible by the extremely high sensitivity
of fluorometric assay. In this study he developed methods for measuring the
activity of individual molecules of the enzyme.

   Tydell and Scheparty [36] measured ß-D-galactosidase activity in cell
cultures by a fluorometric procedure.

   d. N-Acetyl-ß-D-glucosaminidase. Verity et al. [37] have devised a
sensitive assay for N-acetyl-ß-D-glucosaminidase (found in lysosomes, where
it plays a role in the stepwise degradation of hyaluronic acid breakdown
products) by fluorometric measurement of the 2-acetamido-1-naphthol liberated

from the synthetic substrate 1-naphthyl-2-acetamido-2-deoxy-D-gluco-
pyranoside.

e. ß-Glucosidase. ß-Glucosidase is widely distributed in nature and splits
a variety of ß-glucosides. The use of a glucoside of the highly fluorescent
umbelliferone makes for an extremely sensitive procedure for the assay of this
enzyme [38].

By using this procedure Robinson [38] showed that the enzyme is widely
distributed in plants and animals. The fluorescent procedures were also used
in partially successful attempts to separate ß-glucosidase from ß-glucuronidase
by paper electrophoresis. The sensitivity of this fluorometric procedure is
again so high that less than milligram quantities of tissues are sufficient for
assay [38].

f. ß-Glucuronidase. Umbelliferone (7-hydroxycoumarin (110)) and
4-methylumbelliferone are highly fluorescent compounds. In the body they are
converted to the corresponding glucuronides, which are nonfluorescent. In the
presence of ß-glucuronidase the glucuronides are split to release the free
fluorescent products [39].

$$(C_6H_9O_6)O\text{—}\!\!\!\!\xrightarrow{\ \beta\text{-glucuronidase}\ }\ HO\text{—}\ +\ C_6H_{10}O_7 \qquad (112)$$

(110)

The glucuronide of 4-methylumbelliferone can be prepared in excellent
yield by feeding the free compound (commercially available) to rabbits and
isolating the derivative from the urine. The procedure described by Mead et
al. [39] gives a 20 to 25 % yield (isolated) from an administered dose of 2.5 g.

Verity et al. [40] devised a method for ß-glucuronidase assay in which the
glucuronide of 1-naphthol is used as substrate. After incubation of 1-naphthyl-
ß-D-glucuronide at pH 4 to 5 with tissue extract, the solution is chilled, and
made alkaline to about pH 11.0 by the addition of an appropriate volume of
0.5 N NaOH; fluorescence is then assayed. The free 1-naphthol formed during
incubation is excited maximally at 340 nm and fluoresces at 460 nm. The
glucuronide is also fluorescent but is maximally excited at about 310 nm and
emits in the ultraviolet at 345 nm.

Greenberg [41] utilized the α- and ß-isomers of 2-naphthyl-D-glucuronide
as substrates of purified beef-liver ß-glucuronidase. He found that the
ß-isomer was several times more active as substrate and had a K_m of
4×10^{-5} M. Enzyme assay is carried out by measuring the appearance of
fluorescence at 410 nm, with excitation at 350 nm (pH $>$ 11).

g. Hyaluronidase. Still another example of a complex, tedious, and inaccurate procedure that can be easily replaced with a fast, highly sensitive and accurate fluorometric method is that of hyaluronidase assay. Guilbault et al. [42] proposed the use of indoxyl acetate as a substrate for this enzyme. The highly fluorescent indigo white is produced on hydrolysis, and as little as 10^{-3} μg of enzyme can be assayed in 1 to 2 min with an accuracy and precision of 1.8%.

h. Lipase. Guilbault and Kramer [43,44] described a rapid and simple method for the determination of lipase based on its catalysis of the hydrolysis of the nonfluorescent dibutyryl ester of fluorescein:

$$\text{dibutyrylfluorescein} \xrightarrow{\text{lipase}} \text{fluorescein} \qquad (113)$$

This reaction can be monitored by measuring the rate at which the highly fluorescent fluorescein is produced with time. The concentration of enzyme can then be calculated from linear calibration plots of $\Delta F/\text{min}$ versus enzyme concentration.

In a thorough study of fluorometric substrates for lipase Guilbault and Sadar [45] evaluated 12 different compounds from the aspects of stability, spontaneous hydrolysis, enzymatic hydrolysis, Michaelis constant for the enzyme-substrate complex, and total fluorescence of the final product. Optimum conditions of analysis were found for all substrates, and the lowest detectable enzyme concentration was found for each substrate. From all aspects, 4-methylumbelliferone heptanoate was found to be the best substrate for pig-pancreas lipase, and 4-methylumbelliferone octanoate was best for fungal lipase. As little as 1×10^{-5} unit could be determined by a direct reaction-rate method, with an accuracy and precision of about 1.5%.

Guilbault and Hieserman [46] prepared several new fluorometric substrates for the assay of lipase. A study of six N-methylindoxyl esters as substrates for lipase indicated N-methylindoxyl myristate to be best. By using this ester, from 0.0002 to 4.0 unit/ml of pig pancreas can be determined, in the presence of several other esterases, with an accuracy and precision of about 1.5%. Analysis is performed by a direct initial-reaction-rate method in 2 to 3 min.

i. Nucleases. A fluorometric assay for nucleases has been suggested by Stevens [47], who synthesized sulfonyl chlorides of fluorescent compounds and showed that they reacted with the amino groups of deoxyribonuclease to yield nondialyzable fluorescent derivatives. Stevens pointed out that should such a fluorescent derivative be acted on by deoxyribonuclease, it will yield dialyzable fluorescent products as a measure of the nuclease activity.

j. Peptidase. Smith and Hill [48] utilized naphthylamine-containing peptides to demonstrate aminopeptidase activity by fluorescence microscopy. Subsequently Greenberg [41] reported the use of phenylalanyl-ß-naphthylamide for the fluorometric assay of aminopeptidase activity in solution. By carrying out the reaction in a 0.1-ml volume at pH 8.0 (tris buffer) it was possible to measure as little as 10^{-12} mole of aminopeptidase by following the appearance of naphthylamine (λ_{ex} 335, λ_{em} 410 nm). Subsequently Roth [49] utilized L-leucyl-ß-naphthylamide to assay leucine aminopeptidase in kidney extracts, serum, urine, and duodenal juice.

Oxytocinase, an enzyme that cleaves oxytocin at the cystinyl-tyrosine bond, also splits cystine-di-ß-naphthylamide to yield free naphthylamine. Roth [49] adapted this reaction to the fluorometric assay of oxytocinase in the serum of pregnant women.

k. Proteases. Riedel and Wünsch ['50] introduced benzoylarginine-ß-naphthylamide as a substrate for trypsin assay. In their studies the naphthylamine formed as a result of tryptic activity was assayed colorimetrically. More recently Roth [51] used the same substrate to develop a highly sensitive fluorometric procedure for trypsin assay. Assay of trypsin activities equivalent to 1 µg of the crystalline enzyme is relatively simple and can be carried out kinetically using incubation times of 1 to 5 min. The method can be used to measure much smaller amounts of enzyme. Fluorescence is measured at a λ_{ex} of 338 and a λ_{em} of 410 nm.

Trypsin can be assayed by its esterolytic activity with the substrate p-tosylarginine methyl ester hydrochloride (TAME). The methanol that is' released can be assayed by a variety of procedures. Sardesai and Provido [52,53] determined the enzymatically formed methanol by a fluorometric procedure that involves oxidation to formaldehyde and condensation of the latter with acetylacetone and ammonia to a fluorophor. The procedure can be used to determine as little as 100 ng of trypsin and has been applied to measure trypsin activity in serum. Although the substrate is not specific for trypsin and can also be hydrolyzed by plasmin and thrombin, it can be made specific for trypsin by carrying out the assay with and without trypsin inhibitor.

A fluorometric method for the determination of trypsin by using a-benzoyl-L-arginine-ß-naphthylamide (BANA) as substrate was described by Uete, Asahara, and Tsuchikura [54]. The method is reported to be simpler and more sensitive than the Bratton-Marshall reaction.

The specific chrymotrypsin substrates N-acetyl-L-tryptophan ethyl ester (ATrEE) and N-acetyltyrosine ethyl ester (ATEE) are fluorescent, as are the products of chymotryptic digestion, N-acetyltryptophan and N-acetyltyrosine.

Bielski and Freed [55] noted that in both cases fluorescence increased on deesterification. They utilized this finding to develop a sensitive procedure for chymotrypsin assay. One limitation of the method is that chymotrypsin itself is fluorescent at these wavelengths because of its aromatic amino acids. This is no problem as long as the concentration of enzyme is two orders of magnitude less than that of the substrate. Under such conditions there is an almost three-fold increase in the fluorescence of ATrEE on hydrolysis and an almost fivefold increase in the fluorescence of ATEE. Because it fluoresces more intensely with excitation and emission at longer wavelengths, ATrEE is preferred.

When proteins are used as substrates of proteolytic enzymes, it is possible to measure enzyme activity by following the appearance of amino acids in the supernate after precipitation with trichloroacetic acid. Guroff [56] followed proteolytic activity with protein substrates by fluorometric assay of the increase in trichloroaceticacid-soluble tyrosine (free and peptide) during incubation. The method he used involved condensation with nitrosonaphthol.

Lüscher and Käser-Glasszmann [57] have developed a method for the measurement of fibrinolysin with the fluorescent dye Lissamine Rhodamine B 200, which is chemically bound to fibrinogen and is liberated on lysis of the fibrin clot. The fluorescent fibrinogen is prepared by treating the protein with the sulfonyl chloride of the dye. To measure fibrinolytic activity of a tissue sample the dye-labeled protein is incubated with it and other components required for clot formation. The fluorescence liberated into the fluid on clot lysis is proportional to fibrinolytic activity [58].

Fluorescein-labeled fibrin has been used as a substrate for the fluorometric assay of fibrinolysin. Strässle [59] used a similar method for the assay of fibrinolytic activity.

1. Phosphatase. Moss [60] described the use of the naphthyl phosphates as substrates for acid and alkaline phosphatase. The excitation maxima for 1-naphthol are at 250 and 335 nm, with peak emission at 455 nm. The corresponding phosphate ester is maximally excited at 235 and 295 nm, emission being at 365 nm. Thus, even though the ester is fluorescent, it is possible to select appropriate wavelengths to distinguish the free phenol from the ester. This is achieved at 335 nm for excitation and 455 nm for fluorescence. With 2-naphthol, excitation at 350 nm and measurement of fluorescence at 425 nm permits measurement of the free phenol in the presence of the ester. Both 1- and 2-naphthol have maximal and constant fluorescence at pH 10 and higher. It is therefore possible to follow alkaline phosphatase continuously. However, acid phosphatase necessitates alkalinization of the reaction mixture before fluorometric assay.

Campbell and Moss [61] used a similar procedure to assay acid phosphatase. Greenberg [62] compared the 1- and 2-derivatives and showed that
2-naphthyl phosphate yields three times as much fluorescence as the 1-com-
pound when hydrolyzed by alkaline phosphatase. He also noted that the
sensitivity of the method was such that 10^{-14} mole of enzyme could be assayed
in 50 μl volumes.

A histochemical procedure for measuring phosphatase in tissues, using
riboflavin-5'-phosphate as substrate and fluorescence microscopy for detection,
has been reported by Takeuchi and Nagami [63]. The procedure was also
applied to the localization of riboflavin nucleotidase activity.

Tierney [64] used 3-O-methylfluorescein phosphate (**111**) as a substrate for
measuring phosphatase activity in single cells of human fibroblast cultures.

$$(\underline{\text{III}})$$

The advantages of compound (**111**) over naphthyl phosphates are that its
excitation is maximal in the visible region (450-480 nm) and at neutral pH.
The former precludes background fluorescence; the latter means that
fluorescence can be measured at the pH of the enzyme reaction without the
need of an additional step, the addition of alkali.

Westley [65] reported that the phosphate esters of fluorescein derivatives
offer many advantages, including low blanks and higher sensitivity. Among
the compounds he studied was 3-O-methylfluorescein phosphate. In a subsequent paper Shneour [66] reported the use of the reduced form of the dye
3-methoxydihydrofluoran phosphate. The latter requires two enzymatic steps
to yield fluorescence: hydrolysis of the phosphate and peroxidation to the
fluorescein stage. It cannot be used as a test for phosphatase activity alone but
may be useful for detecting living cells, which contain both phosphatase and
peroxidase activities.

The fluorescein phosphates offer quite sensitive assays for phosphatase
activity, but they are not very stable. Land and Jackim [67] found that the
background fluorescence of a 3-O-methylfluorescein phosphate solution
(0.1 mg/ml) at pH 8 increases sevenfold in 24 h at room temperature. Enough
blank fluorescence appears on incubation at 37°C to limit the sensitivity of

phosphatase assays. Land and Jackim [67] synthesized the phosphate ester of
3-hydroxyflavone (flavone 3-phosphate) and found it to be highly stable and an
excellent substrate for phosphatase. The liberated 3-hydroxyflavone is
maximally excited at 360 nm and fluoresces at 510 nm. In 50% ethanol
fluorescence is increased 200 times. The aluminum complex of the dye is even
more highly fluorescent, with excitation at 400 nm and fluorescence at 450 nm.
Picogram quantities (10^{-12} g) of Escherichia coli alkaline phosphatase can be
assayed with this reagent.

Guilbault et al. [68] prepared umbelliferone phosphate as a fluorogenic
substrate for acid and alkaline phosphatase, and compared this substrate with
other substrates listed in the literature. From aspects of stability, rate of
enzymatic hydrolysis, and fluorescence of product formed, umbelliferone
phosphate appears to be an ideal substrate. As little as 10^{-6} unit of alkaline
phosphatase and 10^{-5} unit of acid phosphatase are detectable. This represents
an increase in sensitivity of two to four orders of magnitude over other
techniques.

Guilbault and Vaughan [69,70] described the use of Naphthol AS derivatives
as substrates for phosphatase. The fluorescent properties of a series of
Naphthol AS derivatives (Naphthol AS, AS-BI, AS-D, AS-GR, AS-LC, AS-MX,
and AS-TR) were compared (Table 70). The phosphate esters of these
compounds were investigated as fluorogenic substrates for acid and alkaline
phosphatase. The enzyme was determined by measuring the rate of formation
of the fluorescent Naphthol AS. For example, the reaction by which the nonfluorescent Naphthol AS-BI phosphate (112) is converted to the fluorescent
Naphthol AS-BI (113: λ_{ex} 405, λ_{em} 515 nm) is as follows:

$$(114)$$

(112)   (113)

The amount of enzyme is calculated from calibration plots of initial rate
against the concentration of alkaline phosphatase. In acid solution Naphthol AS
derivatives are nonfluorescent, but calibration curves for acid phosphatase can
be obtained by quenching the enzyme reaction after a fixed time with alkali,
which develops the fluorescence of the liberated Naphthol AS. The amount of
acid phosphatase is calculated from plots of change in fluorescence in the
fixed reaction time against enzyme concentration.

Naphthol AS-BI phosphate (112) was found to be the best substrate for both

TABLE 70

Fluorescence Properties of Naphthol AS Derivatives at pH 9.0

Naphthol	λ_{ex} (nm)	λ_{em} (nm)	Fluorescence coefficient[a] (x 10^5)	Blank fluorescence of phosphate ester[b]
AS	388	516	2.2	0.133
AS-BI	405	515	2.5	0.11
AS-D	383	515	6.6	0.210
AS-GR	388[c]	488[c]		
AS-GR Phosphate	388	488	13	26.1[d]
AS-LC	388	522	2.8	0.09
AS-MX	388	512	2.8	0.076
AS-TR	388	512	3.1	0.014

[a] Fluorescence of hydrolyzed substrate divided by the concentration of original substrate in M. Value for quinine sulfate in 0.1 N H_2SO_4 at λ_{ex} 350 and λ_{em} 450 nm was 2.5 x 10^6.

[b] The net increase in fluorescence of 3.2 x 10^{-4} M Naphthol AS phosphates at pH 9.0 after storage as a 10^{-2} M solution in Methyl Cellosolve at 3°C.

[c] In Methyl Cellosolve.

[d] 3.5 x 10^{-5} M.

enzymes (Tables 71 and 72). It was found to give the highest rates of hydrolysis and the best K_m values of all the substrates investigated. By using this substrate from 5 x 10^{-4} to 0.5 unit of alkaline phosphatase and 2 x 10^{-4} to 2 x 10^{-2} unit of acid phosphatase can be determined.

m. Sulfatases. Mead et al. [39] prepared the ethereal sulfate of 4-methylumbelliferone and investigated its use in the assay of sulfatase. However, they found that, unlike the glucuronide, the sulfate possessed appreciable fluorescence of its own, thereby giving a large "blank" in an enzyme assay. However, it is highly likely that the ethereal sulfate differs from free 4-methylumbelliferone not only in the intensity of fluorescence but also in the location of excitation and fluorescence maxima. With presently available fluorescence spectrometers it should be possible to distinguish the two fluorophors in the same manner as for the naphthols and their phosphates.

Guilbault and Hieserman [71] prepared five new fluorometric substrates for sulfatase: the sulfate esters of indoxyl, 2-naphthol, 4-methylumbellifer-

TABLE 71

Naphthol AS Phosphates as Substrates for Acid Phosphatase

Naphthol phosphate	Rate[a]	Lowest detectable amount[b] (unit x 10^{-4})	K_m (x 10^{-4})
AS	0.01	20	--
AS-BI	0.204	2	1.5
AS-D	Very slow		--
AS-LC	0.03	8	1.45
AS-MX	0.078	2	1.8
AS-TR	0.033	8	2.5

[a] Net change in fluorescence with 2×10^{-3} unit in 3 min.

[b] Lowest detectable amount obtained by using a reaction time of 10 min.

TABLE 72

Naphthol AS Phosphates as Substrates for Alkaline Phosphatase

Naphthol phosphate	Rate[a]	Lowest detectable amount (unit x 10^{-4})	K_m (x 10^{-4})
AS	0.13	8	5
AS-BI	0.155	5	0.22
AS-D	0.25	8	2.3
AS-G	0.075	50	--
AS-LC	0.086	7	0.18
AS-MX	0.069	5	2.8
AS-TR	0.057	5	6

[a] Rate with 5×10^{-3} unit of alkaline phosphatase expressed as $\Delta F/min$. Blank rate of all esters is 0.

one, fluorescein, and resorufin and compared these with each other and with other colorimetric substrates for sulfatase. Of these, 2-naphthyl sulfate and 4-methylumbelliferone sulfate appear to be optimum for the assay of various types of sulfatase. As little as 10^{-4} unit/ml of sulfatase can be determined in 2 to 3 min with a precision and accuracy of about 1.5%.

2. Dehydrogenases and Transaminases

a. Methods Based on the Fluorescence of NADH or NADPH. The dehydrogenases are an important class of enzymes that effect the dehydro-

genation of hydroxy compounds in the presence of a hydrogen acceptor or
coenzyme, such as NAD or NADP. Since NADH (the reduced form of NAD)
is fluorescent and NAD forms a fluorescent derivative on heating in alkali,
many fluorometric methods for enzyme assay with NAD-NADH and NADP-
NADPH systems have been proposed. The initial rate of formation or
disappearance of NADH is measured and related to the concentration of
dehydrogenases, substrate, or coenzyme. For example, lactate is converted
to pyruvate at pH 9 by lactic dehydrogenase (LDH) in the presence of NAD, and
the reverse reaction is effected at pH 7:

$$\text{lactate} + \text{NAD} \xrightarrow{\quad\text{LDH}\quad} \text{pyruvate} + \text{NADH} \tag{115}$$

The rate of NADH formation at pH 9 or the rate of disappearance of this
substance at pH 7 is proportional to the concentration of lactic dehydrogenase,
NAD (or NADH), and substrate [72].

Likewise Ochoa et al. [73] have described fluorometric methods using
NADP for the assay of glucose-6-phosphate (G-6-P) and its dehydrogenase
(G-6-PDH):

$$\text{G-6-P} + \text{NADP} \xrightarrow{\quad\text{G-6-PDH}\quad} \text{6-phosphogluconate} + \text{NADPH} + \text{H}^+ \tag{116}$$

Enzymatically formed NADPH can be measured continuously by its native
fluorescence, and the rate of its formation equated to the concentration of
glucose-6-phosphate or G-6-PDH.

Fluorometric methods for hexokinase have been described by Greengard
[74,75], who used a coupled enzyme system. In the first method hexokinase
converts glucose to glucose-6-phosphate, which is in turn converted to
6-phosphogluconate with simultaneous formation of NADPH. Again the rate
of NADPH production is proportional to the ATP concentration, at non-rate-
limiting excesses of glucose, hexokinase, and NADP.

$$\text{ATP} + \text{glucose} \xrightarrow{\quad\text{hexokinase}\quad} \text{ADP} + \text{G-6-P} \tag{117}$$

$$\text{G-6-P} + \text{NADP} \xrightarrow{\quad\text{G-6-PDH}\quad} \text{6-phosphogluconate} + \text{NADPH} + \text{H}^+ \tag{116}$$

A second method is based on the use of kinase and NADH:

$$\text{ATP} + \text{3-phosphogycerate} \xrightarrow{\quad\text{kinase}\quad} \text{ADP} + \text{1,3-diphosphoglycerate} \tag{118}$$

$$\text{1,3-diphosphoglycerate} + \text{NADH} + \text{H}^+ \xrightarrow{\dfrac{\text{glyceraldehyde}}{\text{phosphate dehydrogenase}}} $$
$$\text{glyceraldehyde-3-phosphate} + \text{P}_i + \text{NAD}^+ \tag{119}$$

The rate of NADH disappearance is proportional to the ATP concentration.

Lowry and co-workers have described methods for glutamic dehydrogenase [76,77], lactic dehydrogenase [77], and 6-phosphogluconate

dehydrogenase [20]. Maitra and Estabrook [78] have developed fluorometric methods for the enzymatic determination of glycolytic intermediates by measuring the fluorescence of NADH and NADPH. The enzymes proposed for these methods are glucose-6-phosphate dehydrogenase, hexokinase, phosphoglucoisomerase, phosphoglucomutase, α-glycerophosphate dehydrogenase, triosephosphate isomerase, aldolase, glyceraldehyde 3-phosphate dehydrogenase, phosphoglycerate mutase, and pyruvate kinase.

Graham and Aprison [79] have described enzymatic fluorometric methods for aspartate, glutamate, and γ-aminobutyrate in nerve tissue. The method is based on the use of glutamic-oxalacetic transaminase (GOT), which in the presence of α-ketoglutarate (α-KG) converts aspartate to oxalacetate. The oxalacetate formed was reduced to malate in the presence of NADH and malic dehydrogenase (MDH):

$$\alpha\text{-KG} + \text{aspartate} \xrightarrow{\text{GOT}} \text{glutamate} + \text{oxalacetate} \tag{120}$$

$$\text{oxalaceate} + \text{NADH} + \text{H}^+ \longrightarrow \text{malate} + \text{NAD}^+ \tag{121}$$

Either the disappearance of NADH or the increase in NAD could be measured, fluorometrically. By this method as little as 5×10^{-12} mole of aspartate, 1×10^{-11} mole of γ-aminobutyrate, and 1×10^{-10} mole of glutamate can be measured.

Laursen and co-workers [80, 81] have developed methods for transaminases by coupling the following reactions:

$$\text{L-alanine} + \alpha\text{-KG} \xrightarrow{\text{transaminase}} \text{pyruvate} + \text{glutamate} \tag{122}$$

$$\text{pyruvate} + \text{NADH} \xrightarrow{\text{LDH}} \text{lactate} + \text{NAD}^+ \tag{123}$$

The increase in NAD^+ is a measure of transaminase activity.

Pitts, Quick, and Robins [82] have coupled the NAD^+-linked succinic semialdehyde dehydrogenase reaction with a transaminase reaction to measure γ-aminobutyric-α-oxoglutaric transaminase. The NADH produced is measured fluorometrically.

Young and Renold [83] have described a fluorometric procedure for the determination of ketone bodies in very small quantities of blood. The enzyme D-ß-hydroxybutyric dehydrogenase is used to determine either acetoacetate or D-ß-hydroxybutyrate; the NADH or NAD is measured fluorometrically:

$$\text{D-ß-hydroxybutyrate} + \text{NAD}^+ \xrightarrow{\text{dehydrogenase}} \text{acetoacetate} + \text{NADH} \tag{124}$$

Cooper [84] assayed acetylcholine by a sodium borohyride reduction to

ethanol, followed by a coupled alcohol dehydrogenase reaction. The rate of
NADH formation was proportional to the concentration of acetylcholine.

b. Resazurin Method. Guilbault and Kramer [85, 86] have devised a simple
and rapid fluorometric method for dehydrogenases, based on the conversion of
the nonfluorescent substance resazurin (114) to the highly fluorescent resorufin
(115: λ_{ex} 540, λ_{em} 580 nm) in conjunction with the NAD-NADH or NADP-NADPH
systems:

$$\text{Substrate + NAD} \xrightarrow{\text{Dehydrogenase}} \text{oxidized substrate + NADH}$$
$$\text{(NADP)} \qquad\qquad\qquad\qquad\qquad\qquad\qquad \text{(NADPH)}$$

(125)

(126)

(114)   (115)

Because of the intense fluorescence of resorufin (10^{-9} M is easily detectable), a twofold increase in sensitivity over the NADH method is obtainable.
The rate of resorufin formation is proportional to the concentration of the
dehydrogenase, substrate, NAD, and diaphorase. Using non-rate-limiting
excesses of any three of these, the fourth can be determined. Approximately
10^{-1} to 10^{-4} unit/ml of lactic dehydrogenase, malic dehydrogenase, alcohol
dehydrogenase, glutamic acid dehydrogenase, glucose-6-phosphate dehydrogenase, L-α-glycerophosphate dehydrogenase, and glycerol dehydrogenase can
be determined with standard deviations of about $\pm 1\%$. Also, diaphorase,
0.00040 to 0.080 unit/ml, and NADH, 2×10^{-7} to 2×10^{-5} M, can be determined with a standard deviation of about $\pm 0.5\%$.

Resorufin has excitation and emission wavelengths of 560 and 580 nm,
respectively, and a fluorescence coefficient of 1.56×10^7 (compared with
1.40×10^6 for quinine sulfate in 0.1 N H_2SO_4).

Resazurin has been successfully used as a substrate for dehydrogenase in
automated systems developed by Technicon [87]. The resorufin formed is
continually monitored.

3. Oxidative Enzymes

a. Xanthine Oxidase. Lowry [88] has shown that 2-amino-4-hydroxypteri-
dine is a good fluorogenic substrate for xanthine oxidase. The product is
isoxanthopterin, which is fluorescent in phosphate buffer. The substrate is
nonfluorescent, and hence the rate of fluorescence formation is proportional to

the enzyme concentration. The fluorophor is assayed directly and continuously in the incubation mixture, and each sample acts as its own standard to correct for blank fluorescence or fluorescence losses.

Guilbault et al. reported homovanillic acid [4, 89] and p-hydroxyphenyl-acetic acid [5] to be ideal substrates for the determination of oxidative enzymes; as little as 10^{-5} unit of amino acid oxidase, peroxidase, glucose oxidase, and xanthine oxidase [4, 89] can be determined fluorometrically:

$$\text{substrate} \xrightarrow{\text{oxidase}} H_2O_2 \tag{127}$$

$$H_2O_2 + \text{p-hydroxyphenylacetate} \xrightarrow{\text{peroxidase}} \text{oxidized product} \tag{128}$$

The rate of formation of the fluorescent oxidized product is followed and equated to the concentration of enzyme.

In attempting to develop methods for galactose oxidase and invertase Guilbault, Brignac, and Juneau [5] tried 25 different substrates for the assay of oxidative enzymes. p-Hydroxyphenylacetic acid was chosen as the best substrate for measuring oxidative enzymes because of its stability, low cost, and the high fluorescence coefficient of its oxidized form.

b. Peroxidase. Keston and Brandt [90] have described a fluorometric method for the analysis of as little as 10^{-8} mole/l (10^{-11} mole/ml) of hydrogen peroxide. The procedure is based on the oxidation, by hydrogen peroxide and peroxidase, of the nonfluorescent diacetyl dichlorofluorescin to the highly fluorescent dichlorofluorescein. The method is also applicable to peroxidase and other enzyme systems that produce hydrogen peroxide. Andreae [91] and Perschke and Broda [92] used scopoletin (6-methoxy-7-hydroxy-1, 2-benzo-pyrone) as a substrate for peroxidase. The disappearance of scopoletin fluorescence was a measure of the peroxidase concentration.

Guilbault, Kramer, and Hackley [89] have shown that homovanillic acid (116) is an excellent substrate for the determination of hydrogen peroxide and peroxidase.

$$\tag{129}$$

(116)   (117)

The nonfluorescent compound (116) is oxidized to the highly fluorescent compound (117) (λ_{ex} 315, λ_{em} 425 nm). The rate of formation of (117) with time is proportional to the concentration of hydrogen peroxide and peroxidase. The solutions of (116) appear to be very stable and thus afford a distinct advantage over the diacetylfluorescin method of Keston and Brandt [90]. As little as 10^{-11} mole/ml of hydrogen peroxide and 10^{-3} unit/ml of peroxidase are determinable.

c.   Aromatic L-Amino Acid Decarboxylase. Decarboxylation of 3,4-di-hydroxyphenylalanine (dopa) and 5-hydroxytryptophan is effected by the same enzyme, aromatic L-amino acid decarboxylase [93]. The substrates and end products of this reaction are aromatic or cyclic compounds, and most of them either possess a high degree of native fluorescence or can be converted to fluorescent derivatives.

In the enzyme reaction dopa (118) is converted to dopamine (119). After incubation the mixture is passed over a permutit column. The amine is adsorbed onto the column, and the amino acid passes through in the effluent. By washing the column with several volumes of water it is possible to remove all unreacted dopa. The dopamine is then eluted with several milliliters of 20% HCl and assayed fluorometrically (λ_{ex} 285, λ_{em} 325 nm).

$$ \text{(118)} \xrightarrow{\text{aromatic L-amino acid decarboxylase}} \text{(119)} + CO_2 \qquad (130) $$

(118)   (119)

Dopa is not a good substrate with which to assay the activity of the enzyme because this amino acid combines nonenzymatically with the coenzyme pyridoxal phosphate, leading to substrate inhibition [94]. 5-Hydroxytryptophan is an excellent substrate and can be used for enzyme assay in the same way as dopa. The resulting amine, 5-hydroxytryptamine (serotonin), is even more highly fluorescent than dopamine, so that with this substrate the method becomes even more sensitive. Fluorescence can be measured at neutral pH (λ_{ex} 295, λ_{em} 330 nm) or in 3 HCl (λ_{ex} 295, λ_{em} 550 nm).

A highly sensitive method for the assay of this enzyme has been described [48] based on the decarboxylation of 5-hydroxytryptophan to 5-hydroxytryptamine in the presence of a monoamine oxidase inhibitor. After incubation proteins were precipitated and the amine was extracted with butanol and returned to an aqueous phase with heptane. The 5-hydroxytryptamine was then treated with ninhydrin as described in Section VI. A. 4 of Chapter 7, and the fluorescence of the ninhydrin derivative was assayed. Excitation is maximal at 380 nm and fluorescence at 500 nm. The ninhydrin procedure is particularly well suited to assay the decarboxylation of 5-hydroxytryptophan since the amino acid yields

only 2% as much fluorescence as does the decarboxylated product, 5-hydroxy-
tryptamine. This in combination with the solvent-extraction procedure ensures
the specificity of the assay. By using microtechniques it was possible to use
the procedure to assay decarboxylase activity in 4 to 20 μg (dry weight) of brain
tissue.

d. Monoamine Oxidase. Monoamine oxidase (MAO) is widely distributed in
animal tissues and is localized largely in the mitochondrial fraction of the cell.
It has received a great deal of attention, and many procedures are available for
its assay--manometric, colorimetric, and spectrophotometric. However, the
most sensitive of all procedures is a fluorometric one involving the conversion
of tryptamine (120) to indoleacetaldehyde (121) and ultimately to indoleacetic
acid (122). In the presence of excess aldehyde dehydrogenase the formation of
indoleacetic acid from tryptamine becomes proportional to the monoamine
oxidase activity [95].

(131)

The sensitivity of the method makes it particularly appropriate for studies
on small amounts of tissue as are available from sympathetic nerves and
ganglia, blood vessels, and various portions of the brain [96]. The fluorescence
is measured at a λ_{ex} of 280 and a λ_{em} of 370 nm.

e. Diamine Oxidase (Histaminase). This enzyme is found in many tissues
but mainly in the liver and kidney. Although not specific for histamine, it is
associated mainly with the metabolism of this active humoral agent. Cohn and
Shore [97] have developed a highly sensitive procedure for assaying this enzyme
by measuring the disappearance of histamine or agmatine with the fluorometric
phthalaldehyde procedure for the amine.

The sensitivity of the fluorometric assay for histamine permits the deter-

mination of diamine oxidase in small amounts of tissue or in tissues with
relatively low enzyme activity. The determination of K_m for the guinea-pig-
kidney enzyme required such a sensitive method since it was found to be
approximately 5×10^{-6} M [97].

f. 3-Hydroxyanthranilic Acid Oxidase. Bokman and Schweigert [98] have
studied the enzyme systems responsible for converting 3-hydroxyanthranilic
acid to quinolinic acid by measuring the disappearance of 3-hydroxyanthranilic
acid fluorometrically. With mammalian enzymes only quinolinic acid is formed.
With less pure systems N'-methylnicotinamide and nicotinic acid also appear.
However, none of these compounds possess native fluorescence comparable to
that of 3-hydroxyanthranilic acid. It was also shown that the substrate can be
recovered quantitatively from heat-inactivated enzyme preparations, so that its
disappearance can be used as a direct measure of enzyme activity. 3-Hydroxy-
anthranilic acid fluoresces maximally at pH 7, excitation maximum at 320 nm
and fluorescence maximum at 415 nm [99].

g. Kynureninase. Jakoby and Bonner [100] used a fluorometric procedure to
assay kynureninase. The assay takes advantage of the differences in
fluorescence characteristics between the substrate kynurenine (123) and the end
product anthranilic acid (124).

$$(132)$$

(123)   (124)

Anthranilic acid fluoresces maximally in neutral or slightly acid solution,
whereas kynurenine fluorescence is maximal at pH 11. The substrate also
fluoresces less intensely and at different wavelength than the product. Excitation and fluorescence maxima for anthranilic acid are 300 and 405 nm,
respectively; for kynurenine they are 370 and 490 nm [99]. After incubation
with enzyme the samples are deproteinized, and aliquots are buffered to pH 5.5,
at which kynurenine fluorescence is minimal. With the monochromators set to
the appropriate wavelengths for anthranilic acid, or with suitable filters, the
fluorescence is a linear function of anthranilic acid concentration. This is in
turn directly proportional to kynureninase activity. Comparable assays have
been carried out with hydroxykynurenine and formylkynurenine as substrates of
kynureninase [100].

III. DETERMINATION OF SUBSTRATES

A. Introduction

At a fixed enzyme concentration the initial rate of an enzymatic reaction increases with increasing substrate concentration until a nonlimiting excess of substrate is reached, after which additional substrate causes no increase in rate. The region in which linearity is achieved and in which an analytical determination of substrate concentration can be made based on the rate of reactions lies below $0.1 \, K_m$. The most important advantage of an enzymatic assay is its specificity. Frequently only one member of a homologous series is active in the enzyme-catalyzed reaction; other members are totally inactive or react at much slower rates. Most enzymes are also specific for one optical isomer of the substrate. Thus in the enzymatic assay of amino acids bacterial amino acid decarboxylase is specific for L-amino acids only. Another advantage in the use of enzymes for substrate analysis lies in the great sensitivity obtained; for example, glucose is oxidized at the rate of a few percent per minute, regardless of concentration. Thus a 10^{-7} M solution can be analyzed as easily as a 10^{-4} M solution.

A complete compilation of enzymatic methods for the assay of carbo-hydrates, amines, amino acids, organic acids, hydroxy compounds, esters, aldehydes, and inorganic substances has been prepared by Guilbault [101, 102] and by Bergmeyer [103]. Some of the more important assays are discussed in this section.

B. Carbohydrates

Enzymes offer advantages of specificity and sensitivity over other nonenzymatic methods for the determination of carbohydrates. Furthermore, by the use of several enzymes, each selective for one carbohydrate, a complex mixture of carbohydrates can be assayed. Guilbault, Sadar, and Peres [104] described fluorometric methods for the assay of mixtures of the carbohydrates, glucose, fructose, maltose, cellobiose, lactose, glycogen, and salicin. Glucose and fructose are determined fluorometrically by using hexokinase and the resazurin-resorufin indicator reaction:

$$\text{glucose} + \text{ATP} \xrightarrow[\text{Mg}^{2+}]{\text{hexokinase}} \text{G-6-P} \tag{117}$$

$$\text{G-6-P} + \text{NADP} \xrightarrow{\text{G-6-PDH}} \text{NADPH} + \text{phospho gluconate} \tag{116}$$

$$\text{NADPH} + \text{resazurin} \xrightarrow{\text{PMS}} \text{NADP} + \text{resorufin} \tag{126}$$

The rate of resorufin formation is measured and is proportional to the glucose present (or to fructose if phosphohexose isomerase is used).

$$\text{Fructose} + \text{ATP} \xrightarrow[\text{Mg}^{2+}]{\text{hexokinase}} \text{fructose-6-phosphate} \tag{133}$$

Maltose, cellobiose, lactose, glycogen, and salicin are enzymatically hydrolyzed to glucose, which is then determined fluorometrically by using the respective enzyme, glucose oxidase, together with p-hydroxyphenylacetic acid and peroxidase to obtain the fluorescent compound ($\underline{125}$) (λ_{ex} 317, λ_{em} 414 nm).

$$\text{glycogen} \xrightarrow{\text{amyloglucosidase}} \text{glucose} \tag{134}$$

$$\text{maltose} \xrightarrow{\text{maltase}} \text{glucose} \tag{135}$$

$$\text{cellobiose} \xrightarrow{\text{ß-glucosidase}} \text{glucose} \tag{136}$$

$$\text{salicin} \xrightarrow{\text{ß-glucosidase}} \text{glucose} \tag{137}$$

$$\text{lactose} \xrightarrow{\text{lactase}} \text{glucose} \tag{138}$$

$$\text{glucose} \xrightarrow{\text{glucose oxidase}} \text{H}_2\text{O}_2 \tag{139}$$

(128), (125)

Some of the enzyme systems, together with the range of concentrations and interferences in the determination of various sugars, are listed in Table 73. All the enzyme systems are highly selective; only cellobiose and salicin, two substrates seldom found together, are substrates for ß-glucosidase; maltase act specifically on maltose; ß-galactosidase (lactase) acts specifically on lactose; and amyloglucosidase specifically on glycogen.

The analysis of a mixture of lactose, maltose, fructose, and glycogen was attempted with the four enzyme systems ß-galactosidase, maltase, hexokinase, and amyloglucosidase, without prior separation. The sample was split into four aliquots, and each was assayed for one of the four components. Table 74 indicates the excellent results obtained for the simultaneous determination of these four carbohydrates.

TABLE 73

Enzyme Systems, Concentration Ranges, and Interferences
in the Determination of Various Sugars

Substrate	Enzyme	Range (μg/ml)	Interference
Glucose	Hexokinase	0.05-50	None
Fructose	Hexokinase	0.05-52	Glucose
Cellobiose	ß-Glucosidase	0.3-35	Glucose, salicin
Maltose	Maltase	3-35	Glucose
Lactose	ß-Galactosidase	3-35	Glucose
Salicin	ß-Glucosidase	0.3-35	Glucose, cellobiose
Glycogen	Amyloglucosidase	3-35	Glucose

TABLE 74

Determination of a Mixture of Sugars[a]

Lactose		Maltose		Fructose		Glycogen	
Added	Found[b]	Added	Found[b]	Added	Found[b]	Added	Found[b]
13.3	13.0	3.3	3.3	0.100	0.102	3.30	3.40
10.0	10.0	26.7	27.0	0.100	0.100	10.0	9.70
26.7	27.3	13.3	13.7	1.00	0.970	20.0	20.3
33.3	33.3	16.7	16.7	1.00	1.02	26.7	26.7
Average error	1.2%		1.0%		1.7%		1.6%

[a]Concentrations in micrograms per milliliter of total solution.

[b]Represents an average of three or more results with a precision of ±1.2%.

C. Organic Acids

The approach outlined for carbohydrates could be applied to the assay of
any complex mixture of organic substances, provided an enzyme were available
for the specific assay of each component of the mixture. Guilbault, Sadar and
McQueen [105] described an assay procedure using six enzyme systems for the
determination of mixtures of 21 organic acids. The dehydrogenases used were
lactate (types II and IV), malate, glutamate, isocitrate, and ß-hydroxybutyrate.
They were coupled with NAD, phenazine methosulfate, and resazurin (114) in a

fluorometric procedure for the determination of acetic, adipic, benzilic, butyric, D-α-hydroxybutyric, D-ß-hydroxybutyric, chloroacetic, citric, formic, L-glutamic, glutaric, glycolic, threo-D-isocitric, L-lactic, L-malic, malonic, oxalic, phthalic, succinic, and L-tartaric acids in the approximate range 0.1 to 500 μg, with an accuracy and precision of about 2%. The rate of production of the highly fluorescent resorufin (115) is proportional to the concentration of the acid (see reactions (125) and (126)).

Some typical results for the analysis of acids are shown in Table 75.

TABLE 75

Typical Determinations of Acids[a]

D-ß-Hydroxybutyric acid[b]			L-Glutamic acid[c]			L-Lactic acid[d]		
Added	Found	R. E. [e] (%)	Added	Found	R. E. [e] (%)	Added	Found	R. E. [e] (%)
1.00	1.01	+1.0	5.00	4.95	-1.0	0.500	0.510	+2.0
5.00	4.90	-2.0	15.0	14.7	-2.0	1.50	1.47	-2.0
10.00	10.0	0.0	25.0	25.0	0.0	3.0	3.08	+2.7
50.0	51.0	+2.0	75.0	77.0	+2.2	5.0	4.85	-3.0
75.0	74.0	-1.5	100.0	102.0	+2.0	10.0	10.0	0.0
Average R. E. (%)		+1.3			+1.4			+1.9

[a]Concentrations in micrograms per milliliter.

[b]Analysis with ß-hydroxybutyric dehydrogenase in the presence of 1 mg/ml each of L-glutamic, L-lactic, L-malic, and D-α-hydroxybutyric acids.

[c]Analysis with glutamate dehydrogenase in the presence of 1 mg/ml each of acetic, D-tartaric, and D-ß-hydroxybutyric acids.

[d]Analysis with lactate dehydrogenase type IV in the presence of 1 mg/ml each of L-malic, L-glutamic, and D-ß-hydroxybutyric acids.

[e]Relative error.

With ß-hydroxybutyrate dehydrogenase, 1 to 75 μg/ml of D-ß-hydroxybutyrate was analyzed in the presence of L-glutamic, L-lactic, L-malic, and D-α-hydroxybutyric acids at concentrations of 1 mg/ml, with an accuracy of +1.3% and a precision of 2%. Citric acid is the only interference being determinable in the range 10 to 110 μg/ml with a precision and accuracy of about 2%.

A four-component acid mixture of citric, D-isocitric, L-lactic, and L-glutamic acids was analyzed (see Table 76). The sample was divided into four equal parts, and analysis of each acid was performed with a different dehydrogenase: ß-hydroxybutyrate dehydrogenase (ß-OH-BuDH) for citric acid, iso-

TABLE 76

Analysis of a Mixture of Citric, Isocitric, Lactic, and Glutamic Acids

Added (μg/ml)			
Citric	D-Isocitric	L-Lactic	L-Glutamic
10.0	1.00	10.0	10.0
10.0	1.00	5.00	50.0
10.0	1.00	10.0	50.0
50.0	0.500	10.0	50.0

Found (μg/ml)			
Citric	D-Isocitric	L-Lactic	L-Glutamic
10.0	1.00	9.90	10.2
10.0	1.01	5.10	49.2
9.9	0.980	10.1	51.0
49.2	0.495	10.0	51.0

Relative Error (%)			
Citric	D-Isocitric	L-Lactic	L-Glutamic
+1.0	0.0	-1.0	+2.0
0.0	+1.0	+2.0	-1.6
-1.0	-2.0	+1.0	+2.0
-1.6	-1.0	0.0	+2.0

citric dehydrogenase (ICDH) for D-isocitric acid, lactate dehydrogenase (LDH) type II for L-lactic acid, and glutamate dehydrogenase (GDH) for L-glutamic acid. A phosphate buffer (pH 6.5) was used for ICDH, a glycine-hydrazine buffer (pH 9.5) for ß-OH-BuDH, and tris buffer (pH 9.5) for LDH and GDH. Essentially the same precisions and accuracies were obtained for the analysis of the mixture (Table 76) as were obtained in the determination of the acid alone (Table 75). Pyruvate can be assayed according to reaction (123). By following the decrease in the native fluorescence of NADH [74] as little as 2.5×10^{-8} mole/l of pyruvate can be assayed.

D. Hydroxy Compounds

Sadar [106] has developed enzymatic methods for the assay of mixtures of hydroxy compounds. Four enzyme systems were used, each for the assay of

one alcohol in the mixture: alcohol dehydrogenase, alcohol oxidase, carbohydrate oxidase, and sorbitol dehydrogenase. In each case a fluorometric
monitoring system was used to follow the reaction.

System 1:

$$\text{alcohol} \xrightarrow{\text{oxidase}} H_2O_2 \tag{140}$$

or

$$\text{carbohydrate} \xrightarrow{\text{oxidase}} H_2O_2 \tag{141}$$

$$HO\!-\!\!\left\langle\!\!\!\!\begin{array}{c}\\ \end{array}\!\!\!\!\right\rangle\!\!-\!CH_2COOH + H_2O_2 \xrightarrow{\text{peroxidase}} \underline{(125)} \tag{128}$$

The rate of formation of the fluorescent product (125) is followed and is
found to be proportional to the alcohol or carbohydrate content.

System 2:

$$\text{alcohol} + \text{NAD} \xrightarrow{\text{dehydrogenase}} \text{NADH} + \alpha\text{-ketoacid} \tag{142}$$

or

$$\text{D-sorbitol} + \text{NAD}^+ \xrightarrow[\text{dehydrogenase}]{\text{sorbitol}} \text{NADH} + \text{D-fructose} + H^+ \tag{143}$$

$$\text{NADH} + \text{resazurin} \xrightarrow{\text{PMS}} \text{NAD} + \text{resorufin} \tag{126}$$

The formation of resorufin is measured and is proportional to the concentration of sorbitol or alcohol present. Sorbitol, methanol, ethanol, allyl alcohol,
cyclohexanol, and sec-butyl alcohol did not interfere in the determination of
glucose, xylose, or sorbose by the use of the appropriate carbohydrate oxidase.
Likewise, neither sugars such as glucose, xylose, and sorbose nor alcohols
such as methanol, ethanol, and allyl alcohol interfered in the assay of sorbitol
with sorbitol dehydrogenase, nor did sorbitol or any sugars interfere in the
determination of alcohols with alcohol oxidase.

Some typical results of the analysis of a mixture of glucose and sorbitol
using carbohydrate oxidase and sorbitol dehydrogenase, respectively, are
shown in Table 77. An average error of about 1.4% and a precision of about
1.5% were obtained.

Finally a three-component mixture of methanol, xylose, and sorbitol was
assayed by using three different enzymes; alcohol oxidase, carbohydrate
oxidase, and sorbitol dehydrogenase. The results given in Table 78, indicated
a precision of about 1.5% and an average error of 1 to 1.5%.

TABLE 77

Assay of a Mixture of Glucose and Sorbitol[a]

Glucose		Sorbitol	
Added	Found[b]	Added	Found[b]
79. 9	79. 0	17. 2	17. 2
70. 0	70. 0	36. 3	36. 0
83. 6	84. 2	29. 6	30. 2
99. 2	99. 2	16. 8	16. 8
Average Error	1. 5 %		1. 4 %

[a]Concentrations in micrograms per milliliter of total solution. Glucose assayed with carbohydrate oxidase, sorbitol with sorbitol dehydrogenase.

[b]Represents an average of three or more results with a precision of $\pm$ 1. 5 %.

TABLE 78

Assay of a Mixture of Methanol, Xylose, and Sorbitol[a,]

Methanol		Xylose		Sorbitol	
Added	Found[b]	Added	Found[b]	Added	Found[b]
10. 3	10. 3	226. 7	226. 7	17. 2	17. 2
15. 6	15. 8	230. 0	232. 1	36. 3	36. 0
25. 1	25. 1	245. 2	245. 2	29. 6	30. 2
33. 2	33. 8	210. 5	218. 0	16. 8	16. 8
Average error	1. 5 %		1 %		1. 4 %

[a]Concentration in micrograms per milliliter of total solution. Methanol assayed with alcohol oxidase, xylose with carbohydrate oxidase, and sorbitol with sorbitol dehydrogenase.

[b]Average of three or more results with a precision of $\pm$1. 5 %.

IV. DETERMINATION OF ACTIVATORS AND COFACTORS

A. Introduction

An enzyme activator or cofactor is a substance that is required for an enzyme to be an active catalyst:

$$\text{E (inactive) + activator} \rightleftharpoons \text{E (active)} \tag{144}$$

The activity of the enzyme will increase until enough activator has been used to

fully activate the enzyme. The initial rate of the enzyme reaction is proportional to the activator concentration at low concentrations, thus providing a
method for its determination. Very little has been done in the analytical
determination of activators. A method for magnesium in plasma is described
by Baum and Czok [107] based on the activation of isocitric dehydrogenase.
With constant amounts of enzyme the rate is dependent on magnesium concentration down to 10^{-6} M. A thorough study of this reaction was made by Adler,
Gunther, and Plass [108], and by Blaedel and Hicks [14], who found that only
Mg^{2+} and Mn^{2+} efficiently activate this enzyme. The useful analytical range
extends up to about 100 ppb for Mn^{2+} and 2×10^{-4} M for Mg^{2+}; Hg^{2+} and Ag^{2+}
at 10^{-5}M; and Ca^{2+} at 10^{-4}M completely inhibited the Mn^{2+} activation.

A number of enzymes require for their activity a specific coenzyme that
participates in the enzymatic reaction. By measuring the amount of activation
of such an enzyme by the coenzyme, a plot of initial rate of reaction against
coenzyme concentration may be constructed. At low concentrations of coenzyme
the degree of activation will be proportional to the concentration of coenzyme
added.

B. Flavine Adenine Dinucleotide

Flavine adenine dinucleotide, which is the coenzyme of D-amino acid
oxidase, can be determined by this method:

$$\text{D-alanine} + \text{FAD} \xrightarrow{\text{D-amino acid oxidase}} \text{pyruvic acid} + \text{FADH} + \text{NH}_3 \qquad (145)$$

$$\text{FADH} + \text{O}_2 \longrightarrow \text{FAD} + \text{H}_2\text{O}_2 \qquad (146)$$

Warburg and Christian [109] and Straub [110] were the first to describe
this method. Unknown samples of FAD containing 0 to 0.3 μg of FAD can be
determined by measuring the increase in the rate of alanine oxidation by
D-amino acid oxidase [111].

C. Adenosine Triphosphate

The firefly reaction has been shown to require ATP and Mg^{2+} in addition to
luciferin (106), luciferase, and oxygen [1]:

$$\text{luciferin} + \text{O}_2 + \text{ATP} \xrightarrow[\text{Mg}^{2+}]{\text{luciferase}} \text{oxyluciferin} + \text{ADP} + \text{Pi} \qquad (104)$$

This reaction has been used as the basis for the most sensitive of all methods
for ATP [112]. This reaction can also be used to assay oxygen at partial
pressures below 10^{-3} mm Hg [113], when the gas is passed through a bacterial
suspension containing all requirements for the luminescent reaction except
oxygen.

Greengard [74, 75] has reported two procedures for the enzyme-fluorometric assay of ATP. The first utilizes the enzyme systems corresponding to those used in the spectrophotometric procedure of Kornberg [114] (see reactions (116) and (117)). Readings of native NADPH fluorescence are made at suitable intervals until the reduction is maximal. The peak reading is proportional to the ATP content. There is sufficient hexokinase activity in most of the nerve extracts to permit completion of the reaction in 20 to 30 min. A reagent blank and a calibration curve are used for calculation.

A second fluorometric procedure for ATP based on the spectrophotometric procedure of Thorn et al. [115], is also available (see reactions (118) and (119) [75]). Modifications of both methods to permit assay of creatine phosphate and of adenosine monophosphate are also described.

D. Organic Phosphates

Greengard and Straub [116] applied these microfluorometric assay procedures to the measurement of the various organic phosphates and other metabolites in normal and stimulated nerves.

Seraydarian et al. [117] have adapted the Slater procedures [118] for the enzymatic determination of hexose phosphate and triose phosphate intermediates to fluorometry, utilizing the disappearance of native NADH fluorescence during oxidation. The methods are based on the following reactions:

Enzyme preparation A:

fructose-1,6-diphosphate $\rightleftharpoons$ glyceraldehyde-3-phosphate

$$\text{dihydroxyacetone phosphate} \qquad (147)$$

$$\text{Dihydroxyacetone phosphate} + \text{NADH} \longrightarrow \text{glycerolphosphate} + \text{NAD}^+ \qquad (148)$$

Enzyme preparation B:

$$\text{glucose-6-phosphate} \longrightarrow \text{fructose-6-phosphate} \qquad (149)$$

$$\text{fructose-6-phosphate} + \text{ATP} \longrightarrow \text{fructose-1,6-diphosphate} \qquad (150)$$

Enzyme preparation A determines the sum of fructose-1,6-diphosphate and triose phosphates. Subsequent addition of enzyme preparation B yields the additional determination of total hexose monophosphates (including glucose-1-phosphate).

E. Enzyme Cycling for Pyridine Nucleotides

Because of the versatility of the coenzymes NAD and NADP, almost every substance of biological interest can be measured by their use. Hundreds of papers have been published on analytical procedures using these substances and either a spectrophotometric or a fluorometric method for performing the analysis. But because of instrumental limitations, one is limited to a sensitivity of about 10^{-8} mole/ml in spectrophotometry, and 10^{-11} mole/ml with the more sensitive fluorometric methods.

Lowry et al. [119] proposed an enzyme-cycling method for measuring pyridine nucleotides, in attempts to increase this sensitivity limit by several orders of magnitude.

The nucleotide to be assayed is made to catalyze an enzymatic reaction between two substrates, which are transferred in amounts far greater than the nucleotide. Thus the measurement of the nucleotide through its catalytic effect increases sensitivity by a factor of 10^3 to 10^4 over a direct measurement. Coenzyme NAD is measured with lactate dehydrogenase and glutamate dehydrogenase:

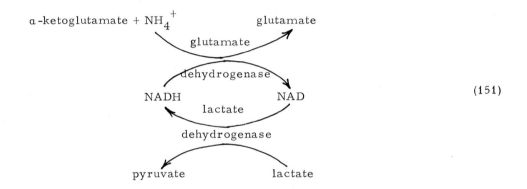

$$(151)$$

Pyruvate is produced in 2500-fold yield in 30 min and is measured in a second cycle with added NADH and lactate dehydrogenase. The rate of transformation is measured by following the change in the fluorescent NAD or NADH. Since the nucleotides are used at concentrations well below their Michaelis constants, the reaction rates are proportional to the nucleotide concentrations. The final product is again a pyridine nucleotide, so the cyclic process can be repeated with an overall multiplication factor of 10^6 to 10^8.

The system for NADP measurement described by Lowry et al. [119] utilizes glucose-6-phosphate dehydrogenase and glutamate dehydrogenase. Each molecule of NADP catalyzes the formation of up to 10,000 molecules of 6-phosphogluconate in 30 min. The 6-phosphogluconate is then measured in a second incubation with 6-phosphogluconate dehydrogenase and extra NADP. The NADPH produced is measured fluorometrically.

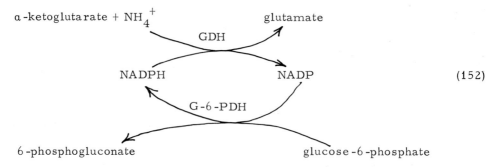

$$\text{(152)}$$

Two-cycle determinations have been performed on as little as 10^{-15} mole of NADP. This detectable concentration represents the amount that would be formed by 1000 molecules of an enzyme with a turnover number of 10^4 per minute, if it could be coupled to a NADP reaction. In principle one could measure as little as one single enzyme molecule by the reduction of sample size or of the blank [119].

V. DETERMINATION OF INHIBITORS

A. Introduction

An inhibitor is a compound that causes a decrease in the rate of enzyme reaction, either by reacting with the enzyme E to form an enzyme-inhibitor complex EI or by reacting with the enzyme-substrate intermediate ES to form a complex:

$$E + I \rightleftharpoons EI$$
$$E + S \rightleftharpoons ES \longrightarrow P + E$$
$$\downarrow I$$
$$E-S-I$$

$$\text{(153)}$$

In general the initial rate of an enzymatic reaction will decrease with increasing inhibitor concentration, linearly at low inhibitor concentrations, then gradually approaching zero.

Analytical working curves for inhibitor assay are generally constructed by plotting percentage of inhibition against concentration of inhibitor. The percentage of inhibition is calculated as follows:

$$\% \text{ inhibition} = \frac{(\text{rate})_{\text{no inhibitor}} - (\text{rate})_{\text{inhibitor}}}{(\text{rate})_{\text{no inhibitor}}} \times 100.$$

$$\text{(154)}$$

B. Assay of Inorganic Substances

Enzymatic methods have been described for the determination of many common inorganic cations and anions: Ag^+, Al^{3+}, Be^{2+}, Bi^{3+}, Ce^{3+}, Cd^{2+}, Co^{2+}, Cu^{2+}, Fe^{2+}, Hg^{2+}, In^{3+}, Mn^{2+}, Ni^{2+}, Pb^{2+}, Zn^{2+}, CN^-, $Cr_2O_7^{2-}$, F^-, and S^{2-}. A complete listing of these procedures was compiled by Guilbault et al. [4].

C. Assay of Pesticides

One of the most interesting newer aspects of enzymatic methods of analysis has been the selective assay of pesticides.

Several analytical methods that have been proposed are based on the inhibition of an enzyme reaction. Keller [120] has reported a fairly specific method for the determination of microgram concentrations of DDT. It is based on the inhibition of carbonic anhydrase, which is inhibited by DDT at concentrations at which other inhibitors (except the sulfonamides) are inactive. The most specific and sensitive method for some organophosphorus compounds is based on the inhibition of cholinesterase, and numerous papers have been published on this system. Kitz [121] has published a review with 158 references on the chemistry of anticholinesterase compounds. Giang and Hall [122] assayed TEPP, paraoxon, and other insecticides that inhibit cholinesterase in vitro, and Kramer and Gamson [123] have developed a colorimetric procedure with compounds related to the indophenyl acetates for the determination of 1 to 10 μg of various organophosphorus compounds.

Guilbault et al. [124] described an electrochemical method for the analysis of sarin, Systox, parathion, and malathion. The decrease in the rate of the cholinesterase-catalyzed hydrolysis of butyrylthiocholine iodide, as measured by dual-polarized platinum indicator electrodes, is linearly related to concentrations of the organophosphorus compounds.

Guilbault and Sadar [125] have developed sensitive methods based on the use of enzyme systems for the determinations of chlorinated insecticides, carbamate insecticides, and herbicides. Fluorescence methods were used because, whenever these have been tried, they have been shown to allow the determination of much lower concentrations of enzymatic inhibitors (for lower enzymatic activities can be measured). The substrate 4-methylumbelliferone heptanoate was used; it is cleaved by as little as 10^{-5} unit of lipase to 4-methylumbelliferone.

The chlorinated insecticides aldrin, heptachlor, lindane, and DDT, the carbamate Sevin, and the herbicide 2,4-D were found to inhibit lipolytic activity and could be determined in the concentration range 0.1 to 30 μg/ml.

The results reported indicate that the method of lipase inhibition is the
most sensitive enzymatic method yet reported for heptachlor, aldrin, lindane,
and 2,4-D.   The proposed inhibition scheme is likewise a good one for DDT.
Although carbonic anhydrase is inhibited by lower concentrations of DDT, the
lipase procedure is easier and more convenient to carry out.

A fluorometric procedure for the assay of aldrin, heptachlor, and methyl
parathion was developed by Guilbault, Sadar, and Zimmer [126].   The method
is based on the inhibition of acid and alkaline phosphatase by these pesticides.

The use of cholinesterase for the detection of pesticides has received the
most interest from analysts.   Reviews on cholinesterase and cholinesterase
inhibitors have been prepared by Cohen et al. [127] and Delga and Foulhoux
[128].

Matousek, Fischer, and Cerman [129] proposed a highly sensitive method
for the determination of organophosphorus compounds by their inhibition of
cholinesterase; indoxyl acetate was used as substrate.   Butygin and
Vyatchannikov [130] evaluated the toxicity of Sevin by its effect on cholinesterase activity in blood, and Davis and Maloney [131] and Zweig and Devine [132]
used anticholinesterase inhibition to study water polluted by organic pesticides
and their residues by an enzymatic cholinesterase method.

Abou-Donia and Menzel [133] studied the inhibition of fish-brain
cholinesterase by carbamates and developed an automatic method for the assay
of this pesticide.   Winteringham and Fowler [134], in a survey of the inhibition
of acetylcholinesterase by carbamates, suggested that the Sevin inhibition is
caused by an enzymatic destruction of the carbamates.

Weiss and Gakstatter [135] detected pesticides in water by the in vitro
inhibition of brain cholinesterase in fish.   Bluegill sunfish were found to be
most sensitive.   Matousek and Cerman [136] have reported a highly sensitive
and simple method for detecting cholinesterase inhibitors.   Paper impregnated
with butyrylthiocholine iodide and Bromthymol Blue as a pH indicator were
used.   The presence of an inhibitor was indicated by the inability of the
cholinesterase used to effect the hydrolysis of the substrate.

Beyman and Stoydin [137] reported a rapid screening test for cholinesterase
inhibition by pesticides making use of agar plates.   As little as 1 ng of DDVP
(Vapona) and other inhibitors could be detected.   Several workers described the
automation of cholinesterase-inhibition determinations with the AutoAnalyzer.
Voss [138] and Ott and Gunther [139] proposed procedures that required prior
extraction and cleaning, and Ott and Gunther in a later publication [140] used
the spots scraped off a thin-layer-chromatography (TLC) plate as input sample
for the AutoAnalyzer.   A test for the detection of organophosphorus pesticides

on TLC plates, using indoxyl acetate, was described by Ortloff and Franz [141].
Ackerman [142] used silica gel TLC plates for the semiquantitative determination of organophosphorus and carbamate pesticides.

A TLC enzymatic inhibition procedure for pesticides in plant extracts
without elaborate cleanup was described by Mendoza et al. [143,144), who
evaluated indoxyl, substituted indoxyl, and naphthyl acetates as substrates for
esterases in a TLC enzymatic detection system. Wales, McLeod, and
McKinley [145] extended these procedures of Mendoza et al. [143,144] to the
separation and detection of carbamates. Winterlin, Walker, and Frank [146]
described a TLC procedure for the separation and detection of 17 anticholinesterase pesticides on alumina and silica gel, and a thin-layer enzymatic
detection scheme for insecticides was outlined by Ackerman [147].

Because various animals and insects are known to be adversely affected
by low concentrations of various pesticides, cholinesterases from these sources
have been isolated and used for assay of pesticides. Guilbault et al. [148]
purified cholinesterases from honey bees and boll weevils and studied the effect
of 12 different pesticides on these enzymes. The boll weevil was used for the
specific assay of Vapona and paraoxon. Zhuravskaya and Bobyreva [149] found
the esterase of the cotton aphid was most sensitive to methylmercaptophos,
whereas the moth Laphygma was most sensitive to Sevin.

Soliman [150] and Goszezynska and Styczynska [151] studied the inhibition
of housefly-head cholinesterase by organophosphorus compounds, and Voss
[152] found peacock plasma to be a useful cholinesterase source for the
analysis of insecticidal carbamates.

The effect of 12 different pesticides on liver enzymes isolated from rabbit,
pigeon, chicken, sheep, and pig was studied by Guilbault, Sadar, and Kuan [153].
The cholinesterases from these sources were found to be inhibited at very low
concentrations by the organophosphorus pesticides DDVP, parathion, and
methyl parathion. None of these enzymes was inhibited by any of the
chlorinated pesticides, and pigeon- and sheep-liver cholinesterase were not
inhibited by Sevin.

Guilbault et al. [154] studied the effect of 15 different pesticides, including
carbamates, chlorinated hydrocarbons, and organophosphorus compounds on
cholinesterases from such insects as the housefly, sugar boll weevil, fire ant,
and German cockroach. Some of the results obtained are shown in Table 79.

Fire-ant cholinesterase is specifically inhibited by DDVP and methyl
parathion. Even parathion had little effect on this enzyme. None of the other
12 inhibitors--paraoxon, aldrin, captan, dalapon, DDT, dieldrin, 2,4-D-acid,
heptachlor, lindane, mirex, methoxychlor, and Sevin--inhibits it at all.
Cholinesterase from both types of housefly (NAIDM and DDT-resistant strains)

TABLE 79

I_{50} Values for Various Cholinesterases[a]

| Enzyme Source | Pesticide concentrations (M X 10^{-8}) | | | | | | |
	Paraoxon	Parathion	Methyl Parathion	DDVP	Sevin	Aldrin	Heptachlor
Housefly DDT- Resistant	x	-	-	6.3	82	-	-
Housefly (NAIDM)	x	-	-	6.5	88	-	-
Sugar boll weevil	x	-	-	530	-	-	-
Fire ant	-	-	150	9.6	-	-	-
Cockroach	3.2	20	73	4.8	93	500	600

[a]Key: a dash (-) indicates that the I_{50} is greater than 1×10^{-3} M or that inhibition does not reach 50%; x indicates that compound was not investigated.

are specific for DDVP and Sevin, and others have little or no effect up to 10^{-3} M concentration of the inhibitor. Although not as sensitive as other cholinesterases, that of the sugar boll weevil is totally specific for DDVP, below 10^{-6} M concentration. Parathion interferes at concentrations greater than 10^{-3} M; although paraoxon was not tried, it is expected to interfere also. The enzyme from American cockroaches is the most sensitive of all but lacks specificity. Although it is most sensitive to DDVP, paraoxon, parathion, and methyl parathion, a good number of chlorinated insecticides, carbamates, and herbicides inhibit it also.

From the results of this study it can be concluded that in addition to the sensitivity (see Table 79) a further specificity and selectivity in pesticide assay by using enzymatic methods have been achieved.

Guilbault and Sadar [155] have discussed a specific enzymatic method for the assay of chlorinated pesticides in the presence of herbicides, organophosphorus pesticides, carbamates, and fungicides. The method is based on the selective inhibition of hexokinase by aldrin, chlordane, DDT, and heptachlor (Table 80). Parts-per-million concentrations of these four pesticides have been assayed in the presence of large amounts of other chlorinated pesticides (DDD, DDE, dieldrin, lindane, methoxychlor, etc.), as well as other types of pesticides, with an accuracy of about 2%.

TABLE 80

Inhibition of Hexokinase by Pesticides[a]

Pesticide	I_{50} $(M \times 10^{-6})$
Aldrin	6. 8
Chlordane	12. 2
DDT	10. 8
Heptachlor	10. 75
Dieldrin	x
Lindane	-
2, 4-D	x
Paraoxon	-
Parathion	-
DDVP	-
Sevin	-

[a]Key: x indicates that inhibition does not excees 20 to 30% at high concentrations; - indicates no inhibition at any concentration.

REFERENCES

1. E. W. White, F. McCapra, and G. F. Field, J. Amer. Chem. Soc., 85, 337 (1963).

2. S. P. Colowick and N. O. Kaplan, eds., Methods of Enzymology, Academic Press, New York, 1957, p. 107.

3. H. Netter, Theoretische Biochemie, Springer-Verlag, Berlin, 1959, 554.

4. G. G. Guilbault, P. Brignac, and M. Zimmer, Anal. Chem., 40, 190 (1968).

5. G. G. Guilbault, P. Brignac, and M. Juneau, Anal. Chem., 40, 1256 (1968).

6. H. Pardue, M. Burke and D. O. Jones, J. Chem. Educ., 44, (11), 684 (1967).

7. G. G. Guilbault, Enzymatic Methods of Analysis, Pergamon Press, Oxford, England, 1970.

8. W. C. Purdy, Electroanalytical Methods in Biochemistry, McGraw-Hill, New York, 1965.

9. G. G. Guilbault, in Fluorescence, Theory, Instrumentation and Practice, (G. G. Guilbault, ed.), Dekker, New York, 1967, p. 297.

10. S. Udenfriend, Fluorescence Assay in Biology and Medicine, Vol. 1, Academic Press, New York, 1962.

11. R. Ruyssen and E. L. Vandenriesche, eds., Enzymes in Clinical Chemistry, Elsevier, Amsterdam, 1965.

12. R. E. Phillips and F. R. Elevitch, in Progress in Clinical Pathology (M. Steffani, ed.), Grune and Stratton, New York, 1966, p. 62.

13. H. U. Bergmeyer, ed., Methods of Enzymatic Analysis, 2nd ed., Verlag Chemie, Weinheim, 1965.

14. W. J. Blaedel and G. P. Hicks, in Advances in Analytical Chemistry and Instrumentation, Vol. 3 (C. N. Reilley, ed.), Interscience, New York, 1964, p. 105.

15. G. G. Guilbault, Anal. Chem., 38, 537R (1966); 40, 459R (1968); 42, 334R (1970).

16. H. U. Bergmeyer, Z. Anal. Chem., 212, 77 (1965).

17. T. M. Devlin, Anal. Chem., 31, 977 (1959).

18. M. Roth, Methods Biochem. Anal., 17, 189 (1969).

19. K. Oldham, U. S. Atomic Energy Commission NP-17585, 1968; Nucl. Sci. Abstr., 22, 44978 (1968).

20. O. H. Lowry, N. R. Roberts, and J. I. Kapphahn, J. Biol. Chem., 224, 1047 (1957).

21. G. G. Guilbault and D. N. Kramer, Anal. Chem., 35, 588 (1963).

22. G. G. Guilbault and D. N. Kramer, Anal. Chem., 37, 120 (1965).

23. G. G. Guilbault and D. N. Kramer, Anal. Chem., 37, 1675 (1965).

24. A. Seligman and R. Barnett, Science, 114, 579 (1959).

25. D. N. Kramer and C. Gelman, CRDL Rep., 1960, p. 541.

26. B. Gehauf and J. Goldenson, Anal. Chem., 29, 276 (1957).

27. A. Seaman and M. Winell, Acta Histochem., 8, 381 (1959).

28. A. Seligman and H. Ravin, Arch. Biochem. Biophys., 42, 337 (1953).

29. R. Tashian, Biochem. Biophys. Res. Commun., 14, 256 (1964).

30. D. M. Hercules and L. B. Rogers, Anal. Chem., 30, 96 (1958).

31. G. G. Guilbault, M. H. Sadar, R. Glazer, and C. Skou, Anal. Letters, 1, 365 (1968).

32. A. K. Prince, Biochem. Pharmacol., 15, 411 (1966).

33. G. G. Guilbault and A. Heyn, Anal. Letters, 1, 163 (1967).

34. B. Rotman, Proc. Natl. Acad. Sci. U.S., 47, 1981 (1961).

35. B. Rotman, J. Zdevic, and M. Edelstein, Proc. Natl. Acad. Sci. U.S., 50, 1 (1963).

36. A. A. Tydell and A. Scheparty, Federation Proc., 21, 161 (1962).

37. M. A. Verity, J. Gambell, and W. J. Brown, Arch. Biochem. Biophys., 118, 310 (1967).

38. D. Robinson, Biochem. J., 63, 39 (1956).

39. J. Mead, J. Smith, and R. T. Williams, Biochem. J., 61, 569 (1955).

40. M. A. Verity, R. Caper, and W. J. Brown, Arch. Biochem. Biophys., 106, 386 (1964).

41. L. S. Greenberg, Anal. Biochem., 14, 265 (1966).

42. G. G. Guilbault, D. N. Kramer, and E. Hackley, Anal. Biochem., 18, 241 (1967).

43. D. N. Kramer and G. G. Guilbault, Anal. Chem., 35, 588 (1963).

44. G. G. Guilbault and D. N. Kramer, Anal. Chem., 36, 409 (1964).

45. G. G. Guilbault and M. H. Sadar, Anal. Letters, 1, 551 (1968).

46. G. G. Guilbault and J. Hieserman, Anal. Chem., 41, 2006 (1969).

47. V. L. Stevens, Tech. Rept. IRL-1029, Dept. of Genetics, Stanford Univ., 1965.

48. E. L. Smith and R. L.•Hill, in The Enzymes (P. D. Boyer, H. Lardy, and K. Myrback, eds.), Vol. 4, Academic Press, New York, 1960, p. 41.

49. M. Roth, in Enzymes in Clinical Chemistry (R. Ruyssen and E. L. Vandenriesche, eds.), Elsevier, Amsterdam, 1965, p. 10.

50. A. Riedel and E. Wünsch, Z. Physiol. Chem., 316, 61 (1959).

51. M. Roth, Clin. Chim. Acta, 8, 574 (1963).

52. V. M. Sardesai and H. S. Provido, J. Lab. Clin. Med., 65, 1023 (1965).

53. V. M. Sardesai and H. S. Provido, J. Lab. Clin. Med., 64, 977 (1964).

54. T. Uete, M. Asahara, and H. Tsuchikura, Clin. Chem., 16, 322 (1970).

55. B. H. J. Bielski and S. Freed, Anal. Biochem., 7, 192 (1964).

56. G. Guroff, J. Biol. Chem., 239, 149 (1964).

57. E. F. Lüscher and R. Käser-Glasszman, Vox Sang., 6, 116 (1961).

58. A. R. Pappenhagen, J. Lab. Clin. Med., 59, 1039 (1962).

59. R. Strässle, Thromb. Diath. Haemorrhag., 8, 112 (1962).

60. D. W. Moss, Clin. Chem. Acta, 5, 283 (1960).

61. D. Campbell and D. W. Moss, Clin. Chim. Acta, 6, 307 (1961).

62.   L. J. Greenberg, Biochem. Biophys. Res. Commun., 9, 430 (1962).

63.   T. Takeuchi and S. Nagami, Acta Pathol. Japan, 4, 277 (1954).

64.   J. H. Tierney, Federation Proc., 23, 2381 (1964).

65.   J. Westley, Tech. Rept. IRL-1010, Dept. of Genetics, Stanford Univ.,
      July 1964.

66.   E. A. Shneour, Tech. Rept. IRL-1015, Dept. of Genetics, Stanford Uiv.,
      May 1965.

67.   D. B. Land and E. Jackim, Anal. Biochem., 16, 481 (1966).

68.   G. G. Guilbault, S. H. Sadar, R. Glazer, and J. Haynes, Anal. Letters,
      1, 333 (1968).

69.   G. G. Guilbault and A. Vaughan, Anal. Letters, 3, 1 (1970).

70.   G. G. Guilbault and A. Vaughan, Anal. Chem., 43, 721 (1971).

71.   G. G. Guilbault and J. Hieserman, Anal. Biochem., 26, 1 (1968).

72.   S. Ochoa, A. H. Mahler, and A. Kornberg, J. Biol. Chem., 174, 979
      (1948).

73.   S. Ochoa, J. B. S. Sales, and P. J. Ortiz, J. Biol. Chem., 187, 863
      (1950).

74.   P. Greengard, Nature, 178, 632 (1956).

75.   P. Greengard, in Methoden der enzymatische Analyse (H. Bergmeyer,
      ed.), Verlag Chemie, Germany, 1962, p. 551.

76.   O. H. Lowry, N. R. Roberts, and M. L. Chang, J. Biol. Chem., 222,
      97 (1956).

77.   O. H. Lowry, N. R. Roberts, and C. Lewis, J. Biol. Chem., 220, 879
      (1956).

78.   P. K. Maitra and R. W. Estabrook, Anal. Biochem., 7, 472 (1964).

79.   L. T. Graham and M. H. Aprison, Anal. Biochem., 15, 487 (1966).

80.   T. Laursen and G. Espersen, Scand. J. Clin. Lab. Invest., 11, 61 (1959).

81.   T. Laursen and P. F. Hansen, Scand. J. Clin. Lab. Invest., 10, 53 (1958).

82.   F. N. Pitts, C. Quick, and E. Robins, J. Neurochem., 12, 93 (1965).

83.   D. A. B. Young and A. E. Renold, Clin. Chim. Acta, 13, 791 (1966).

84.   J. R. Cooper, Biochem. Pharmacol., 13, 795 (1964).

85.   G. G. Guilbault and D. N. Kramer, Anal. Chem., 36, 2497 (1964).

86.   G. G. Guilbault and D. N. Kramer, Anal. Chem., 37, 1219 (1965).

87.   Private communication, Technicon Co., New York, 1971.

88. O. H. Lowry, O. A. Bessey, and E. J. Crawford, J. Biol. Chem., 180, 399 (1949).

89. G. G. Guilbault, D. N. Kramer, and E. Hackley, Anal. Chem., 39, 271 (1967).

90. A. S. Keston and R. Brandt, Anal. Biochem., 11, 1 (1965).

91. W. A. Andreae, Nature, 175, 859 (1955).

92. H. Perschke and E. Broda, Nature, 190, 257 (1961).

93. S. Udenfriend, W. Lavenberg, and H. Weissbach, Federation Proc., 19, 7 (1960).

94. H. F. Schott and W. C. Clark, J. Biol. Chem., 196, 449 (1952).

95. H. Weissbach, T. E. Smith, and S. Udenfriend, Biochemistry, 1, 137 (1962).

96. W. Lovenberg, R. Levine, and A. Sjoerdsma, Federation Proc., 20, 318 (1961).

97. V. Cohn and P. Shore, Federation Proc., 20, 258 (1961).

98. A. Bokman and B. Schweigert, J. Biol. Chem., 186, 153 (1950).

99. D. Duggan, R. Bourman, B. Brodie, and S. Udenfriend, Arch. Biochem. Biophys., 68, 1 (1957).

100. W. R. Jakoby and D. M. Bonner, J. Biol. Chem., 205, 699 (1953).

101. G. G. Guilbault, Anal. Chem. Annual Review, April 1968, 1970.

102. G. G. Guilbault, Enzymatic Methods of Analysis, Pergamon Press, Oxford, 1970.

103. H. Bergmeyer, Enzymatic Methods of Analysis, 2nd ed., Verlag Chemie, Weinheim, 1968.

104. G. G. Guilbault, M. H. Sadar, and K. Peres, Anal. Biochem., 31, 191 (1969).

105. G. G. Guilbault, S. H. Sadar, and R. McQueen, Anal. Chim. Acta, 45, 1 (1969).

106. S. H. Sadar, M. S. dissertation, Louisiana State University, New Orleans, January 1971.

107. P. Baum and R. Czok, Biochem. Z., 332, 121 (1959).

108. E. Adler, G. Gunther, and M. Plass, Biochem. J., 33, 1028 (1939).

109. O. Warburg and W. Christian, Biochem. Z., 298, 150 (1938).

110. F. B. Straub, Biochem. J., 33, 787 (1939).

111. F. M. Huennekens and S. P. Felton, Methods of Enzymology, Academic Press, New York, 1957, p. 950.

112.   R. Wahl and L. Kozloff, J. Biol. Chem., 237, 1953 (1962).

113.   A. M. Chase, in Methods of Biochemical Analysis (D. Glick, ed.),
        Vol. 8, Interscience, New York, 1960, p. 61.

114.   A. Kornberg, J. Biol. Chem., 182, 779 (1950).

115.   W. Thorn, G. Pfleiderer, R. A. Frowein, and I. Pors, Arch. Ges.
        Physiol. Pfluger's, 261, 334 (1955).

116.   P. Greengard and R. W. Straub, J. Physiol. (London), 148, 353 (1959).

117.   K. Seraydarian, W. F. H. M. Mommaerts, and A. Wallner, J. Biol.
        Chem., 235, 2191 (1960).

118.   E. C. Slater, Biochem. J., 53, 157 (1953).

119.   O. H. Lowry, J. V. Passonneau, D. Schulz, and M. K. Rock,
        J. Biol. Chem., 236, 2746 (1961).

120.   H. Keller, Naturwissenschaften, 39, 109 (1965).

121.   R. J. Kitz, Acta Anaesthes. Scand., 8 (4), 197 (1964).

122.   P. A. Giang and S. A. Hall, Anal. Chem., 23, 1830 (1951).

123.   D. N. Kramer and R. M. Gamson, Anal. Chem., 29 (12), 21A (1957).

124.   G. G. Guilbault, B. Tyson, D. N. Kramer, and P. L. Cannon,
        Anal. Chem., 34, 1437 (1960).

125.   G. G. Guilbault and M. H. Sadar, Anal. Chem., 41, 366 (1969).

126.   G. G. Guilbault, M. H. Sadar, and M. Zimmer, Anal. Chim. Acta,
        44, 361 (1969).

127.   J. Cohen, R. Oosterbaan, and F. Berends, Methods Enzymol., 11,
        686 (1967).

128.   J. Delga and P. Foulhoux, Prod. Probl. Pharm., 24, 57 (1969).

129.   J. Matousek, J. Fischer, and J. Cerman, Chem. Zvesti, 22, 184
        (1968).

130.   V. Butygin and K. Vyatchannikov, Zdravookhr. Beloruss., 56 (1968);
        cf. Refer. Zh. Biol. Khim., 1968, Abstr. No. 18F1810.

131.   T. Davis and G. Maloney, Water Sewage Works, 114, 272 (1967).

132.   G. Zweig and J. Devine, Residue Rev., 26, 17 (1969).

133.   M. B. Abou-Donia and D. B. Menzel, Comp. Biochem. Physiol., 21,
        99 (1967).

134.   F. Winteringham and K. S. Fowler, Biochem. J., 99, 6P (1966).

135.   C. M. Weiss and J. H. Galstatter, J. Water Pollut. Contr. Fed., 36,
        240 (1964).

136.   J. Matousek and J. Cerman, Procovni Lekorstvi, 16, 13 (1965).

137.   K. I. Beyman and G. Stoydin, Nature, 208, 748 (1965).

138.   G. Voss, J. Econ. Entomol., 59, 1288 (1966).

139.   D. E. Ott and F. A. Gunther, J. Assoc. Off. Agr. Chem., 49, 662 (1966).

140.   D. E. Ott and F. A. Gunther, J. Assoc. Off. Agr. Chem., 49, 669 (1966).

141.   R. Ortloff and P. Franz, Z. Chem., 5, 388 (1965).

142.   H. Ackerman, Nahrung, 10, 273 (1966); Chem. Abstr., 65, 9657a (1966).

143.   C. Mendoza, P. Wales, H. McLeod, and W. McKinley, Analyst, 93, 173 (1968).

144.   C. Mendoza, P. Wales, H. McLeod, and W. McKinley, Analyst, 93, 34 (1968).

145.   P. Wales, H. McLeod, and W. McKinley, J. Assoc. Off. Anal. Chem., 51, 1239 (1968).

146.   W. Winterlin, G. Walker, and H. Frank, J. Agr. Food Chem., 16, 808 (1968).

147.   H. Ackerman, J. Chromatogr., 36, 309 (1968).

148.   G. G. Guilbault, M. H. Sadar, S. Kuan, and D. Casey, J. Agr. Food Chem., 18, 692 (1970).

149.   S. Zhuravskaya and T. Bobyreva, Uzb. Biol. Zh., 12, 55 (1968).

150.   S. Soliman, Ain Shams Sci. Bull., 9, 127 (1966).

151.   K. Goszezynska and B. Styczynska, Rocz. Panstiv. Zakl. Hig., 19, 491 (1968).

152.   G. Voss, Bull. Environ. Contam. Toxicol., 3, 339 (1968).

153.   G. G. Guilbault, M. H. Sadar, and S. Kuan, Anal. Chim. Acta, 51, 83 (1970).

154.   G. G. Guilbault, M. H. Sadar, and S. Kuan, Anal. Chim. Acta, 52, 75 (1970).

155.   G. G. Guilbault and M. H. Sadar, J. Agr. Food Chem., 19, 357 (1971).

Chapter 9

CHEMILUMINESCENCE

I. INTRODUCTION

Light emission, as we have already discussed, can arise from the reemission of absorbed radiation. In chemiluminescence the emission process is identical, but the necessary excited molecule is produced by a chemical reaction. Chemiluminescence entails a three-step mechanism:

1. A "key intermediate" is formed by a preliminary reaction.

2. The chemical energy of this intermediate is converted to electronic excitation energy by an excitation step.

3. Fluorescence emission occurs from the excited reaction product of the excitation step.

The total quantum yield of chemiluminescence is the product of the yields of these three steps. Hence there are few efficient chemiluminescent reactions. The chemical reactions do not produce significant amounts of excited products. Furthermore, most excited molecules are not very fluorescent.

$$\text{Yield of excited emitter} = \frac{\text{chemiluminescence quantum yield}}{\text{fluorescence quantum yield}}. \qquad (155)$$

If the emitter can be identified (by matching the fluorescence spectrum observed

397

with that of the observed or a suspected reaction product), its quantum yield can be measured independently of the chemiluminescence reaction, and the yield of excited product can be determined by Eq. (155).

Bioluminescence is identical with chemiluminescence except that a biological reaction produces the luminescent species.

II. INSTRUMENTATION

Any conventional fluorometer can be easily converted to do chemiluminescence work. All that is required is that the exciting light source be blocked so that the luminescence can be produced by the chemical reaction. In the case of fast reactions an oscilloscope or a fast-response recorder can be hooked up to the output of the photomultiplier tube.

Du Pont Instruments (Wilmington, Del.) markets a luminescence biometer that can be used to monitor chemiluminescent or bioluminescent reactions. The instrument, shown in Fig. 106, possesses the following characteristics: automatic ranging of readout over five decades; digital read-out of concentration; a unique microsecond injection system with disposable filters; and a signal-storage system that measures the maximum intensity, compares it with a reference calibration, and displays the concentration corresponding to that signal intensity. The instrument sells for $4860, which includes reaction cuvettes and reagent for 500 ATP assays.

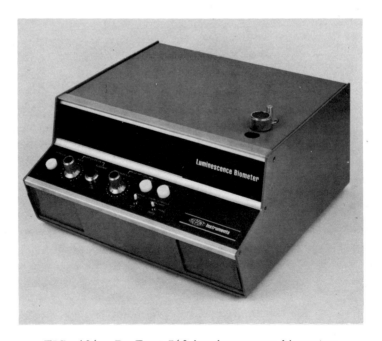

FIG. 106.   Du Pont 760 luminescence biometer.

American Instrument Company (Silver Spring, Md.) manufactures a chemiluminescent instrument, the CHEM-GLOW. The instrument is designed to measure ATP as small as 10^{-14} moles in a 10 μl sample volume and creatine phosphokinase. The basic instrument sells for $800, but an integrator-time accessory costs an additional $1200.

An ATP-Photometer is sold by JRB (La Jolla, California) for $3950. The range of the instrument is 10^{-9} mg ATP/ml to 1 mg ATP/ml. The instrument displays counts per minute for an unknown ATP sample which can be compared with a standard curve to determine the absolute quantity present. The time integral approach, incorporated in the ATP-Photometer, has standardized times for the integral.

III. APPLICATIONS

A. General Remarks

Scientists have been attracted by the luminosity of living organisms for many years. Fireflies are bioluminescence producers, giving off bright luminescence on a dark night; bacteria effect a luminescence of the surface of the sea.

Many chemical reactions have been screened for their luminescence-producing properties, but few have been found to be of analytical usefulness. In addition, the mechanism of the chemiluminescence process has been poorly understood in many cases. Rauhut [1] has investigated chemiluminescent reactions in great detail and has studied conditions for optimizing the quantum efficiency of such reactions. The net result has been the first truly useful luminescent product, a "chemiluminescent stick," sold commercially by American Cyanamid Co., which on breaking exhibits a very intense green chemiluminescence that lasts 3 h. This chemical flare is being sold in gasoline stations and other markets around the United States.

B. Chemiluminescent Indicators

Chemiluminescent compounds have been commonly used as indicators in acid-base and in oxidation-reduction titrations. Their great advantage is that they can be easily observed in colored and turbid solutions. Their use is based on the fact that chemiluminescence is initiated only under definite conditions of oxidation-reduction potential and pH. If the conditions for the start or termination of a chemiluminescence reaction are those obtained at the end point of a titration, the indicator will function properly to yield luminescence or to be quenched of luminescence.

Some typical compounds used as indicators for oxidation-reduction reactions are siloxene, luminol (3-aminophthalhydrazide), lucigenin (N, N'-dimethyl-9, 9'-bisacridinium nitrate), and lopine.

Siloxene (126) has been used as an indicator in the titration of (a) Fe(II) with Ce (IV) [2]; (b) iodide, As(III), oxalate, Sn(II), and Mo(III) with permanganate [3]; (c) Cd(II) with ferricyanide [4]; and (d) Tl(I) iodide with dichromate [3].

Luminol (127) is another useful indicator. It must be used in a basic medium with an oxidizing agent to produce its greenish white fluorescence. It has been used for the hypobromite titration of AsO_3^{3-}, SbO_3^{3-}, SO_3^{2-}, $S_2O_3^{2-}$, CN^-, and SCN^- [5]; in the AsO_3^{3-} titration of ClO^- and BrO^- [6]; and in the hypochlorite titration of AsO_3^{3-}, and N_2H_4 [7].

(126)   (127)

Lucigenin (128) has been used for the titration of H_2O_2 and $Fe(CN)_6^{3-}$, ClO^-, BrO^-, and IO^- [8], and for the titration of ClO^- and BrO^- by N_2H_4 [9]. Similarly lopine (129) has been used.

Chemiluminescence indicators have also been used for the determination of the end point of acid-base neutralization reactions. To accomplish this the indicator (luminol, lucigenin, or lopine), an oxidant (usually peroxide), and a catalyst (ferricyanide or blood hemin for luminol and lopine) are added to the acid, and the solution is titrated against a solution of an alkali. The pH of the solution changes during the reaction, and chemiluminescence occurs at the end point.

(128)   (129)

The pH for the initiation of luminescence is 8 to 8.5 for luminol [10], 9 to 10 for lucigenin [11], and 9 to 10 for lopine [12]. The proper indicator is chosen depending on the end-point pH. Since chemiluminescence reactions are temperature dependent, the pH of the solution should be controlled [13].

C. Chemiluminescent Analytical Reactions

1. Luminol

Probably one of the best known of the chemiluminescent compounds is luminol (127). It is stable indefinitely in the absence of oxygen, but light is produced when a basic solution of luminol is treated with an oxidizing agent [14].

White et al. [15, 16] showed that the mechanism of the chemiluminescence production involves first the formation of a dinegative ion of luminol (130), which reacts with oxygen or an oxidizing agent to yield an excited singlet state of the aminophthalate ion (131) and the emission of light by this species.

$$\quad\quad\quad\quad (130) \quad\quad\quad\quad\quad\quad\quad (131) \quad\quad\quad\quad\quad\quad\quad\quad\quad\quad\quad (156)$$

The luminol reaction has been used for the determination of oxidizing agents like hydrogen peroxide and for metal ions like copper or cobalt, which catalyze this chemiluminescence reaction. Ponomarenko [17] used the luminol reaction for the quantitative determination of Cu, H_2O_2, and other oxidants. Luminescence values were determined for chloro-, amino-, and nitro-substituted phthalhydrazide. He found that both the intensity and the duration of the light emitted by luminol with hydrogen peroxide and iron-hemin complexes were dependent on the concentration of both these substances [18].

Babko and Lukovskaya [19-21] determined hydrogen peroxide, copper, and cobalt by means of the luminol reaction. By holding the luminol and hydrogen peroxide concentration constant, the total fluorescence was found to be proportional to the concentration of copper or cobalt. Alternatively, with the concentration of luminol and copper constant, hydrogen peroxide could be determined. As little as 0.002 and 0.03 µg/ml of cobalt or copper, respectively, and 0.2 µg/ml of hydrogen peroxide could be determined. Iron and copper interfered in the determination of cobalt, but the interference could be eliminated by the use of sodium salicylate. Zinc and nickel did not interfere in either determination. In fact as little as 1×10^{-4} % of copper in metallic zinc can be determined by the luminol reaction [22]. The measurements could be performed either photographically or with a recording photometer.

Armstrong and Humphreys [23] have proposed a dosimeter based on the copper-catalyzed reaction of luminol and hydrogen peroxide. The hydrogen peroxide produced in irradiated water could be determined in the range of 3×10^{-8} to 1.5×10^{-5} M.

Small quantities of ozone can be determined by its effect on the luminescence intensity and color of luminol, fluorescein, and fuchsin, when adsorbed on silica gel [24]. Air is drawn through columns of silica gel impregnated with one of these compounds, and the height of the column where luminescence is quenched is proportional to the concentration of ozone in the air. For example, 0.15 µg of ozone caused an extinguished segment of 1 mm with luminol, whereas 0.34 and 0.39 µg were needed to cause similar quenching on fluorescein and fuchsin columns.

2. Indole

Patrovsky [25] has described a method for the assay of organophosphorus compounds ("nerve gas") in the atmosphere, based on their catalysis of the oxidation of indole by perborate ($NaBO_2 \cdot H_2O_2$) to the fluorescent leuco-indigo.

A detailed study of the factors affecting the progress of this reaction and a description of an apparatus designed to follow this reaction was presented in 1958 [26]. The accelerating effect of these organophosphorus compounds on the chemiluminescent reaction of peroxide or perborate with luminol was described by Bozhevol'nov and Yanishevskaya [27]. The catalytic effect was believed due to the P-F or P-CN bonds:

$$R-P-X + H_2O_2 \longrightarrow R-P-O-O-H + HX \qquad (157)$$

$$R-P-O-O-H + \text{indole} \longrightarrow \text{"luminescence"} \qquad (158)$$
$$(\text{luminol})$$

3. Rhodamine B

A novel method for the assay of ozone in the 0.0003% range in air involves the chemiluminescence of Rhodamine B in an ethanol solution of gallic acid [28]. The ozone produces an activated intermediate from gallic acid, which in turn excites Rhodamine B to an activated state. Luminescence is observed on return to the ground state.

4. Chemiluminescence from Concerted Hydrogen Peroxide Decomposition Reactions

Rauhut [1], in search for chemiluminescence reactions with high quantum efficiencies for use in practical light-producing systems, has reported on a

number of concerted hydrogen peroxide decomposition reactions leading to the
design of several new chemiluminescent reactions.

Rauhut [29] and McCapra and co-workers [30, 31] investigated the hydrogen
peroxide-acridine (132) system:

$$\text{(159)}$$

(132)

This reaction was found capable of producing quantum yields as high as 3%
under suitable conditions.

Chemiluminescence is obtained from the reaction of oxalyl chloride,
hydrogen peroxide, and a fluorescent compound like anthracene [32]. The
reaction was found to be capable of relatively high ($\sim$5%) quantum yields.

$$\text{(160)}$$

$$\text{(161)}$$

Attempts by Rauhut to discover other types of peroxyoxalate luminescent
reactions led to the discovery that some very useful high-quantum-efficiency
reactions can be obtained (Table 81).

In each reaction of Table 81 light is derived from the first singlet excited
state of the fluorescing compound and a variety of fluorescing compounds may be
used to provide a range of colors. Efficient chemiluminescence requires that
the starting oxalate have a high order of reactivity toward hydrogen peroxide;
this in part accounts for the variation in maximum reported quantum yields.
Most of the reactions are base catalyzed. The key chemiluminescent intermediate in each reaction is derived from a 1:1 H_2O_2-oxalate stoichiometry.

Rauhut [1] concluded that it is possible to design new chemiluminescent
reactions and that peroxyoxalate chemiluminescence has demonstrated that
efficient nonenzymatic reactions are possible. Some of such systems have gone
into design of the "chemical light sticks" now available commercially from the
American Cyanamid Co.

TABLE 81

High-Quantum-Efficiency Peroxyoxalate Luminescent Reactions

Reaction	Quantum efficiency (%)	Ref.
(structure) + H$_2$O$_2$ + rubrene $\xrightarrow{\text{light}}$	23	33
(structure) + H$_2$O$_2$ + rubrene $\xrightarrow{\text{light}}$	18	34
(structure) + H$_2$O$_2$ + DPA[a] $\xrightarrow{\text{base}}$ light	8.7	35
(structure) + H$_2$O$_2$ + BPEA[b] $\xrightarrow{\text{base}}$ light	1.8	34
(structure) + H$_2$O$_2$ + rubrene $\xrightarrow{\text{acid}}$ light	16	36
$(C_6H_5)_3$ C-C-O-C-C-O-C-C-C(C$_6$F$_5$)$_3$ + H$_2$O$_2$ + DPA[a] $\xrightarrow{\text{light}}$	13	37

[a] DPA: 9,10-diphenylanthracene.

[b] BPEA: 9,10-bis(phenylethynyl)anthracene.

5. Other Systems

Hydrogen peroxide with sodium hypochlorite produces two fluorescent
peaks that indicate molecular oxygen [38].   Khan and Kasha [39] have described
molecular-oxygen-sensitive chemiluminescent systems that are mixtures of
organic compounds and oxidizing agents.   Others are organic compounds that
can be used for luminescent marking of surfaces [40], such as a mixture of
paraffin, mineral oil, a microcrystalline wax, and silicone with the intensely
chemiluminescent compound tetrakisdimethylaminoethylene (133) [41].   These
oxygen-sensitive compounds are used to provide luminescent night markers.

$$
\begin{array}{cc}
(CH_3)_2N & N(CH_3)_2 \\
\diagdown C = C \diagup \\
\diagup & \diagdown \\
(CH_3)_2N & N(CH_3)_2
\end{array}
$$

(133)

D. Bioluminescence Reactions

The mechanism of the firefly reaction and the composition, structure, and
synthesis of luciferin have been elucidated by White and co-workers at Johns
Hopkins [42-44], and the luminescent reaction has been performed in vitro by
mixing cell-free extracts and even pure reactants.   The firefly reaction requires
ATP and Mg^{2+} in addition to luciferin, luciferase, and oxygen [43]:

$$
luciferin + O_2 + ATP \xrightarrow[\text{Mg}^{2+}]{\text{luciferase}} oxyluciferin + ADP + P_i \tag{162}
$$

Ordinary fluorometric equipment can be easily converted for use with chemiluminescence reactions, so that one can measure the rate of luminescence
formation (or the peak luminescence) and use this to determine luciferase,
Mg^{2+}, O_2, or ATP.   This reaction was found to be more rapid, more reproducible, and more accurate than any previously described method of ATP determination [45].   As little as 10^{-13} mole of ATP is determinable, and the method
is highly specific for ATP.   Only ADP and guanosine triphosphate were found to
give a slight positive interference.   The yellow light of the firefly has a
λ_{em} of 562 and 610 nm at pH 7.0 and 5.4, respectively.

This reaction can also be used to assay oxygen at partial pressures below
10^{-3} mm [46], when the gas is passed through a bacterial suspension containing
all requirements for the luminescent reaction except oxygen.   Coenzyme A is
also determinable, since the oxyluciferin produced in the reaction complexes
and inactivates the luciferase, causing the luminescence to stop.   Coenzyme A

removes the inhibiting oxyluciferin from the luciferase, causing an increase in
the luminescence that can be correlated with coenzyme A concentrations.

The use of the luciferin-luciferase reaction to determine the concentration
of bacteria by the measurement of the total luminescence produced is claimed
to be one of the greatest advances for counting bacteria in the last 100 years
[47].

The Du Pont 760 Luminescence Biometer (see Fig. 106) was originally
designed to measure the firefly luciferin-luciferase reaction for rapid assay of
ATP (10^{-13} g per 10-μl injection) and for measurement of any substance that
can be coupled to a reaction involving ATP.

The luminol-peroxide system has also been used to detect microorganisms,
as was demonstrated by the assay of six types of pathogenic microorganism in
water [48].

The detection of as little as 5 ppb of calcium was described by Shimomura
et al. [49], who utilized the luminescence produced by the protein aequorin,
which is extracted from jellyfish, in the presence of calcium.

IV.   OTHER REFERENCES

Excellent reviews on chemiluminescence are "Chemiluminescence in
Solutions," by Haas [50], which contains 76 references on theory, mechanisms,
and methods; "Chemiluminescent Techniques in Chemical Reactions," by
Shlyapintokh et al. [51]; Chemiluminescence Analysis, by Babko et al. [52],
a 250-page book in Russian; and Luminescent Redox Titrations, a 19-page
review by Erdey and Buzas [53], of analytical chemiluminescent methods.

REFERENCES

1.   M. M. Rauhut, Acc. Chem. Res., 2, 80 (1969).

2.   I. Buzas and L. Erdey, Talanta, 10, 467 (1963).

3.   L. Erdey, I. Buzas, and L. Polos, Z. Anal. Chem., 169, 187, 263 (1959).

4.   F. Kenney and R. Kurtz, Anal. Chem., 36, 529 (1964).

5.   L. Erdey and I. Buzas, Acta Chim. Acad. Sci. Hung., 6, 93 (1955).

6.   L. Erdey and I. Buzas, Acta Chim. Acad. Sci. Hung., 6, 123 (1955).

7.   L. Erdey and I. Buzas, Acta Chim. Acad. Sci. Hung., 6, 115 (1955).

8.   L. Erdey and I. Buzas, Acta Chim. Acad. Sci. Hung., 6, 77 (1955).

9.   L. Erdey and I. Buzas, Acta Chim. Acad. Sci. Hung., 6, 127 (1955).

10. F. Kenney and R. Kurtz, Anal. Chem. , 23, 339 (1951); 24, 1218 (1952).

11. L. Erdey, Acta Chim. Acad. Sci. Hung. , 3, 81 (1953); R. Parizek and L. Moucka, Chem. Listy, 48, 626 (1954).

12. L. Erdey and I. Buzas, Anal. Chim. Acta, 15, 322 (1956).

13. A. Ponomarenko, Dokl. Akad. Nauk SSSR, 102, 539 (1955).

14. J. Goldenson, Anal. Chem. , 29, 877 (1957).

15. E. White, O. Zafirou, H. Kagi, and H. Hill, J. Amer. Chem. Soc. , 86, 940 (1964).

16. E. White and M. Buisey, J. Amer. Chem. Soc. , 86, 941 (1964).

17. A. Ponomarenko, Tr. Mosk. Obshichestva Ispytalelei Pirody, Otd. Biol. , 21, 165 (1965).

18. E. Bovalini and M. Prazyi, Ann. Chim. , 53, 1103 (1963).

19. A. Babko and M. Lukovskaya, J. Anal. Chem. USSR, 17, 47 (1962).

20. A. Babko and N. Lukovskaya, Zavod. Lab. , 29, 404 (1963).

21. A. Babko and N. Lukovskaya, Zavod. Lab. , 29, 409 (1963).

22. A. Babko and L. Dubovenko, Zavod, Lab. , 30, 1325 (1964).

23. W. Armstrong and W. Humphreys, Can. J. Chem. , 43, 2576 (1965).

24. E. Peregud and E. Stephanenko, Zh. Anal. Khim. , 15, 96 (1960).

25. V. Patrovsky, Chem. Listy, 48, 537 (1954).

26. E. A. Bozhevol'nov, Tr. VNII Khim. Reak. , 22, 70 (1958).

27. E. A. Bozhevol'nov and V. Yanishevskaya, Zh. V. Kh. O. , 5, 356 (1960).

28. D. Bersis and E. Vassilow, Analyst, 91, 499 (1966).

29. M. M. Rauhut et al. , Office of Naval Research Tech. , Rept. AD No. 419-212, (1964).

30. F. McCapra and D. Richardson, Tetrahedron Letters, 43, 3167 (1964).

31. F. McCapra, D. Richardson, and Y. Chang, Photochem. Photobiol. , 4, 1111 (1965).

32. E. Chandross, Chem. Commun. , 761 (1963).

33. M. M. Rauhut, J. Amer. Chem. Soc. , 89, 6515 (1967).

34. L. J. Bollyky, R. H. Whitman, B. G. Roberts, and M. M. Rauhut, unpublished work.

35. L. J. Bollyky, R. H. Whitman, and B. G. Roberts, J. Org. Chem. , 33, 4266 (1968).

36. L. J. Bollyky, B. G. Roberts, R. H. Whitman, and J. Lancaster, J. Org. Chem. , in press (1972).

37. L. J. Bollyky, R. H. Whitman, B. G. Roberts, and M. M. Rauhut, J. Amer. Chem. Soc. , 89, 6523 (1967).

38. J. Scott, M. John, and R. Phillips, U. S. Pat. 3, 366, 572 (1968).

39. A. Khan and M. Kasha, J. Chem. Phys. , 39, 2105 (1963).

40. L. E. Huniston, U. S. Pat. 3, 375, 176 (1968).

41. H. Winberg, J. Downing, and D. Coffman, J. Amer. Chem. Soc. , 87, 2054 (1965).

42. E. White, F. McCapra, and G. Field, J. Amer. Chem. Soc. , 83, 2402 (1961).

43. E. White, F. McCapra, and G. Field, J. Amer. Chem. Soc. , 85, 337 (1963).

44. E. White and J. Harding, J. Amer. Chem. Soc. , 86, 944 (1964).

45. G. Winter, Ann. N. Y. Acad. Sci. , 87, 875 (1960).

46. A. Chase, in Methods of Biochemical Analysis, Vol. 8 (D. Glick, ed.), Interscience, New York, 1960, p. 61.

47. D. Eustachio, DuPont Mag. , 62, No. 41 (July 1968).

48. W. Oleniacy, M. Pisano, M. Rosenfeld, and R. Elgart, Environ. Sci. Technol. , 2, 1030 (1968).

49. O. Shimomura, F. Johnson, and Y. Saiga, Science, 140, 13339 (1963).

50. J. Haas, J. Chem. Educ. , 44, 396 (1967).

51. V. Shlyapintokh, O. Karpukhin, L. Postnikov, V. Tsepalov, and A. Vichutinskii, Chemiluminescent Techniques in Chemical Reactions, Consultants Bureau, New York, 1968.

52. A. Babko, L. Dubovenko, and N. Lukovsky, Khemilyumineststentryi Analy. Tekhnika, Kiev, 1966; Chem. Abstr. , 67, 96694g (1967).

53. L. Erdey and I. Buzas, Luminescence Redox Titrations, AD62, 7166 CESTI, 1966.

Chapter 10

ATOMIC FLUORESCENCE FLAME SPECTROMETRY[*]

---

[*]Based in part on Chapter 13, by J. D. Winefordner and J. M. Mansfield, of Fluorescence. Theory, Instrumentation, and Practice (G. G. Guilbault, ed.), Dekker, New York, 1967.

I.   INTRODUCTION

In the preceding chapters I have described the technique of molecular luminescence, i. e. , the emission of radiation by a molecule that has been raised to an excited state by the absorption of energy (radiative, chemical, or electrochemical).

In the same way atoms can be excited to a higher state by the absorption of radiation, and the radiation emitted when a fraction of these atoms lose their energy by a radiational process or processes can be measured. This technique, called atomic fluorescence, was first observed in the early 1900s [1, 2].

The technique of atomic fluorescence flame spectrometry, in which the atomic vapor is in a flame cell, is of more recent origin. This method is similar in principle to atomic emission (AE) and atomic absorption (AA) flame spectrometry, differing only in the mechanism of excitation and the method of measurement of the characteristic signals. In AE flame spectrometry the analyte atoms are excited by collision with flame-gas molecules. Some of the radiation emitted when a fraction of the excited atoms undergoes radiational deactivation is measured. In AA flame spectrometry the analyte atoms are excited by means of an external light source and the fraction of radiation absorbed as a result of radiational excitation is measured. In atomic fluores-cence (AF) flame spectrometry the analyte atoms are excited as in AA, but the

AF radiation resulting when a fraction of the excited atoms undergo radiational deactivation is measured.

The atomic fluorescence of metal atoms (Ca, Sr, Ba, Li, Na) in flames was observed for the first time by Nichols and Howes [3] in 1924. Badget [4] in 1927 studied the fluorescence of thallium, mercury, magnesium, copper, silver, cadmium, and sodium.

Alkemade [5] has reviewed the methods by which atoms are excited in flames. One of these methods is radiational. He described the use of AF flame spectrometry for measuring quantum efficiencies, and he used this method for measuring the quantum efficiency for the sodium D doublet.

Winefordner and Vickers [6] first described the use of AF flame spectrometry for chemical analysis. Winefordner and Staab [7, 8] used AF flame spectrometry for measuring zinc, cadmium, mercury, and thallium in aqueous solutions. Mansfield, Winefordner, and Veillon [9] used a simple experimental system to measure zinc, cadmium, mercury, and thallium, in the parts-per-billion concentration range. Veillon et al. [10] used a xenon-arc source to obtain the analytical curves and limits of detection of 13 elements by AF flame spectrometry. Dagnall, West, and Young [11] have described the determination of cadmium by AF and AA flame spectrometry. Armentrout [12] measured nickel in the parts-per-million concentration range by flame spectrometry, and Prugger [13] has described an experimental system for measuring zinc. Goodfellow [14] studied the fluorescence of zinc, cadmium, copper, gallium, and indium and found no interference from great excesses of other metals. Dagnall, Thompson, and West [15] have studied the influence of varying a number of experimental parameters on the fluorescence of 10 elements. Ellis and Demers [16] obtained excellent sensitivities by AF flame spectrometry for magnesium, silver, zinc, and calcium by using a hydrogen-entrained air flame and a 450-W xenon-arc lamp. West [17, 18] has reviewed spectrochemical methods for inorganic analysis and included a summary of AF research and results.

Excellent recent reviews on atomic fluorescence have been prepared by Winefordner, Svoboda and Chine [19] and Winefordner and Elser [20].

II. TYPES OF ATOMIC FLUORESCENCE RADIATION [1,6]

There are four types of atomic fluorescence, as illustrated in Fig. 107. Resonance fluorescence results when atoms are excited from the ground electronic state to an excited state and then undergo radiational deactivation to the ground state, reemitting radiation of the same energy that was absorbed. The most intense resonance fluorescence generally occurs from the first excited

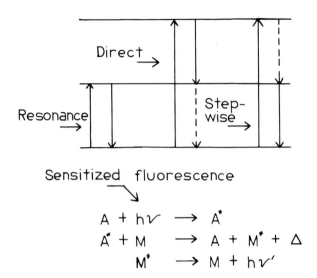

$$A + h\nu \longrightarrow A^{*}$$
$$A^{*} + M \longrightarrow A + M^{*} + \Delta$$
$$M^{*} \longrightarrow M + h\nu'$$

FIG. 107. Types of atomic fluorescence. Symbols: A = atom or molecule in high concentration; M = metal atom in low concentration.

state of an atom. For example, resonance fluorescence occurs when zinc or cadmium atoms absorb and reemit the 2139- and 2288-Å lines, respectively. The transition probabilities for resonance transitions are usually much greater than they are for other transitions. Resonance fluorescence has been and will probably continue to be of the greatest interest to analysts using AF flame spectrometry because resonance-fluorescence intensities are generally significantly greater than the intensities observed with other types of atomic fluorescence. This, of course, is also the basis of the so-called resonance detectors. In addition to resonance fluorescence, two other types of atomic fluorescence (direct line and stepwise line fluorescence) could possibly be used to extend the method's range of application.

Direct line fluorescence results when an atom is raised to an excited state considerably above the ground state and then undergoes a radiational transition to a lower excited state (not the ground state). For example, direct line fluorescence occurs in the emission of the thallium 5350-Å line after excitation of thallium atoms by the thallium 3776-Å line. Stepwise line fluorescence results when an atom is excited to an electronic state above the first excited electronic state, undergoes deactivation (usually radiationless) to a lower excited state, and finally undergoes the fluorescence transition to a lower state (often the ground state). For example, stepwise line fluorescence occurs in the emission of the sodium 5890-Å line after excitation of sodium atoms by the sodium 3303-Å doublet.

The final type of atomic fluorescence, called sensitized fluorescence, occurs when donor atoms (or molecules) are excited by means of an external light source; the excited donor species collides with the sample atom (acceptor), transferring energy and exciting the sample atom; and then the sample atom undergoes radiational deactivation, resulting in atomic fluorescence. For example, if a quartz container has a high concentration of mercury vapor and a low concentration of some metal (e. g. , thallium) vapor, mercury atoms excited by means of a low-pressure mercury arc transfer energy quite effectively to thallium atoms, which then emit their 3776- and 5350-$\overset{\circ}{\text{A}}$ lines. Unfortunately, as long as flame cells are used, sensitized fluorescence will undoubtedly never be of analytical use because the concentration of the donor species (mercury atoms in our example) necessary for energy transfer can never be made sufficiently large to allow an efficient transfer of energy. In addition, the probability of collisional deactivation of excited sensitizer species in flames is much greater than deactivation via energy transfer. Sensitized fluorescence might, however, be used for analytical purposes if a heated quartz cell or other nonflame cell were used to produce the atomic vapor. It should be mentioned that the lack of efficient energy transfer is advantageous to AF flame spectrometry because the so-called interelement effects should therefore be negligible.

III. THEORY

A. The Profile of Atomic Absorption Lines [1, 21]

The absorption of radiation by atoms occurs over a very narrow frequency interval called an atomic absorption line. The absorption-line profile and half-width are determined by broadening parameters. The absorption of radiation results in the electronic activation of an atom from the ground (sometimes from metastable states above the ground) electronic state 1 to an upper electronic state u. The excited atom remains in the upper level u for a short period, for example, 10^{-8} sec, and then undergoes a nonradiational deactivation via collisions with flame-gas molecules such as H_2O, CO_2, O_2, N_2, or CO, or via a radiational deactivation that is called atomic fluorescence and is emitted isotropically.

The intensity of radiation transmitted through an atomic vapor of absorbers, I , varies with frequency and is related to the incident intensity I^0 at the same frequency ν and to the product of the atomic absorption path length L (in cm) times the atomic absorption coefficient k_ν (in cm^{-1}). The expression relating

I_ν to these factors is given by

$$I_\nu = I_\nu^0 \, \exp(-k_\nu L) = I_\nu^0 \, 10^{-0.43 k_\nu L}. \tag{163}$$

The exponent on the right-hand side is more commonly called the absorbance, extinction, or optical density.

The broadening of a spectral line results in the absorption of radiation by atoms over a finite frequency interval $\Delta\nu$ rather than an infinitely narrow frequency interval. The half-width $\Delta\nu$ of an atomic absorption line is defined as the half-width of the plot of k_ν versus ν. Therefore it is the width of the absorption-line profile at $k_{max}/2$, where k_{max} is the maximum (or peak) atomic absorption coefficient.

The major causes of spectral line broadening of atoms in flames and in hot gases in general are Doppler broadening due to thermal agitation of the absorbing or emitting atoms and collisional broadening due to perturbation of the absorbing or emitting atoms by foreign gas atoms or molecules (collisions by foreign species can not only broaden the spectral line but also cause shift in the peak frequency ν_0 and asymmetry of the spectral line shape). In the discussion of this chapter, however, shift and asymmetry effects are assumed to be negligible.

Resonance broadening is generally negligible in atomic absorption studies and need not be considered. It is similar to collisional broadening but is due to collisions between the same type of absorbing or emitting atoms. Natural broadening is due to the finite lifetime of an atom in the excited state. Stark broadening is due to the splitting of an atomic line due to the application of an electric field. It is negligible for atoms in flames and is also generally not significant for atoms in hollow-cathode or electrodeless discharge tubes; it is, of course, important in arcs or sparks of high charge density. Zeeman broadening is a result of the splitting of atomic lines into hyperfine components due to the application of a magnetic field. This effect is negligible unless an extremely strong external magnetic field is applied, and this is not done in analytical studies.

For flame temperatures between 1500 and 3000°K the Doppler half-width for most lines [20] of most elements is on the order of 0.006 to 0.09 Å. The Doppler half-width $\Delta\nu_D$ varies with $\sqrt{T}$, where T is the temperature (in °K) of the atomic vapor--that is, the temperature of the flame or the atomic vapor within a discharge lamp; For most resonance lines (generally resonance lines are used in AF flame spectrometry), of most atoms, the collisional half-widths [20] are between 0.01 and 0.1 Å for analytical flames at 1-atm pressure (i. e., for acetylene-air, hydrogen-oxygen, etc., flames). The collisional half-width $\Delta\nu_C$ varies with $\sqrt{1/T}$. The natural half-width of most resonance lines [22] is about 10^{-4} $\overset{\circ}{A}$. This is a negligible contribution to the total width of a spectral line, which is primarily determined by Doppler and collisional broadening (Doppler and collisional half-widths add quadratically to give the total half-width of the line).

B. The Atomic Absorption Coefficient [1, 21]

The atomic absorption coefficient (in cm^{-1}) as a function of frequency k_ν can be expressed by the well-known Voigt profile expression, which accounts for line broadening due to a Gaussian effect (Doppler broadening) and due to a Lorentzian effect (collisional and natural broadening). Instead of general Voigt profile expressions we give here the simplified expression

$$k_\nu = \frac{k^0 a \rho}{\pi},$$
(164)

where a is the damping constant defined by $\sqrt{\ln 2}$ times the ratio of the collisional half-width to the Doppler half-width, ρ is a complex function of the types of broadening (fortunately tables of ρ values [23], in which ρ is given for a number of values of a and frequency ν are available), and k^0 is the atomic absorption coefficient at frequency ν_0. For the case of pure Doppler broadening k^0 is given by

$$k^0 = \frac{2 \sqrt{\ln 2} \; \lambda_0^2 \, g_u N_1 A_t}{8 \pi^{3/2} \, \Delta \nu_D g_1},$$
(165)

where λ_0 is the wavelength (in centimeters) at the line center, the g terms are statistical weights of the upper and lower energy levels involved in the transition, N_1 is the concentration of atoms in the lower (ground) energy level (in atoms per cubic centimeter) in the hot gases, $\Delta \nu_D$ is the Doppler half-width (in sec^{-1}), and A_t is the probability of the transition u $\longrightarrow$ 1 (in sec^{-1}).

C. The Total Absorption Factor A_T for Radiation
Absorbed from a Continuous Source [1, 21]

The total absorption factor A_T represents the half-width of radiation absorbed by an isolated spectral line from a continuous source and is given by

$$A_T = \int_0^\infty [1 - \exp(-k_\nu L)] \, d\nu \; (\text{sec}^{-1}),$$
(166)

where all terms are as previously defined. The total absorption can be pictorially represented by the area under any of the curves shown in Fig. 108, which also shows the variation in A_T with atomic concentration.

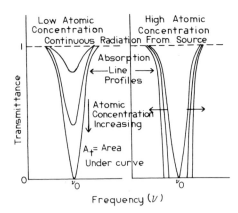

FIG. 108. Pictorial representation of the case where the total absorption-source-line half-width is much greater than the absorption-line half-width.

At small absorbances (i. e. , small values of $k_\nu L$ at all frequencies) A_T is given by

$$A_T = \int_0^\infty k_\nu d\nu, \tag{167}$$

$$A_T = \frac{\lambda_0^2 g_u A_t N_1 L}{8\pi g}, \tag{168}$$

where all terms are as already defined. Equation (168) is of great significance in theoretical spectroscopy because the value of A_T remains constant, whatever processes are responsible for broadening the absorption line, as long as $N_1 L$ is a constant.

At large absorbances near the absorption peak A_T is given by the well-known square-root relationship

$$A_T = \sqrt{\frac{\Delta\nu_D \lambda_0^2 g_u N_1 LA_t a}{4\pi g_1 \sqrt{\ln 2}}}, \tag{169}$$

where a is the classical damping constant and is given approximately by

$$a = \frac{\Delta\nu_C \sqrt{\ln 2}}{\Delta\nu_D}. \tag{170}$$

D. The Total Absorption Factor A_T' for Radiation Absorbed from a Line Source [24]

The total absorption factor A_T' represents the half-width of radiation absorbed by an isolated spectral line from a narrow line source and is given by

$$A_T' = \int_0^{\Delta\nu_S \sqrt{\pi}\, 2\sqrt{\ln 2}} [1 - \exp(-k_\nu L)] d\nu \,, \tag{171}$$

where the limits are over the half-width of the source line $\Delta\nu_S$ (in sec^{-1}). The total absorption can be pictorially represented by the area under any of the curves shown in Fig. 109, which also shows the variation in A_T' with atomic concentration.

At small absorbances A_T' is given by

$$A_T' = \int_0^{\Delta\nu_S \sqrt{\pi}/2 \sqrt{\ln 2}} [1 - \exp(-k_\nu L)] d\nu = \int_0^{\Delta\nu_S \sqrt{\pi}/2 \sqrt{\ln 2}} k_\nu d\nu \tag{172}$$

and, since k_ν is approximately constant over the narrow interval $\Delta\nu_S$,

$$A_T' = \frac{\sqrt{\pi}\, k^0 L \delta \Delta\nu_S}{2\sqrt{\ln 2}} \,, \tag{173}$$

where δ is a factor to correct the broadening of the absorption line by factors other than Doppler broadening.

At large absorbances near the line center A_T' is given by

$$A_T' = \int_0^{\Delta\nu_S \sqrt{\pi}/2 \sqrt{\ln 2}} d\nu = \frac{\Delta\nu_S \sqrt{\pi}}{2\sqrt{\ln 2}} \,. \tag{174}$$

The factor $\sqrt{\pi}/2\sqrt{\ln 2}$ is needed to convert from a Gaussian to a triangular half-width to use in the calculation of A_T' (i. e., the area shown in Fig. 109).

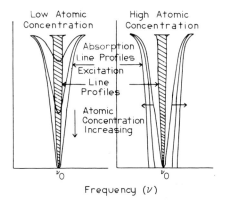

FIG. 109. Pictorial representation of the case where the total absorption-source-line half-width is much smaller than the absorption-line half-width.

E.   Intensity of Atomic Fluorescence [25]

The intensity of atomic fluorescence depends on the intensity of exciting radiation; the ratio of the excitation-source-line half-width $\Delta\nu_S$ (in sec^{-1}) to the absorption-line half-width $\Delta\nu_A$ (in sec^{-1}); the diameter of the absorption cell and the solid angle over which excitation occurs (see Fig. 110); the atomic concentration of the absorbing atoms; the efficiency of converting the absorbed energy to fluorescence energy; and the spectral parameters, such as frequency of the absorption line, the transition probability, and the types of line broadening.   Because the measured photodetector signal and the shape of analytical curves depend directly on the intensity of atomic fluorescence and expressions for intensity are generally derived here.   In this discussion it is assumed for convenience that the source of excitation has an area greater than the flame cell, that the entire flame cell is within the solid angle over which excitation occurs, and that the flame cell is within the solid angle of radiation, Ω_f, subtended by the monochromatic device viewing the fluorescence.

For these conditions the integrated AF expression for an isolated spectral line is given by

$$I_F = \frac{I_A \phi}{4\pi} , \qquad (175)$$

where I_F is the fluorescence intensity (in W/cm^2-ster), I_A is the total intensity absorbed by the spectral line resulting in activation of the fluorescent state (the upper level involved in the fluorescence process), and ϕ is the total power efficiency of the fluorescence transition that is, ratio of watts emitted by the fluorescence process to watts absorbed causing the fluorescence process.

The absorbed intensity I_A is given in general by

$$I_A = \Omega_A \sum_i I_i^0 A_{T_i} , \qquad (176)$$

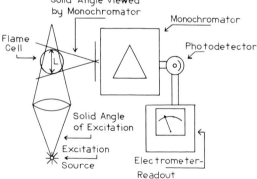

FIG. 110.   Schematic diagram of atomic fluorescence flame spectrometer.

where Ω_A is the solid angle over which excitation occurs and is determined by the entrance optics of the monochromator, I_i^0 and A_{T_i} are the incident intensity of the exciting radiation (in W/cm^2-ster-sec) and the total absorption factor (in sec^{-1}), respectively, for the absorption line causing fluorescence process. If the only process of concern is resonance absorption-resonance fluorescence, such as Zn 2139 $\overset{o}{A}$, and Cd 2288 $\overset{o}{A}$, then

$$I_A = \Omega_A I^0 A_T, \tag{177}$$

where I^0 and A_T are for the resonance-absorption process. The intensity I^0 for a continuous source is merely the watts emitted per square centimeter per steradian per frequency interval. The intensities of line sources are usually listed as integrated intensities (i.e., W/cm^2-ster), and, to obtain I^0 for a line source, the integrated intensity for the line in concern is simply divided by the source-line half-width $\Delta\nu_S \sqrt{\pi}/2 \sqrt{\ln 2}$. The expressions for I_F for small and large absorbances in two extreme cases (where $\Delta\nu_S \gg \Delta\nu_A$, such as in using a continuous source of excitation, and where $\Delta\nu_S \ll \Delta\nu_A$, such as in using a narrow line source, e.g., an electrodeless discharge tube) can be derived by substitution for A_T in Eq. (177). These expressions are tabulated in Table 82.

F. Growth Curves [25]

As can be predicted from the expressions in Table 82, growth curves of $\log I_F$ versus $\log N_1$ for a given element in the case of $\Delta\nu_S \gg \Delta\nu_A$ would be expected to have a slope of unity at low N_1 (low absorbance) and a slope of 0.5 at high N_1 (high absorbance). Furthermore growth curves of $\log I_F$ versus $\log N_1$ for a given element for the case of $\Delta\nu_S \ll \Delta\nu_A$ would be expected to have a slope of unity at low N_1 and a slope of zero at high N_1. This, of course, assumes no change in power efficiency ϕ with change in N_1. It is reasonable to predict that ϕ will decrease due to quenching processes as N_1 increases. A typical growth curve is shown in Fig. 111.

G. Variation in Atomic Concentration with Flame Temperature

The atomic concentration N_u of atoms in state u of energy E_u is related to the atomic concentration N_1 of atoms in state 1 (E = 0.0 if 1 is the ground state) by the Boltzmann equation

$$N_u = \frac{N_1 g_u}{g_1} \exp(-E_u/kT). \tag{178}$$

For most atoms $N_u/N_1 \ll 1.00$; that is, the ratio of excited to ground-state

TABLE 82

Resonance-Fluorescence Equations for Several Limiting Cases[a]

$\Delta \nu_S \gg \Delta \nu_A$:

Small absorbance

$$I_F = \frac{I_C^0 \Omega_A \phi \, \lambda_0^2 g_u A_T N_1 L}{32 \pi^2 g_1}$$

Large absorbance

$$I_F = \frac{I_C^0 \Omega_A \phi}{4\pi} \sqrt{\frac{\lambda_0^2 g_u A_T \Delta \nu_D a N_1 L}{4\pi \sqrt{\ln 2} \, g_1}}$$

$\Delta \nu_S \ll \Delta \nu_A$:

Small absorbance

$$I_F = \frac{2 \sqrt{\ln 2} \, I_L^0 \Omega_A \phi \, \lambda_0^2 g_u A_T N_1 L}{32 \pi^{5/2} g_1 \Delta \nu_D}$$

Large absorbance

$$I_F = \frac{I_L^0 \Omega_A \phi}{4\pi}$$

---

[a]Adapted from Ref. [22].

[b]Symbols: $\Delta \nu_S$ = source-line half-width; $\Delta \nu_A$ = absorption-line half-width; I_C^0 = intensity of a continuous source (W/cm^2-ster-sec); I_L^0 = integrated intensity of a line source (W/cm^2-ster).

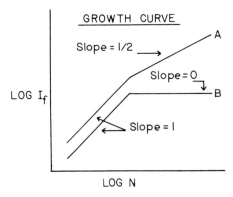

FIG. 111. Growth curves for atomic fluorescence: curve A, source-line half-width much greater than absorption-line half-width; curve B, source-line half-width much smaller than absorption-line half-width.

atoms is very small (see Table 83 for some representative values of N_u/N_1).

TABLE 83

Fraction of Excited Atoms, N_u/N_1, as a Function of
Wavelength and Temperature for Typical Resonance Lines

Resonance-line wavelength (Å)	g_u/g_1	N_u/N_1			
		$T = 2000°K$	$T = 3000°K$	$T = 4000°K$	$T = 5000°K$
2000	2	3.1×10^{-16}	3.5×10^{-11}	6.8×10^{-8}	1.2×10^{-6}
4000	2	6.8×10^{-8}	1.4×10^{-5}	2.4×10^{-4}	1.6×10^{-4}
6000	2	1.5×10^{-5}	6.8×10^{-4}	5.0×10^{-3}	1.6×10^{-2}
8000	2	5.0×10^{-5}	6.6×10^{-4}	9.0×10^{-3}	2.7×10^{-2}

Therefore the atomic concentration in state 1, N_1, is not greatly affected by temperature changes if only excitation is considered. The total atomic concentration N in all electronic states is given by

$$N = N_1 + N_u + \ldots = N_1 \left[1 + \frac{g_u}{g_1} \exp(-E_u/KT) + \ldots\right]. \qquad (179)$$

Because the terms containing $\exp(-E_j/KT)$, where E_j is the energy of any electronic state, are usually much smaller than unity,

$$N = N_1. \qquad (180)$$

Therefore the atomic concentration in all electronic states is essentially the same as in the ground state. This is valid for most elements in most analytical flames.

H. Variation in Atomic Concentration with Sample Concentration Introduced into the Flame Gases

Winefordner and Vickers [26] give an expression relating the atomic concentration N (in atoms per cubic centimeter), of flame gases to solution concentrations C (in moles per liter) introduced into the flame gases:

$$N = \frac{10^{19} F \epsilon \beta C}{e_f Q}, \qquad (181)$$

where F is the solution flow rate into the flame gases (in cubic centimeters per minute), C is the sample concentration (in moles per liter), Q is the flow rate (in cubic centimeters per second) of unburnt gases introduced into the flame at room temperature and 1 atm, e_f is the flame-expansion factor (no units), ϵ is the sample-introduction efficiency (no units), and β is the atomization efficiency (no units). The flame-expansion factor is given by

$$e_f = \frac{n_T T}{n_r T_r} , \qquad (182)$$

where n_T and n_r are the number of moles of combustion products at the flame temperature T and room temperature T_r, respectively.

It is important to note that the parameters ϵ, β, and e_f in Eq. (181) vary with flame temperature T and composition. The variation in e_f with T is approximately linear (see Eq. (182)) since n_T/n_r is approximately constant with change in T; that is, n_T/n_r is 1.0 to 1.2 for many analytical flames.

The variation in ϵ and β with T is much more complex. The efficiency of sample introduction ϵ is defined as the ratio of the moles of salt produced in the flame gases per unit time to the moles of salt actually introduced into the flame gases per unit time. The term ϵ is a complicated function of the type and geometry of aspirator-burner used; of the flow rates and pressures of unburnt gases and solution introduced into the flame; of the viscosity, surface tension, and density of the introduced solution; and of the rate of solvent evaporation. All of these factors influence the size of the droplets produced in a flame when an aspirator (nebulizer) is used and influence the rate of evaporation of the solution droplets to produce a salt mist. Winefordner, Mansfield, and Vickers [27] and later Parsons and Winefordner [28] measured approximate values of ϵ for several turbulent flames (H_2-O_2 and C_2H_2-O_2) produced by total-consumption aspirator-burners and found them to be in the range 0.2 to 0.8 for most conditions used in analytical flame spectrometry. Bernstein [29] measured ϵ values of 0.05 to 0.1 in premixed flames produced by using chamber-type aspirator-burners, which have lower values of ϵ because most of the solution aspirated into the chamber is rejected. In these burners the total of the droplets reaching the flame gases is only a small fraction of the solution volume introduced into the chamber. This fraction is commonly called the yield and is identical with the efficiency of aspiration ϵ for such an aspirator-burner. Of course the efficiency of aspiration of the droplets reaching the flame gases is approximately unity in the case of chamber-type aspirator-burners.

The efficiency of atomization β is simply the ratio of the concentration of atoms of concern to the total concentration of all species containing that same atom; for example, if $BaCl_2$ is aspirated into a flame, then β_{Ba} is defined [26] as

$$\beta_{Ba} = \frac{N_{Ba}}{N_{Ba} + N_{BaO} + N_{BaH} + N_{BaOH} + N_{BaCl} + N_{BaCl_2} + N_{Ba^+}} , \qquad (183)$$

where the N values correspond to the concentration of the species appearing as subscripts, that is, particles per cubic centimeter of flame gases. If a flame is in thermal equilibrium (this assumption is approximately valid for most analytical flames), then the value of ß can be estimated from the total pressure of all species containing Ba from the equilibrium constants of such processes as BaO $\rightleftharpoons$ Ba + O, and BaH $\rightleftharpoons$ Ba + H, and from the partial pressures of flame-gas products (e. g. , O, H, OH). If this is done, it will be found that ß is a very complex function of flame-gas temperature and composition.

De Galan and Winefordner [30] have given a method for measuring ß factors and have measured ß factors for a number of salts of several elements introduced into acetylene-air flames. It is evident from such studies that there is an optimum flame temperature and composition for which the ß factor is maximal (of course, ß = 1. 00 is maximal), and therefore an experimental design should be performed prior to an analysis by AF flame spectrometry in order to obtain optimum conditions of analysis.

It should be stressed that all atomic flame methods (AA, AE, and AF) depend directly on the concentration N of ground-state atoms in the flame. Therefore the change in any parameter causing a variation in N (e. g. , flame temperature and composition, solution- and gas-flow rates) will cause the same type of interference in all three methods.

IV. EXPERIMENTAL SYSTEM

A. General

Figure 110 is a block diagram of an experimental setup originally recommended by Winefordner and Vickers [6] for use in AF flame spectrometry. This arrangement consists of an intense source of radiation focused through a flame cell containing atoms of the element(s) being investigated. The intensity of fluorescent radiation (emitted in all directions) is measured at an angle $\leq 90^{\circ}$ to the excitation beam by some type of monochromator-photodetector-readout system. This optical arrangement has much in common with those found in commercially available molecular-fluorescence spectrometers, which also must measure low intensities of nondirectional radiation emanating from a sample cell through which an intense excitation beam passes. However, the actual instrumental components used in AF flame studies have typically been those developed for, and used in AA and AE spectrometry.

Recent publications [7-16] have described various modifications of the basic arrangement just described and their operation in AF studies. These systems, although of equivalent arrangement, have differed extensively both in the elegance of their design and detail, and in the quality of their instrumental

components. These experimental setups are described according to components
in Table 84.

B. Sources of Excitation

To be analytically useful in AF applications sources of excitation must be
extremely intense (I_L^0 or I_C^0 must be large) over the center of the absorption
line. Although narrow-line sources are preferred in most instances, the
width of the source-line profile is not nearly so critical as it is in AA flame
spectrometry. Other excitation source requirements are: (a) long term (low
drift) and short term (little flicker) stability; (b) long lifetime; (c) low cost;
(d) operation under continuous or pulsed conditions; (e) simple tuning and
focusing; and (f) safety of operation.

Schematic diagrams of line and continuum sources are shown in Fig. 112.

The Osram and Philips metal-vapor arc lamps have been used extensively
for zinc, cadmium, and thallium [7-9, 11, 13-15]. Although these sources are
available for other metals, only the zinc, cadmium, and thallium lamps have

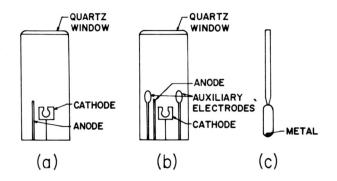

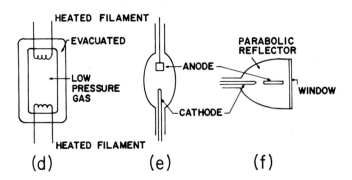

FIG. 112. Schematic diagrams of line and continuum sources: (a) sealed
and demountable hollow cathode discharge lamp-line source, (b) sealed boosted-
output hollow cathode discharge lamp-line source, (c) sealed electrodeless
discharge lamp-line source, (d) metal vapor discharge lamp-line source,
(e) point source xenon arc lamp-continuum source, (f) collimated (Eimac-type)
xenon arc lamp-continuum source.

TABLE 84

Experimental Systems Used for Atomic Fluorescence Flame Spectrometry

Sources	Flame cell	Entrance optics and monochromator	Detector-electrometer-readout system	Refs.
Osram and Philips lamps for Cd and Zn; electrodeless discharge tubes for Tl and Hg	Beckman total-consumption H_2-O_2 and H_2-air; chamber-type natural gas-air	Compact grating monochromator; no additional lenses or mirrors in most studies	Direct-current photometer-recorder readout from RCA 1P28 photomultiplier	7, 8
Osram and Philips lamps for Cd, Zn, and Tl; electrodeless discharge tube for Hg	Beckman total-consumption H_2-O_2	Czerny-Turner type of grating monochromator; single lens used to focus excitation radiation on flame cell; baffle box at monochromator entrance	Direct-current photometer-recorder readout from RCA 1P28 photomultiplier	9
Xenon-arc lamp (150 W; continuous source)	Beckman total-consumption H_2-O_2 and H_2-argon-entrained air	Czerny-Turner type of grating monochromator; single lens used to focus excitation radiation on flame cell; baffle box at monochromator entrance	Mechanically chopped source with phase-sensitive amplifier; RCA 1P28 photomultiplier	10
Osram lamp	Unicam chamber-type burner with acetylene-air	Prism monochromator; modification of existing burner housing to control light scattering	Electrically modulated source and ac amplifier	11

Sources	Flame cell	Entrance optics and monochromator	Detector-electrometer-readout system	Refs.
High-intensity hollow-cathode discharge tube	Beckman total-consumption H_2-air	Prism monochromator; lenses used in both radiation beams; additional parabolic mirror used in excitation beam	Mechanical chopper and tuned ac amplifier	12
Osram or Philips lamp	Zeiss total-consumption	Prism monochromator; no elaborate entrance optics so far as is known	Unknown	13
Philips lamps and electrodeless discharge tubes	Premixed acetylene-air; burner with settling chamber and Beckman total-consumption air-H_2	Prism monochromator; entrance optics similar to those described in Ref. [9]	Direct-current system at lower wavelengths (< 3000 Å); ac amplifier with mechanical chopper at wavelengths above 3000 Å; RCA 1P28 photomultiplier	14
Xenon (450 W; continuous source)	Keyes total-consumption (H_2-O_2 and H_2-entrained air flames)	Ebert-type grating monochromator; entrance optics similar to those described in Ref. [9]	Mechanically chopped phase-sensitive amplifier; EMI 6255B photomultiplier	16
Osram Zn lamp and ac 150-W xenon source (continuous)	Unicam chamber-type burner with propane-air and acetylene-air	Prism monochromator; modification of existing burner housing to control light scattering	Electrically modulated source and ac amplifier; EMI 9601B photomultiplier	15

been found to be analytically usable. Unless some radical improvements are
forthcoming in their design, the application of metal-vapor arc lamps will
probably be limited to these three elements. The mercury and alkali-metal
Osram and Philips lamps are not usable for atomic fluorescence in spite of
their great intensities because they are highly self-reversed when operated
under any practical conditions [7-9]. Other metal-vapor arc lamps (e. g.,
indium and gallium [9]) are extremely weak and are therefore essentially not
usable. It is doubtful that any further research effort will be expended in
attempting to widen the applicability range for this type of source.

Winefordner and co-workers [31] and Goodfellow [14] have used electrodeless discharge tubes for several AF investigations. These sources, which are
operated at radio or microwave frequencies, are simple and inexpensive to
construct [32, 33] and are extremely intense for several elements. Since they
are "cold plasma" sources, they are not plagued with self-reversal problems
to the extent that the metal-vapor arc-type lamps are. The technique of
preparing and operating these lamps is currently limited. Electrodeless
discharge tubes have the advantages of high spectral radiance and low cost of
preparation. However, much time and effort are required in tuning such lamps
to achieve a constant light intensity. These lamps are less stable than hollow-
cathode tubes and have shorter lifetimes.

Electrodeless discharge tubes and their operative equipment are presently
being marketed by at least two commercial firms, and the tubes are available
for virtually any element. A word of caution, however: not all of these are
intense enough for analytically practical AF applications. The commercially
available zinc, copper, thallium, mercury, and cadmium tubes are excellent.
Others are, at best, of moderate intensity. This may at first sound discourag-
ing and contradictory to some of the preceding statements; however, one should
understand that these commercial tubes have not been developed and designed
with high intensities necessarily in mind. This is probably because the most
important applications of electrodeless discharge tubes in the past have been as
narrow-profile line sources for interferometric techniques and spectrometer
calibration, where high intensities are not necessarily needed.

In view of the present state of the art and because of the tremendous need
for intense line sources in future AF applications, workers at the University
of Florida (Winefordner and co-workers) and at the Imperial College, England
(West and co-workers) are presently devoting a large part of their AF research
effort to the development of intense, non-self-reversed electrodeless discharge
tubes.

Dinnin and Helz [34] have used demountable hollow-cathode tubes to excite

gold, silver, lead, and bismuth radiation. Armentrout [12] has used the high-
intensity hollow-cathode discharge tube as an excitation source for nickel
fluorescence. This source, developed by Sullivan and Walsh [35, 36] is certain
to be tried in other AF investigations despite the fact that the lamps are not
really very intense and that two stable dc power supplies plus a heater circuit
are necessary for operation.

There have been three publications [10, 15, 16] by three different groups
regarding the use of continuous excitation sources in AF flame spectrometry.
Veillon et al. [10] first used a 150-W dc xenon-arc continuum and found it
useful for exciting atoms from 2139 to 8000 Å. Similarly Dagnall et al. [15]
used a 150-W ac zenon source whose frequency was keyed to that of the
amplifier. The results obtained in both of these investigations were comparable.
Ellis and Demers [16] reported on a study for which they employed a 450-W
xenon-arc continuum as the excitation source. They obtained very low detection
limits for zinc, silver, magnesium, and cobalt.

The choice of a line versus continuum source must be made on the basis of
analytical requirements: number of elements to be assayed, interferences
expected and radiance required. The continuum source provides better stability,
lifetime and tuning, and has possibilities for multielement analysis; the line
source has a higher radiance over the absorption line, lower interferences and
simpler wavelength selection.

The use of a continuous source requires a fairly high-resolution monochromator to reduce source light-scattering noise. So far only the single-
emission monochromator has been used in the reported experimental studies,
but West [18] has suggested using an additional monochromator in conjunction
with a continuous source to produce a monochromatic excitation beam. However,
there does not seem to be any outstanding advantage (in fact there are a number
of disadvantages) that such a system would offer for the measurement of
resonance fluorescence. On the other hand, such a system would have
advantages for the measurement of stepwise line or direct line fluorescence,
especially if incident-light scattering is a problem.

The 150-W xenon arcs used by Veillon et al. [10] and Dagnall et al. [15] had
a relatively low intensity in the ultraviolet region compared with the line sources
that have been used. It is improbable that, for the case of a continuum source,
sensitivities are linearly or even directly related to source power (source
intensity increases with source power; also source area increases with source
power--the product increases with source power to some exponent greater than
unity). Although Ellis and Demers [16] obtained appreciably greater sensitivities
with a 450-W xenon source, it has not been ascertained that this increased sensitivity was due primarily to the more powerful source. Ellis and Demers feel

that at least part of their success can be attributed to their thorough optimization of optical and flame parameters.

Because of molecular and/or particle scattering in the flame that occurs over the entire spectral bandwidth of the monochromator when a continuum source is used, an increase in source intensity may not increase the sensitivity at all. Although Veillon et al. [10] have shown that this scattering phenomenon can be corrected for when a continuous source is used in conjunction with a scanning monochromator, scattering nevertheless remains a limiting factor in lowering detection limits since the noise factor of this scattering cannot be compensated. A scattering background correction is more difficult to make if line sources are used. However, it should be pointed out that scattering from salt particles causes serious interferences only in samples that are near the limiting detectable concentration of the element being determined and contain a concentrated background matrix. Ellis and Demers [16] have shown that molecular scattering problems can be at least partly eliminated by proper choice of flame conditions.

So far continuum sources have not resulted in limits of detectors comparable with those for line sources in either AA or AF flame spectrometry, although in theory they should be quite useful (see Table 85).

TABLE 85

Experimental Limits of Detection in AF Flame Spectroscopy[a]

Element	Wavelength ($\overset{\circ}{A}$)[b]		Limit of Detection (ppm)	
	LS[c]	CS[d]	LS[c]	CS[d]
Ag	3281	3281	1×10^{-4} [51]	1×10^{-3} [76]
Al	3962[e]	--	0.1 [52,53]	--
As	1937	--	0.1 [54]	--
Au	2676	2676[f]	0.05 [55,56]	4.0 [10]
Be	2349[g]	--	0.01 [57-59]	--
Bi	2231	3068[f]	0.005 [60,61]	2.0 [10]
Ca	4227	4227	0.02 [51]	0.10 [77]
Cd	2288[f]	2288[f]	1×10^{-6} [51]	0.08 [10]
Co	2407	2407	0.005 [62,63]	0.50 [77]
Cr	3579	3579	0.05 [52]	10 [55]
Cu	3247	3247	0.001 [9]	0.02 [76]
Fe	2483	2483	0.008 [52]	1.0 [55]
Ga	4172	4172	0.01 [61]	5.0 [55]
Ge	2652	--	0.1 [64]	--

TABLE 85 --Continued

Element	Wavelength ($\overset{\circ}{A}$)[b]		Limit of Detection (ppm)	
	LS[c]	CS[d]	LS[c]	CS[d]
Hg	2537	--	0. 0002 [65, 66]	--
In	4511	4105	0. 10 [51]	2. 0 [55]
Ir	--	2544	--	100 [55]
Mg	2852[h]	2852	0. 001 [38, 67]	0. 01 [77]
Mn	2795	--	0. 006 [51]	--
Mo	3133	--	0. 50 [53]	--
Ni	2320	2320	0. 003 [68]	1. 0 [76]
Pb	4058	4058	0. 01 [69, 70]	3. 0 [55]
Pd	3405	3405	0. 04 [71]	50 [55]
Pt	--	2660	--	500 [55]
Rh	3692	3692	3. 0 [55]	10. 0 [55]
Ru	--	3728	--	100 [55]
Sb	2311[i]	2311	0. 05 [72]	300 [55]
Se	1960[h]	--	1. 0 [61, 73]	--
Si	2040	--	0. 6 [74]	--
Sn	3034	--	0. 05 [75]	--
Sr	4607	--	0. 03 [51]	--
Te	2143[h]	--	0. 05 [59, 61]	--
Tl	3776	3776	0. 008 [51]	0. 07 [76]
V	3184	--	0. 07 [53]	--
Zn	2138	2138	2×10^{-5} [51, 65]	0. 03 [77]

[a]Data from Refs. [19, 20].

[b]Hydrogen -air flame unless otherwise indicated.   The ratio of fuel to oxidant is not specified.   In several flames the oxidant is entrained air.

[c]Line source.

[d]Continuum source.

[e]Hydrogen -nitrous oxide flame.

[f]Hydrogen -oxygen flame.

[g]Acetylene -nitrous oxide flame.

[h]Acetylene -air flame.

[i]Propane -air flame.

C. Atomizers

The requirements for atomizers in atomic fluorescence are essentially

identical to those for atomic absorption: (1) good atomization efficiency;
(2) low radiation background and background flicker; (3) low concentration of
quenchers (molecular species such as CO, CO_2, N_2, which deactivate excited
atoms via nonradiational means; (4) low residence time of analyte atoms in the
optical path; (5) simplicity of operation; and (6) low cost of initial purchase and
operation.

1. Flame Cells

   A schematic diagram of flame atomizer shapes for AF is shown in
Fig. 113. Total-consumption aspirator-burner and turbulent flames have been
used by most AF investigators [7-10, 12-14, 16] despite their relatively poor
aspiration efficiencies, droplet dispersion, and droplet distribution. These
burners have the advantages of being relatively inexpensive, useful for high-
temperature (turbulent) flames, easily adaptable for any solvent type, and
capable of producing large atomic concentrations of absorbers in the flame
gases. However, such burners produce turbulent flames with a noise component
due to the scattering of exciting light from unevaporated solvent particles in the
flame gases. Organic solvents or organic-aqueous solvent mixtures will
improve aspiration efficiency and therefore will enhance the fluorescence signal.
However, the analyst must be aware that flame background noise increases when
organic solvents are used, and therefore optimization procedures as discussed
in Section III are needed.

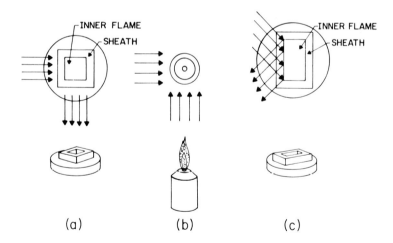

FIG. 113. Schematic diagram of flame atomizer shapes (burner shapes)
for atomic fluorescence flame spectrometry: (a) rectangular flame -- right-
angle illumination -- measurement, (b) round flame -- right-angle illumination--
measurement, (c) rectangular flame -- front-surface illumination -- measurement.

Winefordner and co-workers [7-10] have obtained excellent results with hydrogen-oxygen flames, especially when measurements were made above the highly luminous portion of the flame gases [9,10]. The hydrogen-air flame [8,12] has likewise been found to be relatively noise-free for work in the ultraviolet region and very applicable to AF work. Ellis and Demers [16] have obtained some interesting and excellent results with a hydrogen-entrained air flame with total-consumption burners. They found that not only were fluorescence-line intensities enhanced with this flame as compared with the hydrogen-oxygen flame but also scattering phenomena were subdued.

Chamber-type aspirator-burners and laminar flames have also been used in AF flame spectrometry in order to minimize light scattering from large droplets and particles in the flame gases and to produce a larger flame area. Unfortunately some of the advantages listed for total-consumption aspirator-burners and turbulent flames are not obtained in using the chamber-type aspirator-burners and premixed flames. Winefordner and Staab[8] used a natural gas-air flame for zinc, cadmium, mercury, and thallium. Dagnall, West, and Young [11] used an acetylene-air flame and concluded that the air-propane flame was most suitable for fluorescence measurements in the ultraviolet region (< 3200 $\overset{\text{o}}{\text{A}}$). Goodfellow [14] has used a chamber-type acetylene-air burner as well as a hydrogen-air total-consumption burner, but he did not mention making any direct comparisons between burners. Reported detection limits have been of the same order of magnitude for both burner types.

The best flame-burner system in AF flame spectrometry is probably a combination of acetylene-nitrous oxide and hydrogen-oxygen-air using a rectangular flame with a Pyrex burner, an efficient nebulizer, and complete illumination of the flame [19].

2. Non-Flame Cells

Nonflame atomizers are shown in Fig. 114. The Massmann furnace [37] in Fig. 114a and the West filament [38-40] in Fig. 114b were used for both atomic fluorescence and atomic absorption spectrometric studies. All nonflame cells are electrically heated. An argon atmosphere is also maintained in most nonflame cells by either passing a stream of argon through them or around them. However, the graphite filament and the metal loop (Fig. 114d) can be sheathed either by an argon stream or by a hydrogen diffusion flame. The metal tube furnace (Fig. 114c) of Shull and Winefordner [41] is similar to the Massmann graphite tube furnace except the sample is introduced continuously with an efficient nebulizer. Metal tubes of platinum are simple to fabricate and are not porous compared to graphite tubes, but the upper temperature limit for platinum elements is only 1800°K compared to about 3000°K for graphite.

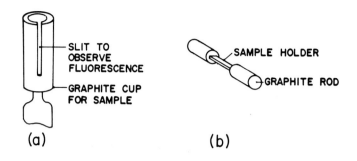

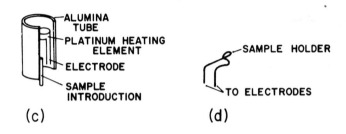

FIG. 114. Schematic diagram of nonflame atomizers for atomic fluorescence spectrometry: (a) Massmann graphite furnace, (b)West graphite filament, (c)Shull metal tube furnace, (d) Bratzel metal loop.

Refractory metals can provide higher temperatures but are highly susceptible
to air oxidation and become brittle after being heated. Tantalum tubes have been
used allowing an increase in temperature of the atomizer to about $3000^{\circ}K$ but
oxidation problems, even in the presence of high flow rates of argon past the
element, are appreciable at present. Sampling for the wire loop (platinum or
tungsten, Fig. 114d) is much more tedious and less precise than for any of the
above nonflame atomizers. The sample is either applied to the loop by dipping
the loop into the solution to be analyzed, or is applied by means of a hypodermic
syringe. For all of the nonflame cells except the metal tube and the metal loop,
discrete sampling of solutions via a hypodermic syringe is utilized. Therefore,
in systems employing atomizers of these types, the detector-readout system
must respond reliably to the transient signal which results when the free atoms
pass through the analytical detection zone of the optical system. Because of the
requirement of short time constants in the detection system, the frequency
response band-width of the measurement system is increased, and background
noise may become a problem.

D.   Entrance Optics and Monochromator

The various entrance-optics arrangements used in AF investigations, as in the other flame-spectrometry areas, have covered a wide range of refinement. Some surprisingly excellent results have been obtained [7, 8, 13] without any focusing arrangements whatsoever in either the exciting or the fluorescence beam axes.   Other experimental setups have included focusing lenses as well as parabolic mirrors for producing a multiple-pass effect through the flame cell with the excitation beam [12] and for collecting fluorescent radiation, otherwise lost, at the monochromator entrance slit [17].   Baffle systems and slits have been utilized to reduce the scattered and stray radiation entering the mono-chromator entrance slit [9-11].   Selection and optimization of entrance optics have a twofold purpose: to enhance the fluorescence signal and to control unwanted (stray) radiation.   The latter reason has been highly important in AF investigations where large slit widths have been employed [9, 10, 12, 14-16].

Any ultraviolet-visible monochromator adequate for emission and absorp-tion methods will be suitable for AF flame spectrometry.   Winefordner and co-workers [7-10] and Ellis[16] have used grating monochromators in all their studies because of the large apertures afforded.   Other workers [11-15] have used quartz-prism monochromators.   The setup used by Dagnall and co-workers [11, 15] was a modified commercially available AA instrument.

E.   Detector-Electrometer-Readout Systems

Photodetectors for studies so far performed have been the RCA IP28 or IP21 and the EMI 6255B and 9601B multiplier phototubes.   Mansfield, Winefordner, and Veillon [9] used a dc amplifier-recorder readout system because all measurements were made above the high-background portion of the flame gases.   Goodfellow [14] also used a dc amplifier in his work below $3000^{\circ}A$, where flame background was low.   Other systems have been more elaborate. Veillon et al. [10] as well as Ellis and Demers [16] used a mechanical chopper to modulate the exciting radiation in conjunction with a phase-sensitive amplifier with a narrow frequency-response bandwidth.   Armentrout [12] used a mechanical chopper and a tuned ac amplifier, as did Goodfellow [14] when working at wavelengths above $3000 \overset{\circ}{A}$.   Dagnall et al. [11, 15] modulated their source electrically at a frequency keyed to their ac amplifier.   So far the ac-amplifier systems have been found to be superior to dc-amplifier systems only for the case of elements fluorescing in the visible region, where the flame background-emission signal is appreciable.   Of course, assuming the same frequency response bandwidth Δf, a dc amplifier should always result in

lower limits of detection than a chopper and ac amplifier since at least half
the exciting radiation is lost when a chopper is used.

F. Multielement Analysis

If a continuum source were used in AF flame spectrometry for only these
elements with resonance lines below 300 nm, the same multidetector spectrometric measurement system could be used and multielement analysis is
possible.   Alternatively different hollow-cathode lamps could be used for AF
assay of different elements.

Instrumentation automated for multielement analysis is available from
Technicon (New York) [44]. The system AFS-6, costing $26,000 has provision
for assaying up to six different elements simultaneously. The instrument has
six prealigned hollow-cathode lamps; six optical filters; a photomultiplier
detector; a multielement prealigned reflective optical system; air-acetylene,
nitrous oxide-acetylene, and air-hydrogen flames; and an electronics control
module for measuring six elements in the fluorescence or emission modes and
directing a four-digit display in concentration units.

V. CHOICE OF EXPERIMENTAL CONDITIONS; USE OF SIGNAL-TO-NOISE RATIO [45]

A. Variation in Photodetector Signal with Sample Concentration

An analytical curve in any analytical method consists of a plot of the
measured signal i_F versus sample concentration C. From the following
approach it is possible to predict the shapes of analytical curves if spectral,
flame-composition and instrumental parameters are known.

For an experimental system, as shown in Fig. 110, the photoanodic current
i_F obtained from the photodetector (usually a photomultiplier tube) is given by

$$i_F = mUk_M WI_F, \tag{184}$$

where m accounts for the loss of light when a mechanical chopper is used
(m = 0.5 when using most choppers and m = 1 if no chopper is used); U is given
by $W/(W + W_c)$, W being the monochromator-slit width and W_c being the
minimum-resolving-power slit width of a monochromator (i. e. , the slit widen-
ing due to diffraction, coma, etc. , that results in a decrease in the measured
signal); and k_M is a constant characteristic of the monochromator-detector
system and is given by the product of the monochromator slit height H, the
transmission t of entrance optics and monochromator, the aperture Ω_F of the
monochromator, and the detector sensitivity γ. For a given photomultiplier and
voltage to the photomultiplier and a given monochromator, k_M is approximately
constant.

By substituting appropriate expressions from Table 82 and by applying the relationship (Eq. (181)) between atomic concentration in the flame gases, N, and solution concentration introduced into the flame gases, C, it is possible to obtain an expression for i_F in terms of solution concentration C, spectral parameters characteristic of the element being studied, instrumental parameters, and flame-gas composition and temperature. If such parameters are known or can be estimated, it is then possible to predict the shapes of analytical curves measured experimentally. The complete expression for i_F will not be given here but can be obtained by the above substitutions.

B. The Signal-to-Noise Ratio in AF Flame Spectrometry and Its Use in the Optimization of Experimental Conditions

In optimizing experimental conditions for analysis in any spectrochemical method the signal-to-noise ratio at the anode of the photodetector or at the readout must be considered. It is generally necessary, or at least desirable, to obtain the maximum signal-to-noise ratio with the existing experimental equipment because the maximum signal-to-noise at any given concentration of the species being determined yields the most precise results. These same experimental conditions will also generally give the lowest limit of detection.

The signal-to-noise ratio can be represented as $i_F/\overline{\Delta i_T}$, where $\overline{\Delta i_T}$ is the total root-mean-square (rms) noise due to all sources. In AF flame spectrometry $\overline{\Delta i_T}$ is essentially a result of six factors:

$$\Delta i_T = (\overline{\Delta i_P}^2 + \overline{\Delta i_B}^2 + \overline{\Delta i_S}^2 + \overline{\Delta i_F}^2 + \overline{\Delta i_E}^2 + \overline{\Delta i_A}^2)^{1/2} \tag{185}$$

where $\overline{\Delta i_P}$ is the rms photomultiplier-shot-noise photoanodic current; $\overline{\Delta i_B}$ is the rms noise photoanodic current due to flame background flicker; $\overline{\Delta i_S}$ is the rms noise photoanodic current due to flicker in the incident radiation scattered off unevaporated solvent droplets in the flame gases; $\overline{\Delta i_F}$ is the rms noise photoanodic current due to flicker in the fluorescence radiation; $\overline{\Delta i_E}$ is the rms noise photoanodic current due to flicker in the thermal radiation emitted by the atoms being studied at the wavelength λ_0; and $\overline{\Delta i_A}$ is the rms noise photoanodic current due to the electrometer-readout system but referred to the electrometer input. These noises add quadratically because they are essentially independent.

A rigorous derivation of an expression for $i_F/\overline{\Delta i_T}$ will not be given here. However, the general approach [36] to the selection of optimum experimental conditions will be indicated. If a general expression for $i_F/\overline{\Delta i_T}$ is known, the optimum value of any parameter, say X, can be found from theory by differentiating $i_F/\overline{\Delta i}_T$ with respect to X, maximizing, and solving for the optimum value of X:

$$\frac{d(i_F/\overline{\Delta i}_T)}{dX} = 0 \text{ (solve for } X_{opt}). \tag{186}$$

However, more information is obtained if the optimum value of X is obtained from the maximum in the plot of $i_F/\overline{\Delta i}_T$ versus X (see curve A, Fig. 115). For example, the variation of $i_F/\overline{\Delta i}_T$ versus monochromator-slit width W results in such a curve, which indicates that it would be worthwhile to determine (theoretically to obtain an estimate of W or experimentally to obtain the value of W_{opt} for use) the optimum monochromator-slit width.

If no optimum value of X exists, then Eq. (186) has little value; but a plot of $i_F/\overline{\Delta i}_T$ versus X is still quite useful. For example, the variation of $i_F/\overline{\Delta i}_T$ with X can also result in a curve in which $i_F/\overline{\Delta i}_T$ continues to increase as X increases (see curve B, Fig. 115). Such a plot results for $i_F/\overline{\Delta i}_T$ versus Δf^{-1}, where Δf is the frequency-response bandwidth over which the electrometer-readout system responds. In other words, if there is no drift in the fluorescence signal i_F, the signal-to-noise ratio will continue to increase without limit as Δf is decreased.

The frequency-response bandwidth Δf can be decreased without limit by integrating devices. However, the smaller the value of Δf, the longer the time needed to take a signal reading. In addition, it is seldom practical experimentally to make Δf smaller than about 0.2 sec^{-1} because of drift in the fluorescence signal due to changes in the aspiration rate.

A curve similar to $i_F/\overline{\Delta i}_T$ versus Δf^{-1} results for $i_F/\overline{\Delta i}_T$ versus $\overline{\Delta I}_B^{-1}$, where $\overline{\Delta I}_B$ is the rms flame background intensity flicker. However, it is less convenient to vary $\overline{\Delta I}_B$ than Δf, since the flame-gas composition and temper-

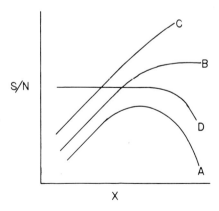

FIG. 115. Typical plots of signal-to-noise ratio S/N versus values of a variable parameter X (see text for discussion of curves A-D).

ature must be changed to obtain a variation in $\overline{\Delta I}_B$. Similar plots also result with other types of flicker in the fluorescence intensity and flicker in the scattered light, etc. These latter types of flicker can frequently be reduced by a more stable flame and source of excitation, and a more uniform means of aspiration of the sample solution.

The variation in $i_F/\overline{\Delta i}_T$ with X can in other cases result in a flat curve (see curve C, Fig. 115), in which an increase or decrease of X results in little change in $i_F/\overline{\Delta i}_T$. Such a plot results for $i_F/\overline{\Delta i}_T$ versus voltage applied to the photomultiplier. Therefore the photomultiplier voltage is unimportant as long as it is below the point of regenerative ionization and yet sufficient to obtain an accurately measurable signal i_F.

The variation in $i_F/\overline{\Delta i}_T$ versus X can in still other cases result in a plot similar to the combination of curves B and C in Fig. 115, in which $i_F/\overline{\Delta i}_T$ increases as X increases, but at high X values $i_F/\overline{\Delta i}_T$ reaches a plateau (see curve D, Fig. 115). For example, this type of curve results for $i_F/\overline{\Delta i}_T$ as a function of source intensity. As the source intensity increases, i_F increases faster than $\overline{\Delta i}_T$ until the major source of noise is the noise due to scattering of incident radiation and/or the noise due to flicker in the fluorescence radiation.

Another type of curve, in which $i_F/\overline{\Delta i}_T$ versus X increases linearly with X but reaches a maximum value results if $i_F/\overline{\Delta i}_T$ versus ϕ is plotted. For resonance fluorescence, Φ is identical with ϕ, the quantum efficiency of the fluorescence transition. The maximum value of ϕ in this case is therefore unity; that is, for every photon absorbed, one photon is emitted. McCarthy, Parsons, and Winefordner [47] have given a general kinetic approach to the evaluation of quantum and power efficiencies for resonance fluorescence as well as stepwise line, direct line, and a combination of these fluorescence types. They considered means of increasing the value of ϕ. For example, dilution of the flame gases with a poor quenching species (e. g. , an inert gas) results in an increase in the ϕ values. Alkemade [5] has measured ϕ values for the sodium 5890- to 5896-Å doublet and has shown that ϕ can be increased significantly (e. g. , 10 times) by dilution with an inert gas like argon. The reader is referred to these more specialized treatments of the factors influencing quantum and power efficiencies.

It should be stressed that optimum values of experimental parameters can be found theoretically if all spectral, flame-composition, and instrumental parameters are known or can be approximated. Also, by means of proper statistical experimental designs [46,47] interdependent experimental variables can be experimentally optimized (e. g. , several parameters can be simultaneously optimized by a factorial design). A simpler but less time-consuming statistical approach is to use a central composite rotatable experimental design [48]. However, even the latter is time consuming, and so optimum values should be estimated theoretically by using graphical methods or Eq. (186) and then determined more accurately experimentally by using a statistical design method.

It should be pointed out that the same signal-to-noise ratio results whether one measures the ratio of photoanodic current i_F to photoanodic current noise $\overline{\Delta i_T}$ or the ratio of readout voltage E_F to readout voltage noise $\overline{\Delta E_T}$; that is,

$$\frac{i_F}{\overline{\Delta i_T}} = \frac{E_F}{\overline{\Delta E_T}} , \qquad (187)$$

since E_F is related to i_F by $E_F = GRi_F$, and $\overline{\Delta E_T}$ is related to $\overline{\Delta i_T}$ by $GR\overline{\Delta E_T}$. The value of the load resistor connected to the anode of the photodetector is R ohms and the voltage gain of the electrometer is G. Therefore these expressions for the signal-to-noise ratio are identical whether given in terms of readout voltage or photoanodic current.

C. Limiting Detectable Concentrations

The limiting detectable sample concentration C_m is defined as the concentration resulting in a signal-to-noise ratio of

$$\frac{i_F}{\overline{\Delta i_T}} = \frac{t \sqrt{2}}{\sqrt{n}} , \qquad (188)$$

where n is the number of signal and background measurements and t is the Student "t" value, which can be found from statistical tables for the desired confidence level (99. 5% is often used) and the degrees of freedom (2n - 2) associated with the experiment.

Equation (188) is very useful from an experimental standpoint since it is based on a statistical treatment and has no personal bias. It is also useful from a theoretical standpoint since it should be possible to predict the value of C_m that will give a certain $t \sqrt{2} / \sqrt{n}$ value, assuming accurate values of all spectral, flame-composition and instrumental parameters are available to substitute into the $i_F / \overline{\Delta i_T}$ expression.

The theoretical calculation of C_m should be of great aid to the analyst in choosing the best method for a given analysis, the best instrument (if a choice is to be made) and instrumental conditions, and the best flame type and temperature for measuring the specified atomic system. Even if all spectral, flame composition, and instrumental parameters are not accurately known, it is still possible to obtain ranges of variables to use in a statistical design for experimentally optimizing the experimental system. Also, theoretical expressions for C_m are quite useful to indicate to the analyst what parameters affect C_m and in what direction these parameters should be varied to lower the value of C_m.

D. Conclusions Concerning the Theoretical Limits of Detection [19]

If the noise, as well as signal, is considered, the limits of detection should be lower for AF than for AA flame spectrometry. With high-temperature flames the limits of detection for the two techniques should be comparable. In theory, AE should be used for all analyte atoms whose resonance lines have wavelengths above 300 nm, and AF or AA for those below 300 nm.

Because the same measurement system can be used with AF and AE flame spectrometry, this is an ideal combination of techniques with either low or high temperature flames for analysis.

VI. ANALYTICAL RESULTS

A. Limits of Detectability

The lowest currently available experimental limits of detection by AF flame spectrometry using a line or a continuum source are listed in Table 85. It is evident from this table that at present the limits of detection are much lower with line sources than with continuum sources (the same observation is true in AA flame spectrometry). This is contrary to theoretical calculations, which show that continuum sources in AF or AA flame spectrometry should give better limits of detection.

Winefordner et al [19] have compared the experimental limits of detection in AA, AF, and AE flame spectrometry and concluded that AE and AF-line source is the most ideal combination for the assay of metal ions at present. They did point out that if the experimental limits of detection in AF-continuum source ever approached the theoretical values, the ideal combination would be AE and AF-continuum source, with a single scanning spectrometric system with dc readout that could be used for all elements between 200 and 800 nm.

For all the elements that have been reliably studied by AE, AA, and AF flame spectrometry, AF is more sensitive than AA and AE methods for Ag, Cd, Cu, Hg, and Zn; AA is more sensitive than AF or AE for Be, Mg, and Rh; AE is more sensitive than AA or AF for Cs, Ga, In, K, Li, Na, Rb, and Sr; AA and AF are more sensitive than AE for As, Au, Bi, Co, Fe, Ni, Pb, Sb, Se, Sn, and Te; AA and AE are more sensitive than AF for Ca and Cr; and AA, AF, and AE are comparable in sensitivity for the assay of Ge, Mn, and Tl. In all cases the use of a continuum source in AF and AA gave lower limits of detection than the use of a line source.

B. Analytical Curves

Analytical curves (see Fig. 116) in AF flame spectrometry are generally analytically useful over a 10^3 to 10^5 concentration range, which is similar to the range of analytical curves obtained in AE flame spectrometry but greater than the range normally found in AA working curves.

C. Selectivity

From theoretical considerations it would appear that essentially the same interferences that occur in AA should be present in AF flame spectrometry. For example, radiation interferences should be small in AF if the radiation from a line source is chopped and a tuned ac amplifier is used or if AF measurements are taken above the flame gases and a line source of excitation and a dc amplifier-readout system are used. Chemical interferences are the most important interferences in all flame-spectrometry studies; that is, the analyte forms a less or a more volatile compound with flame-gas products than the analyte itself or the analyte is occluded within a more or a less volatile matrix of a concomitant (interferent) than the analyte itself [49].

The influence of variation in flame temperature on the measured signal in AF, AA, and AE flame spectrometry is quite similar despite many reports to the contrary. It can be shown that a temperature variation results in about the same error in all three flame-spectrometry methods if the analyte forms a stable molecule with flame-gas products [50]. Of course, if the analyte does not undergo appreciable compound formation with flame gases and if ionization is not appreciable, then AE flame spectrometry is influenced much more

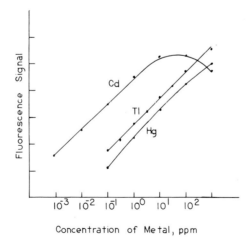

FIG. 116. Typical analytical curves for cadmium, thallium, and mercury.

severely   by variation in flame temperature than the other two spectrometric
methods.

Experimental investigations of interferences were performed by Dagnall
et al. [11, 15], who studied the effects of hundredfold concentration excess of
41 cations and 18 anions on 10^{-5} M cadmium and zinc and in no case found
variation in the photodetector signal i_F greater than $\pm 5\%$. These results are in
agreement with the theoretical predictions concerning the influence of interferences.

The most extensive experimental study of interferences in AF flame
spectrometry to date has been carried out by Goodfellow [14], who concluded
that "the technique can be as free from interelement effects as atomic absorption spectrometry." He noted especially the lack of interference in the presence
of alkali metals, whose low excitation potentials often cause considerable
suppression of normal AE spectra.

VII.   COMPARISON WITH OTHER FLAME-SPECTROMETRY METHODS

Several advantages of AF over AA and AE flame spectrometry should be
mentioned. These are more ways in AF to vary experimental conditions to
lower limits of detection than in AA. For example, the limits of detection in
AF can generally be lowered by increasing the source intensity, the flame
stability, and the power efficiency ϕ . The power efficiency can be increased
by proper choice of gas composition [5, 47]. Source instability in AF is much
less important than it is in AA for signals near the limit of detection. Source
fluctuations have no significant influence in AF when no or little metal is
present in the flame, but in AA they have a significant effect near the limit of
detection since the fluctuation signal from the source is usually the predominant
limiting noise. In AF it is necessary only that the excitation line be intense
over the absorption-line half-width. If the power efficiency for the transition
is large, atomic fluorescence can be usable even if atomic absorption is small.
Usable analytical curves over a concentration range of 10,000 are frequently
obtained in AF compared with a range of 100 for most AA studies. The AF
method has essentially the same advantages over AE as AA has over AE and the
reader is referred to the many review articles and books on AA.

There are also several disadvantages of AF as compared with AA and AE.
Incident-light scattering by solvent droplets and salt particles in the flame
gases, which is negligible in AA and nonexistent in AE, may be a significant
problem in some AF applications. So far AF has been useful only for a limited
number of elements, namely the so-called soft metals (e. g. , Zn, Cd, Hg, Cu,
Pb, Tl) because of the lack of intense line sources. However, recent research

by several workers indicates that it may be possible in the future to obtain intense electrodeless discharge tubes for most of the elements now determined conveniently by AA. In addition, the use of intense continuous sources (e. g., a 1000-W xenon arc) and pulsed-light sources may also increase the number of elements that can be measured more sensitively by AF than by AA.

Although there are as yet few applications of AF flame spectrometry, it should be useful for applications similar to those that now use AA or AE.

REFERENCES

1. A. C. G. Mitchell and M. W. Zemansky, Resonance Radiation and Excited Atoms, University Press, Cambridge, 1961.

2. P. Pringsheim, Fluorescence and Phosphorescence, Interscience, New York, 1949.

3. E. L. Nichols and H. L. Howes, Phys. Rev., 23, 472 (1924).

4. R. M. Badget, Z. Physik, 55, 56 (1929).

5. C. T. J. Alkemade, in Proceedings of the Xth Colloquium Spectroscopicum Internationale (E. R. Lippincott and M. Margoshes, eds.), Spartan Books, Washington, D. C., 1963.

6. J. D. Winefordner and T. J. Vickers, Anal. Chem., 36, 161 (1964).

7. J. D. Winefordner and R. A. Staab, Anal. Chem., 36, 165 (1964).

8. J. D. Winefordner and R. A. Staab, Anal. Chem., 36, 1367 (1964).

9. J. M. Mansfield, J. D. Winefordner, and C. Veillon, Anal. Chem., 37, 1049 (1965).

10. C. Veillon, J. M. Mansfield, M. L. Parsons, and J. D. Winefordner, Anal. Chem., 38, 204 (1966).

11. R. M. Dagnall, T. S. West, and P. Young, Talanta, 13, 803 (1966).

12. D. N. Armentrout, Anal. Chem., 38, 1235 (1966).

13. H. Prugger, Zeiss Information, No. 56, p. 54 (1960) (work by R. Klaus).

14. G. I. Goodfellow, Anal. Chim. Acta, 35, 132 (1966).

15. R. M. Dagnall, K. C. Thompson, and T. S. West, Anal. Chim. Acta, 36, 269 (1966).

16. D. W. Ellis and D. R. Demers, Anal. Chim. Acta, 38, 1945 (1966).

17. T. S. West, Chem. Ind. (London), 1005 (1966).

18. T. S. West, Analyst, 91, 1079 (1966).

19. J. D. Winefordner, V. Svoboda, and L. J. Chine, CRC Critical Reviews in Anal. Chem. , 1, 233 (1970).

20. J. D. Winefordner and R. C. Elser, Anal. Chem. , 43, 24A (April, 1971).

21. A. Unsöld, Physik der Sternatmosphären, Springer, Berlin, 1955.

22. M. L. Parsons, W. J. McCarthy, and J. D. Winefordner, Appl. Spectry. , 20, 223 (1966).

23. C. Young, Tables for Calculating the Voigt Profile, University-Microfilms, Ann Arbor, Michigan, 1965.

24. J. D. Winefordner, W. W. McGee, M. L. Parsons, J. M. Mansfield, and K. E. Zacha, Anal. Chim. Acta, 36, 25 (1966).

25. J. D. Winefordner, M. L. Parsons, J. M. Mansfield, and W. J. McCarthy, Spectrochim. Acta, 23B, 37 (1967).

26. J. D. Winefordner and T. J. Vickers, Anal. Chem. , 36, 1939 (1964).

27. J. D. Winefordner, C. T. Mansfield, and T. J. Vickers, Anal. Chem. , 35, 1607 (1963).

28. M. L. Parsons and J. D. Winefordner, Anal. Chem. , 38, 1593 (1966).

29. R. E. Bernstein, S. Afr. J. Med. Sci. , 20, 57 (1956).

30. L. de Galan and J. D. Winefordner, J. Mol. Spect. Radiat. Transf. , 7, 251 (1967).

31. J. D. Winefordner and co-workers, unpublished data, University of Florida, Gainesville.

32. E. F. Worden, R. G. Gutmacher, and J. G. Conway, Appl. Opt. , 2, 707 (1963).

33. F. C. Fehsenfeld, K. M. Evenson, and H. P. Broida, Rev. Sci. Instr. , 36, 294 (1965).

34. J. I. Dinnin and A. W. Helz, paper presented at 1967 Pittsburgh Conference on Analytical Chemistry and Applied Spectroscopy.

35. J. V. Sullivan and A. Walsh, Spectrochim. Acta, 21, 721 (1965).

36. J. V. Sullivan and A. Walsh, Spectrochim. Acta, 21, 727 (1965).

37. H. Massmann, Spectrochim. Acta, 23B, 215 (1968).

38. T. S. West and X. K. Williams, Anal. Chim. Acta, 45, 27 (1969).

39. R. G. Anderson, I. S. Maines and T. S. West, Anal. Chim. Acta, 51, 355 (1970).

40. J. F. Alder and T. S. West, Anal. Chim. Acta, 51, 365 (1970).

41. M. Shull and J. D. Winefordner, unpublished results, University of Florida, Gainesville (1970).

42. M. P. Bratzel, R. M. Dagnall,and J. D. Winefordner, Anal. Chim. Acta, 48, 197 (1969).

43. M. P. Bratzel, R. M. Dagnall, and J. D. Winefordner, Appl. Spectrosc. , 24, 518 (1970).

44. D. G. Mitchell and A. Johansson, Spectrochim. Acta, 25B, 175 (1970).

45. J. D. Winefordner, M. L. Parsons, J. M. Mansfield, and W. J. McCarthy, Anal. Chem. , 39, 436 (1967).

46. J. D. Winefordner, W. J. McCarthy, and P. A. St. John, J. Chem. Educ. , 44, 80 (1967).

47. W. J. McCarthy, M. L. Parsons, and J. D. Winefordner, Spectrochim. Acta, 23B, 25 (1967).

48. W. G. Cochran and G. M. Cox, Experimental Designs, 2nd ed. , Wiley, New York, 1957.

49. C. T. J. Alkemade, Anal. Chem. , 38, 1252 (1966).

50. L. de Galan and J. D. Winefordner, Anal. Chem. , 38, 1412 (1966).

51. K. E. Zacha, M. P. Bratzel, J. D. Winefordner, and J. M. Mansfield, Anal. Chem. , 40, 1733 (1968).

52. R. M. Dagnall, M. R. Taylor, and T. S. West, Spectry. Letters, 1, 397 (1968).

53. R. M. Dagnall, G. F. Kirkbright, T. S. West, and R. Wood, Anal. Chem. , 42, 1029 (1970).

54. R. M. Dagnall, M. R. Taylor, and T. S. West, Spectry. Letters, 4, 147 (1971).

55. D. C. Manning and P. Heneage, Atomic Absorption Newsletter, 7, 80 (1968).

56. J. P. Matousek and V. Sychra, Anal. Chim. Acta, 49, 175 (1970).

57. D. N. Hingle, G. F. Kirkbright, and T. S. West, Analyst, 93, 522 (1968).

58. M. P. Bratzel, R. M. Dagnall, and J. D. Winefordner, Anal. Chem. , 41, 1527 (1969).

59. R. M. Dagnall, K. C. Thompson, and T. S. West, Talanta, 14, 557 (1967).

60. R. M. Dagnall, K. C. Thompson, and T. S. West, Talanta, 14, 1467 (1967).

61. A. Hell and S. Ricchio, paper presented at 21st Pittsburgh Conference, Cleveland, March, 1970.

62. D. G. Mitchell and A. Johansson, Spectrochim. Acta, 25B, 175 (1970).

63. B. Flett, K. Liberty, and T. S. West, Anal. Chim. Acta, 45, 205 (1969).

64. R. M. Dagnall, G. F. Kirkbright, T. S. West, and R. Wood, Analyst, 95, 425 (1970).

65. P. D. Warr, Talanta, 17, 543 (1970).

66. T. J. Vickers and S. P. Merrick, Talanta, 15, 873 (1968).

67. T. S. West and X. K. Williams, Anal. Chim. Acta, 42, 29 (1968).

68. J. Matousek and V. Sychra, Anal. Chem., 41, 518 (1969).

69. R. F. Browner, R. M. Dagnall, and T. S. West, Anal. Chim. Acta, 50, 375 (1970).

70. V. Sychra and J. P. Matousek, Talanta, 17, 363 (1970).

71. V. Sychra, P. J. Slevin, J. P. Matousek, and F. Bek, Anal. Chim. Acta, 52, 259 (1970).

72. R. M. Dagnall, K. C. Thompson, and T. S. West, Talanta, 14, 1151 (1967).

73. M. S. Cresser and T. S. West, Spectry. Letters, 2, 9 (1969).

74. R. M. Dagnall, G. F. Kirkbright, T. S. West, and R. Wood, Anal. Chim. Acta, 47, 407 (1969).

75. R. F. Browner, R. M. Dagnall, and T. S. West, Anal. Chim. Acta, 46, 207 (1969).

76. D. W. Ellis and D. R. Demers, paper presented at Miami Beach ACS Meeting, April 1967.

77. D. W. Ellis and D. R. Demers, Anal. Chem., 38, 1943 (1966).

Chapter 11

ELECTROGENERATED LUMINESCENCE[*]

[*]Based in part on a discussion by A. J. Bard and co-workers, in
Fluorescence. Theory, Instrumentation, and Practice (G. G. Guilbault, ed.),
Dekker, New York, 1967.

I. INTRODUCTION

In preceding chapters we have discussed the generation of luminescence by
light absorption and chemical reactions. Still another source of luminescent
species is electrochemical reaction. There have been scattered reports in the
literature on the emission of light from solutions undergoing electrolysis
[1-9]. This emission has been found to be due to the generation of oxygen at an
anode, followed by the usual chemiluminescent reaction, or a chemiluminescence resulting from the electrolysis of aromatic hydrocarbons in aprotic media.

In this chapter we consider some of the research in this area, with
emphasis on some of the analytical procedures that are possible.

Bard and co-workers have written a chapter on electrogenerated chemiluminescence in Guilbault's Fluorescence. Theory, Instrumentation, and
Practice [10]. Much of this work is reviewed here. Cruser [11] has written a
thesis on the analytical aspects of electrogenerated chemiluminescence under
the direction of Prof. Bard. A very good recent review of this field was
prepared by Hercules [12].

II. EXPERIMENTAL SETUP

A typical electrochemical cell for electrochemiluminescence (ECL) experiments is shown in Fig. 117. The working electrode is usually platinum or
mercury, and the auxiliary electrode is separated from the main chamber by a
sintered-glass disk. The reference electrode is either an aqueous saturated
calomel electrode (SCE) connected through a salt bridge of the appropriate

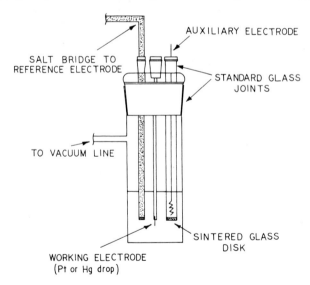

FIG. 117. Electrolytic cell for chemiluminescence.

solvent or sometimes simply a silver wire contained in a separate compartment, which does not behave as a truly poised reference electrode. It does, however, maintain a constant potential throughout the experiment, which can later be related to an SCE and which avoids the problems associated with preparing and connecting the aqueous SCE to the nonaqueous solution. The cell is usually provided with a joint, so that it can be connected to a vacuum line.

The supporting electrolyte, usually tetra-n-butylammonium perchlorate and the compound of interest are added to the cell, the cell is evacuated, and fresh solvent, such as dimethylformamide (DMF) or acetonitrile, is distilled into it. This is usually followed by several freeze-and-thaw cycles to remove dissolved gases [13]. A potential program--either potential steps, potential scans, or sinusoidally varying potentials--is applied to the working electrode, and the variation in light intensity with potential or the emission spectrum is recorded.

In some cases, when only the emission spectrum is desired, a simpler arrangement, such as that shown in Fig. 118, can be employed.

Maloy, Prater and Bard [14] described the use of a rotating-ring-disk electrode (rrde) to generate the radical ion precursors of electrogenerated chemiluminescence (ecl). A cell assembly was designed to allow solutions to be degassed under vacuum and spectroscopic experiments. Digital simulation techniques have been employed to treat ecl at the rrde. The simulations predict the effect of rrde rotation rate and the kinetics of the radical ion annihilation reaction on the intensity of ecl light. When this annihilation reaction is very fast, i.e., greater than 10^7 M^{-1} sec^{-1} for the rrde used, the ecl is seen as a sharp ring of light at the inner edge of the ring electrode, and its intensity, like the disk current, is proportional to the square root of rotation rate. The simultaneous measurement of steady-state disk current and ecl intensity as functions of the disk potential gives direct information about the role of any

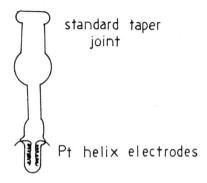

standard taper
joint

Pt helix electrodes

FIG. 118.   Cell used for recording emission spectra.

disk-generated species in the light-producing process. An evaluation of the technique using the ecl of 9,10-diphenylanthracene in N,N-dimethylformamide solution was described.

III. FORMATION OF LUMINESCENT SPECIES

A. Annihilation Reaction

The first experiments in electrochemiluminescence involved the electrolysis of solutions of some aromatic hydrocarbon (e. g., anthracene, diphenylanthra-cene, tetracene, or rubrene), in DMF or acetonitrile, at a platinum electrode with an alternating current. It was observed, visually and spectroscopically, that the emitted radiation was similar to that obtained during fluorescence measurements on the hydrocarbon, which suggested that the emitting species was the excited singlet state of the hydrocarbon. The mechanism suggested for the emission was as follows: The aromatic hydrocarbon R is reduced to the anion radical:

$$R + e \longrightarrow R^-. \tag{189}$$

During the anodic cycle the cation radical is produced:

$$R - e \longrightarrow R^+ \tag{190}$$

or

$$R^- - 2e \longrightarrow R^+. \tag{191}$$

The R^+ diffusing away from the electrode surface encounters R^- diffusing toward the electrode, and the following reaction, which had already been suggested for oxidations of R^- produced by alkali-metal reduction [15], occurs:

$$R^- + R^+ \longrightarrow R^* + R, \tag{192}$$

where R^* denotes the excited singlet of the parent hydrocarbon. This process can be pictured in terms of the molecular orbitals of R as follows:

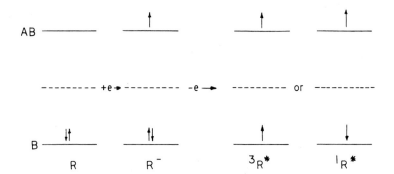

B. Other Oxidants

Further experiments also showed that the oxidant was not necessarily R^+, but could be one produced by oxidation of the solvent, supporting electrolyte, or electrode material [6, 7], so that the reaction

$$R^- + \text{oxidant} \longrightarrow R^* + \text{products} \tag{193}$$

was also possible. For example, cyclic potential-step experiments showed that when the 9, 10-diphenylanthracene anion radical was generated at an electrode and the potential of the electrode was switched to positive potentials resulting in the formation of bromine at a platinum electrode in tetra-n-butylammonium bromide, or mercury(II) at a mercury electrode in tetra-n-butylammonium perchlorate, the same emission was observed [7]. Similarly the reaction of the 9, 10-diphenylanthracene anion radical with lead tetraacetate, ceric ammonium sulfate, or nitrogen dioxide has been reported to result in chemiluminescence [8].

Hercules [5] proposed that the reaction of bromine and R^- did not lead to the emission of light, because flowing a bromine solution over the cathode during electrolysis did not produce light. However, under these conditions it is unlikely that any R^- was produced, since bromine would be preferentially reduced at the cathode. Bader and Kuwana [9] suggested that oxygen and its reduction products may also be implicated in electrochemiluminescence in these media.

C. Excimer Formation

Chandross, Longworth, and Visco [16] performed cyclic potential-step experiments on a number of hydrocarbons at platinum electrodes and recorded the emission spectra during electrolysis. They observed that during the electrolysis of anthracene, 9, 10-dimethylanthracene, phenanthrene, perylene, and 3, 4-benzopyrene the emission spectrum contained a broad structureless band shifted toward the red region from the normal fluorescence spectrum of the parent hydrocarbon in the same solution. They ascribed this longer wave-length band to emission from the excimer, R_2^*, formed by the reaction

$$R^+ + R^- \longrightarrow (R_2)^*. \tag{194}$$

The spectra of 9, 10-diphenylanthracene, tetracene, and rubrene showed a band at the usual parent-hydrocarbon fluorescence wavelengths. The failure to obtain excimer spectra in these cases was ascribed to steric hindrance to the formation of excimer (9, 10-diphenylanthracene and rubrene) or self-quenching (tetracene).

D. Preannihilation Electrochemiluminescence

Reports have also been presented of electrochemiluminescence occurring on generation of either of R^+ or R^- alone, without the need of switching the potential to electrogenerate a reductant or oxidant to react with these. Hercules, Lansbury, and Roe [17] showed that simply oxidizing rubrene to the cation radical in DMF resulted in light emission. This emitted light was about 5% as intense as that resulting from the cation-anion annihilation reaction under similar conditions. This emission on oxidation did not occur when acetonitrile was used as a solvent, but did occur when DMF, triethylamine, or n-butylamine was added to the acetonitrile. In all cases the observed emission spectrum was identical with the fluorescence spectrum. The results suggest that DMF and the amines, or intermediates produced in chemical reaction of these, can act as reductants and produce R^*.

A different type of preannihilation electrochemiluminescence has been described by Maricle, Maurer, and co-workers [8, 18-20]. In these studies the radical is generated, and then the potential is switched to values at which initially no apparent electron-transfer process occurs. After a certain threshold potential has been attained, light emission is observed. For example, in DMF solutions rubrene is reduced to R^- at -1.37 V versus SCE and oxidized to R^+ at +0.95 V. When R^- is generated at -1.6 V and the potential is switched back to potentials more positive than -0.2 V, luminescence is observed [18]. Apparently oxidation current flows at potentials in the range of -0.2 to +0.9 V initially, and the luminescence was ascribed to the reaction of R^- with unspecified impurities yielding products that are oxidized at about -0.2 V and participate in a chemiluminescent reaction, eventually resulting in $^1R^*$. In a similar experiment involving phenanthrene, when the phenanthrene anion radical was produced in DMF at -2.47 V versus SCE and the potential was switched to potentials more positive than +0.15 V but less than the decomposition potential of DMF (+1.5 V), luminescence was seen. In this case, however, the emission spectrum was at considerably longer wavelengths than the fluorescence spectrum and was ascribed to solution phosphorescence [20]. The suggested mechanism therefore involves oxidation of R^- directly to the triplet state $^3R^*$, which is the emitter.

$$R^- - e \longrightarrow {}^3R^*. \tag{195}$$

A direct formation of a rubrene triplet at an electrode by reduction of the cation radical was also proposed, in this case followed by a triplet-triplet annihilation reaction to produce an excited singlet [8], but alternative explanations for the preannihilation emission have since been given [18, 19].

In the case of 9,10-diphenylanthracene luminescence, however, it has been clearly shown that at either platinum or mercury electrodes in DMF, emission occurs only when the potential is scanned into regions where initial anodic and cathodic currents flow [13].

E.   Energy Considerations

1.   General

It has been pointed out that the energy released in a chemiluminescent reaction must be equal to or greater than $h\nu$, where ν is the frequency of the emitted light. This implies that for emission in the visible region (380-780 nm) energy of 76 to 34 kcal/mole or 3.3 to 1.5 ev must be available. The annihilation reaction almost certainly provides sufficient energy for the formation of the excited singlet. For example, for anthracene $E_{R,R^-} = -2.0$ V, $E_{R^+,R} = +1.2$V, and the energy of the lowest excited singlet state is about 3.2 V.

On the other hand, in experiments in which the electrogenerated oxidant is not the radical cation, and the potential of the oxidant can be taken as that required for its formation at the electrode, insufficient energy is available from the simple oxidation-reduction reaction for the formation of the excited singlet. For example, when the radical anion of 9,10-diphenylanthracene is formed at a mercury electrode (at -1.9 V) in a DMF solution containing tetra-n-butylammonium perchlorate and the potential is switched to values more positive than 0.51 V emission characteristic of the excited singlet of 9,10-diphenylanthracene is observed [13], although the difference in potential is less than that required for the formation of this state. Explanations for this "deficient energy" behavior have been based on the initial formation of triplets that then undergo a triplet-triplet annihilation reaction, producing an excited singlet, or on more complex processes in which the initially produced oxidant and reductant undergo a series of reaction steps, eventually leading to the accumulation of sufficient energy to produce the excited singlet. In other cases, such as the preannihilation luminescence observed with rubrene or the case of carbazole described in Section IV. B, where light emission occurs simply by oxidation or reduction of the electroactive species, followed by reaction with solvent or impurities, the energy considerations involved are less clear.

2.   Use of Magnetic Field Effects to Study Electrogenerated Chemiluminescence

Faulkner and Bard [21] have examined the magnetic field dependence of the chemiluminescence intensity for two reactions in this category, i.e., the reactions of the cation radical of N,N,N',N'-tetramethyl-p-phenylenediamine

(WB) with the anion radicals of anthracene and 9,10-diphenylanthracene (DPA) in N,N-dimethylformamide (DMF). For comparison, they also studied the field effects on one reaction which is not energy deficient, the mutual annihilation of the anion and cation radicals of DPA.

It was found that the energy of the ion-radical annihilation between WB^+ (the Wurster's Blue cation) and a hydrocarbon anion radical is indeed insufficient to produce the hydrocarbon in its first excited singlet. In contrast, this state is attainable in the reaction between $DPA \cdot^+$ and $DPA \cdot^-$. Since the emission spectrum is always identical with the fluorescence spectrum of the hydrocarbon, one must classify the hydrocarbon-WB systems as energy-deficient.

For solutions containing WB, the emission intensity increases with the applied magnetic field. This behavior contrasts markedly with that of the solution containing only DPA, where the field had no effect upon the emission intensity.

Two conclusions may be drawn immediately from these observations. First, paramagnetic species are involved in at least one rate-controlling step for light emission in the energy-deficient case, and the rate of that step is field dependent. Secondly, either no paramagnetic species are involved in the rate-controlling steps for the light-producing process in the case of DPA alone, or, more likely, paramagnetic species are involved, but their behavior is unaffected by the field.

Faulkner and Bard [21] indicated that the magnetic field provides a novel tool with which to probe the mechanism of the ECL process.

Faulkner and Bard [22] studied the effect of a magnetic field on the anthracene triplet-triplet annihilation in liquid solution. The fluorescent intensity of anthracene in DMF was found to decrease with increasing field strength. No indication of a low-field enhancement of the intensity was observed. The solvent was also found to affect the intensity-field strength curve.

F. Theory of Controlled-Potential Generation of Chemiluminescence

Several workers have reported chemiluminescence produced by electrogeneration at a single electrode of anion and cation radicals of a large number of aromatic hydrocarbons. Light emissions probably result from a radical-annihilation reaction followed by radiation decay:

$$R^+ + R^- \xrightarrow{k} R + R^*, \tag{196}$$

$$R^* \longrightarrow R + h \tag{197}$$

Feldberg [23] attempted to relate quantitatively the light intensity produced in electrogenerated chemiluminescence to the current, time, and kinetic parameters. He calculated these relationships for a double potential step mode of generation. In a solution containing only the organic species R the electrode potential is initially set so that zero current flows. A sufficiently positive-potential pulse is applied to the electrode so that the concentration of species R at the electrode surface instantaneously becomes zero as R is oxidized to the cation radical R^+. The positive pulse is immediately followed by a negative-potential pulse so that the surface concentrations of both species R and R^+ are zero as the anion radical R^- is generated. During the negative pulse the annihilation and light-emission reactions (see Eqs. (196) and (197)) occur in the diffusion layer.

G. Emission Spectra During Electrochemiluminescence

One type of experiment that has been useful in the study of ECL phenomena is the observation of the emission spectrum during electrolysis and a comparison of this spectrum to the fluorescence spectrum of the parent molecule. Some recent results on the ECL spectra of a number of aromatic compounds and a comparison of these spectra with fluorescence spectra recorded both before and after the ECL experiment were studied by Bard [24].

The fluorescence and ECL spectra of the compounds were recorded in the same cell in DMF containing 0.1 M tetra-n-butylammonium perchlorate with an Aminco-Bowman spectrophotofluorometer. The cell, consisting of two coiled platinum electrodes sealed in to the walls of the cell, was filled with test solution, attached to a vacuum line, evacuated, and sealed off before the experiment. Alternating sinusoidal voltages of various magnitudes were applied to the cell to obtain the ECL spectra. In some experiments (e. g. , with phenanthrene and stilbene) the applied voltages were increased continuously until luminescence was observed, and then the ECL spectra were studied as a function of applied voltage. Although the intensity increased with increasing applied voltage, the locations of the ECL bands were unchanged. The experimental procedure consisted of recording the fluorescence spectrum of the solution, recording the ECL spectra during continuous electrolysis, and finally recording the fluorescence spectrum of the electrolyzed solution.

Three general classes of behavior were noted (Fig. 119 and Table 86). Such hydrocarbons as 9, 10-diphenylanthracene and tetracene showed the preelectrolysis and postelectrolysis fluorescence bands and the ECL bands at the same wavelengths (type I behavior).

Phenanthrene, trans-stilbene, and others show what we call type II beha-

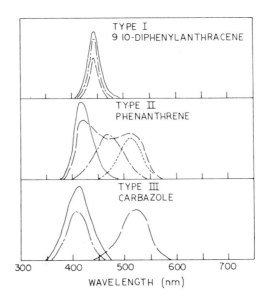

FIG. 119. Types of behavior encountered in ECL systems: preelectrolysis fluorescence (——); postelectrolysis fluorescence (— —); initial ECL (— – — -); final position of ECL spectrum for type II (– – –).

vior. During electrochemiluminescence two emission bands occur: one at the position of the preelectrolysis fluorescence band, and one at longer wavelengths. As the electrolysis is continued the first band decreases, and the longer wavelength band increases, in intensity, until finally (after about 30 min of electrolysis) only the longer wavelength ECL band is observed. Examination of the solution after electrolysis showed that the original, preelectrolysis, fluorescence band was absent, and a new, longer wavelength band was present. Fluorescence spectra of the solution taken at various points during the electrolysis showed that this new fluorescent peak increased in intensity as the long-wavelength ECL band increased in intensity.

Carbazole demonstrated type III behavior. The preelectrolysis and postelectrolysis fluorescence bands were at the same wavelengths, but the ECL spectra were at longer wavelengths, generally decreasing in intensity as electrolysis continued.

The behavior of the type II compounds is particularly interesting, since it showed that decomposition of the electrolysis products of the parent compound occurs during electrolysis and that the long-wavelength ECL band persists even when the parent compound has disappeared. In all cases a postelectrolysis fluorescence band, probably due to a new species, is observed. These results suggest that previous explanations of the long-wavelength emissions of

TABLE 86

Fluorescence and ECL Spectra[a]

Compound	Preelectrolysis fluorescence maximum (nm)	ECL maximum (nm)	Postelectrolysis new fluorescence maximum (nm)
Type I:			
9,10-Diphenyl-anthracene	440	440	None[b]
Rubrene	560	560	None[b]
Tetracene	520	520	None[b]
Type II:			
Phenanthrene	412	412 516	486
Trans-Stilbene	398	398 506	506
Anthracene	412	412 455 569	480
1,2-Benzanthracene	408 420	420 538	452
Type III:			
Bifluorenyl	384	526	None[b]
Carbazole	406	538	None[b]
Dibenzo[c,g]-carbazole	440	530	None[b]

[a]The solution contained 0.1 M tetra-n-butylammonium perchlorate and was 10^{-3} M in the organic compound.

[b]Only original, preelectrolysis, fluorescence band found.

substances showing type II behavior based on the formation of, and emission from, the triplet state or from excimers of the parent compound may have alternative explanations based on ECL reactions of new substances formed during electrolysis.

In experiments of this kind one must be aware of the important effect of trace impurities on the observed behavior. Because energy transfer in the excited state may be very efficient, the observed spectrum may be that of a trace constituent that emits at a longer wavelength. For example, in experiments with fluorene the observed behavior turned out to be caused by a small quantity of carbazole contained in the sample of the hydrocarbon as an impurity.

Decomposition of the radical ions to form new species must also be considered in the interpretation of results.   For example, the anthracene cation radical is known to be very unstable, even in methylene chloride [25], so that the luminescence of products formed on decomposition of the radical cation may be involved in its type II behavior.   One may also note that "purging" of impurities by titration with electrogenerated radical ions may also lead to the production of new species by the decomposition of these ions.

H.   Electrochemiluminescence on Reduction of Carbazole

Substances other than aromatic hydrocarbons have also been shown to undergo electrogenerated chemiluminescence; these include the arylisobenzo-furans and related molecules [19, 26], and carbazole, described in the preceding section.   Chemiluminescence also occurs during the electrochemical reduction of carbazole in DMF at a mercury cathode without the necessity of an electro-chemical oxidation step.

The cyclic voltammetric reduction of carbazole at a hanging-mercury-drop electrode is shown in Fig. 120.   The initial reduction wave at -2. 76 V versus SCE shows no corresponding wave on reversal, probably indicating that the initial reduction step is followed by a fast chemical reaction.

The reduction process is accompanied by the emission of a pale-blue light, which was monitored by a Dumont 6467 photomultiplier tube with an output to either a sensitive galvanometer or an oscilloscope.   If the sweep is stopped on the peak potential, the intensity of the light emission diminishes gradually and reaches a steady state.   The intensity obtained by stepping the potential from -2. 20 V (where no oxidation-reduction process occurs) to potentials on the rising portions of the peak is shown in Fig. 121.   The new peaks found on the reversal oxidation sweep and second and subsequent reduction sweeps indicate that the products of the reaction following the initial reduction step are electro-active.

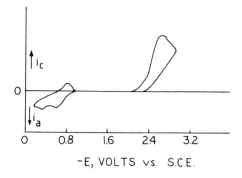

FIG. 120.   Cyclic voltammetric reduction curve of carbazole (2 mM) in DMF containing 0. 1 M tetra-n-butylammonium perchlorate on a hanging-mercury-drop electrode.   Sweep rate 153 mV/sec.

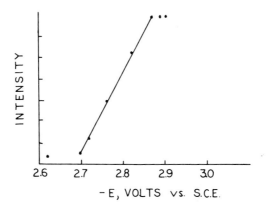

FIG. 121. Intensity of light emission (in arbitrary units) versus potential on the rising portion of the carbazole-reduction wave obtained by stepping the potential from -2. 20 V.

The emission of light is also observed on polarographic reduction of carbazole, which occurs in a drawn-out wave with an $E_{1/2}$ of -2. 68 V versus SCE. The intensity of emitted light increases with increasing polarographic current and remains constant in the diffusion-plateau region. Oscillations in the intensity with the growth and fall of the drop are also observed. During controlled-potential electrolysis at -2. 80 V versus SCE at a mercury-pool cathode with the platinum-anode auxiliary electrode contained in a compartment separated from the test solution by a sintered-glass disk, a bright emission is observed across the whole mercury surface. As the electrolysis continues, the intensity decreases and the cessation of light emission can be used to signal the completion of electrolysis.

The fluorescence spectra of the original solution and the emission spectra during reduction on mercury were recorded with an Aminco-Bowman spectro-photofluorometer. The emission spectra have maxima at 464 nm, shifted from the carbazole fluorescence maxima at 406 nm. After prolonged electrolysis in the spectrophotofluorometer cell, a new fluorescence peak at 414 nm is observed.

Similar results have been obtained for the reduction of dibenzo[c, g]carba-zole at a mercury electrode.

The evidence presented suggests that chemiluminescence results from the reaction of the product of carbazole electroreduction with DMF or supporting electrolyte. The emitting species may be the product of this reaction or possibly carbazole itself emitting at a longer wavelength, for example, via a triplet.

Related experiments that might be mentioned here include those of Hercules and Lytle [27], who recently reported that certain metal complexes and amine radical cations show luminescence on chemical reduction, and those of Chandross and Sonntag [28] on the reaction of the phenylcarbazole anion with oxidants in dimethoxyethane.

I.   Electrochemistry of Perylene

Werner, Cheng, and Hercules [29] investigated the electrochemilumines- cence of perylene with an image-intensifier spectrograph using single-pulse excitation.   The component of perylene ECL spectra at wavelengths longer than perylene fluorescence was found to be dependent on solvent and on the direction of polarity change in the potential pulse.   Such behavior was not consistent with the previous assignment of this long-wavelength component to a perylene excimer.   In acetonitrile the amount of the longer wavelength component is greater than it is in benzonitrile.   Since cation-radical stability appears to be greater in the latter solvent, the long-wavelength components in the perylene ECL spectra can be attributed to cation-decomposition products.   At higher concentrations ($>$ 10^{-3} M) the behavior of the perylene-cation-reduction wave is abnormal.   This behavior is reflected in the difference in ECL intensity versus time curves for cation-to-anion and anion-to-cation excitation.

J.   Anthracene Chemiluminescence in DMF

Faulkner and Bard [30] found that the electrochemiluminescence of anthra- cene is characterized by emission at the frequency of anthracene fluorescence and also at longer wavelengths.   One longer wavelength component is shown to be caused by emission from anthranol, produced by the decomposition of the cation radical of anthracene and probably excited by energy transfer from excited anthracene.   Another component arising from the ECL spectrum of anthranol itself was also observed.   The authors concluded that energy transfer from anthracene to anthranol is an effective mechanism in DMF solution and will be responsible for at least part of the emission observed at 460 nm.

Anthranol was also believed to be implicated in the mechanism producing light in the 565-nm region.   An investigation of the ECL spectra of 1 mM anthracene plus anthranol showed the emission to be maximum at 565 nm.   The emission intensity rose steadily with increasing applied voltage after the threshold of about 4.5 V peak to peak.

K.   Effect of Alkyl Halides on the Chemiluminescence

Siegel and Mark [31] have reported an unexpectedly intense chemilumin- escent reaction that resulted from the electroreduction of solutions containing aromatic hydrocarbons and either 9,10-dichloro-9,10-dihydro-9,10-diphenyl-anthracene (DPACl$_2$) or 1,2-dibromo-1,2-diphenylethane (DBDPE) at platinum electrodes. In most previously reported examples of electrochemically generated chemiluminescence (ecl) of aromatic hydrocarbons, the excited state of the hydrocarbon is produced from the annihilation reaction of the radical anion, $R \cdot ^-$, of the hydrocarbon with either the radical cation, $R \cdot ^+$, or another electrogenerated oxidizing agent.

It was found that this cl, in a variety of nonaqueous solvents, is about two orders of magnitude greater in luminescence intensity than 9,10-diphenyl-anthracene annihilation ecl. This indicates that the mechanism by which the excited state is produced during the electrode reaction is different and probably more efficient than $R \cdot ^-$ and $R \cdot ^+$ annihilation under similar experimental conditions.

A variety of aryl and alkyl halides were found to increase the luminescence yield, and the authors are currently studying the reaction further.

IV. DETERMINATION OF ELECTROGENERATED CHEMILUMINESCENCE EFFICIENCY (Φ ecl)

It is of considerable interest in studies of electrogenerated chemilumin-escence (ecl) to determine the overall ecl efficiency (ϕ ecl), which represents the fraction of radical ion encounters which ultimately result in the emission of a photon. Studies of ecl at a rotating-ring-disk electrode (rrde) are ideally suited for such efficiency measurements, since the radical ion precursers are generated continuously under steady state conditions [12].

Maloy and Bard [32] studied the efficiency and mechanisms of 9,10-diphenylanthracene, rubrene and pyrene systems at the rrde. The efficiency for 9,10-diphenylanthracene (DPA) ecl in dimethylformamide solutions was below 0.1%. Studies performed on the DPA-tetramethyl-p-phenylenediamine (TMPD) system showed that ecl occurs during reaction of $DPA \cdot ^-$ with either $TMPD \cdot ^+$ or $TMPD^{2+}$; the efficiency for this ecl is about an order of magnitude smaller than for DPA alone. TMPD was also shown to be an effective quencher of DPA ecl in these systems. Rrde studies of the rubrene (R) system demon-strated that ϕ_{ecl} was much smaller for it than for the DPA system and that reaction of R^{2+} with $R \cdot ^-$ or R^{2-} with $R \cdot ^+$ produced light at high rotation rates. A detailed study of the pyrene (P)-TMPD system was undertaken. The ϕ_{ecl} for the P-TMPD system was smaller than that for the DPA-TMPD system. A study of the relative pyrene excimer to monomer emission in both ecl and fluorescence demonstrated the existence of a direct path to excimer in ecl, consistent with the triplet-triplet annihilation mechanism. Quenching of pyrene ecl by TMPD was also investigated.

V. ELECTROCHEMILUMINESCENCE ANALYSIS

Cruser and Bard [33] have performed some preliminary experiments suggesting possible analytical applications of electrochemiluminescence. In these experiments rather simple apparatus was employed, so that a practical method might be developed without resorting to complicated electrochemical cells or circuits. The cell employed (see Fig. 118) consisted of two coiled platinum electrodes sealed into the walls of a closed standard-taper joint. This cell was placed, in a reproducible fashion, inside an Aminco-Bowman spectrophotofluorometer, and the emission spectrum was recorded on passage of an ac voltage between the two electrodes. The experiments described here were on solutions containing different concentrations of 9,10-diphenylanthracene in DMF, with 0.1 M tetra-n-butylammonium perchlorate as the supporting electrolyte. Light intensity was measured at a wavelength of 429 nm.

A typical intensity-time curve resulting from the application of ac to the cell is shown in Fig. 122; zero time here corresponds to the moment of switching on the ac. The light intensity is seen to go through a peak value and then decay to a steady value. The effect of varying the magnitude of the ac voltage to the cell is shown in Fig. 123. As the ac voltage is increased, the peak intensity increases, until a constant value is obtained, probably corresponding to the application of a voltage sweep of sufficient magnitude to cause the potential to step to the limiting-current-regions of the current-potential curves.

The effect of concentration on peak intensity, with the voltage adjusted to the limiting-region value for each concentration, is shown in Fig. 124. The peak intensity is seen to be a linear function of concentration, suggesting that this form of emission spectroscopy could find application as an analytical technique. Obviously there are many limitations to this kind of method,

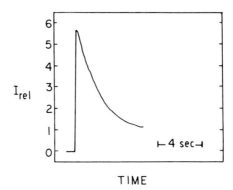

TIME

FIG. 122. Intensity (in arbitrary units) versus time in the 9,10-diphenylanthracene system.

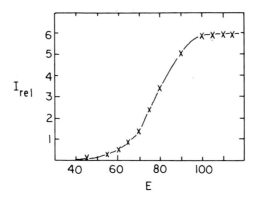

FIG. 123. Intensity of light emission (in arbitrary units) versus applied
ac voltage.   The solution contained 0. 1 M tetra-n-butylammonium perchlorate
and 9, 10-diphenylanthracene in DMF.

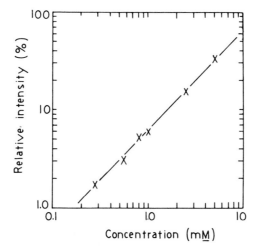

FIG. 124. Intensity of light emission (in arbitrary units) obtained at
various concentrations of 9, 10-diphenylanthracene.

including the fairly specific nature of the phenomenon, the possible variable
effect of quenchers, and the effect of mixtures of chemiluminescent substances.

Cruser and Bard [34] studied the intensity-time and concentration-intensity
relationships and the lifetime of radical cations of aromatic hydrocarbons in
DMF solution. Working curves, obtained by digital-computer simulation,
allowed the determination of the rate constants for the decomposition of the
radical ions. The half-lives of the cation radicals of 9, 10-diphenylanthracene,
1, 3, 6, 8-tetraphenylpyrene, and rubrene were determined experimentally. The

maximum ECL intensity, obtained by passing an alternating current through the cell, was in all cases a linear function of the concentration of the electroactive species. Thus an analysis for these three compounds by the ECL method is possible.

REFERENCES

1. T. Kuwana, in Electroanalytical Chemistry, Vol. I (A. J. Bard, ed.), Dekker, New York, 1966, Chapter 3.

2. V. Ya. Shlyapintokh, L. M. Postnikov, O. N. Karpukhin, and A. Ya. Veretil'nyi, Zh. Fiz. Khim., 37, 2374 (1963).

3. V. P. Kazakov, Zh. Fiz. Khim., 39, 1567 (1965) and references contained therein.

4. R. T. Dufford, D. Nightingale, and L. W. Gaddum, J. Amer. Chem. Soc., 49, 1858 (1927) and references contained therein.

5. D. M. Hercules, Science, 143, 808 (1964).

6. E. A. Chandross and R. E. Visco, J. Amer. Chem. Soc., 86, 5350 (1964).

7. K. S. V. Santhanam and A. J. Bard, J. Amer. Chem. Soc., 87, 139 (1965).

8. M. M. Rauhut, D. L. Maricle, et al., Chemiluminescent Materials, American Cyanamid Co. Tech. Rept. No. 5 to ONR and ARPA, AD 606989; Defense Documentation Center, October 19, 1964, pp. 30-38.

9. J. M. Bader and T. Kuwana, J. Electroanal. Chem., 10, 104 (1965).

10. A. Bard, K. S. V. Santhanam, S. A. Cruser, and L. Faulkner, in Fluorescence. Theory, Instrumentation, and Practice (G. G. Guilbault, ed.), Dekker, New York, 1967.

11. S. A. Cruser, Analytical Aspects of Electrogenerated Chemiluminescence, available from University Microfilms, Ann Arbor, Mich., Order No. 68-16, 060, 1968.

12. D. Hercules, Organic Electroluminescence in Physical Methods of Chemistry (A. Wiesburger and B. Rossiter, eds.), Vol. I, Part 2B, Wiley-Interscience, New York, 1971.

13. K. S. V. Santhanam and A. J. Bard, J. Amer. Chem. Soc., 88, 2669 (1966).

14. J. T. Maloy, K. B. Prater and A. J. Bard, J. Amer. Chem. Soc., 93, 5959 (1971).

15. E. A. Chandross and F. I. Sonntag, J. Amer. Chem. Soc., 86, 3179 (1964).

16. E. A. Chandross, J. W. Longworth, and R. E. Visco, J. Amer. Chem. Soc., 87, 3259 (1965).

17.   D. M. Hercules, R. C. Lansbury, and D. K. Roe, J. Amer. Chem. Soc., 88, 4578 (1966).

18.   D. L. Maricle and A. Maurer, J. Amer. Chem. Soc., 89, 188 (1967).

19.   A. Zweig, A. K. Hoffman, D. L. Maricle, and A. H. Maurer, Chem. Commun., 106 (1967).

20.   A. Zweig, D. L. Maricle, J. S. Brinen, and A. H. Maurer, J. Amer. Chem. Soc., 89, 473 (1967).

21.   L. R. Faulkner and A. J. Bard. J. Amer. Chem. Soc., 91, 209 (1969).

22.   L. R. Faulkner and A. J. Bard, J. Amer. Chem. Soc., 91, 6495 (1969).

23.   S. W. Feldberg, J. Amer. Chem. Soc., 88, 390 (1966).

24.   A. J. Bard, K. S. V. Santhanam, S. A. Cruser, and L. R. Faulkner, in Fluorescence. Theory, Instrumentation, and Practice (G. G. Guilbault, ed.), Dekker, New York, 1967, Chapter 14.

25.   J. Phelps, K. S. V. Santhanam, and A. J. Bard, J. Amer. Chem. Soc., 89, 1752 (1967).

26.   A. Zweig, G. Metzler, A. Maurer, and B. G. Roberts, J. Amer. Chem. Soc., 88, 2864 (1966).

27.   D. M. Hercules and F. E. Lytle, J. Amer. Chem. Soc., 88, 4745 (1966).

28.   E. A. Chandross and F. I. Sonntag, J. Amer. Chem. Soc., 88, 1089 (1966).

29.   T. Werner, J. Cheng, and D. Hercules, J. Amer. Chem. Soc., 92, 5560 (1970).

30.   L. Faulkner and A. J. Bard, J. Amer. Chem. Soc., 90, 6284 (1968).

31.   T. M. Siegel and H. B. Mark, J. Amer. Chem. Soc., 93, 6281 (1971).

32.   J. T. Maloy and A. J. Bard, J. Amer. Chem. Soc., 93, 5968 (1971).

33.   S. A. Cruser and A. J. Bard, Anal. Letters, 1, 11 (1967).

34.   S. A. Cruser and A. J. Bard, J. Amer. Chem. Soc., 91, 267 (1969).

Chapter 12

EXTRINSIC AND INTRINSIC FLUORESCENCE OF PROTEINS

Raymond F. Chen
Laboratory of Technical Development
National Heart and Lung Institute
Bethesda, Maryland

I. INTRODUCTION

Aromatic amino acids present in proteins endow them with an intrinsic ultraviolet fluorescence. Proteins in complex with other fluorescent molecules may be said to have an extrinsic fluorescence. The terms "intrinsic" and "extrinsic" have been applied in an analogous way to Cotton effects observed in the optical rotatory dispersion of proteins [1]. These terms, as applied to fluorescence, were first used in the version of this review chapter that appeared in another volume [2]. The chapter has now been rewritten to include many new studies that reflect the rapid growth of this field.

The characteristics of protein fluorescence are dependent on structure and cannot be fully described by specifying the fluorophor itself. Fluorescence spectroscopy has become widely accepted as a modern method for the study of protein structure, a field of great relevance to control mechanisms, protein biosynthesis, and other biochemical problems. This chapter reviews the principles behind the use of protein fluorescence for configurational studies and describes some of the interesting applications that have been reported in the literature.

II. EXTRINSIC FLUORESCENCE

A. Fluorescence Depolarization

1. Outline of the Method

The technique of fluorescence depolarization was introduced into protein chemistry by Weber [3] in 1952 and is still perhaps the most direct fluorescence method for investigating structural changes in proteins. Briefly, a protein is covalently labeled by reaction with a fluorescent dye. When this dye moiety is excited with polarized light, the fluorescence will also be polarized to a degree that is inversely related to the amount of Brownian motion occurring during the interval between absorption and emission of light. In turn knowledge of the amount of motion gives information on the size and shape of the macromolecule to which the dye is attached. Although the measured quantity is the degree of polarization, the technique is aptly described by the term "depolarization."

Polarized fluorescence is also observed with unpolarized excitation, provided that the direction of observation is normal to that of excitation; this is the case in most instruments used for such measurements. For simplicity, however, the formulas in this section refer to excitation polarized linearly in a direction that is normal to the plane defined by the axes of excitation and observation. P is given by

$$P = \frac{I_V - I_H}{I_V + I_H}, \tag{198}$$

where I_V and I_H are the fluorescence intensities observed with the analyzer oriented parallel and perpendicular, respectively, to the direction of polarization of the incident light.

Perrin [4, 5] and Levshin [6] derived the classical equations relating the fluorescence parameters P and τ, the fluorescence lifetime, to the rotational factors--viscosity η , temperature T, and molecular volume V. For a randomly labeled, rigid, ellipsoidal macromolecule Weber [7, 8] obtained

$$\frac{1}{P} - \frac{1}{3} = \left(\frac{1}{P_o} - \frac{1}{3} \right) \left(1 + \frac{3 \tau}{\rho_h} \right), \tag{199}$$

where P_o is the polarization in the absence of rotation and ρ_h is the harmonic mean of the rotational relaxation times corresponding to the three principal axes of the ellipsoid:

$$\frac{1}{\rho_h} = \frac{1}{3} \left(\frac{1}{\rho_1} + \frac{1}{\rho_2} + \frac{1}{\rho_3} \right). \tag{200}$$

For a sphere $\rho_h = \rho_o$, which is also given by $\rho_o = 3 \eta V / RT$, where R is the gas constant.

In order to obtain ρ_h experimentally one measures polarization as a function of temperature or viscosity and plots $1/P$ versus T/η (a Perrin plot). In accord with theory [4], a straight line that gives $1/P_0$ at T/η = 0 is usually obtained. If τ is known, ρ_h can be calculated.

The relaxation time found in this way is of interest since it should be realated to the size and shape of the protein. The ratio ρ_h / ρ_0 is often used as a quick guide to the degree of the protein's departure from a spherical shape. The value of ρ_0 can be calculated in nanoseconds (assuming a partial specific volume of 0.74) from

$$\rho_0 = \frac{MW}{1.23} \times 10^{-3}, \tag{201}$$

where MW is the molecular weight.

One might ask why the depolarization method requires using protein-dye conjugates rather than exploiting the intrinsic fluorescence of the protein. Actually there is no particular reason why the intrinsic fluorescence cannot be so used in some cases, and indeed it has been employed. At the time of Weber's classical paper [3] describing the technique in 1952, intrinsic protein fluorescence had not yet been discovered. Another consideration is that the accuracy of ρ_h determination by the use of Eq. (199) requires that 3 τ and ρ_h be of similar magnitude, so that differences between $1/P - 1/3$ and $1/P_0 - 1/3$ can be instrumentally defined. After the discovery of intrinsic protein fluorescence it was considered that the lifetime of tryptophan fluorescence was probably too low to be useful. However, many proteins with ρ_h values above 100 nsec, and some even above 200 nsec, have been studied by labeling with dyes whose lifetimes are in the range of 13 nsec. With accurate instrumentation intrinsic fluorescence polarization could yield ρ_h values at least as large as 30 to 100 nsec if the intrinsic lifetimes are in the range 2 to 7 nsec. These values would correspond to rigid protein ellipsoids with molecular weights as high as 15,000 to 50,000.

2. Applications of the Depolarization Method

Weber [8] labeled bovine-serum albumin (BSA) and ovalbumin by reaction with 1-dimethylaminonaphthalene-5-sulfonyl (DNS) chloride. From Oncley's [9] value of 75.5 nsec for the rotational relaxation time of ovalbumin determined by dielectric dispersion Weber concluded from polarization measurements that the DNS moiety had a fluorescence-decay time of 14 nsec. This value of τ was employed to calculate ρ_h for DNS-BSA, and a value of 127 nsec was obtained.

Steiner and McAlister [10] measured the lifetime of DNS directly in conjugates of lysozyme, ovalbumin, and BSA. They obtained 11 to 13.3 nsec, in good agreement with Weber's value [8]. Steiner and McAlister [10] obtained constant ρ_h values for the proteins in experiments using not only DNS but also fluorescein and anthracene sulfonamide conjugates. The values were independent of the degree of conjugation or the wavelength of excitation, and these workers therefore considered that the data supported the assumptions that labeling was random and that the proteins were rigid. In all of the proteins first studied by this method the Perrin plots were linear, and the ratios ρ_h/ρ_0 for the native proteins were uniformly high, in the range 1.7 to 2.

Weber's study [8] showed that DNS-BSA underwent structural changes as indicated by a decrease in polarization below pH 4 and above pH 9.5. At pH 2 the decreased value of the DNS polarization was interpreted by dissociation of the molecule into subunits [8]. Though the changes in BSA at acid pH are now known to be due instead to unfolding and expansion [11], the changes in polarization clearly occurred in the same pH ranges where other physical parameters had indicated the occurrence of structural changes. Harrington et al. [12] confirmed Weber's data on DNS-BSA and obtained similar pH-dependent changes in polarization by using an anthracene carbamide conjugate of BSA. They calculated that the dye moiety had a decay time of 44 nsec.

Although other dyes have been used, DNS chloride is still the dye of choice in most polarization studies. Aside from Steiner and McAlister's study [10], Chadwick and Johnson [13] labeled BSA with a number of dyes and were able to calculate τ for each from polarization data, assuming a rotational relaxation time of 125 nsec for BSA. A pyrene-containing dye had a calculated lifetime of 90 nsec. The interest in long-lifetime dyes was due to a desire to extend the usefulness of the depolarization technique to large rigid proteins with long relaxation times. Though DNS was useful for obtaining information on proteins as small as insulin [14], it was pointed out by Weber [7] that as ρ_h becomes very large with respect to τ, P approaches P_0, so that calculation of the relaxation time from Eq. (199) becomes very imprecise. Therefore the longer lifetimes of the anthracene and pyrene derivatives introduced by Johnson and co-workers [12, 13] would seem to be advantageous, but this is partly offset by their inherently lower P_0 values. In addition to its favorable chemical and fluorescence characteristics DNS has had its usefulness enhanced by more sensitive photoelectric methods of obtaining P and by the fact, to be discussed, that conjugates of very large proteins frequently exhibit measurable changes in polarization with change in T/η.

Now commercially available from many sources, DNS chloride can even be purchased with either ^{14}C or ^{3}H labeling. Although the synthesis described by Weber [8] is simple, Mendel [15] has recently published an improved prepara- tive procedure. Rinderknecht [16] has indicated that labeling of proteins with DNS chloride or other dyes can proceed apace if Celite-adsorbed dyes are used. Apparently the Celite surface has affinity for both protein and dye, which are thus brought into close proximity. The actual labeling conditions in conventional solutions in which DNS chloride is suspended have been studied in detail by Gros and Labouesse [17].

In spite of its advantages, DNS chloride cannot be considered the only suitable labeling agent in every case, and sometimes it is desirable to check the results obtained with DNS conjugates by using another dye as well. Gill et al. [18] compared the fluorescence properties of DNS- and fluorescein- labeled synthetic polypeptides and proteins. They decided that the macro- molecules labeled by reaction with fluorescein isothiocyanate (FITC) gave ρ_h values more in keeping with the known structure than did the DNS-labeled compounds. This finding was interpreted as indicative of nonrigid attachment of DNS. For example, lysozyme labeled with DNS and FITC was found to have relaxation times of 11 and 28 nsec, respectively, at $25^{\circ}C$. It was found that DNS was suitable over a wide pH range, whereas the fluorescein emission changed markedly outside the pH range of 6 to 8. Klotz and co-workers [19] had established that the protonation of the dimethylamino group of DNS occurred in proteins at lower pH than in the free DNS acid. The pK_a in proteins was typically between 2 and 3.

Gill [20] evaluated the internal rigidity of crosslinked polymers by the fluorescence polarization of conjugates labeled with FITC, where the dye had been adsorbed onto Celite. Churchich [21] used FITC-labeled aspartate amino- transferase to study the rotational relaxation time. He assumed a lifetime of 5 nsec for the label and obtained biphasic Perrin plots. This finding was interpreted to indicate separate relaxation times of 130 and 9 nsec, correspond-

ing to the relaxation of the macromolecule and to the free rotation of the dye
about its site of attachment.

Although FITC is easily available and easy to use, it too has some disadvantages. It is much more stable than the old fluorescein isocyanate that was
originally used by Coons et al. [22] to label antibodies, but the usual synthesis
of FITC [23] results in two isomers, which may have different properties.
Commercial preparations of FITC were examined by Cherry and colleages [24]
and found to be highly variable in characteristics. It is also known from direct
measurements, starting with those of Steiner and McAlister [10] that FITC
conjugates have short lifetimes between 3 and 5 nsec, so that depolarization
should be observable only with relatively fast rotations unless very precise
polarization instrumentation is used.

The emission spectra of a number of dyes used to label proteins for
immunochemical studies have been published by Hansen [25]. Some of these
may eventually turn out to be useful for depolarization studies. Knopp and
Weber [26] synthesized a derivative of pyrenebutyric acid that would attach to
BSA. Assuming a ρ_h value of 120 nsec for the conjugates, the authors
calculated from Perrin plots that the extrinsic fluorescence lifetimes ranged
from 75 to 126 nsec, depending on the degree of labeling. The long lifetimes
were desirable if the dye was to be used on very large proteins. However, the
Perrin-Weber relation (Eq. (199)) indicates that for such purposes not only a
high τ but also a high P_0 would be desirable. Unfortunately Knopp and Weber
[26] found the P_0 for pyrenebutyryl conjugates to be on the order of 0.1 compared
with the value of about 0.4 usually found for DNS conjugates.

Knopp and Weber [27] attached pyrenebutyric acid to a human macroglobulin of molecular weight 890,000. When Perrin plots were obtained
isothermally (i.e., by changing the viscosity rather than the temperature),
relaxation times of about 1000 nsec were obtained. In contrast, ρ_0 for the
equivalent sphere was calculated to be 720 nsec. The ρ_h/ρ_0 ratio was smaller
than that found for BSA, indicating that there was greater symmetry in the
macroglobulin. Interestingly, in this case, the same experimental conditions
applied to a DNS-labeled preparation of the macroglobulin gave no detectable
change in P over a wide range of viscosities.

Rawitch et al. [28] improved the activation method for making pyrenebutyric
acid reactive with amino groups. They synthesized the sulfuric acid anhydride
and used it to label thyroglobulin, a protein with a molecular weight of 670,000.
Various preparations had lifetimes measured at 90 to 125 nsec. Perrin plots
obtained isothermally yielded relaxation times of 1500 to 1600 nsec, whereas
plots obtained by heating the solutions gave values of 500 nsec, a finding indica-

tive of thermally activated rotations. The authors contrasted these results with
those obtained with DNS-thyroglobulin conjugates, which were found to have lifetimes of only about 8 nsec and gave horizontal (unusable) Perrin plots. Perrin
plots obtained by heating the DNS-thyroglobulin gave anomalously small relaxation times, confirming the results of Metzger et al. [29].

Several new dyes have been suggested, and no doubt more will appear in the
future. Evaluation of these dyes is desirable, because they might prove to have
some specific favorable characteristics. Maddy [30] described the synthesis of
4-acetamido-4'-isothiocyanatostilbene-2, 2'-disulfonic acid (SITS), which labels
the lipoprotein part of the erythrocyte membrane. However, SITS conjugates of
γ-globulin were found to exhibit photosensitivity and to have fluorescence-decay
times on the order of 1 nsec [31].

Kanaoka et al. [32] synthesized N-[p-(2-benzimidazolyl)phenyl]maleimide
(BIPM) and found it to be virtually nonfluorescent. However, reaction with
sulfhydryl compounds resulted in bright fluorescence. Typically the excitation
and emission maxima were 307 and 370 nm, respectively. Taka-amylase A
was completely inhibited by reaction of BIPM with one of the sulfhydryl groups.

Ghosh and Whitehouse [33] synthesized NBD chloride (7-chloro-4-nitro-
benzo-2-oxa-1, 3-diazole), which they indicated could react with amines, such as
amino acids and proteins. Excitation and emission maxima of typical derivatives
were at 464 and 512 nm, respectively. Conjugates of NBD chloride-labeled
BSA seem to be quite stable and have fluorescence-decay times of about 6 nsec
(Chen, unpublished data).

Chen [31] recorded the absorption and emission spectra, lifetimes, polariza-
tions, and quantum yields of γ-globulin labeled with DNS chloride, FITC,
Rhodamine B isothiocyanate, Rhodamine Lissamine, anthracene-2-isocyanate,
and SITS and found differences in properties depending on the degree of labeling.

Another dye, "mansyl chloride" (N-methyl-2-anilino-6-naphthalenesulfonyl
chloride), was synthesized by Cory et al. [34]. Although the compound was
designed to be used as a covalent fluorescent probe for configurational changes,
it could also be used for depolarization studies.

Fluorescein mercuricacetate was used [35] to label the sulfhydryl groups
of yeast alcohol dehydrogenase. Quenching of fluorescein emission was noted
on reaction with the enzyme, and disruption of the quaternary structure of the
enzyme occurred.

A convenient source of extrinsic fluorescence is a natural fluorescent
cofactor. Churchich [36] has used the pyridoxamine 5-phosphate of alanine
aminotransferase to study the effect of sulfhydryl reagents on the enzyme. The
cofactor was found, by fluorescence-depolarization techniques, to have rotational

mobility at the active site, and the cofactor was released by p-chloromercuri-
benzoate. Pyridoxal derivatives can also be attached to proteins by formation of
a Schiff base and subsequent borohydride reduction. Irwin and Churchich [37]
used this technique for affixing pyridoxal-5-phosphate to lysozyme. Directly
measured, the lifetime was found to be 2.5 nsec, and the Perrin plot indicated
two relaxation times of 0.16 and 8.7 nsec. Both of these lifetimes corresponded
to modes of rotational freedom independent of that of the molecule as a whole.
For FITC-lysozyme the Perrin plot yielded a ρ_h of 25 nsec, which was more
consonant with rotation of the molecule as a whole. In this study Irwin and
Churchich [37] used an unusual alternative method for obtaining the relaxation
with the FITC conjugate. It was found that the fluorescein emission was
quenched by the iodide ion, with a concomitant proportional decrease in
fluorescence lifetime. As can be seen from Eq. (199), a plot of $1/P - 1/3$
versus τ will give a straight line whose slope can be used to calculate ρ_h.
The relaxation time obtained in this way agreed with that found by heating the
FITC conjugate.

3. Structural Studies Utilizing the Depolarization Method

The depolarization technique lends itself naturally to studies of associating-
dissociating systems because the degree of polarization is usually very sensitive
to changes in molecular weight. The study by Laurence [38] in 1952 on the
adsorption of dyes onto BSA was one of the first instances of the use of fluorescence polarization in biochemistry. Weber [8] noted that denaturation of ovalbumin was accompanied by an eightfold increase in ρ_h that could be explained
only by aggregation. Steiner [39] applied fluorescence polarization to the study
of trypsin binding to DNS-labeled soybean-trypsin inhibitor; both of these
proteins have molecular weights of about 20,000. Massey et al. [40] showed
the aggregation of chymotrypsin and chymotrypsinogen from the concentration
dependence of the polarization. The binding of a small DNS-labeled antigen to
a large antibody molecule (molecular weight 160,000) would be expected to
increase the polarization markedly, and this fact forms the basis of studies by
Haber and Bennett [41]. The antibody binding of labeled haptens, such as DNS-
penicillin, and the study of binding equilibria by the polarization technique have
been reviewed by Dandliker and co-workers [42-44].

In contrast, the binding of other molecules can sometimes lead to lower
ρ_h values. Green [45] found that the relaxation time of avidin decreased from
56 to 50 nsec on binding of one molecule of biotin. This showed that the complex
had a more compact structure than the free protein. Similarly Brewer and
Weber [46] found a decreased relaxation time for DNS-enolase on addition of

magnesium ions, which are known to activate the yeast enzyme. Other work by
Brewer and co-workers [47] has utilized fluorescence polarization to show that
yeast enolase is actually made up of two subunits with molecular weights of
34,000.

Fluorescence-polarization measurements were useful in studies by Jonas
and Weber [48,49] on the properties of BSA modified by reaction with dicar-
boxylic acid anhydrides both before and after limited digestion with chymotrypsin. In attempting to use electron-spin-resonance (ESR) measurements on
spin-labeled haptens to determine the rotational motion in an antigen-binding
site on an antibody, Stryer and Griffith [50] resorted to fluorescence-polarization measurements to interpret the ESR results. On binding with antibody the
hapten (DNP-nitroxide) showed distinct line broadening in the ESR spectrum.
The same degree of broadening was observed with DNS-nitroxide in glycerol-
water-ethanol (90:5:5) at 35°. By fluorescence-depolarization measurements
the relaxation time of DNS-nitroxide under these conditions was observed to be
36 nsec. By inference, therefore, DNP-nitroxide in complex with antibody
also had this relaxation time, which is short enough to indicate that there is a
degree of flexibility and motion at the binding site.

A rather surprising finding has emerged as more and more relaxation
times have been reported for native proteins. The evidence seems to indicate
that some proteins have such low ρ_h / ρ_0 values (sometimes less than unity) that
the only possible explanation is internal flexibility. Johnson and co-workers
were among the first in pointing out the existence of abnormally low "relaxation
times" obtained by fluorescence depolarization; the report by Chadwick and
Johnson [13] that DNS-γ-globulin had an unusually low relaxation time was
perhaps the first instance. Johnson and Richards [51] found that legumin, a
seed protein with a molecular weight of 398,000, had a ρ_h of only 200 nsec,
a value they obtained by polarizing the fluorescence of either a DNS or an
anthracene carbamide label. The equivalent oblate ellipsoid was expected to
have a relaxation time of 1160 to 1410 nsec.

Chowdhury and Johnson [52] performed further experiments on bovine-γ-
globulin, which had been found to have a relaxation time of 420 nsec calculated
from sedimentation-diffusion measurements but only 90 nsec from depolarization data. The low relaxation time for γ-globulin has been confirmed by
Steiner and Edelhoch [53] as well as Winkler [54].

Johnson and Massey [55] found that thiocyanate produced only a minor
change in the sedimentation rate of fumarase (molecular weight 220,000)
whereas the fluorescence polarization dropped markedly. Although no relaxation times were calculated, the results suggest that thiocyanate decreases
molecular rigidity without altering the size.

Metzger et al. [29] have reported that a human macroglobulin of molecular weight 890,000 exhibited a relaxation time of only 30 nsec compared with 730 nsec expected for the relaxation time of a sphere of equivalent molecular weight.

Thyroglobulin, a protein with a molecular weight of 670,000, was found to have a relaxation time of only 140 nsec [56], which is only slightly higher than that of serum albumin, a molecule only a tenth as large. In the case of thyroglobulin ρ_h / ρ_0 was only 0.27 [56].

Johnson and Mihalyi [57] found ρ_h to be 195 nsec for DNS-fibrinogen, although the molecular weight of fibrinogen is 340,000.

These observations suggest that the occurrence of flexible segments may be the rule rather than the exception, especially when DNS conjugates are studied by Perrin plots utilizing data obtained by heating [58].

In many cases data from isothermal Perrin plots give very different results. This fact is consistent with side-chain flexibility or other subunit rotations independent of the motion of the entire protein. In studies of synthetic polymers by Wahl et al. [59] and Omenn and Gill [60] the presence of such thermally activated rotations was noted. These polymers can thus serve as models for the behavior of proteins.

Gottlieb and Wahl [61] have considered theoretically the case of a label with complete rotational freedom about one axis and have derived equations relating the diffusion constants of the label and the macromolecule to which it is attached. The presence of internal freedom in proteins means that the DNS conjugates of even very large proteins should show changes in polarization as a result of the mild type of heating generally used to obtain Perrin-plot data. Structural changes in such conjugates can be expected to be reflected by changes in the depolarization data. However, true ρ_h values will obviously not be obtained.

Teale and Badley [62] performed depolarization experiments on pepsin and pepsinogen in various ways in order to evaluate aspects of the method. Perrin plots obtained by heating gave smaller relaxation times than those obtained isothermally by changing the viscosity. The latter method was considered the method of choice for obtaining relaxation times. Heating may have changed the lifetime as well as the ratio of T/η. Varying the polarization by changing the lifetime through collisional quenching with potassium iodide was also studied. Though the method was found suitable by Irwin and Churchich [37], Teale and Badley [62] found some complications in this method due to intertryptophan energy transfer causing changes in the polarization.

One of the most useful applications of fluorescence polarization is in the study of changes in molecular rigidity produced by various perturbants. The coil-to-helix transition was studied by Gill [63] using DNS-labeled polylysine. The relaxation time increased the ratio of ρ_h/ρ_0 going from 0.04 to 0.15. The technique has been employed to examine the degree of disorganization caused by reducing agents that open the disulfide bonds of proteins. The topic is of interest since Anfinsen and co-workers [64] have formed the hypothesis that the primary structure of a protein may determine the secondary and tertiary structure as well.

The reduction of the disulfide bonds of soybean-trypsin inhibitor was studied by Steiner [65], who found that reduced protein had a high polarization value due to extensive aggregation; on reoxidation, polarization fell to normal values and antitrypsin activity was recovered. Churchich [66] reported that reduction of the disulfide bonds in BSA caused an approximate doubling of the rotational relaxation time. Air oxidation of the reduced protein returned ρ_h to a lower level, which was, however, still higher than it was originally.

Such polarization studies have generally shown that reduction and alkylation do not in themselves cause complete loss of protein secondary and tertiary structure. Thus pepsinogen could be reduced and reoxidized with return of potential pepsin activity [67], but polarization studies showed that there was no extensive loss of rigidity on scission of the disulfide bonds [68].

Winkler et al. [69] showed by the depolarization method that reduction and alkylation of rabbit γ-globulin and its papain fragments at neutral pH did not significantly alter molecular rigidity.

A study by Young and Potts [70] on DNS-ribonuclease indicated that conjugation with DNS prevented correct pairing of sulfhydryl groups in the reduced protein, and enzymatic activity could not be regained by reoxidation. White [71] confirmed that attachment of 2 moles of DNS markedly inhibited correct pairing. The reduced and alkylated ribonuclease had a markedly lower relaxation time, but was obviously not in a completely random-coil configuration since further decrease in polarization was observed on addition of urea or guanidine [70].

Similarly Edelhoch and De Crumbrugghe [72] found that reduced, alkylated DNS-thyroglobulin had a high ρ_h value, indicating that the denatured protein had considerable structure. At a single temperature growth hormone was found to undergo a structural transition on acidification from pH 8 to pH 5 [73]. There was an increase in the polarization of the DNS conjugate in this range. Edelhoch and Steiner [74] have reviewed some of their early studies showing the use of the depolarization method to follow structural transitions induced by changes in pH, addition of denaturants, or heat. Denaturation has been monitored by polarization changes by Anufrieva et al. [75].

More recent fluorescence-polarization experiments with immunoglobulins

have been performed, aside from studies already cited. It will be recalled
that Perrin-plot data resulted in relaxation times too low to correspond with
rigid attachment of the DNS label to the protein [53, 54]. Weltman and Edelman
[76] were able to obtain Perrin plots from isothermal data by using DNS-labeled
protein. They obtained a relaxation time of 221 nsec for both human γG-
immunoglobulin and γG-myeloma protein. The data obtained on heating were
very different.

Wahl and Weber [77] performed similar isothermal experiments using two
samples of DNS-immunoglobulin and obtained apparent relaxation times of 170
and 223 nsec. Clearly the short relaxation times obtained on heating showed
that the DNS label was on flexible segments on the protein. Low relaxation
times were also bound by labeling with FITC [31, 78].

Zagyanski et al. [79] found that the fluorescence lifetime of DNS on human
and rabbit γ-globulins was about 7.3 nsec. This figure is considerably smaller
than that obtained by Chen [31] and Wahl and Weber [77]. However, the low
apparent relaxation times obtained by Zagyanski et al. [79] are consonant with
the work cited here.

Nezlin et al. [80] also studied the F_{ab} and F_c papain-produced fragments
after labeling with DNS chloride and found them too to have thermally activated
modes of rotation as determined by fluorescence-depolarization data and direct
lifetime measurements. Wahl [81], using the new technique of time-resolved
fluorescence polarization, determined that DNS-rabbit immunoglobulin at pH 8
exhibited two relaxation times of 370 and 23 nsec corresponding to the rotation
of the entire molecule and of a subunit. However, the same technique applied
to an IgA myeloma protein showed no segmental flexibility [82].

Fluorescence depolarization has continued to be used to study the interaction of antibodies with haptens or antigens. Quantitation of antibody to
conalbumin was obtained with FITC-labeled conalbumin, whose polarization
increased on binding [83]. Binding constants as well as the kinetics of antigen-
antibody combination were obtained by fluorometry [84-86]. With a novel
stopped-flow fluorescence polarimeter Portman et al. [87] obtained kinetic data
on the combination of antifluorescein antibody with haptens.

4. Problems in the Interpretation of ρ_h from Perrin Plots

The use of the Perrin-Weber relation (Eq. (199)) to determine the mean
harmonic rotational relaxation time of macromolecules was based on several
assumptions. Dye moieties attached to the macromolecule were assumed to
be not only randomly distributed but also rigidly attached. Although these
assumptions were seemingly confirmed by Steiner and McAlister's early
studies [10, 14], the preceding section cited many instances of subunit rotat-

ional freedom. Similarly the assumption of random labeling has been question-
ed in many cases. "Randomness" includes both chemical and orientational
aspects; that is, the question is: Are all potential sites of dye attachment,
such as amino groups, equally prone to react, and will the dye axes have no
preferential alignment with respect to the protein axes? Even where chemical
randomness seems approximately valid, the fluorescence may be heterogenous
due to the fluorophors being in different microscopic environments. Also, the
act of labeling a protein may change its properties so much that the results of
fluorescence studies may have little relevance to the native strucutre. Finally,
a complication that may arise is interaction between dyes attached to a
molecule. Such dye-dye interactions may complicate the interpretation of
fluorescence data. The studies to be cited illustrate the reality of these
problems.

Much evidence indicates that fluorescent-labeling reagents attach
preferentially to certain sites on native proteins. Gros and Labouesse [17]
find that only under denaturing conditions can all the potentially labeled sites
be attacked by DNS chloride. The labeling pattern of DNS-ribonuclease was
examined by White [88]. Residues 37 (lysine) and 48 (histidine) were preferentially labeled; residue 31, another lysine, was labeled to some degree.
Specific reaction with DNS chloride was also indicated by Hartley and Massey's
observation [89] of stoichiometric inhibition of chymotrypsin by a single DNS
residue. Likewise attachment of 1 mole of DNS per mole of myosin A markedly
inhibited the ATPase and actin-binding activity [90].

The site of interaction was further investigated by Takashina [91], who
found that a single DNS molecule on heavy meromyosin changes the ATPase K_m
and V_{max} and also masks several sulfhydryl groups. Hill and Laing [92, 93]
found rennin to be 80% inactivated when labeled with one DNS group. They
identified ϵ-DNS-lysine by thin-layer chromatography. However, Rickert [94]
disputes these results, finding the same number of lysines in both native and
labeled rennin. The lability of the DNS group on rennin was thought to be
evidence that it was actually attached to a histidine. Chen [95] found that
lysozyme reacted with only one DNS chloride molecule even when the reagent
was in large excess. The lysozyme had a pronounced tendency to aggregate,
especially in the cold; but its activity in lysing M. lysodeikticus cells was
intact.

Labeling of insulin with FITC resulted in preferential labeling of the amino
terminus [96]. A detailed study of the reaction of fibrinogen with DNS chloride
was performed by Mihalyi and Albert [97, 98]. The protein was labeled with 3
to 36 DNS molecules. With more than six to eight DNS groups present, there
were noticeable changes in the sedimentation rate, salting-out behavior, and

optical rotatory properties. Mihalyi and Albert identified O-DNS-tyrosine,
O, N-di-DNS-tyrosine, and ϵ-DNS-lysine after acid hydrolysis. At low degrees
of labeling tyrosine appeared to be labeled preferentially by a factor of 3 to 4.
Marked heterogeneity of fluorescence was shown by asymmetrical emission
spectra, which changed with degree of labeling. The fluorescence decay was
found to be heterogeneous, and lifetimes in the range of 12 to 15 nsec on the
average were measured.

The heterogeneity of fluorescence of DNS-BSA complexes was documented
by Chen [99]. Spectra, polarization, quantum yields, and lifetimes measured
as a function of the degree of labeling showed that the first DNS moiety bound
to BSA was in a protected environment and had an emission that was more
shifted to the blue than that of subsequently attached DNS groups and was
associated with a long lifetime of about 22 nsec. The heterogeneity of
fluorescence was also noted by Frey and Wahl [100], who found at least two
components of the fluorescence decay, as determined by a nanosecond-flash
technique; the relative amounts of the different components seemed to vary
with the degree of labeling. Figure 125 shows the dependence of fluorescence
parameters on the degree of substitution.

Until recently it was the custom to assume a DNS-fluorescence lifetime of
12 to 13 nsec regardless of the protein labeled. This procedure appeared safe
in view of some studies [10, 14] that seemed to show not much difference in the
lifetimes of various conjugates. However, DNS lifetimes ranging from about
8 to 22 nsec have now been reported, and instrumentation for direct measurements has become available. The variation appears to be due not only to
differences in the amino acid labeled (tyrosine versus lysine or histidine) but
also to the microenvironment of the fluorophor. Klotz et al. [19] found that the
differing environments were reflected in the different pK_a values for the dimethylamino group of DNS in conjugates of different proteins. In BSA the pK_a
was 1.55, whereas in lysozyme it was 2.39. Wang (cited in Ref. [101]) found
that in BSA labeled with six DNS groups solvent perturbation with dimethyl
sulfoxide and propylene glycol showed a fractional exposure of only 0.35,
whereas DNS on lysozyme was completely exposed. It is not unlikely that some
conjugates can contain both exposed and buried fluorescent groups, and thus
exhibit heterogeneous fluorescence.

The presence of several fluorescence lifetimes in a conjugate renders use
of the Perrin-Weber relation (Eq. (199)) invalid, since it is derived on the
assumption of exponential decay. Also, depending on the wavelength of
observation of the fluorescence, the results could be weighted toward either
the long-or the short-decay components. Heavy labeling tends to increase
the validity of random labeling but may increase the chances that the fluores-

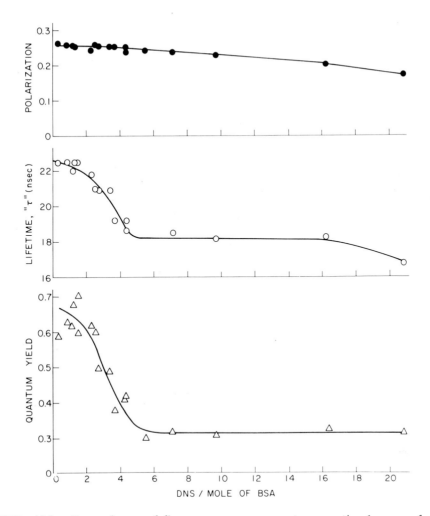

FIG. 125.   Dependence of fluorescence parameters on the degree of labeling of bovine-serum albumin by DNS chloride.   All measurements were made at $23^{\circ}C$, pH 7.   Lifetimes were approximate mean   values in some cases because the decay contained more than one component.   Polarization was measured with 340- and 500-nm excitation and emission.   Reprinted from Ref. [99].

cence is heterogeneous.   Very light labeling may result in a homogeneous fluorescence where the bound dye is nonrandomly attached and oriented.   Figure 126 shows that the Perrin plots obtained for DNS-BSA conjugates differ, depending on the region of the emission spectrum monitored by the phototube. This fact is explained by the difference in the average decay time over the spectrum, the longer lifetimes being associated with the blue side of the emission spectrum.

   Lifetimes should not be assumed to have any given value but should be directly measured if possible.   Wahl et al. [100,102] have reported lifetimes

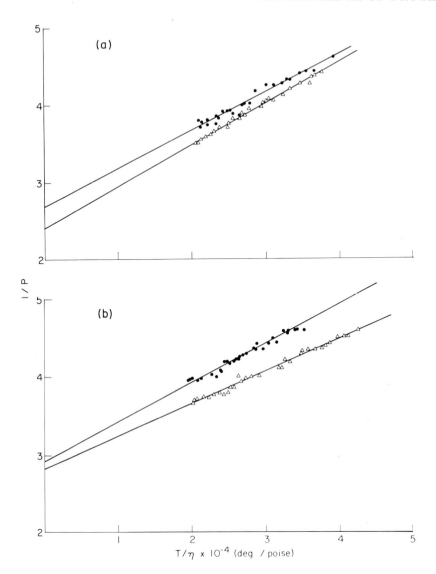

FIG. 126.   Perrin plots for DNS conjugates of BSA containing 0. 31 (A) and 5. 5 (B) DNS groups per mole of protein.   Fluorescence excited at 345 nm was monitored at either 500 nm (-△-) or 550 nm (-◉-).   The curves differ because the two wavelengths monitor fluorescence emissions that have different average decay times. T/η was varied by changing the temperature.

for DNS conjugates of BSA and lysozyme that are at variance with early values [10,14], and their results have been generally confirmed [95,99].   Apparently DNS has a fluorescence lifetime of about 8 nsec when it is attached to a protein at an exposed site, as on lysozyme [95] or a protein in urea [80], but as high as 22 nsec in very hydrophobic pockets [99,100].

Jonas and Weber [48] studied DNS-BSA that had been partly modified chemically, and phase fluorometric measurements of the decay gave clear evidence of fluorescence heterogeneity. The evidence consisted of a disparity between the lifetime values obtained by phase-shift and demodulation measurements. Wahl and Lami [102] reported that several preparations of DNS-lysozyme, each with an average of less than 0.4 dye molecules per mole of protein, had heterogeneous fluorescence decays. Analysis of the decays resulted in postulation of two components having lifetimes of 5.3 and 17 nsec. However, Chen [95] found the DNS group to be very labile in such conjugates. Solutions that were slightly aged showed heterogeneous fluorescence, presumable due to free dye, which could also be identified by chromatography. Fresh DNS-lysozyme seemed to have a single lifetime of 7.8 nsec and a symmetrical emission spectrum; it was postulated that one group on the enzyme reacted preferentially.

The question of randomness of orientation is interesting, and studies have shown that fluorescence-polarization data can indicate whether dye molecules are oriented randomly on a protein, and if not, the technique can well indicate which way the dye molecules are oriented. The fluorescence-polarization spectrum of the DNS moiety [103] shows that there are several electronic transitions responsible for the absorption spectrum, including at least two in the long-wavelength band. Nevertheless, with random labeling, the same relaxation time should be obtained from Perrin plots regardless of the exciting wavelength. Winkler [54], however, reported that fluorescence-depolarization data gave different values of ρ_h for DNS-γ-globulin, depending on exciting wavelength. The relaxation times found by exciting at 320, 340, and 360 nm were 61.5, 65, and 108 nsec, respectively. Winkler postulated that his results reflected nonrandom distribution of the dye molecules on the protein, and that the different exciting wavelengths photoselected DNS moieties at sites with different rotational mobilities. Speyer and Winkler [104] adduced support for this theory in that the polarization values found by 350-nm excitation seemed to be virtually independent of changes in viscosity, whereas P changed with viscosity when excitation was at 320 nm.

An alternative explanation for the difference in results obtained by exciting at different wavelengths was given by Witholt and L. Brand [105], who point out that there are different absorption oscillators in all the dyes used for fluorescence labeling and analyzing the types of Perrin plots one should get with oriented dye molecules. Conjugates were divided into three classes as follows:

1. Fixed, oriented dye labels.
2. Fixed, randomly oriented labels.
3. Loosely affixed labels.

It was shown that Perrin plots obtained by using different excitation wavelengths could distinguish between these classes. Examples of the different
classes were BSA labeled with anthracene-2-isocyanate (class 1), BSA with
adsorbed 8-anilinonaphthalene-1-sulfonate (class 2), and liver alcohol dehydrogenase labeled with anthracene-2-isocyanate (class 3).

Weber and Anderson [106] have mathematically analyzed the effect of
preferential dye orientation on the depolarization data. In addition, they
consider the effect of dye-dye interaction due to resonance energy transfer.
Anderson and Weber [107] have applied these considerations to data obtained
with adsorbates of 8-anilinonaphthalene-1-sulfonate (ANS) on BSA, and conclude
that ANS was preferentially oriented along the equatorial plane of the protein
ellipsoid. Note, however, that this conclusion is contrary to that of Witholt
and L. Brand [105], who found ANS to be randomly adsorbed on BSA.

5. Time-Resolved Fluorescence Polarization

Recent advances in instrumentation have permitted some laboratories to
perform experiments involving a technique known as time-resolved fluorescence
polarization, or "nanosecond polarization." When a sample is excited with a
short pulse of light, the fluorescence that initially appears will have a high
degree of polarization since the molecules have not had time to undergo
Brownian rotation. The polarization diminishes with time in a manner predicted by the relaxation time of the molecule. Measurements of the polarization
decay as a function of time in the nanosecond region therefore permit one to
obtain relaxation times without Perrin plots. For a spherical molecule,

$$A = A_0 \exp(-3t/\rho_0),\qquad(202)$$

where A is the emission anisotropy at time t and A_0 is the limiting anisotropy.
The term A differs from the polarization P in that

$$A = \frac{I_V - 2I_H}{I_V + 2I_H}.\qquad(203)$$

For an ellipsoidal molecule ρ_h is obtained from a modified form of Eq. (202),
which, one should note, does not contain the decay time τ.

Not long ago measurements of A as a function of time on a nanosecond
scale would have been unthinkable, but the application of high-speed electronics
and single-photon-counting techniques for fluorescence-decay studies [108]
have made this possible. The single-photon method is eminently suited for
following fluorescence-decay functions either in the presence or absence of
polarizers. Basically, a sample is irradiated by very short repetitive pulses
of light having a width at half-maximum intensity of about 1 nsec. A photomultiplier is used as a photon counter to determine the probability of light

emission from the sample as a function of time after the flash. The optical system is attenuated so that no more than one photon is received by the photo-tube for each exciting flash. Under these conditions the number of counts recorded at various times after the flash can be stored in a multichannel analyzer and plotted as a histogram whose envelope represents the decay function. More detailed descriptions of the complex electronics involved may be found in several references [108, 109].

Stryer [110] has discussed the advantages of the nanosecond-polarization method over the "steady-state" classical depolarization method described here. In theory several relaxations can be resolved by the nanosecond-polarization method. Wahl [81], for instance, found that the decline of A for DNS-γ-globulin was biphasic. On the other hand, DNS-BSA was found to have only one relaxation time [111]. Tao [109] presented a detailed theoretical analysis of the method and some results on nanosecond-polarization measurements on ANS complexes with apomyoglobin and horseradish peroxidase apoprotein. Yguerabide et al. [112] used this method to show that a DNS-labeled hapten bound to a specific antibody has a certain amount of freedom of rotation in the nanosecond range.

An extension of the single-photon-counting method is the measurement of time-resolved emission spectra. Ware and co-workers [113] have found that the spectra of some aminophthalimides can be seen to shift with time after an excitation flash. The spectral red shifts occur on a nanosecond scale and are direct confirmation of the theory that the emission spectrum is largely determined by dipole interactions with solvent molecules during the excited lifetime. It has been shown [114] that the spectrum of 6-toluidinonaphthalene-2-sulfonate in glycerol at $4^{\circ}C$ or in complex with serum albumin shifts contin-uously toward the red for a few nanoseconds after excitation. Time-resolved spectra also reveal shifts for luminol fluorescence [115].

The advent of single-photon-counting techniques has provided some interesting new possibilities in the measurement of fluorescence parameters. Because of the complex instrumentation involved, it remains to be seen whether or not the techniques will be accessible to only a few laboratories.

B. Fluorescence of Bound Dyes: "Fluorescent Probes"

Changes in the fluorescence properties of dyes adsorbed or covalently attached to macromolecules may signal changes in the microenvironment of the dyes due to configurational changes in the larger molecule. Dyes used in this way have been called "fluorescent probes" [116], a term that is analogous to "reporter group" [117], which is defined as a chromophore that undergoes changes in absorption characteristics as a function of environmental perturb-ations.

Lippert [118] has explained the spectral shift of the fluorescence band, compared with the absorption band, in terms of dipole-dipole interactions in the excited state. According to this generally accepted theory the dipole moment of most molecules increases on excitation; this in turn requires a reorientation of the solvent dipoles forming a shell about the fluorophor. Highly polar solvents require the expenditure of relatively large amounts of energy to accomplish the reorientation, with the result that the emission that finally appears is quite shifted to wavelengths of lower energy as compared with the absorption band. In less polar solvents the solvent reorientation is much less, and hence emission is at shorter wavelengths (higher energy).

It has long been known that covalently bound fluorescent dyes change their characteristics in response to configurational changes. Steiner and Edelhoch [119] found that the lifetime of DNS attached to γ globulin decreased to 8 nsec on denaturation with urea. They also found that the fluorescence of a fluorescein label on BSA diminished in the alkaline pH region where the protein unfolds [120], although fluorescein itself is intensely fluorescent at high pH. Frattali et al. [68] found changes in DNS-fluorescence intensity in a pepsinogen conjugate between pH 8 and 10; this showed the presence of a structural transition, although no change was noted in this range in optical rotation, intrinsic viscosity, or rotational relaxation time. The yield of DNS fluorescence in trypsin conjugates changed on combination with various trypsin inhibitors; the stoichiometry obtained by following the fluorescence agreed with that found by other physical parameters [121]. Edelhoch et al. [122] have used DNS-fluorescence intensity to follow the denaturation of bovine growth hormone.

The use of noncovalently bound dyes as fluorescent probes of protein structure is based on the finding that certain dyes are only weakly fluorescent when free in aqueous solution but strongly fluorescent when adsorbed onto certain proteins [123-126]. The dyes ANS (8-anilinonaphthalene-1-sulfonate) and TNS (6-p-toluidinonaphthalene-2-sulfonate) were reported by Weber and Laurence [123] to be nearly nonfluorescent in water but highly fluorescent when bound to BSA or denatured proteins. Only a few proteins in the native state caused an enhancement in fluorescence, and none to the extent that BSA did. Rees and co-workers [127] have used ANS in an assay for serum albumin.

The anilinonaphthalenesulfonates were studied by Förster [128], who postulated that planarity of the two ring systems was necessary for augmentation of fluorescence intensity. This idea was, however, largely discarded when it was realized that a change in solvent polarity alone could cause the characteristic high quantum yield and blue shift [126, 129]. Because the fluorescence of protein-bound ANS could be mimicked by placing the dye in

nonpolar solvents, the concept arose that spectral analysis of the fluorescence
could give information on the "polarity of the binding site."

Turner and L. Brand [130] studied the influence of solvent polarity on the
fluorescence of the 5,1-7,1- and 8,1-isomers of ANS and TNS. There was a
strong correlation between the position of the fluorescence peak and the polarity
of the solvent. However, Ainsworth and Flanagan [131] found that fluorescent
dyes were unreliable reporters of binding-site polarity. They showed that ANS,
DNS acid, Rhodamine B, and Phenol Blue all bound to the same site on BSA.
The binding-site "polarity" reported by the first two dyes was less than that
reported by the others. These authors postulate that the rigidity and solvation
of the dye were as important as the site polarity in determining the fluorescence
characteristics.

Some of the problems in interpreting the spectra of fluorescent probes seem
to stem from overemphasis on polarity in applying Lippert's [118] theory of
spectral shifts. The shifts are determined by the energy expenditure for
solvent reorientation during the excited state. The dielectric constant, a
measure of polarity, is sometimes assumed to be the main determinant of the
fluorescence characteristics; but there are times when solvent polarity and
solvent-reorientation energy do not parallel each other. Winkler [132]
examined the effect on fluorescence of changes in viscosity rather than in the
dielectric constant. Using mixtures of methanol and 1,2-propanediol, which
both have dielectric constants of 33, he was able to vary the viscosity from
0.6 to 33 cP. The fluorescence of p-aminohippurate (PAH) increased in
intensity by a factor of 2.5 over this range, and there was a blue shift. More
recently it was shown by Seliskar and L. Brand [133] that a fluorescent probe,
2,6-MANS (the N-methyl derivative of TNS), placed in a highly polar environment at low temperature gave spectra similar to those obtained from nonpolar
organic solvents. Clearly the results of Winkler and of Seliskar and L. Brand
are due to the inhibition of solvent relaxation by high viscosity and low
temperature.

Similarly a fluorescent probe on a protein could be near very polar groups,
yet would not be affected by them if they were immobilized and unable to
interact with the excited-state probe. The original idea of Förster [128] that
planarity of the probe molecule may be important has been revived to some
extent by the work of Camerman and Jensen [134,135], who determined the
crystal structure of TNS by X-ray diffraction. They find that electron
delocalization over the two ring systems is possible when the rings are nearly
planar. Slight deformation of the structure of the probe molecule would therefore greatly perturb the fluorescence, and the authors conclude that it is not

safe to conclude that these probes are sensitive only to binding-site polarity.

When adsorbed onto calf-thymus histone, ANS gives a strong fluorescence [136, 137]. Laurence [137] finds that the fluorescence is like that of ANS bound onto BSA and therefore characteristic of nonpolar binding sites. However, no strong binding sites are actually found. It seems possible that the ANS could be bound electrostatically to the many positively charged basic groups on histones, in an environment unconducive to solvent relaxation.

In spite of questions on the interpretation of probe data, these dyes have been useful in studying alterations in protein structure. Callaghan and Martin [124] detected the denaturation of serum albumin by the loss of its ability to form the characteristically fluorescent complex with ANS. Similarly Chen [138] used ANS to show that a defatting procedure using charcoal did not denature serum albumin; this was shown by retention of several properties of the ANS-BSA complex. Gally and Edelman [139] found that Bence-Jones protein could enhance ANS fluorescence after denaturation. Green [140] found that avidin had a hydrophobic binding site for ANS, presumably the same site that binds biotin; however, the quantum yield of ANS fluorescence was much less than in the ANS-BSA complex.

Botts, Morales, and co-workers [141-143] have used ANS as a probe for structural changes in myosin. Generally manipulations or reagents that inhibit the native ATPase activity of the protein also diminish the fluorescence of the ANS-myosin complex. Dissociation of lactic dehydrogenases by acid was correlated by Anderson and Weber [144] with an increased ability to augment ANS fluorescence. On the other hand, Weber and Young [145] followed the peptic digestion of BSA by the decrease in affinity for ANS. Jonas and Weber [146] isolated certain small peptides from digests of BSA and found that the peptides contained the hydrophobic binding sites characteristic of the intact protein. Covalent attachment of the peptides to lysozyme endowed the latter with the ability to bind ANS.

Stryer [129] presented evidence that ANS can bind to apomyoglobin and apohemoglobin in the heme crevice. Whereas it had been reported that ANS quantum efficiency was 0.004 and 0.75 in water and on BSA, respectively [145], Stryer found that ANS on apomyoglobin had a quantum yield of 0.98 [129]. Hsu and Woody [147] showed that ANS and other dyes had optical activity when bound to apohemoglobin. Indeed the binding was so rigid in a preferred orientation that absorption and circular-dichroism spectra could be used to resolve individual vibrational transitions.

Variations with pH of the fluorescence intensity and the strength of association of the ANS-BSA complex(es) have been reported [138, 148] and are

probably related to structural changes. Smekal and Sprindrich [149] investigated ANS binding by human-serum albumin. Among the findings were the
characteristically strong enhancement on binding, the presence of five strong
binding sites, and a pH optimum of about 7 for fluorescence. The work of
Stryer [129] on the binding of ANS to apoproteins of the heme type has been
extended by Anderson et al. [150] to studies of Aplysia apomyoglobin. The
emission spectrum of bound ANS, as well as its lifetime and quantum yield,
were different from that of ANS bound to sperm-whale apomyoglobin. It was
shown by L. Brand et al. [151] that ANS and Rose Bengal bound to liver alcohol
dehydrogenase (LADH) at the active site. Coenzyme would release the bound
dye, which had a high quantum yield. In contrast to the liver enzyme, yeast
alcohol dehydrogenase seemed to retain ANS in the presence of NADH, although
an apparent structural change was signaled by the dye [152].

A novel use of TNS was made by Lynn and Fasman [153], who found it to be
a useful probe for the ß structure of polylysine. This polymer can also exist
in two other forms: as the α-helix or as a random coil. The ß-structure is
obtained by heating or treating with sodium dodecylsulfate in acid and has large
hydrophobic areas. Enhanced TNS fluorescence occurred only with the ß-
structure, and not with the other two forms.

Winkler [154] prepared antibodies against a naphthalene-type hepten and
found that they would combine with TNS with concomitant fluorescence enhancement. McClure and Edelman [155,156] have used TNS to study chymotrypsin.
They found that the dye is a competitive inhibitor that has a much stronger
fluorescence on chymotrypsin than on chymotrypsinogen. Therefore activation
of the zymogen in the presence of TNS led to an enhancement of the fluorescence
of the latter. Although not bound at the active site, the dye was influenced by
the binding of substrates.

The dye ANS was found by K. Brand [157] to bind to transaldolase, with a
marked enhancement of fluorescence. A substrate, fructose-6-phosphate,
reduced the fluorescence without displacing ANS from the complex. A configurational change induced by substrate was postulated. Recently Holler et al.
[158] have found that TNS serves as a reported group on isoleucine-tRNA from
Escherichia coli. The adsorption of the dye onto the enzyme gives a strong
fluorescence, which is largely quenched when the protein binds isoleucine.
No quenching is produced in the presence of amino acids not recognized by the
enzyme. Christian and Janetzko [159] have found that ANS binds and non-
competitively inhibits cholinesterases from horse serum and the electric eel.
ß-Lactoglobulin is a protein known to have several hydrophobic binding sites.

Lovrien and Anderson [160] studied the binding of the N-methyl derivative
of ANS by fluorescence enhancement and equilibrium dialysis. One tight bind-

ing site and two weak sites were discovered. Porcine elastase was reported
[161] to bind ANS and to undergo, at pH 3. 4, a configurational change that was
accompanied by an increase in the ANS quantum yield. Hart et al. [162] found
that a serum high-density lipoprotein bound some 300 moles of ANS, whereas
the apoprotein had only three binding sites. This finding indicates that the
holoprotein held the ANS in the hydrophobic lipid moiety.

It has also been pointed out that the DNS-amino acids, also called "dansyl
amino acids," are similar in properties to ANS and TNS [103]. These compounds have a low quantum efficiency in water (less than 0. 1) and fluoresce
maximally near 580 nm. In apolar solvents the emission becomes more
intense (quantum yield near 0. 7 in dioxane) and blue. A similar change in
emission occurs when the dansyl amino acids are bound to BSA and to certain
other proteins, such as the apomyoglobins. Differences in dye-binding sites
in albumins of different species were shown by the differences in emission
spectra of bound DNS-proline [103].

Himel and co-workers [163-165] have utilized these properties of DNS
derivatives for the design and synthesis of fluorescent molecules that bind
specifically to cholinesterase at the active site. 5-Dimethylaminonapthalene-
1-sulfonamido-N, N-dimethyl-n-propylamine and an aliphatic quaternary
derivative were both found to bind strongly to serum cholinesterase and to
inhibit the enzyme competitively. The binding region of the enzyme includes
an anionic site for which the dye probes compete with cations. 1-Dimethyl-
aminonaphthalene-5-sulfonamide (DNSA) was found to be a strong inhibitor of
bovine-erythrocyte carbonic anhydrase [166]. The binding was signaled by an
increase in the DNSA quantum yield from 0. 005 to 0. 84 and by a marked blue
shift. This finding is illustrated in Fig. 127. The marked increase in yield
was typical of fluorescent probes that find a protected crevice in a protein.
However, the amount of blue shift was anomalously large, larger even than
that observed for DNSA adsorbed onto BSA. Part of the blue shift was explain-
able on the basis of a proton dissociation from the sulfonamide group on
complexation with Zn^{2+} at the active site of the enzyme.

Fluorescent molecules other than naphthalene derivatives have been used
as probes. Cortijo et al. [167,168] found that the pyridoxamine phosphate
fluorescence in sodium borohydride-treated glycogen phosphorylase served as
a built-in probe. Fluorometric titration revealed a possible structural change
in the region of pH 5. 3, where no change in absorption or protein fluorescence
was seen. The native protein had a quantum yield of 0. 12, whereas the
fluorescent prosthetic group had a yield of only 0. 012. The general rule seems
to be that vitamin B_6 derivatives have lower quantum yields when bound to
proteins than when they are free in solution. The quenching of pyridoxamine

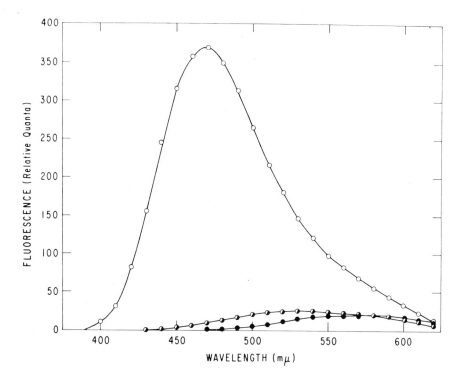

FIG. 127.   Emission spectra of DNSA bound to bovine carbonic anhydrase
(-O-), not bound at pH 7. 4 (-●-), and not bound at pH 12 (-◑-).   Excitation was
at 310 nm; areas under the curves are proportional to quantum yields.
Reprinted from Ref. [166].

5-phosphate that occurs on interaction with aspartate transaminase was used
to determine the strong binding constant of $1.5 \times 10^7 \text{ M}^{-1}$ [169].   Churchich and
Harpring [170] used the enhancement of the B_6 cofactor fluorescence to follow
the course of urea denaturation of aspartate aminotransferase.

The fluorescence of pyridoxal 5-phosphate (PLP) in enzymes may be
influenced by neighboring groups that can accept protons from the cofactor in
its excited state [171].   The fluorescence of PLP in certain enzymes appears
similar to that of the ionized form, although there is no evidence of ionization
in the ground state.

Auramine O, a dye with no detectable fluorescence when free in aqueous
solution, was found by Conrad et al. [172] to form a fluorescent complex with
liver alcohol dehydrogenase.   The quantum yield in the complex was 0. 05, with
a lifetime of 3 nsec.   Excitation and emission peaks were 450 and 523 nm,
respectively.   Sixteen other proteins, including yeast alcohol dehydrogenase,
did   not enhance Auramine O fluorescence.

Takenaka and co-workers [173] have found that 4-methyl-7-diethylaminocoumarin has the features of a probe molecule. The fluorescence in water is excited at 390 nm and peaks at 480 nm. In nonpolar organic solvents the peak is at 420 nm, and when bound to proteins, the fluorescent molecule emits maximally at 450 nm. Albumin, ß-lactoglobulin, and insulin gave the characteristic spectral shifts and are known to have hydrophobic binding sites. Neither lysozyme nor ovalbumin affected the fluorescence of the coumarin derivative.

Takenaka and Shibata [174] found a stilbene dye that was excited at 350 nm and had peak emission at 450 nm to show fluorescence enhancement when bound to a number of proteins. Lasser and Feitelson [175] used a fluorescent substrate, which was hydrolyzed to a fluorescent product, to study the catalytic process of carboxypeptidase A. 2-Carbonaphthoxy-L-phenylalanine was hydrolyzed to the fluorescent 2-naphtholate anion, and the binding of the substrate could be followed separately by its own fluorescence. Michaelis and catalytic constants were obtained.

Since the work of Velick [176] it has been known that NADH bound to enzymes has an enhanced fluorescence. The fluorescence has permitted the determination of binding parameters for NADH and too many enzyme systems to be mentioned here. Ternary complexes may have fluorescence characteristics different from those of binary complexes. Perlman and Wolff [177] found that dimethy sulfoxide was an inhibitor of LDH and formed a ternary complex with the enzyme and NADH, whose fluorescence was enhanced still further. McCarthy et al. [178] found that thyroxine, an inhibitor of horse-liver alcohol dehydrogenase, interfered with the fluorescence enhancement that is normally shown by NADH on binding to the protein. The NADH was not displaced, however. It thus served as a probe that showed interaction of the coenzyme and thyroxine-binding sites.

Suh and Barker [179] synthesized a number of alkyl and aryl phosphates that had the right shapes for binding to the active site of rat-muscle aldolase. Several isomeric naphthalenediol phosphates exhibited a blue shift and a small increase in quantum yield on binding to the enzyme. Certain enzyme inhibitors, such as ribitol-1,5-diphosphate, then quenched this fluorescence on addition to the enzyme-fluorophor complex. Another example of an active-site-directed fluorescent probe is thiochrome pyrophosphate, which was used by Wittorf and Gubler [180] to study the apoprotein of pyruvate decarboxylase. The normal cofactor is thiamine pyrophosphate, but the authors found that thiochrome pyrophosphate, a competitive inhibitor, would bind with a concomitant blue shift of fluorescence. It was shown that thiochrome itself does not bind, and Mg^{2+} was needed for inhibitor binding.

There are other examples where the extrinsic fluorescence of compounds

other than dyes can be used to study protein structure. Changes in NADH
emission observed by Kalckar and co-workers [181] indicated that certain
nucleotides produced structural changes in UDPG-epimerase. An example of
the use of flavin fluorescence is the work of Massey and Curti [182], who
showed that the quenching of FAD on binding to D-amino acid oxidase apoprotein
proceeded in two steps, corresponding to a rapid binding followed by a slow
protein-configuration change. The combination of the Straub diaphorase with a
specific antibody was followed by Stein et al. [183] by means of the quenching
of the visible fluorescence of diaphorase.

Deranleau and Neurath [184] prepared a number of fluorescent substrates
in a study of the active site of chymotrypsin. Their study of fluorescence
spectra, polarization, and activity included binding studies, and they confirmed
that the substrate-binding site was already present in chymotrypsinogen.

Chignell [185] found that the fluorescence of warfarin increased sixfold when
it was bound to human-serum albumin. The binding of DNS-glycine was non-
competitive with warfarin binding. This conclusion was amplified by energy-
transfer calculations that showed the tryptophan-warfarin and tryptophan-DNS-
glycine distances to be 35 and 24 $\overset{\circ}{A}$, respectively, indicating different binding
sites.

In other studies the nonfluorescent drugs phenylbutazone and flufenamic
acid were found to displace DNS-glycine [186, 187]. The binding constants were
calculated from the displacement data, since the association constant for DNS-
glycine was independently known.

Bilirubin-albumin complexes contain an extrinsic fluorescence apparently
due to a constrained configuration of bilirubin [188]. The spectra, quantum
yields, as well as the absorption and circular-dichroism parameters were
different depending on the species of albumin used. The binding constants for
bilirubin-albumin complexes were determined by fluorescence-quenching titrations at different temperatures; the data indicate a large entropy decrease on
complex formation [188].

There is no way to predict the direction of spectral shifts undergone by
fluorescent probes in specific systems. Two examples may be cited where a
DNS label exhibits shifts in different directions. Takagi et al. [189] found that
DNS-Taka-amylase fluorescence was markedly enhanced fourfold by reduction
of disulfide bonds. The blue shift indicated that the DNS groups was transferred
to a hydrophobic region. Kenner and Neurath [190, 191] introduced an amino
group on a tyrosyl residue in trypsin and trypsinogen, and labeled that group
specifically by reaction with DNS chloride at pH 5. The fluorescence of DNS in
the trypsinogen derivative decreased and shifted toward the red when the
zymogen was activated.

Fluorescent-probe molecules have been used in conjunction with polarization measurements to estimate the viscosity of solutions. Because the Perrin equation relates the polarization, lifetime, molecular volume, temperature, and viscosity, any of these parameters can be calculated if the others are known. Burns [192,193] has estimated the intracellular viscosity of yeast and Euglena cells by measuring the polarization of the fluorescein, which is released enzymatically after the cells take up fluorescein diacetate, a non-fluorescent, lipid-soluble substrate. Shinitzky et al. [194] estimated the microviscosity of the hydrophobic parts of membranes and micelles. They measured the polarization and lifetimes of 2-methylanthracene and 9-vinyl-anthracene adsorbed on the materials of interest.

Because of the sensitivity of fluorescent probes to configurational changes, it is not surprising that glutamic dehydrogenase (GDH), an allosteric enzyme par excellence, has been extensively studied with fluorescent dyes. The enzyme is activated and inhibited by a variety of compounds, including nucleotides, amino acids, hormones, and sulfhydryl reagents [195]. Recent estimates of size by light scattering suggest that the active protomer has a molecular weight of 316,000 [196], although the enzyme at high concentrations can associate into multimers whose molecular weight is as high as 2×10^6. The protomer can be dissociated by denaturants into identical subunits with molecular weights of 50,000 [197].

Studies have shown that GDH may be modified by acetylation [198] or dinitrophenylation [199], with the result that there are changes in kinetic and allosteric properties. When GDH was labeled with DNS chloride in the presence of NAD^1, the enzyme was completely active but had lost its ability to be activated by ADP, though it could still be inhibited by GTP [2]. In the absence of NAD, labeling of GDH led to some inactivation. Fluorescence spectra of DNS-GDH were found to differ depending on the presence or absence of co-enzyme during the labeling procedure. Perrin plots for the two types of conjugates also differed. Since ADP had an effect on the Perrin plots, a configurational change is suggested. Structural changes were also suggested by Perrin plots obtained with an anthracene-labeled GDH preparation in the presence and absence of ADP.

Malcolm and Radda [200] reacted GDH with 4-iodoacetamidosalicylic acid. Analysis of the fluorescence of the attached salicylyl moiety showed that there was a 1:1 equivalence of modification and inactivation. The denaturation of GDH by dodecylsulfate was also followed by fluorescence, using both the intrinsic protein emission and that of bound NADH [201].

Adsorbed dyes, especially ANS, have also been used as a probe of GDH [202-204]. The ANS fluorescence was greatly affected by allosteric effectors

[202, 204]. Yielding [205] showed that GDH acylated in the absence of regulatory nucleotides was catalytically inactive, yet responded to allosteric effectors as shown by changes in the fluorescence of adsorbed ANS. Dodd and Radda [204] found that on binding to GDH the fluorescence of ANS increases by a factor of about 100 and increases further on adding GTP or NADH. Monitoring of the ANS fluorescence in a stoppedflow device allowed determination of the rate constants for the structural transitions effected by these nucleotides [204]. Acetylation of a specific tyrosinase was found to reduce the GTP-dependent enhancement of ANS fluorescence [206]. Simple fluorescence-depolarization measurements were used by Churchich [207] to study the binding of NADH by GDH under a variety of conditions.

The immunoglobulins constitute another class of proteins that are uniquely suited for study with fluorescent probes. Parker and co-workers [208, 209] prepared antibody against the DNS group. When antibody combined with a hapten, ϵ-DNS-lysine, the DNS fluorescence yield was greatly enhanced, and the emission was shifted to the blue. Rabbit antibodies obtained at different days after immunization gave different amounts of shift to the blue in the ϵ-DNS-lysine emission. The heterogeneity of the antibodies was studied by ion-exchange chromatography, and it was found that the more positively charged fractions of the immunoglobulins gave a greater shift to the blue than did the more negatively charged molecules.

Lopatin and Voss [210] found that antibody against fluorescein caused a quenching of fluorescence on combination. The group(s) responsible for the quenching were not identified.

Pecht et al. [211] found that the emission of a DNS-labeled antigen could be excited at 280 nm only if the hapten were combined with antibody. The sensitized fluorescence was due to energy transfer from tryptophan in the antibody, and permitted the authors to calculate an association constant of $3.2 \times 10^6 \, M^{-1}$.

Energy transfer also occurred in the complex containing ϵ-DNP-lysine and antibody directed against the DNP group [212]. Here the quenching of antibody fluorescence by the DNP group was stronger in the presence of 4 M guanidine hydrochloride when the hapten was added before, rather than after, addition of the denaturant. The results showed that the hapten stabilized the antibody structure.

Yoo and co-workers [213, 214] have performed extensive studies on antibodies that react with several isomeric ANS probes in an effort to define the binding-site geometry. The antibodies were obtained by coupling 4-azonaphthalene-1-sulfonate and the corresponding 6,2-isomer to a carrier protein. Reaction of antibody with 4,1-ANS and 6,2-ANS was signaled by a large fluorescence enhancement and a blue shift, but the enhancement was greater with antibody against the same positional isomer of azonaphthalenesulfonate. When light chains of specific antibodies were isolated, it could be shown by fluorescence enhancement that specific ANS binding still ocurred [215]. The antibody activity inherent in light chains was also shown in direct immunofluorescent staining, using fluorescently labeled light chains with antistreptococcal activity [216].

Yguerabide et al. [112] used nanosecond-fluorescence-polarization methods to investigate the rotational mobility of the antibody combining site. With ϵ-DNS-lysine as the hapten, they found that combination with protein gave a high quantum yield and a lifetime of 23.1 nsec. The decay of the fluorescence polarization after pulse excitation with polarized light was indicative of two rotational relaxation times corresponding to rotation of the whole molecule plus local segmental flexion. The F_{ab} fragment exhibited only the shorter relaxation time. The experiment showed in a direct way that the localized motion at the combining site was occurring in the nanosecond scale.

C. Interactions of Fluorescent Probes with Membrane-Bound Components

Because proteins are part of the structure of biological membranes, it is not surprising that fluorescent probes have been added to membranes in order to detect configurational changes that might account for the properties of these particulate systems. Azzi, Chance, and co-workers [217, 218] and Packer et al. [219] have added ANS to mitochondria and submitochondrial particles and described the changes in fluorescence enhancement as a function of the excitation state of the system.

Erythrocyte ghosts also apparently are able to enhance probe fluorescence [220, 221). Feinstein et al. [222] found that erythrocyte membranes enhanced ANS fluorescence, which was further increased in the presence of butacaine or calcium chloride. The fluorescence of ANS was increased in the presence of such lipids as sphingomyelin, ganglioside, or phosphatidylserine. Apparent configurational changes could also be followed on erythrocyte membranes with a cationic dye, ethidium bromide [224].

When fluorescent probes interact with membranes, it cannot be assumed that the interaction involves only protein, since lipid material is also present. Bornet and Edelhoch [225] found that various dyes could be used to follow the micellization of lipidlike detergent molecules by monitoring the enhancement of fluorescence. Muesing and Nishida [226] have shown that ANS causes disruption of phospholipid dispersions and changes the structure of lipoproteins.

Vanderkooi and Martonosi [227] studied skeletal-muscle microsomes (sarcosomes) by the change in the fluorescence of adsorbed ANS. The intensity of emission was increased by cations important for muscular activity, and the enhancement was highest at an acid pH. The authors noted that ANS fluorescence appeared in the presence of phospholipids alone.

The cell membrane of polymorphonuclear leukocytes is thought to respond in some way to surfaces that are "foreign" to an organism. Romeo et al. [228] used ANS as a probe for configurational changes in leukocyte-membrane components when placed in contact with polystyrene spherules [228]. The fluorescence increase that occurred on ANS binding to the membrane was further enhanced in the presence of spherules.

The rotational mobility of dyes attached to membrane fragments from electric-eel electroplaxes was studied by Wahl et al. [229] with nanosecond-polarization methods. Both ANS and DNS were bound to membrane fragments and were found to be essentially immobilized in the membrane. Detergents resulted in marked rotational freedom of the dyes.

Brocklehurst et al. [230] used ANS and its N-methyl derivative to probe the membrane structure of electron-transport particles of ox-heart heavy mitochondria. The changes in probe fluorescence were correlated with the energy state of the particles. Pyrene-3-sulfonate was proposed as a "volume indicator" of subcellular organelles, because this compound shows concentration-dependent excimer formation.

Probe fluorescence has even been extended to studies of possible structural changes in nerve protein components [231, 232]. Nerves stained supravitally with any of about a dozen different fluorescent dyes responded to electrical stimulation with changes in the extrinsic fluorescence. The changes were very small: approximately 3×10^{-4} that of the signal from the dye on the nerve in the resting state [232]. These changes were resolved with repetitive stimulation and a computer of average transients. Tasaki et al. [233] have measured the fluorescence polarization of a fluorescent probe, TNS, injected into squid axon. They found the overall polarization to be 0.22, but the portion of the emission that changed on stimulation was completely polarized, suggesting that those dye molecules were oriented along the nerve. The use of optical techniques for detecting structural changes in nerves during stimulation has been reviewed by Tasaki [234].

Nagai et al. [235] have reported that ANS is bound to an Na^+, K^+-sensitive ATPase preparation of brain synaptic membranes. The membrane-bound fluorescence was sensitive to K^+ concentration and to the presence of ouabain, an inhibitor of the enzyme activity.

III. INTRINSIC PROTEIN FLUORESCENCE

A. General Considerations

The fact that proteins have an intrinsic ultraviolet fluorescence has been

known only since about 1957, and such fluorescence has been used as a tool
to study protein structure for approximately one decade. The discovery of
intrinsic protein fluorescence has been reviewed by Udenfriend [236] and
Konev [237].

Shore and Pardee [238] were apparently the first to describe the phenomenon, deducing that the emission maximum must occur between 300 and 400
nm on the basis of the characteristics of the filters they used. Duggan and
Udenfriend [239] noted ultraviolet fluorescence from protein hydrolyzates,
and Konev [240] presented the spectra of protein emission. Earlier, in 1953,
Weber [7] in discussing extrinsic protein fluorescence indicated that proteins
might possess an ultraviolet fluorescence, since their absorption spectra were
in the ultraviolet. Corrected emission spectra of the fluorescent aromatic
amino acids phenylalanine, tyrosine, and tryptophan were reported by Teale
and Weber [241] in 1959. Corrected spectra of some proteins were given by
Teale [242] in 1960, although, according to Weber and Young [145], some of
Teale's data may be in question due to incorrect compensation for scatter.

The quantum yields of phenylalanine, tyrosine, and tryptophan were
reported by Teale and Weber [241] to be 0.04, 0.21, and 0.20. Ten years
later these values were challenged as being uniformly about 50% too high [243].
For instance, the quantum yield of tryptophan was reported as 0.13 and 0.12
by Chen [243] and Börresen [244], respectively, both of whom used a quinine
standard for comparison. Eisinger [245] obtained a tryptophan yield of 0.14
by comparison with p-terphenyl, a standard with a yield of 0.87. As a
corollary, this result shows that the original value of 0.20 for tryptophan must
be too high; otherwise p-terphenyl would have an impossibly high quantum
efficiency of 1.3.

Most studies on proteins do not rely on absolute quantum yields, but
concern over the accuracy of the quantum-yield data is far from academic.
For instance, Bishai et al. [246] measured the quantum yields of phosphorescence and fluorescence of indole derivatives at 77°K and came to the important
conclusion that the sum of yields was always 1.0. However, the quantum-
yield standard used was tryptophan, with an assumed yield of 0.20. Thus it
appears that the data of Bishai et al. [246] indicate that the total luminescence
yield of indoles under the conditions studied was 0.65, and a completely
different interpretation of the results is required.

The corrected fluorescence spectra of the aromatic amino acids, phenol,
and quinine are given in Fig. 128.

The intrinsic fluorescence of proteins can be discussed in terms of
contributions from tyrosine and tryptophan. The fluorescence of phenylalanine and its peptides is of interest in itself, but it has relatively little

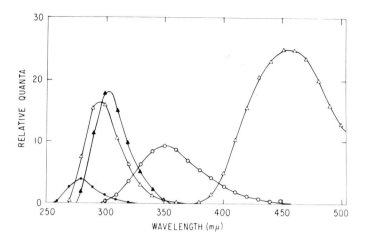

FIG. 128. Corrected emission spectra of phenol (-△-), tyrosine (-▲-), phenylalanine (-●-), tryptophan (-o-), and quinine (-△-). Reprinted from Ref. [243].

importance in proteins in which phenylalanine fluorescence has not been unequivocally detected. Leroy et al. [247] have measured phenylalanine-fluorescence characteristics, especially the lifetimes at different temperatures; they obtained a lifetime of 6.8 nsec at 20°C. Reasons for the non-detectability of phenylalanine fluorescence in proteins relate to the low quantum yield, quenching by energy transfer to other chromophores, and the intrinsically low molar absorptivity. For these reasons the following discussion of intrinsic fluorescence is limited to tyrosine and tryptophan.

B. Tyrosine

1. Characteristics of Tyrosine Fluorescence

In one of the first studies of protein emission spectra Teale [242] was able to demonstrate tyrosine fluorescence only in those proteins that contained tyrosine but no tryptophan. These "class A" proteins included insulin, ribonuclease, zein, and ovomucoid and all had emission maxima at 304 nm and quantum yields of 0.01 to 0.03. No evidence of tyrosine fluorescence could be found in "class B" proteins--that is, those that contained both tryptophan and tyrosine. These results conflicted with those of Konev [240], who reported peaks corresponding to tyrosine fluorescence in a number of proteins having tryptophan as well. The presence of tyrosine fluorescence in a class B protein was later recognized by Weber [248, 249] in the spectrum of human-serum albumin, which has 17 to 18 tyrosine residues but only 1 tryptophan residue. Excitation of this protein with 275-nm light clearly gave more emission in the 304-nm region than did excitation with 290-nm light, which was absorbed only

by tryptophan [248]. The same technique was used to show tyrosine fluorescence
in BSA [249], ovalbumin [250], and pyruvate kinase [251]. Most class B proteins
examined in this way can be shown to possess tyrosine fluorescence.

The early difficulties encountered in identifying tyrosine fluorescence in
proteins stem from the low quantum efficiency of the residues, the low detector
sensitivity of some instruments in this region of the ultraviolet, and the
scattered-light problem arising from the nearness of the tyrosine peak at 303
nm and the excitation peak at 280 nm. Scattered light is an especially trouble-
some obstacle, but it can be markedly diminished in emission spectra by the
use of horizontally polarized excitation [252]. Various combinations of gratings
and phototubes are available for commercial spectrofluorometers now, and the
resultant emission spectra need very little correction [253]. Examples of
protein emission spectra are shown for serum albumins in Fig. 129. The
examples demonstrate the obvious tyrosine contribution as a shoulder near 300
nm in human and rabbit albumins. The emission spectra themselves also show
the species differences in albumins, which has also been shown by dye spectral
differences [103, 254].

Tyrosine seems to have a very low quantum yield in most proteins (0. 01 -
0. 03) so that emission from tyrosine is normally greatly overshadowed by
tryptophan fluorescence. However, tyrosine fluorescence was detected as a
shoulder in the corrected emission spectrum of lysozyme by Lehrer and
Fasman [255] even though this protein has six tryptophan and only three
tyrosine residues. Edelhoch et al. [122] have published corrected emission
spectra of bovine growth hormone and reported the surprising result that the
native protein has an emission maximum at 315 nm when excited at 280 nm.
They attribute the short wavelength of the emission peak to a large contribution
from tyrosine residues.

A considerable amount of effort has been concentrated on the problem of
why the quantum yield of tyrosine fluorescence is usually so low in native
proteins. White [256] studied the pH dependence of the fluorescence of phenol
and its derivatives including tyrosine. From her results it was clear that loss
of the phenolic proton resulted in loss of fluorescence. Quenching occurred not
only at alkaline pH, where the phenolic group ionizes, but also under conditions
where the hydrogen can be transferred to an acceptor such as the ionized
carboxyl or the amine group of tyrosine.

It has been found that the un-ionized carboxyl group quenches phenolic
emission [257, 258]. In the case of quenching of phenol and anisole fluorescence
by un-ionized carboxylic acids, Weber and Rosenheck [257] suggested that a
dark complex was formed in the ground state. Feitelson [258], noting that the
fluorescence of both tyrosine and phenylalanine is quenched by interaction with

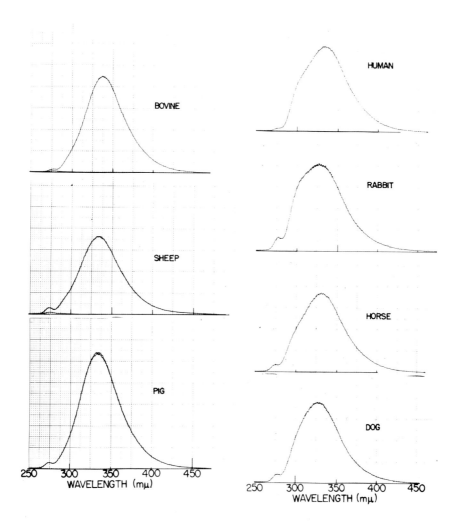

FIG. 129. Emission spectra of serum albumins from different species, excited at 276 nm. Detector response was essentially linear from 260 to 400 nm.

uncharged carboxyl groups, postulated that quenching occurred in the excited state by electron donation from the benzenoid ring to the quencher. Feitelson [258] calculated the rate constants for the quenching of tyrosine fluorescence by a number of bases and deduced that excited-state ionization of the phenolic hydroxyl did not occur to a significant extent.

The fluorescence lifetime of tyrosine was reported to be about 3 nsec in water at room temperature [259]. This is in fair agreement with the more recent 3.6-nsec value of Fayet and Wahl [260]. No lifetimes for tyrosine

fluorescence in proteins have been reported. They would be expected to be
about 0. 3 nsec or so if they are proportional to the quantum yield.

Edelhoch et al. [261] confirmed the strong quenching effect of the peptide
bond on tyrosine fluorescence by studying a group of peptides containing
tyrosine and glycine. They found that the peptide group was a stronger
quencher when the carboxyl group of tyrosine was part of the group then if the
amino group was involved. In these peptides they could also show quenching
by energy transfer between tyrosines, between tyrosine and ionized tyrosine,
and between tyrosine and tryptophan if the latter were also incorporated into
the peptide.

Quenching by peptide bonds was also present in diketopiperazines containing tyrosine [262]. In these compounds there is considerable steric restriction,
and fluorescence-depolarization studies indicated a compact, folded shape; yet
the peptide bonds were still able to quench the phenolic group fluorescence.
Despite the steric factors, the quenching probably resulted because the peptide
groups were still able to come into contact with the tyrosine groups.

Feitelson [263] has stated that direct interaction was needed between
-CONH- or -COOH groups and excited aromatic rings for tyrosine-fluorescence
quenching to occur. The hydroxyl group was felt to enhance the charge density
of the phenyl ring, making tyrosine more easily quenched than phenylalanine.

Weber and Rosenheck [257] used copolymers containing a small percentage
of tyrosine as models for the study of tyrosine fluorescence in proteins. By
comparing the quantum yields and pH dependence of fluorescence of the polymers
before and after O-methylation, they came to the conclusion that proton transfer
accounted for at least 80% of the quenching of tyrosine fluorescence in these
models.

Another mechanism of quenching, energy transfer from tyrosine to
tryptophan, was shown long ago by Weber's polarization studies [264] to be in
principle feasible over the distances encountered in proteins. However,
Teale's study [242] failed to elicit evidence of such energy transfer. Wada and
Ueno [265] found support for energy transfer from tyrosine to tryptophan in a
study of copolymers containing these two amino acids in different proportions.
By comparing the excitation spectra of tryptophan and alcohol dehydrogenase,
Cassen and Kearns [266] found evidence for energy transfer from tyrosine to
tryptophan. Such transfer was also present in glucanohydrolase but absent
from chymotrypsin, papain, pepsin, and carboxypeptidase A. Eisinger's study
[267] on energy migration in peptide horomones also includes evidence for
tyrosine-tryptophan energy transfer. This type of transfer was shown by
Longworth [268] in the dipeptide tryptophyl-tyrosine and in the enzyme ribonuclease T_1, which has but one tryptophan residue. Steinberg [269] reviewed

further instances of tyrosine-tryptophan energy transfer, which now has been
conclusively shown to occur. The transfer probability is relatively low,
however, because of the unfavorably short lifetime of tyrosine fluorescence
which probably prevails in proteins.

Because of the overlap of the absorption and emission spectra of tyrosine,
energy transfer between tyrosine residues may occur. Longworth, Rahn, and
co-workers [270-272] have shown such transfer in poly-L-tyrosine. The
ionization of only 1 tyrosine group causes the quenching of 50 others.

Fasman et al. [273, 274] have investigated the possibility that tyrosine
fluorescence can be influenced by the amount of α-helix present. By using
such copolymers of tyrosine and glutamic acid as copoly-L-Glu-L-Tyr (95:5),
which exist in either random-coil or helical form, depending on pH, they found
an increase of about 25% in tyrosine quantum yield in the helical form [274].
Comparison of random and block copolymers containing tyrosine and tryptophan
was made by Fasman et al. [275] at various pH values. Although the change
from helix to coil did not alter the fluorescence per se, it was considered that
the fluorophors were less accessible to other quenching groups in the helix.
Copolymers of tyrosine with lysine gave quantum yields of 0.02 to 0.04
compared with 0.06 to 0.08 for copolymers with glutamic acid; this reflected
the greater quenching efficiency of the amino group as compared with the
carboxyl group.

Block tyrosine copolymers were found to have very low yields, near 0.02,
and it was reported [274, 275] that a new band suggestive of excimer emission
occurred in these polymers at 400 nm. A putative excimer band was also
reported for polytryptophan [276], and excimer formation was suggested as yet
another pathway for radiationless deactivation in proteins. However, subsequent
work revealed that the polytyrosine "excimer" band was due to a photodegradation artifact [277, 278]. A study by Moon et al. [279] on the model system
phenol-acetate suggests that quenching by acetate has both collisional and static
components and that hydrogen bonding plays a role in deactivation.

A series of investigations by Cowgill have also dealt with tyrosine fluorescence. The quantum yields of peptides containing tyrosine or tryptophan were
much lower than those of the free amino acids [280], observations that extended
White's results for glycyltryptophan and glycyltyrosine [256]. By studying the
effect of substituents on phenol and indole compounds, Cowgill [281] came to the
conclusion that electronegative residues quench by an inductive effect, and
postulated that quenching in peptides occurs via the electronegative peptide
linkage itself.

Because of the possibility of tyrosine-tyrosine energy transfer proposed by
Weber [264], it seemed possible that a nonfluorescent tyrosine could serve as an

"energy sink" for the dissipation of the excitation energy of other residues. From fluorometric titration data on insulin, RNase, and peptides, Cowgill [282] concluded that ionized tyrosines could serve as a sink for energy of un-ionized tyrosines. Furthermore iodination of tyrosines quenched their fluorescence, and the iodinated residues can serve as an energy sink for uniodinated tyrosines [283]. Cowgill [284] has presented evidence that phenolic compounds form hydrogenbonded complexes with amides in nonpolar solvents. He suggested that such complexes may be accurate models for the nonfluorescent, abnormal tyrosines buried in the interior of the RNase molecule [285]. Furthermore synthetic model compounds were examined, and it was concluded that hydrogen-bonded phenolic groups could serve as energy sinks for the fluorescence of groups such as methoxyphenyl, which cannot form hydrogen bonds [286].

Russell and Cowgill [287] have titrated some tyrosyl peptides and in this way demonstrated that (a) protonation of a histidine has no effect on tyrosine fluorescence, and (b) the γ-carboxyl group of glutamyltyrosine is too far away to quench fluorescence.

An important quenching mechanism for tyrosine fluorescence in proteins was uncovered by Cowgill [288], who found that both disulfide and sulfhydryl groups were quenchers. The quenching by disulfide groups was stronger. Arian et al. [289] confirmed thia and found that the sulfur of methionine also has a quenching effect.

The major mechanisms of tyrosine-fluorescence quenching have been reviewed [290, 291]. The principal groups involved are peptide bonds, disulfides, ionized tyrosine, hydrogen-bond-accepting carbonyl groups in hydrophobic environments, ionized carboxyl groups, quenched tyrosines acting as energy sinks, and tryptophans acting as energy acceptors.

2.   Tyrosine Fluorescence in the Study of Protein Structure

Since a variety of mechanisms exist for the quenching of tyrosine fluores-cence, changes in protein structure can be expected to be reflected in the nature of tyrosine emission. Teale [292] found that acid, urea, and detergents greatly increased the tyrosine fluorescence of the albumins as well as of other proteins and suggested that this was due to changes in secondary and tertiary structure. Insulin and ribonuclease fluorescence had been found earlier to double in intensity on transfer from water to propanediol [242]. Gally and Edelman [293] showed that there was a discontinuity in the fluorescence-versus-temperature profile of ribonuclease at precisely the temperature at which a structural transi-tion was known to occur.

A study of the quenching of tyrosine fluorescence in insulin, ribonuclease, and human-serum albumin by phosphate showed that the degree of quenching was different in the three proteins [294]. Phosphate could thus be used as a probe of the relative accessibility of tyrosine residues in proteins. Although human albumin contains tryptophan, it is possible to thoroughly unmask the tyrosine fluorescence by quenching tryptophan fluorescence by energy transfer to an attached DNS group [294] or by destruction of tryptophan by N-bromosuccinimide [292].

Cowgill has performed a number of studies relevant to tyrosine fluorescence in specific proteins. Pancreatic trypsin inhibitor, which has four tyrosines and no tryptophans, was completely quenched by the iodination of only two of the tyrosines [295]. The curve of fluorescence versus degree of iodination suggested that only two of the tyrosines were fluorescent in the native protein and that their iodination resulted in the quenching. No energy transfer was detected between tyrosines and iodotyrosines. Disruption of the native structure by detergents caused a marked increase in fluorescence.

The proteins paramyosin and tropomyosin have tyrosine and no tryptophan, and a high degree of helicity. The native fluorescence efficiency of the tyrosines was 0.4 and 0.26, respectively, times that of free tyrosine; this is much higher than in the case of most proteins. It was demonstrated that the helical structure held the phenolic groups away from the peptide bonds, but, when disrupted into the random-coil forms, the proteins both exhibited the normal degree of tyrosine-fluorescence quenching [296]. Sodium dodecylsulfate and cetyldimethylethylammonium bromide were found to increase the fluorescence of paramyosin and tropomyosin [297]. The mechanism here was thought to be a stabilization of the helix, plus close binding of the detergents to tyrosine groups, thus providing a hydrophobic environment. This reviewer has the impression that the effect of sodium dodecylsulfate on protein fluorescence is unpredictable, because the detergent unfolds some proteins but stabilizes helicity in others.

Cowgill and Lang [298] followed the acetylation of the three exposed tyrosine groups of ribonuclease. The first two tyrosine groups were easily acetylated, and the third one was acetylated less easily and had the most fluorescence. Consideration of the X-ray diffraction model indicated that the first two tyrosines acetylated were probably those in positions 73 and 115, and were initially of low quantum yield as they were only 5 to 6 Å from disulfide bonds.

Donovan [299] showed that the class A protein ovomucoid had "abnormal" titration behavior showing that all the tyrosines were buried. The native quantum yield was very low, only 0.03 that of free tyrosine, but it was enhanced markedly by acidifcation or esterification. Deyl et al. [300] studied the tyrosine fluorescence of calf-skin collagen. The fluorescence was markedly enhanced on

pronase or pepsin digestion of the protein, a procedure that removed end
sequences containing many acidic residues. Both collagen and another class A
protein, elastin, have a visible blue fluorescence that possibly could be di- or
tri-tyrosyl fluorescence [300, 301].

The phosphorescence of tyrosine in proteins has been studied (see, for
example, Refs. [302] through [305]). Although the fluorescence of tyrosine
is weak, it may dominate protein phosphorescence even in class B proteins
[302]. Although of considerable theoretical interest, the phosphorescence of
proteins can normally be observed only at liquid-nitrogen temperatures or in
the solid state. For these reasons phosphorescence has not been used extensively for structural studies and is not treated here.

C.   Tryptophan

1.   Characteristics of Tryptophan Fluorescence

Unlike tyrosine emission, tryptophan fluorescence in proteins was found by
Teale [242] to have quantum yields that could be either higher, lower, or about
the same as that of the free amino acid. Denaturants such as urea tended to
equalize the tryptophan efficiencies of different proteins at a level characteristic
of free tryptophan [241]. In addition, the emission maxima of tryptophan
fluorescence in native proteins were observed to vary from 332 to 342 nm
depending on the protein [242] but shifted to about 350 nm on denaturation. A
recent study by Kronman and Holmes [306] indicates that complete denaturation
of proteins in guanidine hydrochloride results in tryptophan quantum yields
that are similar but still not quite identical. Emission-spectra differences also
persist, and it is concluded that these differences are due to influences of near-
neighbor residues in the polypeptide chains.

Teale [242] reported that the tryptophan quantum yield varied from 0.05 for
γ-globulin to nearly 0.48 for BSA. To understand tryptophan-fluorescence
quenching processes occurring in proteins, White [256] performed fluorometric
titrations of tryptophan; similar titrations were also done by Gally and Edelman
[293]. The results indicate that tryptophan fluorescence is rather constant
between pH 3.5 and 7.5, decreases at lower pH due to the formation of an
un-ionized carboxyl group, which is a quencher, and increases at higher pH
due to deprotonation of the NH_3^+ group, also a strong quencher. At pH values
higher than 10 quenching again occurred, presumably due to abstraction of a
proton from the indole imino nitrogen in the excited state [256, 307].

Proton transfer during the excited state has been postulated to be a major
quenching mechanism for indoles [308]. Conceivably loss of a proton could be
facilitated by acceptor groups in close proximity in proteins. Stryer [308]

showed that indoles and a number of other fluorescent molecules exhibited a
marked deuterium-isotope effect, the deuterated derivatives having quantum
yields of up to twice that of the hydrogen compounds. These results were taken
as confirmation of the theory that proton transfer was the major quenching
mechanism in the indole series. However, other studies [130, 309] have shown
that compounds not containing a dissociable proton also exhibit marked
deuterium-isotope effects. Förster and Rokos [309] emphasized that the isotope
effect is at least partly due to the greater weight of the solvent molecule, which
then is less effective in deactivation processes. Nevertheless, Stryer's
conclusion cannot be completely discounted for indoles, as other studies
[310-312] have shown that the deuterium-isotope effect is much reduced in
N-substituted indoles.

Cowgill [280] has found that the quantum yields of tryptophan peptides in
water are only half as great as that of free tryptophan and postulates that the
peptide bond is a quencher. Tryptophan fluorescence is markedly sensitive to
temperature, decreasing by about 2% per degree near room temperature
[313-315]. Cowgill [316] has recently postulated that tryptophan fluorescence in
peptides is mainly quenched by the carbonyl group, but that the quenching is
enhanced by amino groups. Organic solvents tend to eliminate the quenching of
carbonyl groups [316]. Kronman et al. [317] studied the fluorescence of
copolymers of glutamic acid and lysine containing small amounts of tryptophan.
They came to the conclusion that the transition from random coil to α-helix had
no effect on tryptophan fluorescence. On the basis of fluorometric titrations
and the addition of ethanolamine, they concluded that the amino group was a
strong quenching factor. However, the authors find that tryptophan fluorescence
in the absence of specific quenching groups is constant from pH 2 to 12.5.
Copolymers containing blocks of contiguous tryptophans had very low quantum
yields, near 0.04, attributed to indole-indole interaction, as evidenced also by
a red shift of the emission.

The tryptophan fluorescence of the highly helical protein light meromyosin
was found by Cowgill [318] to have a quantum yield 1.7 times that of free tryptophan. The high yield was attributed to the lack of quenching by peptides involved
in the hydrogen bonding that stabilized the helix. Denaturation to the random
coil was accompanied by reduction of fluorescence yield, as expected. The
peptide-bond quenching was further studied by Cowgill [319]. Large concentrations of amides added to indole solutions did not quench, which suggests that
only the proximity of the peptide groups to the indole in proteins permits it to
quench. It was also confirmed that quenching at high pH by the hydroxide ion
involves the proton on the ring nitrogen, since N-methyl derivatives of indoles
were immune to such quenching.

The question of whether the fluorescence behavior of indoles is basically altered at the electronic-structure level by substituent groups was touched on by Weinryb and Steiner [320]. Measurements of the fluorescence lifetimes and quantum yields of indole derivatives at various temperatures permitted the conclusion that the lifetimes were proportional to the yields, and all the indole compounds would have the same lifetime under conditions where the quantum yield was unity. In other words, the innate-fluorescence rate constant of the indole moiety was not affected by the substituent groups. Those groups therefore affect the observed quantum yields by changing the rate constants for radiationless processes.

By studying the pH dependence of tryptophan fluorescence in compounds also containing tyrosine, Edelhoch and co-workers [321-322] and Cowgill [280] concluded that energy transfer from tryptophan to ionized tyrosine is a pathway for the deactivation of tryptophan. A recent study by Edelhoch et al. [322] with a series of peptides, $Trp-(Gly)_n-Tyr$, where n varied from 0 to 4, showed a decrease in quenching efficiency of ionized tyrosine with chain length. When n = 2, quenching was 50% of maximum, a finding that could be used to test the Förster theory [323] for energy transfer. The distance between indole and phenol rings was found from models to be 14.5 $\overset{o}{A}$; the distance calculated from the Förster theory was 13.3 $\overset{o}{A}$. The energy transfer occurs because the absorption maximum of ionized tyrosine at 295 nm means that there is good overlap of the band with the tryptophan emission spectrum, which in proteins is usually shifted to the blue from that of the free amino acid.

Within the last 5 years two important groups in proteins have been found to quench tryptophan fluorescence. Cowgill [288] showed that disulfides quench both indole and tyrosyl fluorescence. In model peptides containing tryptophan and disulfide bonds, reduction to the sulfhydryl compounds could be followed kinetically by the increase in fluorescence. Sulfhydryl groups were also quenchers, but weaker than disulfides. It was also found [324] that disulfide compounds quenched indole fluorescence by a collisional mechanism. The spatial requirement for quenching seems to be that the disulfide must come into contact with the indole. The cyclic peptide formed by the oxidation of Cys-Trp-Cys is highly constrained, and there is no more tryptophan quenching there than in peptides not containing disulfides [324].

Stryer et al. [325] found that the fluorescence of thioredoxin, an enzyme, was enhanced by a factor of 2.5 when a disulfide bridge was reduced with dithioerythritol. Although they concluded that this enhancement was due to a localized configurational change not reflected in any other physical measurement, in view of Cowgill's findings [288], the enhancement is probably due simply to release from disulfide quenching. Figure 130 illustrates the increase

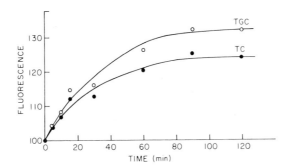

FIG. 130. Effect of reduction of disulfide bonds on the fluorescence of tryptophan in peptides containing cystine. The oxidized forms of N-t-butyloxy-carbonyl-tryptophanyl-cysteine (TC) and N-t-butyloxycarbonyl-tryptophanyl-glycyl-cysteine (TGC), 10^{-4} M, were treated with 6×10^{-4} M dithioerythritol at pH 9. The fluorescence was excited at 290 nm and monitored at 340 nm.

in tryptophan fluorescence obtained by reducing disulfide bonds, which are close to the indole group in small peptides.

Shinitzky and Goldman [326] found that protonated histidine quenches tryptophan fluorescence. Peptides containing the two amino acids were synthesized and titrated; the titration of histidine corresponded exactly with tryptophan-fluorescence quenching of 40 to 80%. It was considered possible that fluorometric titration curves of proteins showing quenching on acidification from pH 8.5 to 5 are due to the close proximity of histidine and tryptophan moieties. Papain, ficin, bromelain, and chymopapain showed this behavior, whereas glucagon, pepsin, and lysozyme did not. The mechanism of quenching was suggested to be charge-transfer interaction, for which some evidence was found in the absorption spectra of model cyclic and linear peptides [327].

The tryptophan-fluorescence spectrum has been the subject of much discussion. Vladimirov and Li [328] showed that the zwitterionic form of tryptophan has an emission-spectra peak that is shifted 2 to 3 nm to the red in comparison with the peak of the cationic form, and the basic form of tryptophan has a further red shift of 2 nm from that of the zwitterionic form. The influence of the charged groups was not explained, but possibly they could affect solvent dipole relaxation after Franck-Condon excitation. Longworth [329] observed the emission spectrum of tryptophan and some of its peptides in various solvents. The emission peaks were quite similar, although the quantum yields of the peptides were frequently much lower; this suggests that the factors responsible for the spectral shifts and the quantum yields are not necessarily related.

The free amino acid at neutral pH in water has an emission peak at 348 nm

[241, 243], but the structureless form bears little resemblance to the fluorescence spectrum of indole derivatives in such apolar solvents as cyclohexane. Van Duuren [330] found marked loss of structure and a red shift as small amounts of alcohol (1 % v/v) were added to indole solutions in cyclohexane. Mataga [331] calculated that indole in the excited state has a much higher dipole moment than in the ground state and attempted to explain the spectral changes with solvent on the basis of Lippert's theory of solvent dipole relaxation. Konev and co-workers [332] also had made the observations reported by Van Duuren [330], but in addition showed that there were marked changes in the absorption spectrum when small amounts of alcohols were added to indole in apolar solvents. The presence of an isobestic point indicated a 1:1 hydrogen-bonded solvated form of indole in the ground state. A large part of the spectral change observed with polar solvents must therefore be due to solvation, or, more accurately, complexation with solvent in the ground state.

The question of whether indoles like tryptophan form "exciplexes" has been brought up by Lumry and co-workers [311, 333, 334]. As defined by Birks [335], an exciplex is an excited-state complex made up of two different molecules:

$$A^* + B \longrightarrow (AB)^* \tag{204}$$

A special case of an exciplex is the complex formed when A = B; the excited complex then is called an excimer. Walker et al. [334] found that the marked red shift and loss of structure of the fluorescence spectra of indoles in polar solvents were too great to be explained on the basis of Lippert's theory alone. They showed data indicating that indole formed a 1:2 exciplex with the polar solvent molecules. However, no attention seems to have been paid to the ground-state solvation effects. Longworth [268, 336] was not able to confirm the 1:2 stoichiometry, although he found examples where there was little if any absorption change, but a marked fluorescence-spectra change, on addition of polar solvents.

Eisinger and Navon [337] measured tryptophan spectra as a function of temperature in polar glasses from $80°K$ to room temperature. The highly structured spectrum at low temperatures gave way to a red shift and loss of vibrational detail in the region of 170 to $230°K$, where the glass softens. The authors concluded that shifts in the solvation shell in the excited state were responsible for the spectral changes. The shifts occurred only when the viscosity was low enough to permit solvent mobility. This view differs subtly from that of Walker et al. [333, 334] in that no stoichiometric exciplexes are postulated.

Possibly related to indole solvation and the possibility of exciplex formation is the observation by Galley and Purkey [338] that the emission spectrum of indole at low temperatures is dependent on the excitation wavelength. This is taken as evidence for several populations of solvated molecules, which have slightly different excitation and emission spectra. At higher temperatures dipole-relaxation events occur with facility, and Franck-Condon states with different solvation configurations become homogenized before emission occurs. The finding lends some support to the idea that one discrete stoichiometric exciplex of indole and solvent does not predominate in polar solvents.

Feitelson [339] has observed the formation of hydrated electrons after excitation of indole derivatives. A chemical trapping system was used to quantitate hydrated electrons produced at various pH values. Electron scavengers are frequently good quenchers of indole fluorescence, but Feitelson felt that prevention of ion recombination did not play a major role in indole quenching, so that electron ejection was an interesting, but not directly related, phenomenon.

Konev [237] has emphasized another facet of tryptophan fluorescence, the electronic structure of the indole group. Weber's studies [340] had shown that the long-wavelength band of indoles is made up of two electronic transitions whose moments are nearly perpendicular. Zimmermann and Joop [341] using highly sensitive equipment, obtained evidence from low-temperature fluorescence emission and polarization data that the fluorescence of indole arises from two singlet states simultaneously. The evidence was based on the vibrational structure of the spectra and the structure in the polarization spectra. Konev et al. [332] found that the polarization in the emission band of proteins is not constant and suggested that protein emission due to tryptophan groups arises from two singlet states of the indole moiety. Konev [237] discussed the possibility that energy transfer between tryptophan residues may involve one transition more than the other, that hydrogen bonding may enhance one transition rather than the other, and that the balance between the amount of emission from the two states may influence the fluorescence spectrum. The novelty of emission from two states in the same molecule is perhaps hard to accept, although such a situation was postulated for quinine and 6-methoxyquinoline [342].

Another possibly neglected source of tryptophan-fluorescence quenching is dissolved oxygen. It is known that the fluorescence of most organic molecules in apolar solvents is strongly quenched by oxygen [343], although the effect in water is usually negligible. Tryptophans in the interior of a protein would, however, be in an essentially apolar environment and could be quenched by

oxygen. Indeed, Barenboim [344] has evidence that oxygen accounts for some of
the quenching of the intrinsic fluorescence of proteins.

2. Applications of Tryptophan Fluorescence Spectroscopy

Tryptophan fluorescence has come into wide use as a tool for studying
protein structure. Teale [242], Konev [240], and others had noted that complete
denaturation of proteins caused marked changes in fluorescence. That such
fluorescence would yield information about intermediate, unfolded states of
proteins was suggested by the work of Steiner and Edelhoch [345]. In fluorometric titrations of BSA, lysozyme, and pepsin these workers found marked
changes in the intensity of fluorescence emission at certain pH values. The
fluorescence of BSA was quenched below pH 5 and also between pH 7 and 9;
these regions corresponded to transitions known from other physical methods.
Lysozyme and pepsin seemed relatively insensitive to acid down to pH 2, a
finding in agreement with their known stability to acids.

Steiner and Edelhoch [345] found that tryptophan fluorescence decreased
linearly with temperature in the case of pepsin, pepsinogen, and γ-globulin
except when the temperature of known structural transitions was reached. In
the case of pepsin and pepsinogen there was a marked quenching of fluorescence
as the thermal transition point was reached; in the case of γ-globulin at alkaline
pH an increase in fluorescence occurred.

At about the same time Gally and Edelman [313] noted that the fluorescence
of proteins decreased with temperature at a rate that differed for various
proteins. In a study of the L-chains of Bence-Jones proteins Gally and Edelman [139] found a sharp rise in tryptophan fluorescence at the temperature at
which a structural transition was also indicated by fluorescence-polarization
and fluorescent-probe methods. At 68°C Bence-Jones protein had a tryptophan
fluorescence that could be seen in the uncorrected emission spectrum to be
distinctly shifted to the red from its position at 25°C [139]. Early studies such
as these established the usefulness of fluorescence spectroscopy for studying
changes in the structure of proteins.

Many proteins have now been studied with regard to their tryptophan
fluorescence. For example, L. Brand et al. [346] observed that the fluorescence of lactic dehydrogenase decreased in the range of 2 to 4 M urea. They
also found that the fluorescence of yeast and liver alcohol dehydrogenases were
markedly influenced by pH and urea. Heat inactivation was noted to cause a red
shift of the intrinsic fluorescence of the yeast enzyme.

Gerstein et al. [347] correlated the quenching of tryptophan fluorescence
in pepsinogen at pH 9.9 with the loss in potential peptic activity and antigenicity.
Pepsin fluorescence did not fall until a higher pH value was reached. The

denaturation of pepsinogen was studied by various fluorescence methods by
Edelhoch et al. [348] who found that 8 M urea caused a time-dependent fall in
tryptophan fluorescence; kinetic analysis showed that the drop had the same
first-order rate constant as was found by changes in optical rotation, difference
spectra, or by use of the pH-stat. On the other hand, Perlmann [349] has
observed that when pepsinogen solutions were cooled from a high temperature,
the levorotation decreased at a slower rate than the rate of recovery of
tryptophan fluorescence and potential pepsin activity.

In a study of trypsin Edelhoch and Steiner [350] found that the tryptophan
fluorescence was quenched and shifted to the blue on combination with the
inhibitor ovomucoid. Neither inhibitor contains tryptophan, and the results
were interpreted as due to an environmental change near the trypsin tryptophan
residues on complexation. These authors have also studied the soybean trypsin
inhibitor [351, 352]. The protein is remarkably resistant to 9 M urea, but by
raising the temperature or pH in this solvent, they produced a fall in fluorescence intensity. The rate of decrease was the same as that found by spectrophotometry, viscometry, or fluorescence polarization. At high pH quenching
of tryptophan fluorescence was noted and attributed to the transfer of energy
to ionized tyrosine residues [351].

Some proteins contain only a single trypothan residue; the fluorescence
thus can be precisely attributed to environmental factors at a particular site,
in contrast to the situation in most proteins, which have several tryptophan
residues. Human-serum albumin, and probably albumins of some other
species (see Fig. 124), have only one tryptophan. The enzymes ribonuclease
T_1 and staphylococcal nuclease are interesting in this regard because they have
only one tryptophan and their three-dimensional structures have been largely
worked out by X-ray crystallography.

Longworth [268, 353] found that tryptophan fluorescence in ribonuclease T_1
was very unusual in that there was distinct structure present even in aqueous
solutions at room temperature. Similarity to indole fluorescence spectra in
apolar solvents suggested that the tryptophan in this protein was in a very hydrophobic pocket. Energy transfer from tyrosine to tryptophan was detected, and
the nonequivalence of absorption and excitation spectra indicated that transfer
efficiency was less than unity, although no tyrosine fluorescence could be
elicited. Irie [354] and Pongs [355] noted the blue-shifted emission peak of
ribonuclease T_1 and the shift from 320 to 350 nm on urea denaturation. Binding
of AMP and CMP resulted in the quenching of tryptophan, which served as a tool
for studying association stoichiometry.

In contrast, staphylococcal nuclease binds substrates and inhibitors with no
change in tryptophan fluorescence [356]. This enzyme, with a molecular weight

of 16, 800, has a tryptophan residue at position 140 at the end of a run of α-helix;
The apolar nature of the tryptophan environment is clearly seen on the model
based on the 2-Å-resolution electron-density map obtained by Arnone et al.
[357]. On the basis of fluorescence spectra at various low temperatures
Longworth [336] concluded that tryptophan in the enzyme forms an exciplex with
a solvent molecule under room-temperature conditions, but not when the
viscosity is high enough to prevent relaxation in the excited state. However, it
is not clear that there is sufficient room to permit the exciplex partner to be a
solvent molecule rather than a neighboring side-chain group.

Cuatrecasas et al. [356] have studied the tyrosine fluorescence of the
enzyme extensively, because there are marked changes on binding of nucleotide
inhibitors. This correlates well with the finding of tyrosine residues in the site
for the binding of Ca^{2+}, which is needed for activity. Although the tryptophan
residue is a considerable distance from the active site, its fluorescence has
been used to follow the kinetics of configurational changes by stopped-flow
fluorometry [358, 359]. The emission of the native enzyme is about 2.3 times
as intense as that of free tryptophan, and acid denaturation brings the yield down
dramatically. Renaturation kinetics were found to follow a biphasic curve,
suggesting two processes, one of which was essentially independent of temperature [359]. The two processes were postulated to be helix formation, followed
by refolding of the protein.

Fluorescence was used by Shifrin and Steers [360] to show structural
changes in the subunits of ß-galactosidase on urea denaturation. No distinction
could be made between the dissociation of the subunits of the tetramer and the
denaturation of the subunits. Schechter and Epstein [361] similarly followed the
denaturation of myoglobin and apomyoglobin by their intrinsic fluorescence and
could not distinguish denaturation of myoglobin from dissociation of the heme
moiety; the two processes were simultaneous. As shown by curves of fluorescence versus concentration of denaturant, the myoglobins were more stable than
the apoproteins.

In proteins that contain more than one tryptophan residue it is not surprising
that the fluorescence from each of the tryptophans may not be identical; that is,
various tests can reveal the heterogeneity of the intrinsic fluorescence. When
carbonic anhydrase fluorescence was quenched by a sulfonamide bound to the
active site, a comparison of the degree of sensitized fluorescence and the degree
of quenching led to the conclusion that energy transfer was occurring primarily
from less fluorescent tryptophans [166]. By using luminescence spectra
obtained at $77^{\circ}K$, Purkey and Galley [362] actually obtained emission spectra
that were dependent on exciting wavelength. Apparently in the proteins studied
(alcohol dehydrogenases, papain, and trypsin) the buried and exposed trypto-

phans have significantly different absorption spectra at low temperatures and can be selectively excited. Elkana [363] showed the heterogeneity of protein fluorescence by the use of quenching by salicylamide, an energy-transfer acceptor with a peak absorption near 300 to 310 nm. In the presence of large amounts of salicylamide the emission from tryptophan residues emitting at short wavelengths is preferentially quenched, leading to an apparent red shift in emission. Such a shift was observed in lysozyme, but not in human-serum albumin, which has only one tryptophan residue.

A number of other workers have utilized fluorescence to study lysozyme. Edelhoch and Steiner [364] showed early that urea denaturation or extremes of pH and temperature could induce fluorescence increases that corresponded to structural changes in the protein. The increase of up to 33% in quantum yield was consistent with the finding by Churchich [365, 366] that the native enzyme has a low quantum yield, 0.07. Reduction and carboxymethylation caused a red shift in emission and an increase in yield to 0.13; much of the original fluorescence characteristics were regained on air oxidation of the reduced protein. Ultraviolet irradiation of the enzyme caused marked loss of activity, destruction of some tryptophans, and a red shift in the fluorescence [367].

Lehrer and Fasman [255] also noted the low quantum yield of lysozyme, although their corrected lysozyme spectrum peaked at a longer wavelength than reported [366]. It was found that the addition of various poor substrates, such as tri(N-acetyl-D-glucosamine), enhanced the fluorescence yield and shifted the spectrum toward the blue. Similar observations were made by Shinitsky et al. [368]; both sets of authors attributed the fluorescence changes to the influence of substrate on the three tryptophans known by X-ray crystallography to be at the active site.

The preferential quenching of tryptophans near the binding crevice was further studied by Lehrer and Fasman [369]. Differential emission spectra obtained at various pH values in the presence and absence of substrates allowed the authors to calculate the spectra of the three tryptophans at the binding site. This clear evidence of heterogeneity in protein emission was also shown by selective modification: Teichberg and Sharon [370] were able to oxidize tryptophan 108 of lysozyme with iodine, leaving tryptophans 62 and 63 to be studied by their behavior with varying pH.

In a study of the mechanism of action of the Δ^5-3-ketosteroid isomerase Wang et al. [371] made use of changes in the ultraviolet fluorescence spectrum on addition of substrates and inhibitors. The changes were not all explicable in terms of quenching by energy transfer and pointed to the occurrence of configurational changes.

Kronman and co-workers [372, 373] have studied α-lactalbumin by emission spectroscopy. The well-defined configurational change occurring at pH 6 as the temperature is lowered from 25 to $1^{\circ}C$ is accompanied by a 10-nm blue shift in the emission, whereas denaturation by dilute acid caused an increase in quantum yield and a red shift. Above pH 10 tryptophan emission was shifted about 10 nm toward the red. Acylation studies [374] showed that the tyrosine residues on the surface of the protein were near enough to tryptophan residues to affect the fluorescence of the latter. Acetylation of amino groups in the interior of the protein was accompanied by changes in the fluorescence of buried tryptophans.

The structural transitions of both human- [375] and bovine-serum albumins [345, 376] have their counterparts in changes in the fluorescence spectrum or intensity. In the so-called N-F transition region of pH 4 human-serum albumin shows reciprocal changes in the tryptophan and tyrosine fluorescences.

Edelhoch and Lippoldt [377] found that glucagon had an emission spectrum showing that the fluorescent groups were exposed to solvent, and pH titrations of the hormone gave no inflection points. These and other criteria indicated that there was no tertiary structure in the enzyme in solution, in spite of the X-ray evidence of helicity in glucagon crystals. Horse-liver alcohol dehydrogenase has an overall tryptophan quantum yield that is higher than that of free tryptophan. Blomquist [378] showed that acid denaturation of the protein could be followed by a decrease in tryptophan fluorescence.

One application of tryptophan fluorescence spectroscopy that is perhaps mundane and obvious, but very effective, is the detection of tryptophan in proteins. Bannister et al. [379] recently have resolved the question of whether tryptophan is present in human erythrocuprein by showing that 280-nm excitation gives a characteristic tryptophan fluorescence that is sensitive to N-bromosuccinimide, an oxidizer of indole rings. Conversely, photodegradation of tryptophan residues is accompanied by loss of both fluorescence and enzymatic activity in glutamic dehydrogenase [380].

Gabel and co-workers [381] showed that trypsin and chymotrypsin attached to a solid support could be studied by both extrinsic and intrinsic fluorescence. Configurational changes induced by denaturants are reflected in the tryptophan fluorescence and in the emission from bound TNS.

In a number of proteins the binding of ligands that do not absorb in the region of tryptophan emission nevertheless results in a quenching of fluorescence. Helene et al. [382, 383] studied the valyl-tRNA synthetase of E. coli. Binding of various substrates (valyl-tRNA, valine, ATP, and Mg^{2+}) led to varying degrees of quenching, which in turn could be used to obtain binding constants. Kuczenski and Suelter [384] found that the tryptophan fluorescence of yeast pyruvate kinase was severely quenched in the presence of fructose-1,6-

diphosphate, and less so by K^+ or Mg^{2+}. In the case of bovine- and pig-serum albumins Spector and John [385] found that the binding of fatty acids resulted in fluorescence quenching. As much as 50% quenching was noted on binding 8 to 9 moles of fatty acid. No effect was noted with human, rabbit, horse, or dog albumins. Attalah and Lata [386] found that testosterone binding also caused quenching of BSA fluorescence. In these examples the ligands cannot quench by Forster's resonance-energy-transfer mechanism, and it is tempting to ascribe the effects on fluorescence to either configurational changes induced by ligand or to binding of the ligand in the immediate environment of the emitting tryptophans.

Protein fluorescence may also be influenced by the binding of various metal ions. The quenching that is observed in colored metal-ion complexes is clearly due to radiationless energy transfer. This was the case in the binding of copper and iron to transferrin, which was studied by Lehrer [387]. Shaklai and Daniel [388] studied the hemocyanin from a mollusk and found that the fluorescence of the tryptophan residues would fluctuate with the oxygen content of the molecule, since the quenching was stronger in the presence of a $Cu \cdots O$ absorption band. From the degree of quenching and the critical transfer distance calculated from Förster's theory the authors arrived at a tryptophan-Cu^{2+} average distance of 25 Å. Bannister and Wood [389] performed similar studies of <u>Murex trunculus</u> hemocyanin. Preparation of the copper-free apoprotein increased the tryptophan quantum yield fivefold. Strong quenching was noted in the holoprotein when oxygenated; thiocyanate reduction resulted in a threefold fluorescence enhancement.

Metal ions that have no visible color also affect protein fluorescence. Rogers [390, 391] reported that Ag^+ quenched the fluorescence of glutamate and lactate dehydrogenases by about 70%. The quenching was correlated with loss of enzymatic activity and with the reaction of sulfhydryl compounds to form mercaptide absorption bands, which are most intense below 250 nm but extend with sufficient intensity up to 340 nm to cause quenching of tryptophan fluorescence via the energy-transfer mechanism. Liver alcohol dehydrogenase and ovalbumin were strongly quenched by Ag^+.

The mercuric ion was also found to quench the fluorescence of many proteins [393]. It had been shown by Ramachandran and Witkop [394] that in simple indoles Hg^{2+} can substitute directly in the ring system, causing marked ultraviolet-spectral changes. It was found [393] that these complexes are nonfluorescent. However, the fluorescence quenching by Hg^{2+} in proteins could not be correlated with the occurrence of such complexes. Instead, the development of mercuric mercaptide absorption bands resulted in energy transfer from excited tryptophans. Calculation of the transfer distances for the quenching by Ag^+ and Hg^{2+} showed that the transfer operated over relatively short distances of 11 to 17 $\overset{o}{A}$ [393, 394]. The possibility was raised that metal-ion quenching could be used as a probe of tryptophan-sulfhydryl group distances in proteins.

Metal-ion effects on fluorescence have not been extensively studied, although there are many metal-protein complexes of interest, and often heavy-metal derivatives are made for crystallographic studies. An example is that of papain, which can be prepared in several Hg^{2+}-containing forms. Arnon and Shapira [395] compared the fluorescence of papain, mercuripapain (a form with a mercuric ion complexed to the active-site sulfhydryl), and papain with mercuric ion inserted in the disulfide bridge connecting positions 43 and 152. Metal-free papain had a much higher fluorescence yield than did the mercury-containing derivatives, and the degree of quenching in the mercuripapain was greatest. No speculations on the mechanism of quenching were offered, and the absorption of the mercuric mercaptides of papain has not been reported.

Papain fluorescence has been studied aside from metal-ion effects. From titration data Barel and Glazer [396] find that the activated enzyme has groups with pK_a values of 6.6 and 8.5 near the emitting tryptophans. In contrast, the unactivated enzyme has only a group with a pK_a of 6.6. Barel and Glazer assign the two groups to histidine and an essential sulfhydryl group in the activated enzyme. However, Sluyterman and De Graaf [397] found that papain preparations could be subfractionated by affinity chromatography, and fluorometric titration of the subfractions indicated that the activated papain has only the titratable group with a pK_a of 8.6, whereas the nonactivatable fraction has a group with a pK_a of 6.6. Weinryb and Steiner [398, 399] have studied papain luminescence, including phosphorescence of tryptophan and tyrosine, in water and D_2O. Among their findings was that tyrosine fluorescence was very weak, although tyrosine phosphorescence was relatively strong, and one particular tryptophan residue had an especially high quantum yield.

Steiner et al. [400] have considered the possibility of using small molecules to perturb tryptophan fluorescence in a manner analogous to the solvent-perturbation technique of Herskovits and Laskowski [401]. The fluorescence of the model compound N-acetyltryptophanamide was increased by such perturbants as propylene glycol and urea, both of which decreased the solvent dielectric constant; however, the fluorescence was diminished by glycine, which increased solvent polarity. The effects of perturbants on the tryptophan fluorescence of various proteins differed markedly, and in general the effects were greater in denatured proteins whose tryptophans are exposed.

Edelhoch and Lippoldt [402] used the same technique in a study of thyroglobulin. It was found that whereas in the native state thyroglobulin was little affected by propylene glycol, in 7 M urea the tryptophan-fluorescence intensity increased by 30% in 20% propylene glycol. This result showed that all the tryptophans were essentially exposed, since the same degree of fluorescence enhancement was observed for N-acetyltryptophanamide under these conditions.

A variation of the perturbation technique was used by Lehrer [403], who showed that the iodide ion could be used to quench selectively the tryptophans that were more exposed. Lysozyme fluorescence was shown in this way to consist of components from different classes of tryptophans [404].

Although not related to intrinsic protein fluorescence, the use of iodide as a quenching probe by Vernotte [405] can be mentioned here. She employed fluorescein-labeled conjugates and showed that the apparent Stern-Volmer constants for potassium iodide quenching varied with different proteins, showing the variation in degree of exposure of the dye. Using this technique, Arrio et al. [406] studied fluorescein-labeled trypsin. The sensitivity of the fluorescein to quenching by potassium iodide was altered by binding the substrate N-acetyltyrosine ethyl ester, and it was postulated that the configuration of the enzyme had changed around the dye site. Winkler [407] employed the borate ion as a quenching probe for dyes adsorbed onto proteins. Blomquist [408] found that D_2O could be used as a perturbant to demonstrate NADH binding in a protected environment on an enzyme. When NADH was free in solution 75% D_2O enhanced its fluorescence by 20%, but there was no effect when the NADH was bound to alcohol dehydrogenase.

Radiationless energy transfer from excited tryptophan groups results in fluorescence quenching by a large number of ligands. Such quenching has been of utility in the study of many proteins; only a few examples can be mentioned here. Perlman et al. [409] studied the effect of iodination on thyroglobulin fluorescence. It was shown that quenching of tryptophan fluorescence by energy transfer to various iodine-containing tyrosine species, including thyroxine, was feasible. The transfer probability increased with ionization of the iodotyrosines. Steiner et al. [410] and Tabachnick et al. [411] utilized fluorescence quenching to study the binding of thyroxine and similar compounds to serum albumin.

D'Anna and Tollin [412] studied the fluorescence of the "Shethna flavo-protein," an enzyme containing flavine mononucleotide (FMN). The coenzyme quenches the protein fluorescence by 95%, although the calculated critical transfer distance is 20 Å. Therefore the average distance between tryptophan and FMN must be considerably closer than 20 Å. The second-order kinetics of the binding of FMN to apoprotein were obtained by the quenching of either FMN or protein fluorescence.

Luisi and Favilla [413] found that NADH quenched only 50% of the fluor-

escence of horse-liver alcohol dehydrogenase and concluded that only one of
the two tryptophan residues was affected. This seems to be only one of
several possible interpretations, yet the conclusion is consonant with the
tryptophan heterogeneity noted by Purkey and Galley [362].

Montanaro and Sperti [414] obtained the dissociation constants for the
NADH-diphtheria toxin complexes from fluorescence-quenching measurements.
Although NADH fluorescence is usually enhanced on binding to proteins, in
this case there was little enhancement.

Protein components of the erythrocyte membrane have been isolated and
shown by fluorescence quenching to bind to hemoglobin [415]. The heme
protein did not quench the fluorescence of proteins to which it did not bind, nor
did myoglobin quench the fluorescence of hemoglobin-binding components.

The quenching of intrinsic protein fluorescence is a very convenient
optical handle for following fast reactions. Geraci and Gibson [416] found that
NADH binds to LDH in a simple one-step reaction; stopped-flow fluorometry
was used. Noble et al. [417] studied the rate of combination of heme proteins
with their specific antibodies. For antihemoglobin-hemoglobin, the second-
order rate constant was $10^6 M^{-1} sec^{-1}$. The ternary complexes formed by
lactate dehydrogenases with pyruvate and NAD have an absorption above 300 nm
that results in fluorescence quenching. With stopped-flow measurements of
fluorescence quenching Wuntch et al. [418-420] studied the rate of ternary-
complex formation by various isozymes of lactate dehydrogenase at different
concentrations. Gutfreund et al. [421] studied the pH dependence of the ternary
complex formation by fluorescence quenching and correlated the substrate
inhibition by high concentrations of pyruvate to the formation of the ternary
complex. Everse et al. [422], using fluorescence quenching and stopped-flow
methods, concluded that the phenomenon of isozymes might be explained on the
basis of kinetic differences with regard to substrate inhibition of isozymes from
different tissues, but this theory is still controversial [420].

Weber [264, 307, 423] has proposed that the excitation polarization
spectrum of proteins may be a good indicator of configurational changes. He
has shown that the polarization corresponding to the absorption differs greatly
for different proteins. In proteins where there is evidence for structural
changes the polarization spectrum changes in amplitude and somewhat in shape.
The polarization depends on the complex geometrical relationships between the
absorption oscillators and the multiple emission oscillators in a protein. In the
instruments used for such studies [423-425] no monochromator is used for the
emission, which is instead isolated by a sharp-cutoff filter. The observed
polarization spectrum is thus a function also of the emission spectrum and
detector-response characteristics, since Konev [332] has shown that the polari-

zation in the emission band is also not constant. Whether these "protein polarization spectra" supply more information than simple polarization measurements at one wavelength, in combination with emission spectral data, remains to be seen. Lynn and Fasman [426] used polarization spectra to show changes in the configuration of poly(Glu-Trp) (99:1) on changing the pH. The looser configuration gave a generally lower polarization spectrum.

IV. ENERGY TRANSFER

No review of protein fluorescence can be complete without mention of energy transfer. Many of the papers already cited have made rich use of energy-transfer phenomena, such as fluorescence quenching or sensitized fluorescence, and it is clear that analysis of energy transfer can give much additional information, especially about distances between groups on a protein. The commonest type of energy transfer is singlet-singlet transfer, which involves a normally fluorescent donor whose excitation energy passes to an acceptor that may or may not be fluorescent. In order for energy to pass from one molecule to another, certain spatial and steric requirements must be met, and the emission spectrum of the donor must overlap the absorption spectrum of the acceptor. Förster [323] provided a means of quantitating some of these factors when he produced an equation for calculating R_0, the critical transfer distance between donor and acceptor where the probability of transfer was equal to the probability of emission:

$$R_0 = \sqrt[6]{\frac{1.66 \times 10^{-33} \times \tau J_\nu}{n^2 \times \bar{\nu}_0^{\;2}}} \, , \tag{205}$$

where τ is the donor-fluorescence decay time in the absence of energy transfer, ν_0 is the mean of the positions of donor emission and absorption spectral peaks in terms of wavenumbers, J_ν is the overlap integral, and n is the refractive index. The subject of energy transfer has been reviewed from time to time (see, for example, Refs. [269] and [427] through [430]). The Förster equation has been tested with specially synthesized compounds containing donor and acceptor groups separated by known distances. The observed and calculated transfer distances are then compared. For a given separation R of donor and acceptor the fraction X of photons transferred is given by [323]:

$$1 - X = \frac{1}{(R_0/R)^6 + 1} \, . \tag{206}$$

Latt et al. [431] synthesized compounds containing both donor and acceptor groups attached via a bisteroid skeleton. The donor-acceptor pairs were

separated by about 20 $\overset{o}{A}$, and the distances calculated from Förster's theory
were in agreement with those derived from molecular models. Similar confirmation of the Förster equation was produced by energy-transfer measurements on model compounds by Lamola et al. [432]. Rauh et al. [433] obtained
much higher transfer rates in some indole alkaloid model compounds than those
predicted on the basis of the energy-transfer theory; they postulate that some
other transfer mechanism also operated in their system.

Stryer and Haugland [434] synthesized oligomers containing DNS (acceptor)
and 1-naphthyl (donor) groups separated by varying numbers of proline spacer
groups. From theory, R_0 was 27.2 $\overset{o}{A}$, and the observed value was 34.6 $\overset{o}{A}$; the
agreement was fair, considering the uncertainty in several experimental factors.
The data gave support to the sixth-root dependence of energy transfer predicted
by Eq. (205).

A very similar series of compounds was used by Gabor [435], who reached
the same conclusion, namely, that the Förster theory was confirmed. Conrad
and L. Brand [436] prepared a series of compounds containing tryptophan
(donor) and DNS (acceptor) groups separated by methylene groups. The degree
of energy transfer diminished with chain length, but it was clear that the chains
were not fully extended. Nevertheless, the data were said to be confirmation of
Eq. (205).

The feasibility of energy transfer in proteins was shown quite early by
calculations based on the Förster theory (see, for example, Refs. [427] and
[437]). Various refinements in the theory have been proposed. For instance,
Feitelson [438] has taken into consideration the effect of Brownian motion during
the excited state of the donor. In that case R varies with time during the transfer process. Recently Cantor and Pechukas [439] considered the situation
where the donor-acceptor distances were not homogeneous but consisted of a
distribution of distances among a population of molecules with different con-
formations. Here it was indicated that if data with several donor-acceptor
pairs were available, mathematical analysis could reveal the actual distance
distribution functions.

Edelhoch et al. [261, 409, 440] have calculated R_0 values for many donor-
acceptor pairs of biochemical interest and have examined energy transfer in
tryptophan peptides. Because the fluorescence spectrum of tyrosine overlaps
its absorption spectrum, energy transfer between tyrosines is possible. Knopp
et al. [272, 441] have been able to show energy transfer in polytyrosine in
several ways. Because the tyrosyl residues are not aligned in parallel, the
effect of energy transfer is to cause depolarization of fluorescence excited with
polarized light. Also, ionization of only a few tyrosine residues resulted in

quenching of fluorescence from many un-ionized tyrosines.

In preceding sections of this chapter examples were cited in which energy transfer was important in the study of protein structure; energy transfer may involve both extrinsic and intrinsic fluorescence. For instance, Latt et al. [442] used DNS-labeled glycyltryptophan as a substrate for cobalt-carboxy-peptidase and were able to follow various fluorescence parameters as a function of time with a stopped-flow device. The rate of binding of the substrate could be measured by the quenching of protein fluorescence by the DNS group, and the release of the DNS-labeled product could be followed by the release from quenching of the DNS fluorescence by the cobalt absorption band.

Pepsin active-site topology was studied by Badley and Teale [443], who analyzed the energy transfer from tryptophans to substrate-analog diazoketones. The unexpectedly high degree of energy transfer they observed indicated a clustering of tryptophan residues near the active center, rather than a random distribution over the whole protein. In another study Badley and Teale [444] studied the energy transfer manifested by the quenching of pepsin fluorescence in complexes of pepsin and heme protein substrates. Deductions were made about the relation of the point of contact of the proteins and the region of tryptophans in pepsin.

It should be remembered that energy transfer can almost always be demonstrated when dyes are attached to proteins. Although the dyes may have been put there for other reasons, analysis of the degree of quenching of the intrinsic fluorescence or the amount of sensitized dye fluorescence will give ancillary information on distances between tryptophans and dye.

Other types of energy transfer are possible aside from the singlet-singlet transfer so far discussed. Triplet-singlet transfer also occurs by a resonance mechanism, but the donor molecule is in the triplet, or prephosphorescent, state. The manifestation of energy transfer is thus a sensitized, delayed fluorescence. Such a phenomenon was observed by Galley and Stryer [445] in chymotrypsin to which proflavin was attached at the active site. Tryptophan phosphorescence was quenched, and delayed fluorescence of proflavin appeared.

Triplet-triplet energy transfer has been observed in biochemical systems (see, for example, Refs. [446] and [447]). In this type of transfer the donor and acceptor molecules must be very close (less than about 10-12 $\overset{o}{A}$).

For observation of triplet-singlet or triplet-triplet transfer the experiments should be performed at liquid-nitrogen temperatures. Although one could argue that this has little relevance to physiological temperatures, the spatial relations elucidated by the energy transfer should be as valid at low temperatures as they are at 37°C.

V. CONCLUSIONS

Fluorescence spectroscopy of proteins has become one of the standard techniques in physical biochemistry. The number of studies cited in this review is much larger than that in a review [2] written 5 years ago. Indeed very many more studies could have been cited here, except for limitations of space. The increased use of fluorometry can be ascribed to a better understanding of its advantages and limitations; advances in, and increased availability of, instrumentation; and the development of new methods and techniques. Further advances are taking place at a rapid rate, and it is clear that this exciting field will continue to command much interest.

REFERENCES

1.   E. R. Blout, Biopolymers Symp. , 1, 397 (1964).

2.   R. F. Chen, in Fluorescence. Theory, Instrumentation, and Practice (G. G. Guilbault, ed.), Dekker, New York, 1967, p. 443.

3.   G. Weber, Biochem. J. , 51, 145 (1952).

4.   F. Perrin, J. Phys. , 7, 390 (1926).

5.   F. Perrin, Ann. Phys. , 12, 169 (1929).

6.   V. L. Levshin, Z. Physik, 26, 274 (1924).

7.   G. Weber, Adv. Protein Chem. , 8, 415 (1953).

8.   G. Weber, Biochem. J. , 51, 155 (1952).

9.   J. L. Oncley, Ann. N. Y. Acad. Sci. , 41, 121 (1941).

10.   R. F. Steiner and A. McAlister, J. Polymer Sci. , 24, 105 (1957).

11.   J. F. Foster, in The Plasma Proteins (F. W. Putman, ed.), Academic Press, New York, 1960.

12.   W. Harrington, P. Johnson, and R. Ottewill, Biochem. J. , 63, 349 (1956).

13.   C. S. Chadwick and P. Johnson, Biochim. Biophys. Acta, 53, 482 (1961).

14.   R. F. Steiner and A. McAlister, J. Colloid Sci. , 12, 80 (1957).

15.   A. Mendel, J. Chem. Eng. Data, 15, 340 (1970).

16.   H. Rinderknecht, Experientia, 16, 430 (1960).

17.   C. Gros and B. Labouesse, Eur. J. Biochem. , 7, 463 (1969).

18.   T. J. Gill, III, E. M. McLaughlin, and G. S. Omenn, Biopolymers, 5, 297 (1967).

19. I. M. Klotz, E. C. Stellwagen, and V. H. Stryker, Biochim. Biophys. Acta, 86, 122 (1964).

20. T. J. Gill, III, Biopolymers, 3, 43 (1965).

21. J. E. Churchich, Biochim. Biophys. Acta, 147, 511 (1967).

22. A. H. Coons, H. J. Creech, R. N. Jones, and E. Berliner, J. Immunol., 34, 159 (1942).

23. J. L. Riggs, R. J. Seiwald, J. H. Burckhalter, C. M. Downs, and T. Metcalf, Amer. J. Pathol., 34, 1081 (1958).

24. W. B. Cherry, R. M. McKinney, V. M. Emmel, J. T. Spillane, G. A. Hebert, and B. Pittman, Stain Technol., 44, 179 (1969).

25. P. A. Hansen, Acta Histochem., Suppl. VII, 167 (1967).

26. J. Knopp and G. Weber, J. Biol. Chem., 242, 1353 (1967).

27. J. A. Knopp and G. Weber, J. Biol. Chem., 244, 6309 (1969).

28. A. B. Rawitch, E. Hudson, and G. Weber, J. Biol. Chem., 244, 6543 (1969).

29. H. Metzger, R. L. Perlman, and H. Edelhoch, J. Biol. Chem., 241, 1741 (1966).

30. A. H. Maddy, Biochim. Biophys. Acta, 88, 390 (1964).

31. R. F. Chen, Arch. Biochem. Biophys., 133, 263 (1969).

32. Y. Kanaoka, M. Machida, K. Ando, and T. Sekine, Biochim. Biophys. Acta, 207, 269 (1970).

33. P. B. Ghosh and M. W. Whitehouse, Biochem. J., 108, 155 (1968).

34. R. P. Cory, R. R. Becker, R. Rosenbluth, and I. Isenberg. J. Amer. Chem. Soc., 90, 1643 (1968).

35. J. R. Heitz and B. M. Anderson, Arch. Biochem. Biophys., 127, 637 (1968).

36. J. E. Churchich, Biochim. Biophys. Acta, 178, 480 (1969).

37. R. Irwin and J. E. Churchich, J. Biol. Chem., 246, 5429 (1971).

38. D. J. R. Laurence, Biochem. J., 51, 168 (1952).

39. R. F. Steiner, Arch. Biochem. Biophys., 49, 71 (1954).

40. V. Massey, W. F. Harrington, and B. S. Hartley, Discussions Faraday Soc., 20, 24 (1955).

41. E. Haber and C. S. Bennett, Proc. Natl. Acad. Sci. U.S., 48, 1935, (1962).

42. W. B. Dandliker and G. A. Feigen, Biochem. Biophys. Res. Commun., 5, 299 (1961).

43. W. B. Dandliker, S. P. Halbert, M. C. Florin, R. Alonso, and H. C. Schapiro, J. Exptl. Med., 122, 1029 (1965).

44. W. B. Dandliker, H. C. Schapiro, J. W. Meduski, R. Alonso, G. A. Feigen, and J. R. Hamrick, Jr., Immunochemistry, 1, 165 (1964).

45. N. M. Green, Biochem. J., 89, 609 (1963).

46. J. M. Brewer and G. Weber, J. Biol. Chem., 241, 2550 (1966).

47. J. M. Brewer and T. E. Spencer, Biochem. Biophys. Res. Commun., 31, 960 (1968).

48. A. Jonas and G. Weber, Biochemistry, 9, 4729 (1970).

49. A. Jonas and G. Weber, Biochemistry, 9, 5092 (1970).

50. L. Stryer and O. H. Griffith, Proc. Natl. Acad. Sci. U.S., 54, 1785 (1965).

51. P. Johnson and E. G. Richards, Arch. Biochem. Biophys., 97, 260 (1962).

52. F. H. Chowdhury and P. Johnson, Biochem. Biophys. Acta, 66, 218 (1963).

53. R. F. Steiner and H. Edelhoch, J. Amer. Chem. Soc., 84, 2139 (1962).

54. M. Winkler, Biochim. Biophys. Acta, 102, 459 (1965).

55. P. Johnson and V. Massey, Biochim. Biophys. Acta, 23, 544 (1957).

56. H. Edelhoch and R. F. Steiner, Biopolymers, 4, 999 (1966).

57. P. Johnson and E. Mihalyi, Biochim. Biophys. Acta, 102, 476 (1965).

58. N. M. Green, Biochem. J., 92, 160 (1964).

59. M. Frey, P. Wahl, and H. Benoit, J. Chem. Phys., 61, 1005 (1964).

60. G. S. Omenn and T. J. Gill, III, J. Biol. Chem., 241, 4899 (1966).

61. Y. Y. Gottlieb and P. Wahl, J. Chem. Phys., 60, 840 (1963).

62. F. W. J. Teale and R. A. Badley, Biochem. J., 116, 341 (1970).

63. T. J. Gill, III, J. Amer. Chem. Soc., 87, 4188 (1965).

64. F. H. White, Jr., and C. B. Anfinsen, Ann. N.Y. Acad. Sci., 81, 515 (1959).

65. R. F. Steiner, Biochim. Biophys. Acta, 100, 111 (1965).

66. J. E. Churchich, Arch. Biochem. Biophys., 97, 574 (1962).

67. V. Frattali, R. F. Steiner, D. Millar, and H. Edelhoch, Nature, 199, 1186 (1963).

68. V. Frattali, R. F. Steiner, and H. Edelhoch, J. Biol. Chem., 240, 112 (1965).

69. M. H. Winkler, M. B. Goldman, and E. A. Sweeney, Biochem. Biophys. Acta, 112, 559 (1966).

70. D. M. Young and J. T. Potts, J. Biol. Chem., 238, 1995 (1963).

71. F. H. White, Jr., J. Biol. Chem., 239, 1032 (1964).

72. H. Edelhoch and B. de Crumbrugghe, J. Biol. Chem., 241, 4357 (1966).

73. H. Edelhoch and H. G. Burger, J. Biol. Chem., 241, 458 (1966).

74. H. Edelhoch and R. F. Steiner, in Electronic Aspects of Biochemistry (B. Pullman, ed.), Academic Press, New York, 1964, p. 7.

75. E. V. Anufrieva, M. V. Volkenstein, M. G. Krakoviak, and T. V. Sheveleva, Dokl. Akad. Nauk. USSR, 182, 361 (1968).

76. J. K. Weltman and G. M. Edelman, Biochemistry, 6, 1437 (1967).

77. P. Wahl and G. Weber, J. Mol. Biol., 30, 371 (1967).

78. M. Fayet and P. Wahl, Biochim. Biophys. Acta, 181, 373 (1969).

79. Yu. A. Zagyanski, R. S. Nezlin, and L. A. Tumerman, Immuno-chemistry, 6, 787 (1969).

80. R. S. Nezlin, Yu. A. Zagyansky, and L. A. Tumerman, J. Mol. Biol., 50, 569 (1970).

81. P. Wahl, Biochim. Biophys. Acta, 175, 55 (1969).

82. J. K. Weltman and R. P. Davis, J. Mol. Biol., 54, 177 (1970).

83. R. P. Tengerdy, J. Lab. Clin. Med., 70, 707 (1967).

84. R. P. Tengerdy, J. Immunol., 99, 126 (1967).

85. S. A. Levison, A. N. Jancsi, and W. B. Dandliker, Biochem. Biophys. Res. Commun., 33, 942 (1968).

86. S. A. Levison, A. J. Portmann, F. Kierszenbaum, and W. B. Dandliker, Biochem. Biophys. Res. Commun., 43, 258 (1971).

87. A. J. Portman, S. A. Levison, and W. B. Dandliker, Biochem. Biophys. Res. Commun., 43, 207 (1971).

88. F. H. White, Jr., Federation Proc., 23, 692 (1964).

89. B. S. Hartley and V. Massey, Biochim. Biophys. Acta, 21, 58 (1956).

90. M. Kasuya and H. Takashina, Biochim. Biophys. Acta, 99, 452 (1965).

91. H. Takashina, Biochim. Biophys. Acta, 200, 319 (1970).

92. R. D. Hill and R. R. Laing, Nature, 210, 1160 (1966).

93. R. D. Hill and R. R. Laing, Biochim. Biophys. Acta, 132, 188 (1967).

94. W. Rickert, Biochim. Biophys. Acta, 220, 628 (1970).

95.   R. F. Chen, Biochem. Biophys. Res. Commun. , 40, 1117 (1970).

96.   W. W. Bromer, S. K. Sheehan, A. W. Berns, and E. R. Arquilla, Biochemistry, 6, 2378 (1967).

97.   E. Mihalyi and A. Albert, Biochemistry, 10, 237 (1971).

98.   E. Mihalyi and A. Albert, Biochemistry, 10, 243 (1971).

99.   R. F. Chen, Arch. Biochem. Biophys. , 128, 163 (1968).

100.   M. Frey and P. Wahl, Compt. Rend. , 262, 2653 (1966).

101.   M. Laskowski, Jr. , Federation Proc. , 25, 20 (1966).

102.   P. Wahl and H. Lami, Biochim. Biophys. Acta, 133, 233 (1967).

103.   R. F. Chen, Arch. Biochem. Biophys. , 120, 609 (1967).

104.   J. L. Speyer and M. H. Winkler, Biochim. Biophys. Acta, 188, 345 (1969).

105.   B. Witholt and L. Brand, Biochemistry, 9, 1948 (1970).

106.   G. Weber and S. R. Anderson, Biochemistry, 8, 361 (1969).

107.   S. R. Anderson and G. Weber, Biochemistry, 8, 371 (1969).

108.   G. Laustriat, A. Coche, H. Lami, and G. Pfeffer, Compt. Rend. , 247, 434 (1963).

109.   T. Tao, Biopolymers, 8, 609 (1969).

110.   L. Stryer, Science, 162, 526 (1968).

111.   P. Wahl, Compt. Rend. , 263, 1525 (1966).

112.   Y. Yguerabide, H. F. Epstein, and L. Stryer, J. Mol. Biol. , 51, 573 (1970).

113.   W. Ware, S. K. Lee, G. J. Brandt, and P. P. Chow, J. Chem. Phys. , 54, 4729 (1971).

114.   L. Brand and J. R. Gohlke, J. Biol. Chem. , 246, 2317 (1971).

115.   M. DeLuca, L. Brand, T. A. Cebula, H. H. Seliger, and A. F. Makula, J. Biol. Chem. , 246, 6702 (1971).

116.   G. M. Edelman and W. O. McClure, Accounts Chem. Res. , 1, 65 (1968).

117.   M. Burr and D. E. Koshland, Jr. , Proc. Natl. Acad. Sci. U. S. , 52, 1017 (1964).

118.   E. Lippert, Z. Elektrochem. , 61, 962 (1957).

119.   R. F. Steiner and H. Edelhoch, J. Amer. Chem. Soc. , 83, 1435 (1961).

120.   R. F. Steiner and H. Edelhoch, Chem. Rev. , 62, 457 (1962).

121.   R. F. Steiner and H. Edelhoch, J. Biol. Chem., 240, 2877 (1965).

122.   H. Edelhoch, P. G. Condliffe, R. E. Lippoldt, and H. G. Burger, J. Biol. Chem., 241, 5205 (1966).

123.   G. Weber and D. J. R. Laurence, Biochem. J., 51, xxi (1954).

124.   P. Callaghan and N. H. Martin, Biochem. J., 83, 144 (1962).

125.   B. Alexander and G. M. Edelman, Federation Proc., 24, 413 (1965).

126.   W. O. McClure and G. M. Edelman, Biochemistry, 5, 1908 (1965).

127.   V. H. Rees, J. E. Fildes, and D. J. R. Laurence, J. Clin. Pathol., 7, 336 (1954).

128.   T. Förster, Naturwissenschaften, 33, 220 (1946).

129.   L. Stryer, J. Mol. Biol., 13, 482 (1965).

130.   D. C. Turner and L. Brand, Biochemistry, 7, 3381 (1968).

131.   S. Ainsworth and M. T. Flanagan, Biochim. Biophys. Acta, 194, 213 (1969).

132.   M. H. Winkler, Biophys. J., 7, 719 (1967).

133.   C. J. Seliskar and L. Brand, Science, 171, 799 (1971).

134.   A. Camerman and L. H. Jensen, Science, 165, 493 (1969).

135.   A. Camerman and L. H. Jensen, J. Amer. Chem. Soc., 92, 4200 (1970).

136.   G. R. Shepherd and B. J. Noland, Anal. Biochem., 11, 443 (1965).

137.   D. J. R. Laurence, Biochem. J., 99, 419 (1966).

138.   R. F. Chen, J. Biol. Chem., 242, 173 (1967).

139.   J. A. Gally and G. M. Edelman, Biochim. Biophys. Acta, 94, 175 (1965).

140.   N. M. Green, Biochem. J., 94, 23C (1965).

141.   J. A. Duke, R. McKay, J. Botts, and M. F. Morales, Biochim. Biophys. Acta, 126, 600 (1966).

142.   S. T. Lim and J. Botts, Arch. Biochem. Biophys., 122, 153 (1967).

143.   H. C. Cheung and M. F. Morales, Biochemistry, 8, 2177 (1969).

144.   S. R. Anderson and G. Weber, Arch. Biochem. Biophys., 116, 207 (1966).

145.   G. Weber and L. B. Young, J. Biol. Chem., 239, 1424 (1964).

146.   A. Jonas and G. Weber, Biochemistry, 10, 4492 (1971).

147. M. -C. Hsu and R. W. Woody, Biopolymers, 9, 1421 (1971).

148. E. Daniel and G. Weber, Biochemistry, 5, 1893 (1966).

149. E. Smekal and J. Sprindrich, Studia Biophys. (Berlin), 2, 133 (1967).

150. S. R. Anderson, M. Brunori, and G. Weber, Biochemistry, 9, 4723 (1970).

151. L. Brand, J. R. Gohlke, and D. S. Rao, Biochemistry, 6, 3510 (1967).

152. F. M. Dickinson, FEBS Letters, 15, 17 (1971).

153. J. Lynn and G. D. Fasman, Biochem. Biophys. Res. Commun., 33, 327 (1968).

154. M. H. Winkler, J. Mol. Biol., 4, 118 (1962).

155. W. O. McClure and G. M. Edelman, Biochemistry, 6, 559 (1967).

156. W. O. McClure and G. M. Edelman, Biochemistry, 6, 567 (1967).

157. K. Brand, FEBS Letters, 7, 235 (1970).

158. E. Holler, E. L. Bennett, and M. Calvin, Biochem. Biophys. Res. Commun., 45, 409 (1971).

159. S. T. Christian and R. Janetzko, Arch. Biochem. Biophys., 145, 169 (1971).

160. R. Lovrien and W. Anderson, Arch. Biochem. Biophys., 131, 139 (1969).

161. S. Wasi and T. Hofmann, Biochem. J., 106, 926 (1968).

162. C. J. Hart, R. B. Leslie, and A. M. Scanu, Chem. Phys. Lipids, 4, 367 (1970).

163. C. M. Himel, R. T. Mayer, and L. L. Cook, J. Polymer Sci., A-I, 8, 2219 (1970).

164. C. M. Himel, W. G. Aboul-Saad, and S. Uk, J. Agr. Food Chem., 19, 1175 (1971).

165. R. T. Mayer and C. M. Himel, Biochemistry, 11, 2082 (1972).

166. R. F. Chen and J. C. Kernohan, J. Biol. Chem., 242, 5813 (1967).

167. M. Cortijo and S. Shaltiel, Biochem. Biophys. Res. Commun., 39, 212 (1970).

168. M. Cortijo, I. Z. Steinberg, and S. Shaltiel, J. Biol. Chem., 246, 933 (1971).

169. J. E. Churchich and J. G. Farrelly, J. Biol. Chem., 244, 72 (1969).

170. J. E. Churchich and L. Harpring, Biochim. Biophys. Acta, 105, 574 (1965).

171. M. Arrio-Dupont, Biochem. Biophys. Res. Commun., 44, 653 (1971).

172.   R. H. Conrad, J. R. Heitz, and L. Brand, Biochemistry, 9, 1540 (1970).

173.   O. Takenaka, Y. Nishimura, A. Takenaka, and K. Shibata, Biochim. Biophys. Acta, 207, 1 (1970).

174.   O. Takenaka and K. Shibata, J. Biochem., 66, 805 (1969).

175.   N. Lasser and J. Feitelson, Biochemistry, 10, 307 (1971).

176.   S. F. Velick, J. Biol. Chem., 233, 1455 (1958).

177.   R. L. Perlman and J. Wolff, Science, 160, 317 (1968).

178.   K. McCarthy, W. Lovenberg, and A. Sjoerdsma, J. Biol. Chem., 243, 2754 (1968).

179.   B. Suh and R. Barker, J. Biol. Chem., 246, 7041 (1971).

180.   J. H. Wittorf and C. J. Gubler, Ehr. J. Biochem., 14, 53 (1970).

181.   C. R. Creveling, A. Bhaduri, A. Christensen, and H. M. Kalckar, Biochem. Biophys. Res. Commun., 21, 624 (1965).

182.   V. Massey and B. Curti, J. Biol. Chem., 241, 3417 (1966).

183.   A. M. Stein, B. Wold, and J. H. Stein, Biochemistry, 4, 1500 (1965).

184.   D. A. Deranleau and H. Neurath, Biochemistry, 5, 1413 (1966).

185.   C. F. Chignell, Mol. Pharmacol., 6, 1 (1970).

186.   C. F. Chignell, Mol. Pharmacol., 5, 244 (1969).

187.   C. F. Chignell, Mol. Pharmacol., 5, 455 (1969).

188.   R. F. Chen, Abstr. 162nd Natl. Amer. Chem. Soc. Meeting, September 1971.

189.   T. Takagi, Y. Nakanishi, N. Okabe, and T. Isemura, Biopolymers, 5, 627 (1967).

190.   R. A. Kenner and H. Neurath, Biochemistry, 10, 551 (1971).

191.   R. A. Kenner, Biochemistry, 10, 545 (1971).

192.   V. W. Burns, Biochem. Biophys. Res. Commun., 37, 1008 (1969).

193.   V. W. Burns, Exptl. Cell Res., 64, 35 (1971).

194.   M. Shinitzky, A. -C. Dianoux, C. Gitler, and G. Weber, Biochemistry, 10, 2106 (1971).

195.   C. Frieden, in The Enzymes (P. D. Boyer, H. A. Lardy, and K. Myrback, ed.), 2nd ed., Vol. 7, Academic Press, New York, 1963, p. 7.

196.   H. Eisenberg and G. M. Tomkins, J. Mol. Biol., 31, 37 (1968).

197. E. Apella and G. M. Tomkins, J. Mol. Biol., 18, 77 (1966).

198. R. F. Colman and C. Frieden, J. Biol. Chem., 241, 3652 (1966).

199. G. diPrisco, Biochem. Biophys. Res. Commun., 26, 148 (1967).

200. A. D. B. Malcolm and G. K. Radda, Eur. J. Biochem., 15, 555 (1970).

201. K. S. Rogers and S. C. Yusko, J. Biol. Chem., 244, 6690 (1969).

202. W. Thompson and K. L. Yielding, Arch. Biochem. Biophys., 126, 399 (1968).

203. G. H. Dodd and G. K. Radda, Biochem. Biophys. Res. Commun., 27, 500 (1967).

204. G. H. Dodd and G. K. Radda, Biochem. J., 144, 407 (1969).

205. K. L. Yielding, Biochem. Biophys. Res. Commun., 29, 424 (1967).

206. N. C. Price and G. K. Radda, Biochem. J., 114, 419 (1969).

207. J. E. Churchich, Biochim. Biophys. Acta, 147, 32 (1967).

208. C. W. Parker, T. J. Yoo, M. C. Johnson, and S. M. Godt, Biochemistry, 6, 3408 (1967).

209. C. W. Parker, S. M. Godt, and M. C. Johnson, Biochemistry, 6, 3417 (1967).

210. D. E. Lopatin and E. W. Voss, Jr., Biochemistry, 10, 208 (1971).

211. I. Pecht, E. Maron, R. Arnon, and M. Sela, Eur. J. Biochem., 19, 368 (1971).

212. R. E. Cathou and T. C. Werner, Biochemistry, 9, 3149 (1970).

213. T. J. Yoo, H. Nakamura, A. L. Grosberg, and D. Pressman, Immunochemistry, 7, 627 (1970).

214. H. Nakamura, T. J. Yoo, A. L. Grossberg, and D. Pressman, Immunochemistry, 7, 637 (1970).

215. T. J. Yoo, O. A. Roholt, and D. Pressman, Science, 157, 707 (1967).

216. C. H. Evans, S. B. Herron, and G. Goldstein, J. Immunol., 101, 915 (1968).

217. A. Azzi, B. Chance, G. K. Radda, and C. P. Lee, Proc. Natl. Acad. Sci. U.S., 62, 612 (1969); B. Chance, A. Azzi, L. Mela, G. Radda, and H. Vainio, FEBS Letters, 3, 10 (1969).

218. A. Azzi, P. Gherardini, and M. Santato, J. Biol. Chem., 246, 2035 (1971).

219. L. Packer, M. P. Donovan, and J. M. Wrigglesworth, Biochim. Biophys. Acta, 35, 832 (1969).

220.   D. F. H. Wallach, E. Ferber, D. Selin, E. Weidekamm, and H. Fischer, Biochim. Biophys. Acta, 203, 67 (1970).

221.   R. B. Freedman and G. K. Radda, FEBS Letters, 3, 150 (1969).

222.   M. B. Feinstein, L. Spero, and H. Felsenfeld, FEBS Letters, 6, 245 (1970).

223.   B. Rubalcava, D. M. de Munoz, and C. Gitler, Biochemistry, 8, 2742 (1969).

224.   C. Gitler, B. Rubalcava, and A. Caswell, Biochim. Biophy. Acta, 193, 479 (1969).

225.   H. Bornet and H. Edelhoch, J. Biol. Chem., 246, 1785 (1971).

226.   R. A. Muesing and T. Nishida, Biochemistry, 10, 2952 (1971).

227.   J. Vanderkooi and A. Martonosi, Arch. Biochem. Biophys., 133, 153 (1969).

228.   D. Romeo, R. Cramer, and F. Rossi, Biochem. Biophys. Res. Commun., 41, 582 (1970).

229.   P. Wahl, M. Kasai, and J.-P. Changeux, Eur. J. Biochem., 18, 332 (1971).

230.   J. R. Brocklehurst, R. B. Freedman, D. J. Hancock, and G. K. Radda, Biochem. J., 116, 721 (1970).

231.   I. Tasaki, L. Carnay, and A. Watanabe, Proc. Natl. Acad. Sci. U. S., 64, 1362 (1969).

232.   L. D. Carnay and W. H. Barry, Science, 165, 608 (1969).

233.   I. Tasaki, A. Watanabe, and M. Hallett, Proc. Natl. Acad. Sci. U. S., 68, 938 (1971).

234.   I. Tasaki, Adv. Biol. Med. Phys., 13, 307 (1970).

235.   K. Nagai, G. E. Lindenmayer, and A. Schwartz, Arch. Biochem. Biophys., 139, 252 (1970).

236.   S. Udenfriend, Fluorescence Assay in Biology and Medicine, Vol. 1, Academic Press, New York, 1962; also Vol. 2, 1969.

237.   S. V. Konev, Excited Electronic States of Biopolymers (in Russian), Nauka i Tekhnika, Minsk, 1965.

238.   V. G. Shore and A. B. Pardee, Arch. Biochem. Biophys., 60, 100 (1956).

239.   D. E. Duggan and S. Udenfriend, J. Biol. Chem., 223, 313 (1956).

240.   S. V. Konev, Dokl. Akad. Nauk SSSR, 116, 594 (1957).

241.   F. W. J. Teale and G. Weber, Biochem. J., 65, 476 (1957).

242.   F. W. J. Teale, Biochem. J., 76, 381 (1960).

243.   R. F. Chen, Anal. Letters, 1, 35 (1967).

244.   H. C. Börreson, Acta Chem. Scand., 21, 920 (1967).

245.   J. Eisinger, Photochem. Photobiol., 9, 247 (1969).

246.   F. Bishai, E. Kuntz, and L. Augenstein, Biochim. Biophys. Acta, 140, 381 (1967).

247.   E. Leroy, H. Lami, and G. Laustriat, Photochem. Photobiol., 13, 411 (1971).

248.   G. Weber, Nature, 190, 27 (1961).

249.   G. Weber, Biochem. J., 79, 29P (1961).

250.   Yu. A. Vladimirov and G. M. Zimina, Biokhimiya, 30, 1105 (1965).

251.   C. H. Suelter, Biochemistry, 6, 418 (1967).

252.   R. F. Chen, Anal. Biochem., 14, 497 (1966).

253.   R. F. Chen, Anal. Biochem., 20, 339 (1967).

254.   J. H. Baxter, Arch. Biochem. Biophys., 108, 375 (1964).

255.   S. S. Lehrer and G. D. Fasman, Biochem. Biophys. Res. Commun., 23, 133 (1966).

256.   A. White, Biochem. J., 71, 217 (1959).

257.   G. Weber and K. Rosenheck, Biopolymers Symp., 1, 333, (1964).

258.   J. Feitelson, J. Phys. Chem., 68, 391 (1964).

259.   R. F. Chen, G. G. Vurek, and N. Alexander, Science, 156, 949 (1967).

260.   M. Fayet and P. Wahl, Biochim. Biophys. Acta, 229, 102 (1971).

261.   H. Edelhoch, R. L. Perlman, and M. Wilcheck, Biochemistry, 7, 3893 (1968).

262.   H. Edelhoch, R. S. Bernstein, and M. Wilchek, J. Biol. Chem., 243, 5985 (1968).

263.   J. Feitelson, Photochem. Photobiol., 9, 401 (1969).

264.   G. Weber, Biochem. J., 75, 345 (1960).

265.   A. Wada and Y. Ueno, Biopolymers Symp., 1, 343 (1964).

266.   T. Cassen and D. R. Kearns, Biochim. Biophys. Acta, 194, 203 (1969).

267.   J. Eisinger, Biochemistry, 8, 3902 (1969).

268.   J. W. Longworth, Photochem. Photobiol., 7, 587 (1968).

269.   I. Z. Steinberg, Ann. Rev. Biochem., 40, 83 (1971).

270.   J. W. Longworth and R. O. Rahn, Biochim. Biophys. Acta, 147, 526 (1967).

271. J. J. ten Bosch, R. O. Rahn, J. W. Longworth, and R. G. Shulman, Proc. Natl. Acad. Sci. , U.S. , 59, 1003 (1968).

272. J. J. ten Bosch and J. A. Knopp, Biochim. Biophys. Acta, 188, 173 (1969).

273. G. D. Fasman, K. Norland, and A. Pasce, Biopolymers Symp. , 1, 325 (1964).

274. A. Pesce, E. Bodenheimer, K. Norland, and G. D. Fasman, J. Amer. Chem. Soc. , 86, 5669 (1964).

275. G. D. Fasman, E. Bodenheimer, and A. Pesce, J. Biol. Chem. , 241, 916 (1966).

276. S. S. Lehrer and G. D. Fasman, Biopolymers, 2, 149 (1964).

277. S. S. Lehrer and G. D. Fasman, J. Amer. Chem. Soc. , 87, 4687 (1965).

278. S. S. Lehrer and G. D. Fasman, Biochemistry, 6, 757 (1967).

279. A. Y. Moon, D. C. Poland, and H. D. Scheraga, J. Phys. Chem. , 69, 2960 (1965).

280. R. W. Cowgill, Biochim. Biophys. Acta, 75, 262 (1963).

281. R. W. Cowgill, Arch. Biochem. Biophys. , 100, 36 (1963).

282. R. W. Cowgill, Biochim. Biophys. Acta, 94, 81 (1965).

283. R. W. Cowgill, Biochim. Biophys. Acta, 94, 74 (1965).

284. R. W. Cowgill, Biochim. Biophys. Acta, 109, 536 (1965).

285. R. W. Cowgill, Biochim. Biophys. Res. Commun. , 16, 332 (1964).

286. R. W. Cowgill, Biochim. Biophys. Acta, 112, 550 (1966).

287. E. C. Russell and R. W. Cowgill, Biochim. Biophys. Acta, 154, 231 (1968).

288. R. W. Cowgill, Biochim. Biophys. Acta, 140, 37 (1967).

289. S. Arian, M. Benjamini, J. Feitelson, and G. Stein, Photochem. Photobiol. , 12, 481 (1970).

290. R. F. Chen, H. Edelhoch, and R. F. Steiner, Physical Principles and Techniques of Protein Chemistry, Part A, Academic Press, New York, 1969, p. 171.

291. R. W. Cowgill, in Molecular Luminescence (E. C. Lim, ed.), Benjamin, New York, 1969, p. 589.

292. F. W. J. Teale, Biochem. J. , 79, 14P (1961).

293. J. A. Gally and G. M. Edelman, Biopolymers Symp. , 1, 367 (1961).

294. R. F. Chen and P. F. Cohen, Arch. Biochem. Biophys. , 114, 514 (1966).

295.   R. W. Cowgill, Biochim. Biophys. Acta, 140, 552 (1967).

296.   R. W. Cowgill, Biochim. Biophys. Acta, 168, 417 (1968).

297.   R. W. Cowgill, Biochim. Biophys. Acta, 168, 439 (1968).

298.   R. W. Cowgill and N. K. Lang, Biochim. Biophys. Acta, 214, 228 (1970).

299.   J. W. Donovan, Biochemistry, 6, 3918 (1967).

300.   Z. Deyl, R. Praus, H. Sulcova, and J. N. Goldman, FEBS Letters, 5, 187 (1969).

301.   J. Blomfield and J. F. Farrar, Biochem. Biophys. Res. Commun. , 28, 346 (1967).

302.   L. Augenstein and J. Nag-Chaudhuri, Nature, 203, 1145 (1964).

303.   M. W. Grimes, D. R. Graber, and A. Haug, Biochem. Biophys. Res. Commun. , 37, 853 (1969).

304.   R. F. Steiner and R. Kolinski, Biochemistry, 7, 1014 (1968).

305.   H. Rau and L. Augenstein, J. Chem. Phys. , 46, 1773 (1967).

306.   M. J. Kronman and L. G. Holmes, Photochem. Photobiol. , 14, 113 (1971).

307.   G. Weber, in Light and Life (W. D. McElroy and B. Glass, eds.), Johns Hopkins Press, Baltimore, 1961.

308.   L. Stryer, J. Amer. Chem. Soc. , 88, 5708 (1966).

309.   T. Förster and K. Rokos, Chem. Phys. Letters, 1, 279 (1967).

310.   R. W. Ricci, Photochem. Photobiol. , 12, 67 (1970).

311.   M. S. Walker, T. W. Bednar, and R. Lumry, in Molecular Luminescence (E. C. Lim, ed.), Benjamin, New York, 1969, p. 135.

312.   M. S. Walker, T. W. Badnar, and R. Lumry, Photochem. Photobiol. , in press.

313.   J. A. Gally and G. M. Edelman, Biochim. Biophys. Acta, 60, 499 (1962).

314.   J. Green, cited in Ref. [236], p. 106.

315.   R. F. Steiner and H. Edelhoch, Biochim. Biophys. Acta, 66, 341 (1963).

316.   R. W. Cowgill, Biochim. Biophys. Acta, 133, 6 (1967).

317.   M. J. Kronman, Biochim. Biophys. Acta, 133, 19 (1967).

318.   R. W. Cowgill, Biochim. Biophys. Acta, 168, 431 (1968).

319.   R. W. Cowgill, Biochim. Biophys. Acta, 200, 18 (1970).

320.   I. Weinryb and R. F. Steiner, Biochemistry, 7, 2488 (1968).

321. H. Edelhoch, L. Brand, and M. Wilchek, Israel J. Chem. , 1, 216 (1963).

322. H. Edelhoch, L. Brand, and M. Wilchek, Biochemistry, 6, 547 (1967).

323. T. Förster, Fluoreszenz organischer Verbindungen, Vandenhoeck u. Rupprecht, Gottingen, 1951.

324. R. W. Cowgill, Biochim. Biophys. Acta, 207, 556 (1970).

325. L. Stryer, A. Holmgren, and P. Reichard, Biochemistry, 6, 1016 (1967).

326. M. Shinitzky and R. Goldman, Eur. J. Biochem. , 3, 139 (1967).

327. M. Shinitzky and M. Fridkin, Eur. J. Biochem. , 9, 176 (1969).

328. Yu. A. Vladimirov and C. -K. Li, Biofizika, 7, 270 (1962).

329. J. W. Longworth, Biopolymers, 4, 1131 (1966).

330. B. L. Van Duuren, Org. Chem. , 26, 2945 (1961).

331. N. Mataga, Y. Torihashi, and K. Ezumi, Theor. Chem. Acta, 2, 158 (1964).

332. S. V. Konev, V. P. Bobrovich, and Y. Chernitskii, Biofizika, 10, 42 (1965).

333. M. S. Walker, T. W. Bednar, and R. Lumry, J. Chem. Phys. , 45, 3455 (1966).

334. M. S. Walker, T. W. Bednar, and R. Lumry, J. Chem. Phys. , 47, 1020 (1967).

335. J. B. Birks, Nature, 214, 1187 (1967).

336. J. W. Longworth and M. Del Carmen Battista, Photochem. Photobiol. , 12, 29 (1970).

337. J. Eisinger and G. Navon, J. Chem. Phys. , 50, 2069 (1969).

338. W. C. Galley and R. M. Purkey, Proc. Natl. Acad. Sci. U.S. , 67, 1116 (1970).

339. J. Feitelson, Photochem. Photobiol. , 13, 87 (1971).

340. G. Weber, Biochem. J. , 75, 335 (1960).

341. H. Zimmermann and N. Joop, Z. Elektrochem. , 65, 61 (1961).

342. R. F. Chen, Anal. Biochem. , 19, 374 (1967).

343. C. A. Parker and W. T. Rees, Analyst, 85, 587 (1960); P. Pringsheim, Fluorescence and Phosphorescence, Wiley, New York, 1949.

344. G. M. Barenboim, Biofizika, 8, 154 (1963).

345. R. F. Steiner and H. Edelhoch, Nature, 193, 375 (1962).

346.   L. Brand, J. Everse, and N. O. Kaplan, Biochemistry, 1, 423 (1962).

347.   J. F. Gerstein, H. Van Vunakis, and L. Levine, Biochemistry, 2, 964 (1963).

348.   H. Edelhoch, V. Frattali, and R. F. Steiner, J. Biol. Chem., 240, 122 (1965).

349.   G. E. Perlmann, Biopolymers Symp., 1, 383 (1964).

350.   H. Edelhoch and R. F. Steiner, J. Biol. Chem., 240, 2877 (1965).

351.   R. F. Steiner and H. Edelhoch, J. Biol. Chem., 238, 925 (1963).

352.   H. Edelhoch and R. F. Steiner, J. Biol. Chem., 238, 931 (1963).

353.   J. W. Longworth, Photochem. Photobiol., 8, 589 (1968).

354.   M. Irie, J. Biochem., 68, 31 (1970).

355.   O. Pongs, Biochemistry, 9, 2316 (1970).

356.   P. Cuatrecasas, H. Edelhoch, and C. B. Anfinsen, Proc. Natl. Acad. Sci. U.S., 58, 2043 (1967).

357.   A. Arnone, C. J. Bier, F. A. Cotton, V. A. Day, E. E. Hazen, D. C. Richardson, J. S. Richardson, and A. Yonath, J. Biol. Chem., 246, 2302 (1971).

358.   A. Schechter, R. F. Chen, and C. B. Anfinsen, Science, 167, 886 (1970).

359.   H. F. Epstein, A. N. Schechter, R. F. Chen, and C. B. Anfinsen, J. Mol. Biol., 60, 499 (1971).

360.   S. Shifrin and E. Steers, Jr., Biochim. Biophys. Acta, 133, 463 (1967).

361.   A. N. Schechter and C. J. Epstein, J. Mol. Biol., 35, 567 (1968).

362.   R. M. Purkey and W. C. Galley, Biochemistry, 9, 3569 (1970).

363.   Y. Elkana, J. Phys. Chem., 72, 3654 (1968).

364.   H. Edelhoch and R. F. Steiner, Biochim. Biophys. Acta, 60, 365 (1962).

365.   J. E. Churchich, Biochim. Biophys, Acta, 92, 194 (1964).

366.   J. E. Churchich, Biochim. Biophys. Acta, 120, 406 (1966).

367.   J. E. Churchich, Biochim. Biophys. Acta, 126, 606 (1966).

368.   M. Shinitzky, V. Grisaro, D. M. Chipman, and N. Sharon, Arch. Biochem. Biophys., 115, 232 (1966).

369.   S. S. Lehrer and G. D. Fasman, J. Biol. Chem., 242, 4644 (1967).

370.   V. I. Teichberg and N. Sharon, FEBS Letters, 7, 171 (1970).

371.   S. F. Wang, F. S. Kawahara, and P. Talalay, J. Biol. Chem., 238, 576 (1963).

372.   M. J Kronman, Biochim. Biophys. Acta, 133, 19 (1967).

373.   M. J. Kronman, L. G. Holmes, and F. M. Robbins, Biochim. Biophys. Acta, 133, 46 (1967).

374.   M. J. Kronman, L. G. Holmes, and F. M. Robbins, J. Biol. Chem., 246, 1909 (1971).

375.   R. F. Chen, Biochim. Biophys. Acta, 120, 169 (1966).

376.   R. F. Steiner and H. Edelhoch, Nature, 193, 375 (1962).

377.   H. Edelhoch and R. E. Lippoldt, J. Biol. Chem., 244, 3876 (1969).

378.   C. H. Blomquist, Arch. Biochem. Biophys., 122, 24 (1967).

379.   W. H. Bannister, C. M. Salisbury, and E. J Wood, Biochim. Biophys. Acta, 168, 392 (1968).

380.   R. F. Chen, Biochem. Biophys. Res. Commun., 17, 141 (1964).

381.   D. Gabel, I. Z. Steinberg, and E. Katchalski, Biochemistry, 10, 4661 (1971).

382.   C. Hélène, F. Brun, and M. Yaniv, Biochem. Biophys. Res. Commun., 37, 393 (1969).

383.   C. Hélène, F. Brun, and M. Yaniv, J. Mol. Biol., 58, 349 (1971).

384.   R. T. Kuczenski and C. H. Suelter, Biochemistry, 10, 2862 (1971).

385.   A. A. Spector and K. M. John, Arch Biochem. Biophys., 127, 65 (1968).

386.   N. A. Attalah and G. F. Lata, Biochim. Biophys. Acta, 168, 321 (1968).

387.   S. S. Lehrer, J. Biol. Chem., 244, 3613 (1969).

388.   N. Shaklai and E. Daniel, Biochemistry, 9, 564 (1970)

389.   W. H. Bannister and E. J. Wood, Comp. Biochem. Physiol., 40B, 7 (1971).

390.   K. S. Rogers, Enzymologia, 36, 153 (1969).

391.   K. S. Rogers, Enzymologia, 37, 174 (1969).

392.   R. F. Chen, manuscript in preparation.

393.   R. F. Chen, Arch. Biochem. Biophys., 142, 552 (1971).

394.   L. K. Ramachandran and B. Witkop, Biochemistry, 3, 1603 (1964).

395.   R. Arnon and E. Shapira, J. Biol. Chem., 244, 1033 (1969).

396.   A. O. Barel and A. N. Glazer, J. Biol. Chem., 244, 268 (1969).

397.   L. A. A. Sluyterman and M. J. M. De Graaf, Biochim, Biophys. Acta, 200, 595 (1970).

398.   I. Weinryb and R. F. Steiner, Biochemistry, 9, 135 (1970).

399.   R. F. Steiner, Biochemistry, 10, 771 (1971).

400.   R. F. Steiner, R. E. Lippoldt, H. Edelhoch, and V. Frattali,
       Biopolymers Symp., 1, 355 (1964).

401.   T. T. Herskovits and M. Laskowski, Jr., J. Biol. Chem., 237, 2481
       (1962).

402.   H. Edelhoch and R. E. Lippoldt, Biochim. Biophys. Acta, 79, 64 (1964).

403.   S. S. Lehrer, Biochemistry, 10, 3254 (1971).

404.   S. S. Lehrer, Biochem. Biophys. Res. Commun., 29, 767 (1967).

405.   C. Vernotte, Compt. Rend., D264, 1892 (1967).

406.   B. Arrio, F. Rodier, C. Boisson, and C. Vernotte, Biochem. Biophys.
       Res. Commun., 39, 589 (1970).

407.   M. H. Winkler, Biochemistry, 8, 2586 (1969).

408.   C. H. Blomquist, J. Biol. Chem., 244, 1605 (1969).

409.   R. L. Perlman, A. van Zyl, and H. Edelhoch, J. Amer. Chem. Soc.,
       90, 2168 (1968).

410.   R. F. Steiner, J. Roth, and J. Robbins, J. Biol. Chem., 241, 560 (1966).

411.   M. Tabachnick, F. J. Downs, and N. A. Giorgio, Jr., Arch. Biochem.
       Biophys., 136, 467 (1970).

412.   J. A. D'Anna, Jr., and G. Tollin, Biochemistry, 10, 57 (1971).

413.   P. L. Luisi and R. Favilla, Eur. J. Biochem., 17, 91 (1970).

414.   L. Montanaro and S. Sperti, Biochem. J., 105, 635 (1967).

415.   G. Wasemiller, A. Abrams, and S. Bakerman, Biochem. Biophys. Res.
       Commun., 30, 178 (1968).

416.   G. Geraci and Q. H. Gibson, J. Biol. Chem., 242, 4275 (1967).

417.   R. W. Noble, M. Reichlin, and Q. H. Gibson, J. Biol. Chem., 244,
       2403 (1969).

418.   T. Wuntch, E. S. Vesell, and R. F. Chen, J. Biol. Chem., 244, 6100
       (1969).

419.   T. Wuntch, R. F. Chen, and E. S. Vesell, Science, 167, 63 (1970).

420.   T. Wuntch, R. F. Chen, and E. S. Vesell, Science, 169, 480 (1970).

421.   H. Gutfreund, R. Cantwell, C. McMurray, R. S. Criddle, and G.
       Hathaway, Biochem. J., 106, 683 (1968).

422.   J. Everse, R. L. Berger, and N. O. Kaplan, Science, 168, 1236 (1970).

423. G. Weber and B. Bablouzian, J. Biol. Chem., 241, 2558 (1966).

424. D. A. Deranleau, Anal. Biochem., 16, 438 (1966).

425. C. -G. Rosen, Acta Chem. Scand., 24, 1849 (1970).

426. J. Lynn and G. D. Fasman, Biopolymers, 6, 159 (1968).

427. G. Karreman, R. H. Steele, and A. Szent-Györgyi, Proc. Natl. Acad. Sci. U.S., 44, 140 (1958).

428. L. Stryer, Rad. Res., Suppl. 2, 432 (1960).

429. A. A. Lamola, Photochem. Photobiol., 8, 601 (1968).

430. R. G. Bennett and R. E. Kellogg, Prog. React. Kinet., 4, 215 (1966).

431. S. A. Latt, H. T. Cheung, and E. R. Blout, J. Amer. Chem. Soc., 85, 998 (1965).

432. A. A. Lamola, P. A. Leermakers, G. W. Byers, and G. S. Hammond, J. Amer. Chem. Soc., 87, 2322 (1965).

433. R. D. Rauh, T. R. Evans, and P. A. Leermakers, J. Amer. Chem. Soc., 90, 6897 (1968).

434. L. Stryer and R. P. Haugland, Proc. Natl. Acad. Sci. U.S., 58, 67 (1967).

435. G. Gabor, Biopolymers, 6, 809 (1968).

436. R. H. Conrad and L. Brand, Biochemistry, 7, 777 (1968).

437. L. Stryer, Biochim. Biophys. Acta, 35, 242 (1959).

438. J. Feitelson, J. Phys. Chem., 44, 1497 (1966).

439. C. R. Cantor and P. Pechukas, Proc. Natl. Acad. Sci. U.S., 68, 2099 (1971).

440. H. Edelhoch, R. L. Perlman, and M. Wilchek, Ann. N.Y. Acad. Sci., 158, 391 (1969).

441. J. A. Knopp, J. J. ten Bosch, and J. W. Longworth, Biochim. Biophys. Acta, 188, 185 (1969).

442. S. A. Latt, D. S. Auld, and B. L. Vallee, Proc. Natl. Acad. Sci. U.S., 67, 1383 (1970).

443. R. A. Badley and F. W. J. Teale, J. Mol. Biol., 58, 567 (1971).

444. R. A. Badley and F. W. J. Teale, J. Mol. Biol., 44, 71 (1969).

445. W. C. Galley and L. Stryer, Biochemistry, 8, 1831 (1969).

446. W. C. Galley and L. Stryer, Proc. Natl. Acad. Sci. U.S., 60, 108 (1968).

447. J. W. Longworth and M. del Carmen Battista, Photochem. Photobiol., 11, 207 (1970).

Chapter 13

CHLOROPHYLL FLUORESCENCE AND PHOTOSYNTHESIS[a]

Govindjee, G. Papageorgiou and E. Rabinowitch

Department of Botany and the Biophysics Division of the
Department of Physiology and Biophysics,
University of Illinois
Urbana, Illinois 61801

---

[a]By Govindjee, G. Papageorgiou and E. Rabinowitch,
University of Illinois, Urbana, Illinois.

I. INTRODUCTION

The physical and chemical processes that lead to the reduction of carbon dioxide and the oxidation of water in plants have been the subject of much study in the last decade [1, 2]. In 1957 Emerson [3, 4] discovered that the rate of photosynthesis was greater when red light (absorbed mainly by chlorophyll b) and far-red light (absorbed mainly by chlorophyll a) were used simultaneously than when they were used separately (the Emerson enhancement effect). To explain this phenomenon, it was suggested that photosynthesis requires two light reactions, sensitized by two pigment systems [5]. It is now widely accepted that two light reactions, arranged in series, are involved in this process [6] (Fig. 131). The first light reaction, arbitrarily called reaction II by Duysens and Amesz [7], leads to the reduction of a cytochrome and the oxidation of water to molecular oxygen; the second light reaction (reaction I) leads to the oxidation of the reduced cytochrome and the reduction of $NADP^+$ (nicotinamide adenine dinucleotide phosphate). Along the electron pathway from water to $NADP^+$ a fraction of light energy is utilized to synthesize ATP molecules from ADP and inorganic phosphate. With sufficient NADPH and ATP available, enzymatic reduction of carbon dioxide to the carbohydrate level (Calvin-Benson cycle) becomes possible.

The experiments of Emerson and Arnold [8] had established that a group of about 2400 chlorophyll a molecules somehow cooperate in evolving one oxygen molecule and reducing one carbon dioxide molecule. Since four hydrogen atoms (or electrons) must be transferred from water to carbon dioxide to achieve this process, about 600 chlorophyll molecules must cooperate in the transfer of one electron. In the two-light-reaction model of photosynthesis each electron must be transferred in two steps, so that the group of 600 chlorophyll molecules has to be divided in two parts, probably of 300 molecules each. These groups are usually referred to as photosynthetic units (PSU). Photosynthetic units of the pigment system that sensitize light reaction I (PSU_I) and those that sensitize light reaction II (PSU_{II}) are spectrally distinguishable.

Each PSU consists of a light-collecting part, comprising the majority of the pigment molecules ("bulk") and a "reaction center," or "trap," that participates directly in the primary photochemical reaction. Light energy absorbed by the bulk pigments is funneled (by a resonance-transfer mechanism) to the "trap," where the primary oxidation-reduction reaction (electron transfer) takes place. The trap in system I has been identified [9] as a chlorophyll a molecule with an absorption peak at 700 nm on the long-wavelength side of the absorption band of the bulk with a peak at 675 nm; it is referred to as "P700." The nature of the trap in system II has not yet been definitely established, but there are indications that it is also a chlorophyll a molecule (P680-P690) [10-12].

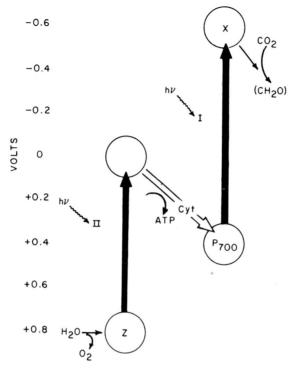

FIG. 131. Hydrogen (or electron) transfer in photosynthesis is now conceived as a two-step process involving two pigment systems. (The scale on the right represents the oxidation-reduction potential E_o of the intermediates in photosynthesis.) The light energy collected by pigment system II, trapped by pigment 680-690 (P680-690), oxidizes the primary H donor ZH to Z and reduces the unknown primary acceptor Q, the quencher of chlorophyll a fluorescence (also known as C550, for an absorbance change at 550 nm; in the text we use the symbol A for it). Molecular oxygen is evolved when water reacts with Z. The electrons are passed downhill to cytochrome f (Cyt), synthesizing ATP in this process. The light energy collected by pigment system I, trapped by pigment 700 (P700), oxidizes cytochrome f and reduxes another unknown acceptor P430 (or X). The reduced X, in turn, can reduce $NADP^+$ to NADPH via an iron protein, ferredoxin. The NADPH and ATP--the end products of the light reactions of photosynthesis--are then utilized to synthesize carbohydrate (CH_2O) from carbon dioxide. A backreaction of reduced entities (XH, NADPH, etc.) with oxidized Cyt (or any component in the intersystem chain) leads to a cyclic flow of electrons that could be coupled to the formation of more ATP.

We now ask: How are the absorbed light quanta distributed equally between the two pigment systems for the efficient operation of photosynthesis? Two different hypotheses have been suggested [13]. In one of them, the "spillover" hypothesis, the excess light energy absorbed in pigment system II can spill over into system I (but not vice versa), leading to balanced excitation of the two systems in the part of the spectrum where more light is absorbed by system II and to a dropoff of the yield where light is absorbed preferentially in system I. The spillover is energetically possible because system I absorbs light of longer wavelengths than system II. This hypothesis explains the high quantum yield of photosynthesis at wavelengths below 680 nm and the "red drop" in the action spectrum of photosynthesis beyond 680 nm (observed in green plants and algae).

In the second, "separate-package," hypothesis, excess light quanta absorbed in pigment system II are not transferred to pigment system I. An approximately balanced excitation of both systems is achieved by the presence in both systems of the same pigments in somewhat different proportions. For example, in the green alga Chlorella system II contains relatively more chlorophyll b, and system I relatively more long-wavelength forms of chlorophyll a. According to this hypothesis, the action spectrum of photosynthesis (and also of chlorophyll a fluorescence as system I is weakly fluorescent) must show drops wherever one of the two systems absorbs more than the other. This must cause a "fine structure" in the $\Phi = f(\lambda)$ curve (where Φ stands for quantum yield). Some such structure has been noted in the action spectrum of photosynthesis [5, 13] and of chlorophyll a fluorescence [14-16]. It has been

suggested [17] that the spillover may occur only between system II and the trap (P700) in system I, but not the bulk of the latter system. Energy transfer can occur only if P700 is in the reduced, colored state. This means that energy transfer from system II to system I is possible only when the electrons flow from system II to P700, keeping it in the reduced state; if P700 is oxidized (i. e., bleached), the energy transfer stops. This hypothesis permits to reconcile certain observations favoring the spillover mechanism with other observations favoring the separate-package hypothesis.

A light quantum absorbed in a pigment molecule in the bulk of one of the two pigment systems faces three alternatives. Within the lifetime of its excited state it may be (a) lost by rapid "radiationless relaxation" into thermal energy; (b) reemitted as fluorescence; or (c) transferred to other pigment molecules. The energy quanta that reach a trap have an additional choice--to be converted into chemical energy by an "uphill" electron transfer. The efficiency of "quantum conversion" in the traps seems to be close to 100% when the electron-transfer chain operates at its highest rate (for recent estimates see Mar et al. [18]).

The quanta reemitted as fluorescence are presumed to originate mainly (or exclusively) in the bulk of the photosynthetic unit. When the trap is less than 100% effective in the conversion of light energy that reaches it into chemical energy, some energy could become available for trap fluorescence. However, since the difference in excitation energy between the bulk and the trap at room temperature is not large compared with kT (k - Boltzmann's constant), the photochemically unused quanta have a good chance of returning from the trap to the bulk to be emitted there. At very low temperatures, however, the energy barrier opposing the diffusion of quanta back into the bulk becomes prohibitive, and emission from the trap can become significant.

Fluorescence, in photosynthetic organisms, represents a waste of light energy; this waste is, however, quite low since the yield of chlorophyll a fluorescence in living cells is only about 3%. In spite of the low yield of fluorescence, and complexity due to overlapping of the absorption and fluores-cence spectra of the various pigments, a systematic analysis of the fluorescence intensity and spectrum has yielded important information concerning the composition of the two pigment systems and the primary mechanisms of photo-synthesis. The intensity of emission is related to the efficiency of the two postulated primary processes, and its spectrum (and the action spectrum of its excitation) is indicative of the properties of the two postulated pigment systems. Fluorescence provides a nondestructive tool for monitoring the transformations of the pigments in photosynthesis.

Prior to the discovery of the two light reactions in photosynthesis, reviews on the relation of chlorophyll fluorescence to photosynthesis were written by Franck [19], Rabinowitch [20], and Weber [21]. More recently Butler [22], Clayton [17], Robinson [23], and Govindjee and Papageorgiou [24] have reviewed the subject in the context of the two-light-reaction model. In this chapter we discuss mainly the complexity of the chlorophyll a fluorescence spectrum in vivo, with emphasis on experiments made in our laboratory in Urbana.

The emission spectrum of chlorophyll a in vivo has a main band with a maximum at 685 ± 2 nm and a vibrational "satellite" at about 740 nm. These bands originate in a $\pi^* \longrightarrow \pi$ transition to the ground state. Excitation in the blue-violet (Soret) absorption band, leading to the second electronic excited state, is followed by radiationless transition to the lowest excited state, which is so fast that no emission originating in the upper state has ever been reported.

II. MULTIPLICITY OF CHLOROPHYLL a FORMS IN VIVO

The red absorption band of chlorophyll a in vivo is broad (half-bandwidth $\sim$30 nm) in comparison with the same band of chlorophyll (Chl) a in organic solvents (18-nm half-bandwidth in diethyl ether). This suggests that chloro-phyll a in vivo is composed of more than one molecular species or complex. Analyses of the absorption bands of chlorophyll a in suspensions of chloroplasts or whole algal cells encounter several difficulties:

1. Other pigments besides chlorophyll a are present.

2. Light is strongly scattered by the suspension (Rayleigh scattering and selective spectral scattering [25]; this may increase the apparent absorption if

measured in an ordinary spectrophotometer, and the absorption spectra may become distorted.

3. In a suspension light that is less absorbed has a larger effective path length than the more strongly absorbed light ("detour" effect [26]).

4. A mutual shading of the pigment molecules takes place in colored particles (chloroplasts and grana), and some light may pass the medium without hitting any particle at all ("sieve effect" [20, 27]).

5. Fluorescence contributes to the "transmitted" light particularly significantly when absorption is strong.

Two methods have been used to remedy some of these difficulties. In one a diffusing glass (an opal glass) is placed between the sample and the detector, so that almost all forward scattering is collected. In the other an integrating sphere (the Ulbright sphere) is used, so that almost all scattered energy is collected and read as "transmitted" light. Neither of the two methods eliminates the sieve effect or the detour effect.

French [29] constructed a "derivative spectrophotometer" in which the first derivative of the absorbance was plotted automatically as function of wavelength. This procedure accentuates the complex structure of a band, since every inflection in the band envelope appears as a crossing of the absicca or as a peak. The derivative absorption spectra of various algae obtained in this way were analyzed by Brown and French [30] (also Ref. [31]) into several Gaussian components. They were identified as Chl a 672, Chl a 683, and Chl a 694. More recently, a new component, which may be the short-wavelength band of long-wavelength aggregated forms, absorbing at 660 nm has been definitely identified by French and co-workers [32].

Cederstrand et al. [33] measured the absorption spectrum of algae in an integrating regular dodecahedron with a photocell on each of the 12 faces: the red band was interpreted as the sum of two components, Chl a 668 and Chl a 683 (again assuming a Gaussian shape of the components). A third band, due to chlorophyll b, was located at 650 nm. The assignment of a Gaussian shape to the bands of individual chlorophyll a components is not completely arbitrary, since a Gaussian curve matches very closely the red band of chlorophyll a in various solvents.

Das et al. [27] (see also Ref. [28]) have shown that elimination of the so-called sieve effect by sonication of Chlorella does not affect the qualitative results obtained with whole cells, but the peaks of the two major chlorophyll a components are moved (as expected) closer together to 670 and 683 nm. The half-width of both components is about 17 nm, very similar to that of the chlorophyll a band in ether.

The action spectrum of Emerson's enhancement effect [34, 35] (the proportional increase in yield when far-red light (680-720 nm) is given together with the short-wavelength light, compared with the sum of the yields in the separate beams) exhibits a peak or a shoulder at 670 nm, apparently attributable to the Chl a 670 component. (Both peaks, at 670 and at 680 nm, are assigned to chlorophyll a because of their location. The Chl a 670 is probably identical with French's Chl a 672, and Chl a 680 with his Chl a 683.) Cederstrand et al. [33] saw no convincing evidence of the existence of a third component in Chlorella, Anacystis, Porphyridium, and Spinacea (equivalent to French's Chl a 694), because in the long-wavelength tail of the band deviation from Gaussian shape occurs also in solution spectra. However, there is other evidence of its existence. The separate existence of a Chl a 670 component is indicated also by the observation of preferential photochemical bleaching in this spectral area [36].

Fractionation of the two (or three) chlorophyll a forms has been attempted by breaking the cells mechanically and solubilizing the pigment complexes selectively by means of detergents (such as digitonin) or by extraction with solvents of varying polarity. The two types of pigment complex identified in this manner are also Chl a 670 and Chl a 683, the latter being the more labile form [37-39].

III. FLUORESCENT AND NONFLUORESCENT (OR WEAKLY FLUORESCENT) FORMS OF CHLOROPHYLL a

Duysens [40] and French and Young [41] observed that the light absorbed by the phycobilins of red algae is more efficient in exciting chlorophyll a fluorescence than light absorbed by chlorophyll a itself. This led Duysens [40] to postulate the existence of two forms of chlorophyll a: one fluorescent and the other nonfluorescent (or weakly fluorescent). The fluorescent form must be associated with the phycobilins much more effectively than the "nonfluorescent" one. It is now considered to be part of pigment system II. The "nonfluorescent" form was first assumed to be inactive in photosynthesis, but we now interpret it as belonging to pigment system I, which can perform only light reaction I.

A second indication of the existence of two forms of chlorophyll a with different fluorescence yields was obtained from measurements of the lifetime, τ, of chlorophyll a fluorescence in vivo and in vitro by a direct flash method [42]. The lifetime was found (in Chlorella) to be 1.5 nsec (assuming a simple exponential decay). If one makes the plausible assumption that the "natural" lifetime τ_0 of chlorophyll a in vivo is the same as it is in vitro (15.2 nsec, calculated by integrating the area under the band envelope), one calculates $\phi = 0.10$ from the relation $\tau = \phi \cdot \tau_0$. However, direct measurements [43] (see also Ref. [44]) of ϕ in Chlorella gave significantly lower values, of about 0.03. This discrepancy can be explained if it is postulated that a significant proportion of chlorophyll a in vivo does not contribute significantly to fluorescence. The exact proportion of nonfluorescent (or weakly fluorescent) chlorophyll a could be estimated only if the measurements of both τ and ϕ were done under identical conditions on the same suspension and if τ_0 could be determined directly for chlorophyll in the living cell. (For recent measurements of, and literature on, chlorophyll a fluorescence lifetime in vivo see Refs. [18] and [45] through [47]).

IV. THE RED DROP IN CHLOROPHYLL FLUORESCENCE YIELD

If the weakly fluorescent or nonfluorescent form of chlorophyll a absorbs more strongly on the long-wavelength side of the red absorption band than the fluorescent form (as we have reason to believe), a decline (red drop) in the quantum yield of fluorescence ϕ_F plotted as a function of wavelength λ can be expected. Indeed, Duysens [40] has noted such a decline beyond 680 nm in green algae. However, Teale [48], who took great precautions to exclude scattering by using a polarizing filter in the path of fluorescence, a suspending medium of high refractive index (concentrated sucrose, ethylene glycol), and a very dilute suspension of Chlorella, could find no red drip up to 690 nm. We reinvestigated the ϕ_F versus λ curve for Chlorella sonicates (in which the scattering is very much reduced [49,50]) and for very dilute suspensions of Chlorella, both by direct measurements and by computing the ϕ_F versus λ curve according to Stepanov's relation between the absorption and the emission spectrum [15,50,51]. Fluorescence was collected from the same surface on which the exciting light fell, thus reducing its reabsorption. As a result of these studies we are convinced that a red drop in the ϕ_F versus λ curve undoubtedly does exist. It begins at about 680 nm in Chlorella.

The fluorescence yield curve $[\phi_F = F(\lambda)]$ in the phycobilin-containing algae is different because almost all the chlorophyll a is present in pigment system I. The system I chlorophyll a, however, is not fluorescent because the long-wavelength-absorbing from (C700, or Chl a 695) drains all the excitation, and the latter is nonfluorescent or weakly fluorescent. Because the absorption by system I chlorophyll a extends to shorter wavelengths in these algae, the red drop begins earlier. Thus in the red alga Porphyridium and in the blue-green alga Anacystis the fluorescence yield begins to decline at about 640 to 660 nm [50,52].

One is naturally tempted to associate the "long-wavelength" nonfluorescent component of chlorophyll a with Cederstrand's Chl a 680, and the "short-wavelength" fluorescent component with Chl a 670. However, quantitative analysis makes this interpretation difficult to maintain. It rather suggests that the red drop is associated with a third, minor, long-wavelength component (French's Chl a 694?). This means that both main forms, Chl a 670 and Chl a 680, are present in both pigment systems I and II, whereas Chl a 694 may be present in pigment system I only, making the whole bulk of chlorophyll a in this system nonfluorescent (or rather weakly fluorescent).

The red drop in the fluorescence yield should disappear if the nonfluorescent (or weakly fluorescent) form of chlorophyll a could be preferentially destroyed.

Under aerobic conditions and at low pH, prolonged sonication of Chlorella cells does lead to preferential bleaching of a long-wavelength form (Chl a 693) of chlorophyll a, as demonstrated by the difference between the absorption spectra of the sonicates prepared by sonication in the absence (pH 7.8) and in the presence (pH 4.5) of air. As expected, the samples deficient in the far-red form of chlorophyll a show a complete absence of the red drop in the action spectrum of chlorophyll fluorescence [49].

V. ANALYSIS OF THE EMISSION SPECTRA AT ROOM TEMPERATURE

As mentioned in Section I, chlorophyll a in vivo fluoresces with a peak at 685 ± 2 nm and a vibrational band at about 740 nm. The main red emission band has been investigated intensively in recent years for evidence of its complexity. Due to experimental difficulties, such as the low quantum efficiency of chlorophyll fluorescence in living tissues, the overlapping of the bands, and the polyphasic structure of the material, special techniques must be applied to obtain meaningful data.

A. Matrix Analysis

The intensity of fluorescence emitted at a certain wavelength by a mixture of several fluorescent forms with different positions of absorption bands will depend on the wavelength of excitation. If intensity data for a sufficient number of excitation-emission wavelength pairs are available, one can determine the number of individual fluorescing species. Weber [53] developed for this purpose a matrix method of analysis. The intensity of emission at different wavelengths is determined for fluorescence excited at different wavelengths, and matrices of increasing order are formed. If the 2 x 2 matrices do not disappear, there must be at least two fluorescence emitters; if the 3 x 3 matrices are significant, at least three, and so on. Weber's method was applied by Brody and Brody [54] to Euglena, and by Govindjee and Yang [55] to spinach chloroplasts. The analyses showed that at room temperature there are (at least) two different fluorescent chlorophyll a species.

More recently Williams et al. [14] confirmed this also for Chlorella; they suggested, from consideration of the sign and magniture of different matrices, that only two fluorescent chlorophyll a species seem to be present, and that the one responsible for strongest emission has an emission peak at about 687 nm, whereas the less or weakly fluorescent component emits at about 700 nm. A band at 700 nm has recently been observed by Govindjee and Briantais [56] in the ratio spectrum of "high" to "low" fluorescence yields (see their paper for details). The two 700-nm bands may not be identical.

B. Polarized and Nonpolarized Fluorescence

The existence of a chlorophyll a component absorbing on the long-wavelength side of the main absorption band was established by Olson et al. [57] and by Lavorel [58]. A dichroism ratio of more than 4 was observed when the analyzer was set parallel and perpendicular to the lamellar plane of the chloroplast at 695 nm; at the shorter wavelengths no dichroism was discernible. The emission spectrum of "polarized" fluorescence showed a maximum at 716 to 720 nm, that of "depolarized" fluorescence, at about 685 nm. The intensity of the former was about $\sim$ 5 to 10% of that of the latter.

Since the red absorption band of chlorophyll a is due to a $\pi \rightarrow \pi^*$ transition, with the oscillator located on the porphyrin plane (and bisecting rings II and IV), the above experiments suggest that at least a part of the chlorophyll a is either lying flat or is tilted at a small angle with respect to the lamellar plane. This orientation can be imposed by the protein-lipid matrix in which the chlorophyll a molecules are embedded. Possibly the weakly fluorescent long-wavelength form of chlorophyll a is simply the oriented fraction. It thus seems that a minor form of chlorophyll a in system I is arranged in a relatively ordered array. Is it the form responsible for the weak, long-wavelength polarized-fluorescence band beyond 700 nm and for the red drop in the overall chlorophyll fluorescence yield in the far-red region? The answer is not yet clear.

Boardman and Anderson [59] found evidence of partial separation of the two postulated pigment systems by differential centrifugation of digitonin-solubilized chloroplast material. Anderson et al. [60] provided further evidence of this separation: the lighter particles, enriched in P700 (i.e., in pigment system I) effectively performed the reduction of $NADP^+$ with reduced 2,6-dichlorophenol indophenol (DCPIPH, a dye with a normal oxidation-reduction potential of about 0.2eV) as an electron donor, whereas the heavier particles, richer in chlorophyll b (i.e., in pigment system II), effectively reduced DCPIP with water as electron donor. The degree of polarization of chlorophyll fluorescence was 2.7% in the heavier, and 5.4% in the lighter, fractions [60,61]. These results are consistent with the conclusion that a fraction of chlorophyll a in pigment system I is ordered and emits more strongly polarized fluorescence. However, such conclusions should be made with caution because the size of the particles may also affect the polarization values.

The excitation spectra [58] of the polarized fluorescence in Chlorella have peaks at about 700, 540, and 400 nm, whereas those of depolarized fluorescence show a red drop beyond 680 nm, again indicating that the oriented chlorophyll a is a part of pigment system I. Our observations on the red alga Porphyridium confirm these findings [62]. The action spectrum of the species fluorescing at 685 nm always follows closely the absorption spectrum of pigment system II, whereas the action spectrum of polarized fluorescence at 720 nm follows the absorption spectrum of system I. Vredenberg [63] also suggested that chlorophyll a_I fluoresces at 720 nm, but he found no correlation between the kinetics of absorbance changes attributable to P700 and the 720-nm fluorescence. (However, if P700 sensitizes the reaction in its triplet state, no correlation between chlorophyll fluorescence and photochemical sensitization can be expected.)

VI. FLUORESCENCE AT LOW TEMPERATURES

The sharpening of the emission bands at low temperatures is a useful means of reducing the difficulties arising from the overlapping of the bands at room temperature (see Section V). In addition, the quantum efficiency of chlorophyll fluorescence is considerably increased at low temperatures (both photochemical and internal quenching is slowed down). The low-temperature spectra are of particular interest for the identification of the photochemical reaction centers (the traps) since the "trap depth" ($\Delta\lambda \cong 10\text{-}15$ nm) increases from 1-2 kT at room temperature, to 5-8 kT at 77°K (liquid-nitrogen temperature), and to 100-150 kT at 4°K (liquid-helium temperature).

S. S. Brody [64] was the first to observe a new emission band, located at 720 nm (F720), in Chlorella cooled to 77°K. Several investigators [5, 54, 55, 65-70] confirmed its existence and observed an additional band at 696 to 698 nm. An extension of this work to liquid-helium temperature (4°K) by Cho et al. [71] and Cho and Govindjee [66, 72] clearly showed the existence of three fluorescence bands, at 689 (F687), 698 (F696), and 725 (F720) nm, respectively. These bands appear at different locations in different organisms. We shall, however, refer to them as F687, F696, and F720.

The overall shape of the red emission band in green cells exhibits a strong

temperature dependence caused by the different behavior of the three components. Govindjee and Yang [55] studied this behavior in spinach chloroplasts,
Krey and Govindjee [65] in Porphyridium, and Cho and Govindjee [73] in
Chlorella, in the range 77 to 293°K. The F696 band appears only in the
temperature range between 77 and 140°K. On warming from 77 to 140°K, both
F696 and F720 decrease in intensity (although at different rates), whereas F687
remains constant. At the lower temperatures (40 to 4°K) Cho et al. [71] found
that the F687 band increased, whereas the F696 band intensity rose only
slightly. These changes can be interpreted as reflecting a reduction in the rate
of energy transfer from the bulk of chlorophyll a (emission peak at 685 nm) to
the trap in system II (emission peak at 698 nm) at these low temperatures.

Furthermore it was found [55, 65, 66, 72] that excitation in chlorophyll b
(in green plants) and in phycobilins (in red or blue-green algae) leads to a lower
F720 but higher F687 and F696 bands; whereas excitation in chlorophyll a leads
to the opposite result. This suggests that F687 and F696 belong mainly to
pigment system II, and F720 belongs mainly to pigment system I, in agreement
with the conclusions reached in a preceding section.

These conclusions are further confirmed by experiments on particles
prepared by solubilizing spinach chloroplasts with detergents (digitonin) and
differential centrifugation, with chloroplasts extracted with acetone and
methanol of different concentrations, and with Chlorella cells sonicated in air
or argon. Particles--prepared by the digitonin method--that perform light
reaction I, according to Boardman and Anderson [59], were found to be richer
in F720-emitting material when cooled down to 77°K, whereas particles that
perform light reaction II were enriched in F696-emitting material [61, 74].
Shimony et al. [75] obtained fractions enriched in F720-emitting material
(Chl a in system I) and fractions poorer in F720-emitting material (Chl a in
system II) from the blue-green alga Anacystis. Thomas and Van der Wal [76]
reported that chloroplasts extracted with methanol of different concentrations
had somewhat different absorption and fluorescence characteristics.

Cederstrand et al. [39] studied the absorption and fluorescence characteristics of (a) chloroplast residues after extraction with aqueous methanol and
acetone of different strength, and (b) the corresponding extracts. They found,
particularly at low temperatures, an apparent enrichment either of system II
or of system I in the residues from extraction with different solvents. For
example, chloroplasts extracted with 50% methanol were enriched in F696-
emitting material (system II). Very dilute aqueous acetone or methanol
apparently extracted whole pigment complexes present in the chloroplasts.
Under aerobic conditions and at acid pH, sonication of Chlorella cells [49] leads
to preferential bleaching of the weakly fluorescent (or nonfluorescent) chloro-

phyll a (Chl a 693); these samples show no F720 band at 77°K, assumed to be due to the long-wavelength form of chlorophyll a (pigment system I).

The excitation spectra of the F720, F696, and F687 fluorescence bands in Chlorella and Anacystis at very low temperatures (4-77°K) provided the following information [66, 72, 73]:

1. There are two excitation peaks, at about 670 and 680 nm, respectively, which must be due to the two forms of chlorophyll a identified in the absorption spectra [27, 33]; the excitation spectra of both F720 and F696 show these peaks, confirming the earlier suggestion that both pigment systems I and II contain both components. Using narrow slits, it was possible [66] to extend the measurement of the excitation spectrum of F687 to the 680-nm region. It appears that Chl a 680 is responsible for F687. If Chl a 680 is preferentially extracted, one can show that Chl a 670 is responsible for a band at 680 nm (F680) that appears under these conditions. The exact location of the 670- and 680-nm peaks in the action spectra may be slightly different for F720 and F696, suggesting some differences in the environment of these chlorophyll a components in the two systems; this, however, needs further study.

2. The ratio of fluorescence intensity excited at 440 nm (Chl a absorption) to that excited at 480 nm (Chl b absorption) is greater for F720 than it is for F687 and F696. This confirms that F720 is excited more effectively by absorption in chlorophyll a (preferentially in pigment system I) and F687 and F696 by absorption in chlorophyll b (preferentially in pigment system II). Near identity of the excitation spectra for F687 and F696 confirms our earlier suggestion [62] that F696 belongs to the same system as F687 (system II). Williams et al. [14] concluded from their analyses that the "F700" band, observed in Chlorella fluorescence in vivo at room temperature, belongs to system I; this shows that it is not identical with the low-temperature F696 band. We now consider the possibility that there are two emission bands in this spectral region--the system I component is present both at room temperature and at 77°K, and the system II component appears at low temperatures (see earlier discussions [77, 78]) and under conditions when photosynthesis is saturated or abolished.

3. The highly efficient energy transfer from chlorophyll b to chlorophyll a appears to be independent of temperature. This is suggested by the constancy of the ratio of chlorophyll fluorescence excited at 440 nm (absorption in Chl a) to that excited at 480 nm (absorption in Chl b) when the temperature was varied from 4 to 77°K. However, the efficiency of energy transfer from phycobilins to chlorophyll a was dependent on temperature--just as for the transfer from Chl a 680 to the trap of system II, noted above.

VII. LIFETIMES OF THE EXCITED STATES

The decay period of chlorophyll a fluorescence in vivo has been estimated from flash fluorometry [42, 79] and phase fluorometry [80, 81] to be in the range of 1 nsec; it was assumed that the decay of chlorophyll a fluorescence in vivo follows a simple first-order exponential function. Under conditions of interrupted photochemical deexcitation, such as low-temperature [18, 81] and 3-(3,4-dichlorophenyl)-1,1-dimethyl urea (DMCU) poisoning [18, 47, 82], the lifetimes were approximately doubled, corresponding to increased quantum yield of fluourescence under these conditions.

Murty and Rabinowitch [83] employed shorter flashes and a faster detecting system than those employed by S. S. Brody and Rabinowitch [42] to trace the decay curve of chlorophyll a fluorescence in various algae. The plot of the logarithm of the fluorescence intensity against time showed [83] not a single straight line but a curve with two linear portions. From the slopes of these curves two time constants were calculated: one corresponding to a decay period τ_1 of 1 to 2 nsec, and the other to a decay period τ_2 of 4 to 5 nsec. This experiment suggested the existence of two components of chlorophyll a fluorescence, decaying at different rates. The authors suggested, as one of the alternatives, that the faster decaying component originates in pigment units with a higher chlorophyll a concentration, in which the excitation is rapidly transferred to the reaction site, lowering the quantum yield of fluorescence. (A similar conclusion has been reached by Pearlstein [84] on theoretical grounds.) This needs further investigation, as Singhal and Rabinowitch [85] (see also Refs. [45] and [46]) could not repeat Murty's earlier observation; it is possible that an "artifact" in the exciting lamp was responsible for the τ_2 observed earlier.

According to Robinson [23], the more rapid photochemical deexcitation of the chlorophyll a singlets in comparison with the slower rates of intersystem crossing renders the participation of the chlorophyll a triplets in the primary reactions unlikely. Assuming that at $77^\circ K$ the singlets can either fluoresce (with a measured lifetime of 3.1 nsec) or cross over to the triplet state, an intersystem-crossing rate of 0.26 $nsec^{-1}$ can be calculated on the basis of the τ_0 value of 15.2 nsec. This rate represents an upper limit since, at room temperature, other competing deexcitation processes will reduce it further. The rate of chemical deexcitation in vivo at room temperature, however, calculated on the basis of the 0.03 quantum yield of fluorescence, is much higher (~ 6 $nsec^{-1}$) than the upper limit of the rate of intersystem crossing, so that the latter cannot compete favorably with the former. The low fluorescence yields indicate a fast chemical turnover, with the participation of the singlet

states of the reaction centers.   This concept is by no means universally
accepted.   For example, Franck and Rosenberg [86] suggested that reaction I
proceeds through the triplet state.   There is also the possibility that triplets
are not involved in the main path of photosynthesis, but they are produced in a
side pathway--either directly at the reaction center by intersystem crossing or
by a backreaction of photoreaction II.   An explanation of this nature has been
recently formulated and analyzed by Stacy et al. [87] and Lavorel [88].   They
explain the delayed fluorescence in algae on the basis of a triplet-triplet fusion
theory.

VIII.   VARIATION IN FLUORESCENCE YIELD WITH LIGHT INTENSITY

There are several pathways for the dissipation of energy by a population
of excited chlorophyll a molecules: radiative deexcitation (fluorescence),
photochemical quenching, nonradiative deexcitation (internal conversion into
heat energy), and transfer to a fluorescent or nonfluorescent species.   The
following equation can be written for the yield ϕ_F of chlorophyll a fluorescence:

$$\phi_F = \frac{k_F}{k_F + k_R + k_C [A]} , \tag{207}$$

where k_F is the rate of radiative deexcitation, k_C is the rate of a bimolecular
quenching process, [A] is the concentration of the quenching partner, and k_R is
the sum of nonradiative and nonphotochemical deexcitation rates.   Since the
availability of the photochemical quencher [A] is limited by a sequence of dark
enzymatic reactions in photosynthesis, it is expected that at saturating light
intensities the quantum yield of fluorescence will reach an upper limit corresponding to [A] = 0, whereas at low intensities a lower limit will be established
corresponding to a constant value [A] = A_0.

In pure dilute chlorophyll a solutions fluorescence intensity F is proportional, within wide limits, to the incident intensity I; that is, the quantum yield
ϕ_F of fluorescence is independent of light intensity.   Chlorophyll a fluorescence in vivo shows, however, a dependence of ϕ_F on I, as expected from
Eq. (207).   This was clearly shown by, among others, Franck [19], Wassink
(see Ref. [20]), and by Brugger [89], who plotted F as a function of I.   They
found that the slope of this curve, at "high" intensities (i. e. , in light strong
enough to "saturate" photosynthesis), is twice that at low intensities (where the
quantum yield of photosynthesis, ϕ_p, is maximal and constant).   Latimer et al
[43], who measured the absolute quantum yields of fluorescence in <u>Chlorella</u>,
confirmed these findings, showing that the quantum yield of chlorophyll a

fluorescence increased from about 0.025 and approached 0.05 with increasing intensity of the exciting light. (The quantum yield of photosynthesis (O_2 evolution) is known to be about 0.12 at low intensities [15, 90] and declines steadily at high intensities).

Krey and Govindjee [65] measured chlorophyll a fluorescence F as a function of exciting light intensity I in the red alga Porphyridium cruentum. They could confirm the nonlinearity of the F = f(I) curve only in the case when excitation took place in the pigment phycoerythrin (i.e., in system II); no dependence of ϕ_F on I could be observed when excitation took place in chlorophyll a itself (i.e., in system I). This difference can be understood if the reaction centers of pigment system I operate in the triplet state, whereas those in pigment system II operate in the singlet state, because only in the second case is there a competition between sensitization and fluorescence.

IX. CHANGES IN FLUORESCENCE YIELD WITH TIME AND THE TWO-LIGHT EFFECT: THE ACTIVATION REACTION

The yield of chlorophyll fluorescence in vivo undergoes complex but reproducible changes with time (the Kautsky effect); not only the intensity but also the spectral composition of the fluorescence is altered. (For literature and other details see a recent review by Govindjee and Papageorgiou [24]).

When dark-adapted Chlorella cells are exposed to strong light, the fluorescence yield rises instantaneously to an initial level (0) that is independent of photochemical processes. It then rises to a level that remains constant (or decreases slightly) for a brief period (ID), in the millisecond range, and then rises again to a peak (P) (reached after 0.25 to 1 sec). Within about 1 to 2 sec it decreases to an almost steady level (S); another peak is observed after about 30 to 50 sec (M), after which the fluorescence declines to its final steady level (T). The exact shape of the F = f(t) curve, and the time of occurrence of the various characteristic points, varies with the conditions of the experiment [91-95].

To explain the induction curve of chlorophyll fluorescence, Kautsky et al. [96] suggested that its yield depends on the presence of the oxidized form of an oxidation-reduction intermediate (A) that quenches the fluorescence. This quencher is reduced by the first light reaction

$$A + e \xrightarrow{h\nu} A^- \qquad (208)$$

and reoxidized by a dark reaction

$$B^+ + A^- \longrightarrow B + A; \qquad (209)$$

a second light reaction

$$B + X \xrightarrow{\ h\nu\ } B^+ + X^- \tag{210}$$

completes the sequence.

According to Kautsky, the fluorescence plateau reached in the millisecond range is due to the removal of the quencher (A) by a first-order photochemical process. However, the rate of the regeneration of the dark reaction becomes significant when the concentration of B^+ is increased by the second light reaction; this causes the first plateau in the induction curve. The renewed rise to the peak reached after about 0.25 sec follows as B^+ is used up. Kautsky al. did not explain the decay of this peak.

Lavorel [97] suggested that the fluorescence yield is a sum of two contributions. One, invariant with time, is independent of any photochemical events; the other, a time-dependent one, is due to a photoactive form of chlorophyll a. With an experimental setup in which an algal suspension flowed at controlled speed through a capillary, a segment of which was illuminated, he succeeded in recording the spectra of the two components of fluorescence. The variable fluorescence contained relatively more of the 685-nm than the 717-nm fluorescence species. This was confirmed by Rosenberg et al. [98] and by Munday [99].

Two light reactions were postulated in Kautsky's scheme, but he reported no attempts to separate them by exciting with light of different wavelengths. The first observation that the yield of fluorescence can be modified by adding far-red light (absorbed in pigment system I) to short-wavelength light (absorbed in system II) was made by Govindjee et al. [100] in experiments analogous to those by which the photosynthetic enhancement effect was demonstrated. They found that fluorescence excited with red or blue light (absorbed in pigment system II) was lowered (quenched) when far-red light (which by itself produced no fluorescence) was added. Butler [101, 102] used strong red (system II) and far-red (system I) actinic light to establish a steady-state concentration of the postulated oxidation-reduction intermediate and followed with a weak exciting beam to observe the fluorescence yield. (The actinic light was eliminated during the fluorescence measurement.) Butler confirmed the antagonistic effects of the two lights on chlorophyll a fluorescence, in whole leaves. The quenching of fluorescence by far-red light was maximal at 705 nm [103]. This maximum depends, however, on the intensity of the actinic light; it is shifted to longer wavelengths at higher light intensities. This band may be due to the same long-wavelength chlorophyll a form that is responsible for the 720-nm emission band at 77°K.

Another method employed to establish a steady photochemical state before measuring fluorescence is to excite fluorescence with a weak modulated beam, superimposed on a strong, constant background light [104]. The two-light effect on fluorescence was confirmed by this method. Duysens and Sweers [104] proposed that in reaction II--sensitized by light absorbed in pigment system II-- the reductant (indirectly, water) reduces an intermediate, which they called Q, we shall continue to use Kautsky's notation, A, for the sake of uniformity in this presentation. The oxidation-reduction potential E_o' of the primary electron acceptor, pigment system II, has been estimated to be +180 mV, over a pH range from 6 to 9 [105]. However, Butler et al. [106] reported two E_o' values for Q: -35 and -270 mV; they considered only the first value to be of significance. Recently Knaff and Arnon [107], and Eriyon and Butler [108] have described an absorbance change at 550 nm due to compound 550, or C550, which has been suggested by the latter authors to be identical with Q.

If A is associated with the trap, and the reaction can occur in the excited-singlet state of the latter, it must cause a quenching of chlorophyll a fluorescence in pigment system II. The steady-state fluorescence yield of pigment system II is governed by the proportion of oxidized "traps," that is, of traps associated with A (as contrasted to those associated with A^-) since only the oxidized form can be utilized for the photochemical reaction. At very high intensities of light absorbed in system II, there is a preponderance of A^- (since light reaction I is unable to reoxidize A^- at a high enough rate). The increase in fluorescence at the higher light intensities is due to this shift in system II. This has been confirmed by observations on the excitation spectra of fluorescence of the green alga Scenedesmus [109] and on the emission spectra of the red alga Porphyridium [65, 110]. On excitation, pigment system I depresses the fluorescence yield because A^- is oxidized to A.

Duysens and Sweers [104] postulated an additional dark backreaction via another intermediate (Q') to account for the inability to observe an increase in fluorescence yield if enough dark period was not provided between light exposures. (We use A' to remain consistent with the terminology used here.) Their scheme is very similar to that of Kautsky et al. [96]:

$$
\begin{array}{c}
H_2O \ \Big| \\
\\
O_2 \ \Big\downarrow
\end{array}
\quad
\begin{array}{c}
A \xrightleftharpoons[hv_I]{hv_{II}} A^- \\
\searrow \quad \swarrow \\
A'
\end{array}
\qquad\qquad (211)
$$

The rise in fluorescence yield is attributed to the removal of A by light reaction II. On poisoning with DCMU, this rise can be observed in its "pure" form,

without subsequent decline. This indicates that DCMU inhibits the reactions
that regenerate A. The brief ID plateau (in the millisecond range) is due to
(as in Kautsky's scheme) the indirect oxidation of A⁻ by system I, involving one
or more additional oxidation-reduction intermediates. Finally, the decay from
the peak (at about 0.25 sec) is due to both system I action and the dark backreaction via the so-called A'. The latter also accounts for the higher steady-
state fluorescence value at the end of the decay period as compared with the
first plateau in the millisecond range. Both A⁻ and A' are photochemically
inactive forms; one can assume that the higher yield at the steady state is due
to the fact that part of the quencher A exists in the nonphotochemical form A'.
Recent experiments (see review [24]) suggest that A' may be nothing else but a
physical state of the system in which there is a high spillover of energy from
pigment system II to the weakly fluorescent pigment system I.

Since A is the primary electron acceptor, the rate of oxygen evolution will
depend on its concentration, and an inverse linear relationship should exist
between it and the fluorescence yield. This prediction has been substantiated
by the experiments of Delosme et al. [111].

Kautsky et al. [96] estimated the relative concentration of the fluorescence
quencher as [A]:[Chl] ≃ 1:400, whereas the estimate of Duysens and Sweers
[104] gave [A]:[Chl a_2] ≃ 1:150. A somewhat higher ratio was found by Malkin
and Kok [112]. These figures indicate that the quencher is present in amounts
approximately equal to the number of reaction centers. In all likelihood
Kautsky's A (or Duysens' Q) is not a usual plastoquinone, because the latter is
known to be present in much larger quantities. Plastoquinone is an intermediate that participates in the electron-transfer chain between the two photosynthetic systems, operating at a site close to system II. (However, there are
suggestions [113] that Q may be a type of quinone, and there may be two
quenchers, Q_1 and Q_2, instead of one.)

After a dark period, simultaneous recordings of the fluorescence kinetics
and oxygen-evolution rate [111,114] showed that the inverse relationship is not
valid over the entire fluorescence-rise curve. During the first phase (referred
to as "activation phase" in the earlier literature) fluorescence and oxygen-
evolution rate increase in parallel. (In weak light there is a lag in O_2
evolution, but not in the fluorescence rise; also, in the presence of DCMU, when
there is no O_2 evolution, the fluorescence yield rises.) During the second
("complementary") phase these processes are antiparallel. (McAllister and
Myers [115] had first observed this phenomenon with the techniques available in
1940.) The duration of the "activation phase" is inversely proportional to the
intensity of the exciting light. In the complementary phase these is a linear

relationship between the fluorescence yield (F/I) and the rate of oxygen evolution (V/I) of the form

$$\frac{F}{I} = a\frac{V}{I} + b. \tag{212}$$

This relationship between the yields is independent of light intensity since the constants a (which is negative) and b are independent of it. To explain his kinetic results Joliot [114] postulated (a) that the initial fluorescence variations originate in pigment system II and (b) that a photochemical quencher designed by him as E at that time (we shall continue to use A for it) is activated during this period. Schematically the reaction sequence given by Joliot can be rewritten as follows:

$$A_iH \underset{k_1}{\overset{k_2I}{\rightleftharpoons}} AH \underset{BH}{\overset{B}{\longrightarrow}} A \underset{}{\overset{k_2I}{\rightleftharpoons}} A^* \left(\begin{array}{c} O_2 \\ \\ H_2O \end{array} \right. \tag{213}$$

In this scheme B is the form of another intermediate and is produced by light reaction I. Both A_iH and A can quench the fluorescence, but AH does not. In the very beginning of the excitation period the photochemical quencher is assumed to exist in the form A_iH. During the activation phase it is transformed to AH, with parallel increase in the fluorescence intensity (due to the removal of A_iH), and in the rate of oxygen evolution (due to the production of $\dot{A}^*$). At the end of the activation phase the form A_iH is consumed, and only the forms A (a quencher) and AH (a nonquencher) are present.

In the "complementary" phase the transformation of AH to A and the subsequent photochemical reaction results in a decrease in fluorescence and increase in oxygen evolution. These initial explanations have now been replaced by more intricate theories in which four oxidizing equivalents accumulate on the donor side (Z) before oxygen evolution occurs, and the "activation" is included in the "main" pathway as the formation of less than four equivalents [116-118].

It has been shown [92, 119] that preillumination with far-red light (pigment system I), even if followed by a brief dark period (0.5 sec), affects the induction of fluorescence by largely eliminating the 0.25-sec peak (P). These experiments indicate the presence of a "pool" of intermediate B formed by far-red light, as suggested earlier by French [120] on the basis of flashing-light experiments on oxygen evolution and as suggested recently by Kok et al. [121] and Witt et al. [122]. Govindjee et al. [119] also found that, after far-red preillumination, the steady-state fluorescence at about 2 to 3 sec is slightly higher than the brief plateau in the millisecond range. Joliot's reaction

sequence can explain these results. Prolonged far-red light treatment converts
BH to B, so that when pigment system II begins to operate, the nonquencher AH
is rapidly transformed to the quencher A, by the dark bimolecular reaction,
thus causing a delay in the rise and a reduction in the height of the 0.25-sec
peak.

Experiments on oxygen evolution in flashing light support the idea of an
activation reaction in photosynthesis. Allen and Franck [123] reported that no
oxygen is produced in algae by a single short light flash (~ 1 msec). However,
when a longer flash (~25 msec) or two brief flashes with a spacing of a few
seconds were given, or when weak preillumination was provided before the
flash, oxygen was evolved. Whittingham and co-workers [124, 125] confirmed
and extended these observations. Joliot [126], who developed a very sensitive
oxygen electrode, provided more precise data on this phenomenon, which can
also be ascribed to the need for an activation reaction (for recent theories see
Refs. [116] through [118]).

The common feature of the three reaction schemes (Kautsky, Duysens, and
Joliot) proposed to account for the cause of induction in chlorophyll fluorescence
in vivo is a photochemical substrate that can be directly reduced by pigment
system II and indirectly reoxidized by pigment system I, both processes pro-
ceeding with a quantum efficiency of unity. This picture permits an explanation
of the antagonistic effect of lights absorbed in system II and system I, respectively, and of complementarity between fluorescence intensity and the rate of
oxygen evolution. However, so far no chemical identification of the postulated
primary photochemical reactants has been possible, and additional complications due to changes in the efficiency of energy spillover for system II to system
I spoil this simple relationship.

X. VARIATIONS IN FLUORESCENCE: CHANGES IN INTENSITY AND SPECTRA

Light-induced variations in the quantum yield of chlorophyll a fluorescence
have been employed in the study of the complexity of chlorophyll a in vivo (see
Sections VIII and IX).

Butler and Bishop [109] compared the excitation spectra of an intact leaf
after red and far-red preillumination and observed that it was the pigment
system II fluorescence that increased at high light intensities, whereas a band at
705 nm was virtually independent of the preillumination of the sample. Difference emission spectra of Porphyridium, constructed by Krey and Govindjee
[110] by subtracting the emission spectrum obtained in low light (below the
saturation of photosynthesis) from the spectrum obtained in high light, exhibit
a band at 693 nm in addition to the main 687-nm band (pigment system II).
Recently we have obtained [56] similar results with Chlorella. The increase

in 687-nm fluorescence confirms Butler and Bishop's conclusions. The 693-to 700-nm band was, however, new. From its spectral location, the expected Stokes' shift, and the consideration that, when photosynthesis is saturated, some emission from the traps may become possible, the 693- to 700-nm band has been tentatively assigned to the trap in system II.

An alternative means of modifying the quentum yield of fluorescence, in living cells, is the interruption of the photosynthetic electron flow by selective poisons, such as certain derivatives of phenylurea (e. g. , DCMU) and 9,10-phenanthroline. Poisoning with DCMU results in the diminution of oxygen evolution (although the possibility of its complete abolition is questionable) and an increase in the fluorescence yield by a factor of more than 2. Under conditions of interrupted photochemical quantum conversion one may expect that previously nonfluorescent species, such as energy traps, may also emit some fluorescence, although most of the excess fluorescence must originate in the bulk, as pointed out in Section I. Indeed, "difference spectra" obtained by comparison of DCMU-poisoned with nonpoisoned Porphyridium or Chlorella cells [56, 65] reveal the existence of a large 685-nm band (bulk) and an additional band at 692 to 700 nm, which for the above-stated reasons may also originate in the trap of system II.

The occurrence of an effective energy transfer from accessory pigments (e. g. , phycocyanin, Chl b) to chlorophyll a is well known [40, 41, 79, 127-129]. M. Brody and Emerson [130] and Brody and Brody [131] reported changes in the rate of oxygen evolution and the quantum yield of fluorescence in Porphyridium induced by prolonged preillumination--changes they ascribed to changes in the efficiency of energy transfer from phycoerythrin to chlorophyll a. Such a modification could be achieved by growing Porphyridium in light of different wavelengths. Ghosh and Govindjee [132] obtained similar results with the blue-green alga Anacystis; they suggested that decreased efficiency of energy transfer from phycocyanin to "fluorescent" chlorophyll a in pigment system II may mean a more effective transfer to the weakly fluorescent form of the same pigment in system I.

Papageorgiou and Govindjee [77, 93] studied the effect of prolonged illumination on normal and DCMU-poisoned Anacystis; they found that not only the quantum efficiency of fluorescence changes in a specific way but also the spectral distribution of fluorescence is altered, the effect being largest at about 695 nm. In order to interpret their kinetic data the authors assumed that light-induced configurational changes may occur and affect both the rate of energy transfer (from phycobilins to Chl a) and the quantum yield of chlorophyll a fluorescence (for more recent ideas see recent reviews [24, 133, 134].

The effects of prolonged illumination on the quantum efficiency of chlorophyll fluorescence were also studied in <u>Chlorella</u> [94]. Under constant exciting-light conditions the fluorescence yield, after the initial induction effects are over in about 1 sec, rises for 30 to 40 sec (S $\longrightarrow$ M phase); this rise is followed by a slower decline to a lower steady level (T) than that observed after 2 to 3 sec. The rates of change, as well as the final established level of the yield, depend on the light intensity, the pH of the medium [135], and the integrity of the electron-transport chain and phosphorylation. In whole cells poisoning by DCMU or 9,10-phenanthroline abolishes the transients and inhibits electron transport in photosynthesis. When phosphorylation is uncoupled from electron transport by ammonium chloride and p-trifluoromethoxyphenylhydrazone of ketomalonyldinitrile (FCCP), the fluorescence transients are also affected. The compound FCCP, which is known to uncouple phosphorylation and increase electron transport in the Hill reaction, may affect the whole cells differently. In whole cells, where carbon dioxide is the oxidant and ATP is needed in the Calvin cycle, uncoupling of phosphorylation would reduce the rate of the Calvin cycle, thus reducing, rather than accelerating, electron transport. Increased fluorescence may then be expected. We found that in whole <u>Chlorella</u> cells FCCP, in a concentration of 5×10^{-5} M, completely eliminates the (M $\longrightarrow$ T) decay after the 30- to 40-sec fluorescence peak, and the fluorescence remains at a constant level. The dependence of the long-time light-induced fluorescence changes on the same parameters (pH, uncouplers of phosphorylation, etc.) that affect the light-induced configurational (volume and scattering) changes in chloroplasts may indicate a close relationship between fluorescence changes and energy-preserving processes like phosphorylation. Recently Mohanty et al. [95] found that glutaraldehyde, which abolishes configurational changes, also abolishes slow fluorescence changes. Closer analysis of this subject cannot be undertaken without further data on short-time fluorescence induction and observations on isolated chloroplasts.

XI. FLUORESCENCE INTENSITY AND PARTIAL REACTIONS

Prior to the discovery of the two light reactions, Lumry et al. [136] studied the relationship of chlorophyll fluorescence and the Hill reaction, which is the reduction of an added oxidant (other than CO_2) and the simultaneous oxidation of water to molecular oxygen. Quenching of chlorophyll a fluorescence was observed when Hill oxidants were added to chloroplasts. Kok [69] observed quenching of chlorophyll a fluorescence in chloroplasts when $NADP^+$ and ferredoxin were added; this quenching was abolished by the addition of DCMU. Further addition of 2,6-dichlorophenol indophenol plus ascorbate did not cause renewed quenching. These experiments show a clear relationship between chlorophyll a fluorescence yield and the efficiency of electron transport involving pigment system II (or both systems I and II, but not system I alone).

Govindjee and Yang [137] observed a strong quenching of chlorophyll a_2 fluorescence (687 nm) when phenazine methosulfate (PMS) was added to DCMU-treated spinach chloroplasts. (Excitation was at 610 nm to avoid absorption by PMS.) It is known that PMS accelerates "cyclic" phosphorylation (i. e., phosphorylation coupled with reversal of light reaction I), which is sensitized by system I. One could suggest that the yield of chlorophyll a fluorescence is also affected by the "physical state" of the chloroplast membrane, which may change during "phosphorylation." These results can also be explained if we suggest that under these conditions spillover of excess energy takes place from the strongly fluorescent pigment system II to the weakly fluorescent pigment system I, resulting in the quenching of fluorescence. However, a direct reaction of PMS with pigment system II cannot be excluded.

XII. CONCLUDING REMARKS

Absorption and fluorescence spectrophotometry has been used successfully in the study of photosynthetic pigments in vivo. Chlorophyll a is the most important pigment since it is present in all photosynthesizing plants and algae, and a special "complex" of chlorophyll a--designated as P700--participates in the primary oxidation-reduction reaction of photosynthesis. Although only one form of chlorophyll a can be identified in organic solvent extracts, the evidence presented in this chapter suggests that several modifications of it exist in vivo; they may be different holochromic forms or different aggregates of chlorophyll a. Figure 132 shows the probable distribution of the different forms of chlorophyll a and b in the two pigment systems of photosynthesis in higher plants and green algae. In the red and blue-green algae chlorophyll b is replaced by the phycobilins and most of the chlorophyll a is in pigment system I.

Some of the chlorophyll a forms are designated as Chl a 670, Chl a 680, Chl a 695, and P700. These forms (see Table 87) have been identified mainly from the absorption spectra, the difference absorption spectra, the action spectra of the light reactions, and the action spectra of chlorophyll a fluorescence. It is generally believed that the two pigment systems (I and II) contain both chlorophyll b and a (Chl a 670 and Chl a 680), but in different proportions. (In phycobilin-containing algae most of the Chl a 670 and Chl a 680 belong to pigment system I.) However, Chl a 695 may be present exclusively in pigment system I. A form of chlorophyll a designated as Chl a 660 may simply be the short-wavelength band of this aggregate form of Chl a 695, as suggested by S. S. Brody.

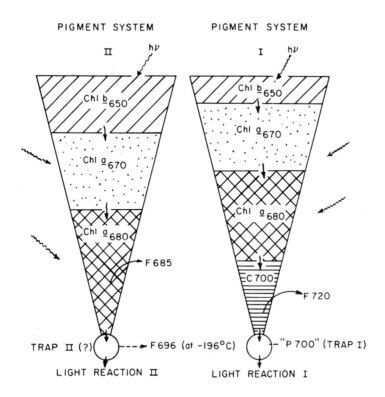

FIG. 132. A working hypothesis for the distribution of the chlorophylls in the two pigment systems (I and II) in higher plants and green algae. The two systems seem to contain both chlorophyll a (Chl a) and chlorophyll b (Chl b 650), but in different proportions. (In red and blue-green algae the phycobilins replace Chl b.) It is suggested that the long-wavelength form of chlorophyll a (C700, Chl a 695) is present only in pigment system I. Chl a 660 may be a sub-band of this aggregate form of chlorophyll a. The two "bulk" chlorophylls (Chl a 670 and Chl a 680) are almost equally distributed in the two systems, pigment system II containing more Chl a 670. (In red and blue-green algae a larger proportion of Chl a is in pigment system I.) The energy trap (trap I) of system I is P700; trap II has not been definitely identified, but it may be identified with the recently discovered P680-P690. It has also been suggested that the new emission band that appears at 696 nm (F696) when plants are cooled to 77°K originates in trap II. At room temperature most (80%) of the main fluorescence band at 685 to 687 nm (and its 740-nm satellite) originates from system II, whereas a band at 700 nm originates mainly in system I; however, a band at 700 nm also appears in system II when photosynthesis is saturated or abolished.

TABLE 87

In vivo Chlorophyll a Absorption Bands

Band maximum (nm)	Method of observation	Refs.
668-672 and 678-683	Derivative absorption spectrophotometry	28-31
	Direct analysis of absorption spectra	27, 28, 33
	Action spectra of Emerson enhancement	34, 35
	Fractionation and separation of pigment complexes	37-39, 59, 74
	Excitation spectra of chlorophyll a fluorescence at low temperatures	66-70, 72
694-705[a] (+ Chl a 660?)	Derivative absorption spectrophotometry	28-31
	Chlorophyll a dichroism	57
	Action spectra of the quenching of fluorescence by far-red light	102
	Action spectra of chlorophyll a fluorescence at 77°K	22, 55, 66
	Action spectra of polarized fluorescence	58
P700	Difference absorption spectrophotometry	9

[a]This band is composed of more than one species.

In higher plants and algae fluorescence measurements (see Table 88) at room temperature have revealed the presence of at least two emission bands at 687 nm (with a satellite band at 740 nm) and at about 700 nm. The former may be mainly present in pigment system II, and the latter in system I. However, a band at 695 to 700 nm also appears in system II when photosynthesis is saturated or abolished. At low temperatures (4 to 140°K), however, three bands are clearly observed in all the organisms so far studied; these bands are at 687, 696, and 720 nm. The exact locations of these bands differ with different organisms. These 687- and 696-nm bands originate mainly in the pigment system II. Portions of the 687- and 696-nm bands may also come from system I. The 720-nm band, however, originates mainly in system I.

TABLE 88

In vivo Chlorophyll a Fluorescence Bands

Band maximum (nm)	Method of observation	Refs.
683-687	Emission spectra at room temperature	Well known
	Emission spectra at low temperature	5, 54, 55, 64-72
	Induced variations in the quantum yield of fluorescence	56, 65, 77, 95, 97, 98, 104
693-696[a]	Emission spectra at low temperatures	5, 54, 55, 64-72
	Induced variations in the quantum yield of fluorescence	56, 65, 77, 110, 119
716-720[a]	Emission spectra at room temperature (by matrix analysis)	14, 54, 55
	Emission spectra at low temperatures	5, 54, 55, 64-72
	Emission spectra of polarized fluorescence	57, 58[b]

[a]These bands are composed of more than one species.

[b]There is some doubt as regards this assignment.

REFERENCES

1. L. P. Vernon and M. Avron, Ann. Rev. Biochem. , 34, 269 (1965); G. Hind and J. M. Olson, Ann. Rev. Plant Physiol. , 19, 249 (1968).

2. G. W. Robinson, Ann. Rev. Phys. Chem. , 15, 311 (1964).

3. R. Emerson, R. Chalmers, and C. Cederstrand, Proc. Natl. Acad. Sci. U.S. , 43, 133 (1957).

4. R. Emerson and E. Rabinowitch, Plant Physiol. , 35, 477 (1960).

5. Govindjee, Natl. Acad. Sci. -Natl. Res. Council, 1145, 318 (1963).

6. E. Rabinowitch and Govindjee, Scientific American, 213, 74 (1965); Photosynthesis, Wiley, New York, 1969.

7. L. N. M. Duysens and J. Amesz, Biochim. Biophys. Acta, 64, 243 (1962).

8.   R. Emerson and W. Arnold, J. Gen. Physiol. , 15, 391 (1932).

9.   B. Kok, Biochim. Biophys. Acta, 21, 399 (1956).

10.   G. Doring, H. H. Stiehl, and H. T. Witt, Z. Naturforschg. , 22b, 639 (1967).

11.   G. Doring, J. L. Bailey, W. Kreutz, and H. T. Witt, Naturwissen- schaften, 55, 220 (1968).

12.   Govindjee, G. Doring, and R. Govindjee, Biochim. Biophys. Acta, 205, 303 (1970).

13.   J. Myers, Natl. Acad. Sci. -Natl. Res. Council, 1145, 301 (1963).

14.   P. Williams, N. R. Murty, and E. Rabinowitch, Photochem. Photobiol. , 9, 455 (1969).

15.   R. Govindjee, E. Rabinowitch, and Govindjee, Biochim. Biophys. Acta, 162, 539 (1968).

16.   M. Bazzaz, Ph. D. thesis, University of Illinois, Urbana, 1972.

17.   R. K. Clayton, in The Chlorophylls, Physical, Chemical, and Biological Properties (L. P. Vernon and G. R. Seely, eds.), Academic Press, New York, 1966, p. 610.

18.   T. Mar, Govindjee, G. S. Singhal, and H. Merkelo, Biophys. J. , in press (1972).

19.   J. Franck, in Photosynthesis in Plants (J. Franck and W. E. Loomis, eds.), Iowa State College Press, Ames, Iowa, 1949, p. 293.

20.   E. I. Rabinowitch, Photosynthesis and Related Processes, Vols. I and II (parts 1 and 2),Interscience, New York, 1945, 1951, 1956, Chapters 23, 24, and 37c.

21.   G. Weber, in Comparative Biochemistry of Photoreactive Systems (M. B. Allen, ed.), Academic Press, New York, 1966, p. 49.

22.   W. L. Butler, in Current Topics in Bioenergetics (D. R. Sanadi, ed.), Vol. I, Academic Press, New York, 1966, p. 49.

23.   G. W. Robinson, Brookhaven Symp. Biol. , 19, 16 (1966).

24.   Govindjee and G. Papageorgiou, in Photophysiology (A. C. Giese, ed.), Vol. 6, Academic Press, New York, 1971, p. 1.

25.   P. Latimer, Ph. D. thesis, University of Illinois, Urbana, 1956.

26.   W. L. Butler, Optical Soc. Amer. , 52, 292 (1962).

27.   M. Das, E. Rabinowitch, L. Szalay, and G. Papageorgiou, J. Phys. Chem. , 71, 3543 (1967).

28.   E. Rabinowitch, L. Szalay, M. Das, N. R. Murty, and Govindjee, Brookhaven Symp. Biol. , 19, 1 (1966).

29. C. S. French, in Photobiology, Proceedings of the 19th Annual Biology Colloquium, Oregon State College, 1958, p. 52.

30. J. Brown and C. S. French, Plant Physiol., 34, 305 (1959).

31. J. Brown, in Methods in Enzymology, Vol. 23, Part A (A. San Pietro, ed.), Academic Press, New York, 1971, p. 477.

32. C. S. French, J. S. Brown, and M. C. Lawrence, Carnegie Inst. Wash. Year Book, 70, 487 (1971).

33. C. N. Cederstrand, E. Rabinowitch, and Govindjee, Biochim. Biophys. Acta, 126, 1 (1966).

34. Govindjee and E. Rabinowitch, Science, 132, 355 (1960); Biophys. J., 1, 73 (1960).

35. C. S. French, J. Myers, and G. C. McCloud, in Comparative Biochemistry of Photoreactive Systems (M. B. Allen, ed.), Academic Press, New York, 1960, p. 361.

36. L. M. Vorobyova and A. A. Krasnovsky, Biokhimiya, 21, 126 (1956).

37. M. B. Allen, J. C. Murchio, S. W. Jeffrey, and S. A. Bendix, in Studies on Microalgae and Photosynthetic Bacteria, Japanese Society of Plant Physiologists, University of Tokyo Press, Tokyo, 1963, p. 407.

38. J. Brown, C. Bril, and W. Urbach, Plant Physiol., 40, 1086 (1965).

39. C. N. Cederstrand, E. Rabinowitch, and Govindjee, Biochim. Biophys. Acta, 120, 247 (1966).

40. L. N. M. Duysens, Ph. D. thesis, University of Utrecht, The Netherlands, 1952.

41. C. S. French and V. M. K. Young, J. Gen. Physiol., 35, 873 (1952).

42. S. S. Brody and E. Rabinowitch, in Proceedings of the First National Biophysics Conference, Yale University Press, 1959, p. 110.

43. P. Latimer, T. T. Bannister, and E. Rabinowitch, in Research in Photosynthesis (H. Gaffron et al., eds.), Interscience, New York, 1957, p. 107.

44. L. Szalay, M. Török, and Govindjee, Acta Biochim. Biophys. Acad. Sci. Hung., 2, 425 (1967).

45. W. J. Nicholson and J. Fortoul, Biochim. Biophys. Acta, 143, 577 (1967).

46. A. Muller, R. Lumry, and M. S. Walker, Photochem. Photobiol., 9, 113 (1969).

47. H. Merkelo, S. Hartman, T. May, G. S. Singhal, and Govindjee, Science, 164, 301 (1969).

48. F. W. J. Teale, Report to U. S. Army Research and Development in Europe (DA-91-508-EUC-281), 1959.

49. M. Das and Govindjee, Biochim. Biophys. Acta, 143, 570 (1967).

50. M. Das, E. Rabinowitch, and L. Szalay, Biophys. J. , 8, 1131 (1968).

51. L. Szalay, E. Rabinowitch, N. R. Murty, and Govindjee, Biophys. J. , 7, 137 (1967).

52. G. Papageorgiou and Govindjee, Biochim. Biophys. Acta, 131, 173 (1967).

53. G. Weber, Nature, 190, 27 (1961).

54. S. S. Brody and M. Brody, Natl. Acad. Sci. -Natl. Res. Council, 1145, 455 (1963).

55. Govindjee and L. Yang, J. Gen. Physiol. , 49, 763 (1966).

56. Govindjee and J. -M. Briantais, Fed. Europ. Biochem. Soc. Letters, 19, 287 (1972).

57. R. A. Olson, W. H. Jennings, and W. L. Butler, Biochim. Biophys. Acta, 88, 318, 331 (1964).

58. J. Lavorel, Biochim. Biophys. Acta, 88, 20 (1964).

59. N. K. Boardman and J. Anderson, Nature, 203, 166 (1964).

60. J. M. Anderson, D. C. Fork, and J. Amesz, Biochem. Biophys. Res. Commun. , 23, 874 (1966).

61. C. N. Cederstrand and Govindjee, Biochim. Biophys. Acta, 120, 177 (1966).

62. Govindjee, in Currents in Photosynthesis (J. B. Thomas and J. H. C. Goedheer, eds.), Ad Donker, Rotterdam, 1965, p. 93.

63. W. Vredenberg, Ph.D. thesis, University of Leiden, 1965.

64. S. S. Brody, Science, 128, 838 (1958).

65. A. Krey and Govindjee, Biochim. Biophys. Acta, 120, 1 (1966).

66. F. Cho and Govindjee, paper presented at the annual meeting of the Biophysical Society held at Houston, Texas, February 1967; Biophys. J. , 9, A119 (1969); Biochim. Biophys. Acta, 216, 139 (1970).

67. F. F. Litvin, A. A. Krasnovsky, and G. T. Rikhireva, Dokl. Akad. Nauk SSSR, 135, 1528 (1960).

68. J. Bergeron, Natl. Acad. Sci. -Natl. Res. Council, 1145, 527 (1963).

69. B. Kok, Natl. Acad. Sci. -Natl. Res. Council, 1145, 45 (1963).

70. J. H. C. Goedheer, Biochim. Biophys. Acta, 88, 304 (1964); ibid. , 102, 73 (1965).

71. F. Cho, J. Spencer, and Govindjee, Biochim. Biophys. Acta, 126, 174 (1966).

72. F. Cho and Govindjee, Biochim. Biophys. Acta, 216, 151 (1970).

73. F. Cho and Govindjee, Biochim. Biophys. Acta, 205, 371 (1970).

74. N. K. Boardman, Ann. Rev. Plant Physiol., 21, 115 (1970).

75. C. Shimony, J. Spencer, and Govindjee, Photosynthetica, 1, 113 (1967).

76. J. B. Thomas and U. P. Van der Wal, Biochim. Biophys. Acta, 79, 490 (1964).

77. G. Papageorgiou and Govindjee, Biophys. J., 7, 375 (1967).

78. P. Mohanty, B. Zilinskas-Braun, Govindjee, and J. P. Thornber, Plant Cell Physiol., 132 (in press, 1972).

79. G. Tomita and E. Rabinowitch, Biophys. J., 2, 483 (1962).

80. O. D. Dmietrievsky, V. L. Ermolaev, and A. N. Terenin, Dokl. Akad. Nauk SSSR, 114, 751 (1957).

81. W. L. Butler and K. Norris, Biochim. Biophys. Acta, 66, 72 (1963).

82. A. Müller and R. Lumry, Proc. Natl. Acad. Sci. U.S., 54, 1479 (1965).

83. N. R. Murty and E. Rabinowitch, Biophys. J., 5, 655 (1965).

84. R. M. Pearlstein, Proc. Natl. Acad. Sci. U.S., 52, 824 (1964).

85. G. S. Singhal and E. Rabinowitch, Biophys. J., 9, 586 (1969).

86. J. Franck and J. Rosenberg, J. Theoret. Biol., 7, 276 (1964).

87. W. T. Stacy, T. Mar, C. E. Swenberg, and Govindjee, Photochem. Photobiol., 14, 197 (1971).

88. J. Lavorel, Progr. Photosynth. Res., 2, 883 (1969).

89. J. E. Brugger, in Research in Photosynthesis (H. Gaffron et al. eds.), Interscience, New York, 1957, p. 113.

90. R. Emerson, Ann. Rev. Plant Physiol., 9, 1 (1958).

91. J. C. Munday and Govindjee, Biophys. J., 9, 1 (1969).

92. J. C. Munday and Govindjee, Biophys. J., 9, 22 (1969).

93. G. Papageorgiou and Govindjee, Biophys. J., 8, 1299 (1968).

94. G. Papageorgiou and Govindjee, Biophys. J., 8, 1316 (1968).

95. P. Mohanty, G. Papageorgiou, and Govindjee, Photochem. Photobiol., 14, 667 (1971).

96. H. Kautsky, W. Appel, and H. Amann, Biochem. Z., 332, 277 (1960).

97. J. Lavorel, in La Photosynthese, Coll. Intern. C.N.R.S., 119, 159 (1963); Biochim. Biophys. Acta, 60, 510 (1962).

98.   J. L. Rosenberg, T. Bigat, and S. Dejaegere, Biochim. Biophys. Acta, 79, 9 (1964); J. L. Rosenberg and T. Bigat, Natl. Acad. Sci. -Natl. Res. Council, 1145, 122 (1963).

99.   J. C. Munday, Ph. D. thesis, University of Illinois, Urbana, 1968.

100. Govindjee, S. Ichimura, C. Cederstrand, and E. Rabinowitch. Arch. Biochem. Biophys. , 89, 322 (1960).

101. W. L. Butler, Plant Physiol. Suppl. , 36, IV (1961).

102. W. L. Butler, Biochim. Biophys. Acta, 64, 309 (1962).

103. W. L. Butler, Biochim. Biophys. Acta, 66, 275 (1963).

104. L. N. M. Duysens and H. E. Sweers, in Studies on Microalgae and Photosynthetic Bacteria, Japanese Society of Plant Physiologists, University of Tokyo Press, Tokyo, 1963, p. 353.

105. B. Kok and O. Owens, cited by B. Kok and G. M. Cheniae, in Current Topics in Bioenergetics (D. R. Sanadi, ed.), Vol. 1, Academic Press, New York, 1966, p. 1.

106. W. Butler, W. A. Cramer, and T. Yamoshita, Biophys. J. , 9, A-28 (1969).

107. D. Knaff and D. Arnon, Proc. Natl. Acad. Sci. U. S. , 63, 963 (1969).

108. K. Eriyon and W. Butler, Biochim. Biophys. Acta, in press, 1971.

109. W. L. Butler and N. I. Bishop, Natl. Acad. Sci. -Natl. Res. Council, 1145, 91 (1963).

110. A. Krey and Govindjee, Proc. Natl. Acad. Sci. U. S. , 52, 1568 (1964).

111. R. Delosme, P. Joliot, and J. Lavorel, Compt. Rend. , 249, 1409 (1959).

112. S. Malkin and B. Kok, Biochim. Biophys. Acta, 126, 413 (1966).

113. R. Govindjee, Govindjee, J. Lavorel, and J. -M. Briantais, Biochim. Biophys. Acta, 205, 361 (1970).

114. P. Joliot, Biochim. Biophys. Acta, 102, 116 (1965).

115. E. D. McAllister and J. Myers, Smith Inst. Misc. Coll. , 99, 1 (1940).

116. G. Forbush, B. Kok, and M. P. McGloin, Photochem. Photobiol. , 14, 307 (1971).

117. P. Joliot, A. Joliot, A. Bouges, and G. Barbieri, Photochem. Photobiol. , 14, 287 (1971).

118. T. Mar and Govindjee, J. Theoret. Biol. , in press, 1972.

119. Govindjee, J. C. Munday, and G. Papageorgiou, Brookhaven Symp. Biol. , 19, 434 (1966).

120.   C. S. French, in Currents in Photosynthesis (J. B. Thomas and J. H. C. Goedheer, eds.), Ad Donker, Rotterdam, 1966, p. 285.

121.   B. Kok, S. Malkin, O. Owens, and B. Forbush, Brookhaven Symp. Biol., 19, 446 (1966).

122.   H. T. Witt, G. Döring, B. Rumberg, P. Schmidt-Mende, U. Siggel, and H. H. Stiehl, Brookhaven Symp. Biol., 19, 161 (1966).

123.   F. L. Allen and J. Franck, Arch. Biochem. Biophys., 58, 510 (1955).

124.   C. P. Whittingham and A. H. Brown, J. Exptl. Botany, 9, 311 (1958).

125.   C. P. Whittingham and P. M. Bishop, Natl. Acad. Sci. -Natl. Res. Council, 1145, 371 (1963).

126.   P. Joliot, Brookhaven Symp. Biol., 19, 418 (1966).

127.   H. J. Dutton, W. M. Manning, and B. B. Duggar, J. Phys. Chem., 47, 308 (1943).

128.   E. C. Wassink and J. A. H. Kersten, Enzymologia, 12, 3 (1946).

129.   W. Arnold and J. R. Oppenheimer, J. Gen. Physiol., 33, 434 (1950).

130.   M. Brody and R. Emerson, J. Gen. Physiol., 43, 251 (1959).

131.   S. S. Brody and M. Brody, Arch. Biochem. Biophys., 82, 161 (1959).

132.   A. K. Ghosh and Govindjee, Biophys. J., 6, 611 (1966).

133.   J. Myers, Ann. Rev. Plant Physiol., 22, 289 (1971).

134.   Govindjee and P. Mohanty, in Fluorescence News (R. A. Passwater and P. Welker, eds.), Vol. 6(2), American Instrument Co., Silver Spring, Md., 1971, p. 1.

135.   G. Papageorgiou and Govindjee, Biochim. Biophys. Acta, 234, 428 (1971).

136.   R. Lumry, B. Mayne, and J. D. Spikes, Discussions Faraday Soc., 27, 149 (1959).

137.   Govindjee and L. Yang, unpublished data, 1965-1966.

Chapter 14

ANALYSIS ON SOLID SURFACES

I.   INTRODUCTION

One of the newer uses of fluorescence has been in the assay of substances directly on solid surfaces, such as paper chromatograms, thin-layer chromatography (TLC) plates, potassium bromide disks, electrophoresis strips, and silicone-rubber pads. Chromatography and electrophoresis are widely used analytical techniques and it is not surprising that fluorometric methods are used in conjunction with them. Advantages are increased sensitivity and speed of analysis.

In this chapter we consider the use of fluorescence in the assay of substances after separation on paper and TLC chromatograms, as well as the direct assay of substances on a solid surface placed directly in a commercial fluorometer.

II. INSTRUMENTATION

In paper and thin-layer chromatography, and electrophoresis the position of the separated components is detected either by their fluorescence under ultraviolet light or by fluorescence quenching. In the first method the substance may either show native fluorescence or be converted to a fluorophor by a spray

reagent. In the second method the paper, thin layer, or electrophoresis strip
is itself rendered fluorescent, and the separated components appear as dark
spots on a bright background by absorbing the ultraviolet light.

The detection of the fluorescent spots in the first case or the dark spots in
the second is effected by exploring the chromatograms under ultraviolet light in
a darkened room. Portable hand lamps that emit the 254- or 360-nm mercury
lines are available for this purpose. Alternatively one could use a portable
laboratory cabinet for this purpose, such as the Chromato-vue (UV Products,
Inc., San Gabriel, Calif.), which can be operated even in lighted rooms. The
cabinet contains lamps, filters, and reflectors for both the 254- and 365-nm
excitation radiation.

Early applications of fluorescence in quantitative measurements in solid-
surface analysis consisted principally of scraping off the substance from the
solid TLC plate or cutting the substance out of the paper chromatogram,
dissolving the substance in an appropriate organic solvent, and then measuring
in solution with a commercial fluorometer. This technique was time consuming
and had limited accuracy. Hence much attention was paid to the development
of instrumentation capable of direct measurement on the solid surface. As a
result several attachments to commercial instruments capable of direct
measurements on TLC plates, electrophoretic strips, or paper chromatograms
are now available.

An accessory for either the Turner 110 or 111 fluorometer capable of
quantitative fluorescence measurements on paper or TLC plates is available.
A holder that is provided can scan strips up to 5 x 35 cm with several selected
exciting wavelengths in the ultraviolet and visible regions. Emitted light can be
measured in the 360- to 650-nm region.

G. K. Turner Associates reported the use of the same instruments for
measuring enzymes developed electrophoretically on gels [1]. To measure
serum lactate dehydrogenase isoenzymes after gel electrophoresis, the gel is
incubated with a dry reagent film containing lactate and NAD. On incubation the
areas on the gel showing lactate dehydrogenase activity yield the highly fluorescent NADH, whose emission is proportional to enzyme activity. Assay of the
lactate isoenzymes with this thin-gel fluorometric procedure is rapid and
requires only 1 µl of serum or plasma.

The American Instrument Company has introduced a motorized scanner
that permits direct absorbance or fluorescence analysis of TLC plates up to
8 x 8 in. in size. The Thin-Film Scanner accessory (Fig. 133) developed for
the American Instrument Company's Aminco-Bowman spectrophotofluorometer
(SPF) and the Fluoro-Microphotometer are useful for qualitative and quantitative
scanning of TLC plates and paper chromatograms. The scanner is capable of

FIG. 133. Motorized thin-film scanner mounted on the Aminco-Bowman
spectrophotofluorometer.

measuring absorption and fluorescence on the same scan. Ultraviolet or visible
light from the excitation monochromator is directed up through interchangeable
slits and onto the layer side of the plate. The spots are excited in the ultraviolet, and the visible fluorescence that passes through the glass plate and
through the fiber optic is measured. Sterols and higher alcohols made into
fluorescent urethanes by reaction with fluoranthenyl-3-isocyanate have been
quantitatively determined on TLC plates at concentrations from 0.005 to 5 μg
by this scanning method [2]. In addition this application has been used in air-
pollution studies [3].

For fluorescence studies with excitation below 350 nm the plate is turned
upside down; the spots are excited in the ultraviolet, and the visible fluorescence that passes through the glass plate and through the fiber optic is
measured. The fiber optic is positioned at a 45-degree angle to the plate for
fluorescence measurements and at a 90-degree angle for absorption measurements. When the fiber-optic pickup is used, the range is from approximately
380 nm up into the visible. By replacing the fiber optic with an available
accessory 1P28 tube the range can be extended to below 380 nm. The scanning
unit travels a distance of 7 in. in 1, 2, or 4 min from left to right or vice versa.

The motorized thin-film scanner, available for use with the Aminco-Bowman spectrophotofluorometer or the Fluoro-Microphotometer costs $1400 plus $155 for the ultraviolet assembly. A cheaper model, the motorized Uniscan, useful for studies of electrophoresis or TLC separations, adapts to the Fluoro-Microphotometer and costs $650.

A similar attachment to the Perkin-Elmer Model MPF-2A fluorescence spectrophotometer is shown in Fig. 134. It permits direct analysis on TLC plates up to 8 x 8 in. and costs $2000. The attachment is kinematically mounted as shown in Fig. 134. The plate can be scanned automatically, forward and reverse along the axis at a rate of 25 or 50 mm/min, and manually along the X and Y axes.

An adapter for scanning fluorescence on paper chromatograms and electrophoretograms with the Farrand spectrofluorometer has been reported by Takemoto [4].

Eisenbrand [5] attached paper-chromatography strips to a rotating drum and by appropriate rotation of the drum in front of a monochromator was able to detect delayed fluorescence and phosphorescence. He was also able to obtain spectra of the emitted luminescence.

Wadman et al [6] utilized the Beckman spectrophotometer to measure the fluorescence of reducing oligosaccharides on paper chromatograms after condensation with an amine reagent. Each spot was cut out and fastened with a rubber band in front of a hole in a wooden block. The block was then placed in the cell-holder compartment with the paper as near the phototube as possible to expose the latter to the maximum fluorescent light. A filter to absorb the exciting light and to transmit the fluorescent light was placed between the paper and the phototube. The slit was opened wide, and the sensitivity set near maximum. Fluorescence was read on the percent-transmission scale. Transmittance without paper was checked frequently to correct for variations in sensitivity of the instrument. The selector switch was set at 1.0 or 0.1, depending on the intensity of fluorescence.

Hamman and Martin [7] described a scanning device that incorporates flexible fiber optics to allow connection to any suitable arrangement of light source and photocell in various spectrophotometers with simple adapters. The details of the scanning arm are shown in Figure 135.

When the arm is used for fluorescence measurement, both fiber optics are placed in the upper head of the scanner. Fluorescence below 400 nm is difficult to detect since the specific fiber optics employed (Flexible Light Guides, 36 in. long and 1/8 in. in diameter, American Optical Co., Inc., Southbridge, Mass.) do not transmit below this wavelength. For quantitative measurements excitation

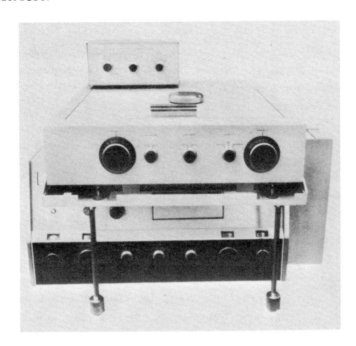

FIG. 134. Scanning attachment for the Perkin-Elmer Model MPF-2A fluorescence spectrophotometer.

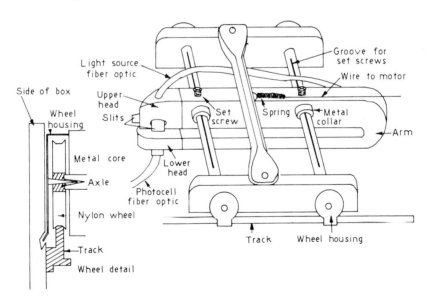

FIG. 135. Scanning arm containing flexible fiber optics. Reprinted from Ref. [7] by courtesy of the Academic Press, Inc.

and emission monochromators can be set at the optimum wavelengths for the substance to be determined.

Solid sample accessories, available for use with the Aminco-Bowman SPF, Farrand, or Perkin-Elmer MPF-2, can be used for fluorescence assay of many types of solid samples, for example, powder, crystal, paper, and plastic. A typical accessory, the Aminco Model C73-62140, is shown in Fig. 136. Monochromatic light from the excitation monochromator is focused on the solid sample. The fluorescence is reflected by an aluminum mirror into the fluorescence monochromator. The sample can be compressed into a potassium bromide pellet for analysis.

III. ANALYSIS OF PAPER CHROMATOGRAMS

Different grades of filter paper vary in their content of fluorescent materials, which may interfere with attempts to measure fluorescence emission from the paper. Also the problems concerning the purity of filter paper depend on the compound under analysis. Van Horst et al. [8] stated that Whatman No. 1 paper is not suitable for measuring the fluorescence of amino acids on paper chromatograms and suggested Whatman No. 4 or 54 for this purpose. Shore and Pardee [9] suggested Whatman No. 52 or S and S No. 57 for amino acid analysis. This same group had reported earlier [6] that Whatman No. 1, Whatman No. 4, and S and S No. 597 were best suited for the fluorometric assay of reducing oligosaccharides on paper chromatograms.

Any one of the commercially available fluorometer or fluorescence spectrometers can be used (with appropriate modification) to measure fluorescence emitted from paper chromatograms. Fluorescence can be excited by shining the exciting light through the paper, as did Wadman et al. [6]. Semm and Fried [10] used a similar arrangement, although complete details were not given. The paper can also be placed at a 45-degree angle across the cell compartment, and the emitted fluorescence can be measured from the surface.

If one sets up to measure fluorescence from paper in a spectrophotometer of in a fluorescence spectrometer, it should be possible to determine the entire fluorescence spectrum of a given compound. Korte and Weitkamp [11] report procedures for measuring the fluorescence spectra of berberine, fluorescein, xanthopterin, and thiochrome on paper chromatograms.

Van Horst et al. [8] carried out a quantitative study of amino acids on paper with a Photovolt Model 525 densitometer with a phototube detector connected to a various reader. A mercury lamp was used for excitation.

Photographs of fluorescence from chromatographic surfaces can be made with an ultraviolet lamp and the usual photographic equipment. Since it is necessary to vary exposure times with each chromatogram in order to produce the best contrast, it is convenient to use a Polaroid camera. Abelson [12] developed a simple procedure for photographing, without the use of a camera,

FIG. 136.   Solid-sample accessory, C73-62140.

spots that emit fluorescence when excited with light at 360 nm.   A Kodak 2A
gelatin filter, which absorbs radiant energy below 405 nm, is interposed
between the chromatogram and the contact paper (Kodabromide F 5').   The
chromatogram is exposed to a 365-nm source (mercury lamp with Wood's filter)
of appropriate intensity and at a suitable distance.   The photographic paper is
then developed in the normal manner.

IV. ANALYSIS OF TLC PLATES

Mixed-adsorbent two-dimensional thin-layer chromatography has proved
useful, as has gas chromatography, when followed by fluorometric and phosphorimetric examination, in characterizing all types of large conjugated
compounds (benzo[a]pyrene, anthracene, carbazole, chrysene, etc.) in complicated organic fractions of airborne particulates and effluents from air-pollution
sources [13].

In paper and thin-layer chromatography the positions of the separated
components are often detected either by their fluorescence under ultraviolet
light or by fluorescence quenching.

Preparation of fluorescent and phosphorescent thin layers can be accomplished by incorporating fluorescent inorganic or organic chemicals in the
adsorbent prior to chromatography.   The technique is an adaptation of that used
by Sease [14,15] for column chromatography.   Kirchner et al. [16] incorporate

0. 15 g of zinc silicate (phosphor No. 609, E. I. du Pont de Nemours and Co. ,
Inc.) and 0. 15 g of zinc cadmium silicate (phosphor No. 1502) in 20 g of the
adsorbent and binder. According to Sease [15], the use of both silicates gives a
mixture whose excitation range extends from 230 to 390 nm. Other phosphors,
such as zinc silicate luminescent material P1, Type 118-2-7 (General Electric
Co. , Cleveland, Ohio) and Ultraphor (Badische Anilin-und Soda-Fabrik,
Ludwigshafen, Germany), have been used. Positions of compounds that absorb
in this range are revealed by a loss or decrease of fluorescence intensity of the
inorganic phosphors under ultraviolet illumination. For illumination a short-
wavelength light, Mineralite Model SL 2537, and a long-wavelength light, Model
SL 3600, are satisfactory for most purposes. General Electric's 15-W Blacklite
F15T8-BLB fluorescent lamp can be easily used in the usual desk-top lamp
fixture as a useful source of long-wavelength ultraviolet light. Alternative
processes incorporate organic fluorescent agents in the slurry during the
preparation. These include the use of 0. 04% aqueous sodium fluorescein
solution instead of water [17], 0. 0011 g of Rhodamine 6G in 30 g of adsorbent
[18,19], and a 0. 02% solution of 2', 7'-dichlorofluorescein [20].

A spray method may be used after chromatography, including such fluorescent reagent sprays as a 0. 02% solution of 2', 7'-dichlorofluorescein in ethanol
[21], a 0. 05 to 0. 004% solution of fluorescein in water [16, 17], a 0. 05% solution
of Rhodamine B in water [22, 23] or a 0. 05% solution of morin in methanol [24].
Compounds containing conjugated double bonds appear as colored spots when
viewed under ultraviolet light. If fluorescein-sprayed layers, while still damp,
are exposed to a small amount of bromine vapor, the fluorescein is converted to
eosin except where compounds are located, with the result that under ultraviolet
light the compounds show up as a greenish yellow spot on a quenched background.

One of the early fluorescent chromatogram scanners used a mercury-vapor
lamp attached to a Beckman Model DU spectrophotometer as the source of
monochromatic light and the usual interference filter-phototube combination for
the determination of the fluorescence intensity [25]. Several methods allow for
direct fluorescence scanning of the chromatogram. Jänchen and Pataki [26]
described the Camag-Turner TLC-Scanner and demonstrated reproducibilities
of 3 to 5% for the fluorescence method and 5 to 8% for the quenching technique
for results on the same chromatogram. A reproducibility of 10 to 15% was
obtained for results compared from different chromatograms. The quantity of
fluorescent compounds chromatographed was between 0. 5 and 10 µg. Determinations of purine and pyrimidine derivatives [27] as well as quantitative
determinations of the dinitrophenyl, dimethylaminonaphthalenesulfonyl, and
phenylthiohydantoin derivatives of amino acids also were accomplished by this
same instrument [28]. Carbohydrates made fluorescent by reaction with

p-aminohippuric acid have been determined on thin layers supported by flexible
stainless-steel plates so that the plate could be attached to a rotating drum in
the chromatography door of the Turner fluorometer [29].

A method developed for the determination of pesticides on chromatograms
by reflectance measurements [30] may prove useful as a further improvement in
fluorescence scanning. In this method both the light-source glass fibers and
monitor glass fibers are located randomly in the same head. A second head
containing both light-source fibers and monitor glass fibers scans the blank
area adjacent to the spots and is used to correct for background differences on
the plate through a bridge circuit with the first scanner head and recorder.
Thin-layer scanners such as the Nester/Faust Uniscan 900 TLC include
fluorescence-monitoring capability.

Applications of fluorescence and thin-layer chromatography in the assay of
environmental pollutants will be discussed in Chapter 16.

V. ANALYSIS BY SOLID-SURFACE ATTACHMENTS

Fluorescence analysis from solid surfaces with attachments to the Aminco-
Bowman SPF, Perkin-Elmer MPF, or Farrard spectrofluorometer (Fig. 136)
has been used by several workers, although the true potential of this technique
is yet to be tapped. Perhaps after researchers realize that this method can be
as accurate as measurements in solution (see Section VI) shall we see a more
widespread use.

Van Duuren and Bardi [31] have measured the reflectance-fluorescence
spectra of hydrocarbons in KBr pellets in the concentration range of 0.02 to 200
μmole/gram of KBr. Spectral measurements were made on KBr disks, 0.7 to
1.0 mm thick, 13.0 mm in diameter, which were prepared with an appropriate
die. The disks were placed on a black sample holder in the cuvette compartment
of a Farrand spectrofluorometer. Van Duuren and Bardi found that excitation
spectra in KBr pellets generally agreed with spectra obtained in cyclohexane.
On the other hand, the fluorescence spectra of many of the hydrocarbons
investigated differed significantly from those obtained in solution. The
significance and utility of KBr fluorescence remain to be determined.

A method used by Sawicki [32] and associates for obtaining the excitation
and emission fluorescence characteristics of spots on chromatograms uses an
Aminco-Bowman spectrophotofluorometer equipped with a solid-sample
attachment. In this method the chromatogram is sprayed with GS chromatogram-
preserving medium (Gallard-Schalesinger Corp., Carle Place, N.Y.) before
being removed from the plate. The chromatogram may then be cut into pieces
for fluorometric examination of each piece in the attachment.

This technique can be a powerful one for assay of pharmaceuticals and in forensic analysis. It is possible to place a tablet or pill on the solid-surface attachment and determine quantitatively the drugs present (provided the drug is fluorescent). Passwater [33] has shown that sugar cubes impregnated with lysergic acid diethylamide ($<$ 10 μg/cube) emit an intense fluorescence that is not observed with untreated sugar cubes.

VI. QUANTITATIVE ANALYSIS ON SOLID SURFACES [34-36]

A. Introduction

Guilbault and co-workers have attempted to develop methods for quantitative analysis on solid surfaces. The basis of this procedure was the adaption of the attachment to an Aminco filter fluorometer to accept, instead of a glass cuvette, a metal slide painted black to lower the background. On the slide is placed a silicone-rubber pad on whose surface are placed all the reagents for a quantitative assay. On addition of the sample to be assayed, a fluorescence is produced on the surface and is measured as in solution analysis. This appears to be the first report of the accurate measurement by a fluorescence method of the rate of a chemical reaction on a solid surface.

B. Experimental Data

In attempting to design a solid-surface device Guilbault et al. started by preparing clear-glass slides with small strips of chromatographic paper (Chrom AR Sheet 500; Mallinckrodt), adding the reagents and solution to be analyzed, and recording the rate of change of the fluorescence. The fluorescence rate of identical slides varied too greatly for the method to be of any analytical use. The next step was to paint a glass slide the complementary color of the fluorescent radiation, apply the reagents directly to the painted surface, add a drop of the solution to be analyzed, and finally record the rate of fluorescence change. The main problem with this procedure was that when the slide was placed into the fluorometer, in a vertical position, the drop of enzyme solution would sometimes roll off the plate. Since the first two methods were not satisfactory a silicone-rubber pad on a clear-glass slide was used as the solid matrix on which the substrate and enzyme could be applied. This method proved to be acceptable.

In this device pads of silicone rubber (Dow Corning Glass and Ceramic Adhesive, Dow Corning Co. , Midland, Mich.) were prepared by pressing uncured silicone rubber between two pieces of Glassine paper (Eli Lilly and Co.) into the cross-sectional form shown in Fig. 137. The pads were prepared in a strip about 12 cm long and 15 mm wide. Usually 15 to 20 individual pads were obtained from such a strip.

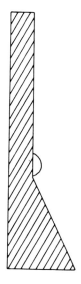

Pad support

FIG. 137.   Pad support.

Two methods were used to apply the substrate to the pad.   The batch
method involved putting a solution of the substrate onto the strip of silicone
rubber, allowing the solution to evaporate, and then cutting the strip into the
desired pad dimensions.   The prepared pads were then placed on their respective
glass slides.   The second method involved cutting the individual pad from the
strip, placing it on the glass slide, and then applying the substrate to the pad.
This isolated-pad method was found to be preferable.

The pads were placed approximately 1 to 2 cm from the bottom of ordinary
3 x 1 in. microscope slides 0.96 to 1.06 mm thick.   It was essential that the
entire pad be in the optical path.

A special cell was constructed to hold the glass slides in the optical path
of the fluorometer (Fig. 138).   The cell was constructed of ordinary sheet metal
with portholes to allow radiation to enter and leave the cell cavity.   The guides
for the glass slides were 7.5 cm long by 0.4 cm wide, and the space between
the guides was 1.5 mm.   In order to obtain as low a background as possible the
cell was painted with a dull-black optical paint.

Difficulties were encountered with this apparatus due to the vertical
positioning of the pad; gravity caused the drop to fall off the surface.   Guilbault
and Vaughn circumvented this difficulty by turning the fluorometer on its end,
supported by two wooden blocks placed parallel to the primary filter holder
(Fig. 139).   This prevented instability of the drop on the pad since the pad could

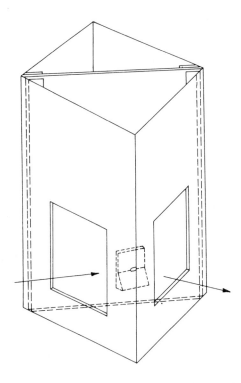

FIG. 138.   Cell for holding glass slides in the optical path of the fluorometer.

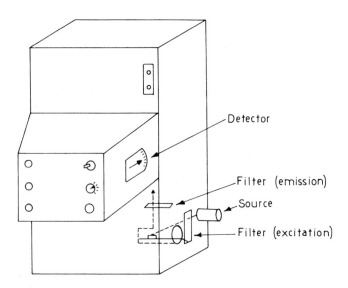

FIG. 139.   Fluorometer placed on end for horizontal pad studies.

now be held horizontally. The base of the instrument was supported with a piece of wood to prevent electronic noise.

An Aminco-Bowman cell adapter of black Delrin, catalog number A6-63019, was modified to hold a metal plate with the dimensions shown in Fig. 140, view A. The metal strip was painted black.

Two grooves were cut in the cell adapter in such a way that the plate could be slid into the cell adapter with the plate lying flush to the base of the two opposing portholes. The position of the plate is shown in Fig. 140, view B. The groove was cut parallel to the lower side of the portholes with reference to the porthole on the top of the cell holder.

View C of Fig. 140 shows the top of the cell holder. The groove does not pass centrally through the cell holder.

The top of the cell is shown in view D of Fig. 140. The porthole was filed so that the light emitted from the sample on the silicone-rubber pad would not be blocked out by the sides of the cell holder. The pad and its holder are shown in Fig. 141.

1. Drop Shape and Volume

The shape of the drop is critical. Because of the geometry of the system, the drop can act as a lens, and the excitation light passing into the drop can be internally reflected many times. This implies that the background and rate of change of fluorescence will vary from sample to sample. In order to prevent

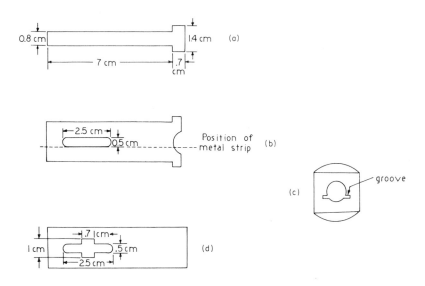

FIG. 140. Construction of the cell holder and metal plate. (A) Dimensions of metal plate; (B) side view of cell holder showing position of metal plate; (C) end view of cell holder showing position of groove; (D) top view of cell holder showing the filed aperture.

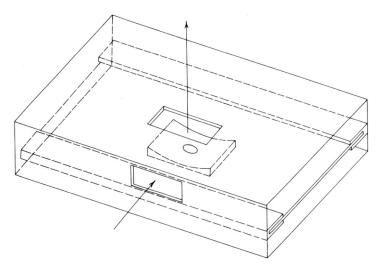

FIG. 141.   Pad and holder.

this variation the drop was spread over half the surface of the pad.   This also
helps to mix the sample.   Without this precaution reproducible results are not
obtained.

2. Pad Shape

The shape of the pad should be kept constant.   Differences between pads
change the background and rates of change of fluorescence.   Both these factors
increase as the pad size increases.   Only pads that gave identical backgrounds
(no substrate or solution present) were used in the determinations.

3.   Color of Pads

The color of the silicone rubber affects both the background and rate of
change of fluorescence.   With any of the filter systems, the background and rate
of change of fluorescence increase in the order black ⟨ gray ⟨ clear ⟨ white.

TABLE 89

Comparison of Pad Colors[a]

Color	Background	Rate (ΔF/min)
Translucent	1. 29	4. 6
Gray	0. 30	4. 3
Black	0. 31	1. 3

[a]Pad composition was 20 µl of both ß-NAD and lithium lactate solution;
also 20 µl of glycine buffer and 30 µl of the 3850 I. U. /ml commercially
available enzyme solution were used in the analysis.

Each possible combination of pad color and filter was examined, and it was found that the most accurate results could be obtained if a gray silicone-rubber pad was used (Table 89).

C. Results

1. Assay of Cholinesterase in Blood [34]

Pads for the assay of cholinesterase were prepared by placing 10 μl of a 10^{-2} M solution of N-methylindoxyl acetate (Isolabs, Akron, Ohio) onto a silicone-rubber pad.

(nonfluorescent)   (fluorescent)

The slide was then removed from the fluorometer, and 20 μl of an enzyme solution was applied to the slide, being careful to get the enzyme solution to cover as much of the substrate film as possible. The recorder was started immediately, the slide was placed back into the instrument, and the rate was recorded. A calibration plot of the change in fluorescence units per minute versus enzyme concentration was prepared and used for subsequent analyses.

Linear calibration plots were obtained for cholinesterase concentrations from 10^{-6} to 10^{-2} units/ml with a precision of about 2% and an accuracy of 2.2%.

Preliminary studies have indicated that pads made were stable for at least 30 days (Table 90) if kept in a cold, dark place. Further research is under way.

This stability compares very favorably with that of the solution methods: N-methylindoxyl acetate solutions are very unstable and must be prepared fresh daily. In contrast, the substrate on the pad, since it is present in a dry state, is quite stable for at least 30 days.

2. Alkaline Phosphatase in Blood [35]

Onto a series of gray silicone-rubber pads (adhesive sealant 3145 RTV) was placed 10^{-2} M Naphthol AS-BI phosphate (0.01 ml). The ethanol was allowed to evaporate off so that the Naphthol AS-BI phosphate was deposited on the

TABLE 90

Pad-Stability Data[a]

Day[b]	Rate $(\Delta F/\min)$[c]
0	3.10
10	3.04
15	3.08
20	3.20
30	3.00

[a]A 500-μg/ml solution of cholinesterase was freshly prepared for each day; 20 μl of a 0.1 M substrate solution were applied to the pad and allowed to evaporate to dryness.

[b]Days after preparation of pad.

[c]Rates were an average of at least three runs.

surface of the pad. These pads can be used immediately or stored below 0°C for at least 2 months without any deterioration.

$$\text{Naphthol AS-BI phosphate} \xrightarrow{\text{alkaline phosphatase}} \text{Naphthol AS-BI} \qquad (215)$$
$$\text{(nonfluorescent)} \qquad\qquad\qquad\qquad \text{(fluorescent)}$$

The pad was placed on the blackened metal strip, which was then slid into the cell holder, in such a way that the pad was centrally positioned below the filed aperture of the cell holder. Onto the pad was placed from syringes pH 9.8, 1 M, 2-amino-2-methylpropanol-HCl buffer (0.02 ml) and then serum (0.01 ml). The drop formed was mixed with the needle of a syringe so that the drop covered half the area of the pad.

The cell holder was placed in the fluorometer, and the rate of reaction was recorded. It may be necessary to wait 15 sec before the reaction starts; this is particularly noticeable with serum samples having low alkaline phosphatase values. The serum alkaline phosphatase value can be obtained from the initial rate determined in two ways.

A plot of serum alkaline phosphatase values obtained from an alternative procedure versus initial rate of reaction gives a straight line. In the study carried out the rate of reaction was plotted against units alkaline phosphatase obtained from a standard laboratory procedure using phenolphthalein phosphate. The results, plotted in Fig. 142, indicate good agreement between the two methods over a wide range of serum values.

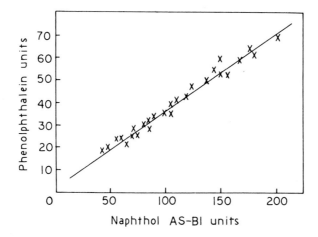

FIG. 142. Correlation between results obtained with the Naphthol AS-BI method and the phenolphthalein method.

3. Lactate Dehydrogenase in Serum [36]

The reaction system used was

$$\text{Lithium lactate} + \text{NAD} \xrightarrow{\text{LDH}} \text{NADH} + \text{pyruvic acid} \qquad (216)$$

The rate of production of the fluorescent product NADH was monitored.

A 50-μl aliquot of the NAD solution was applied to the pads, and evaporated over silica gel under reduced pressure, to produce a thin film of solid NAD on the surface of the pad. Then a 20-μl aliquot of the lactate solution was added to the pad and evaporated similarly, to produce a thin film of solid lactate. Thus a thin film of solid NAD and lithium lactate on the silicone-rubber surface was the final pad composition.

The fluorometer was calibrated before work each day by putting a 10-μl sample of a 10-μg/ml solution of quinine sulfate on a black pad and setting the sensitivity control of the instrument so that a reading of 60% transmittance on the 0.03 scale was obtained. The choice of this particular transmittance value was purely arbitrary. The pad was permanently attached to one of the metal slides so that reproducible results were easier to obtain.

A pad was placed in the proper position on the metal slide. A 10-μl aliquot of the buffer solution was put onto the pad and spread out so that the entire substrate film was dissolved in the drop of solution. The metal slide was then set in the fluorometer, and a background rate was recorded.

After recording the background for about 1 min, the slide was removed from the instrument and a 20-μl aliquot of either the commercially available enzyme solution or the serum solution, whichever was to be analyzed, was put onto the pad. The slide was immediately placed back into the instrument, and the fluorescence rate was recorded.

A calibration plot of the change in fluorescence units per minute versus the enzyme concentration could then be constructed and used for future assay of unknown solutions.

From 160 to 1000 units of lactate dehydrogenase in serum were assayed with an average relative error of 2.3%.

This method is more convenient than most normal solution methods because only one or two solutions are used during the actual analyses and there are fewer manipulations, such as pipetting, to be performed. Substrate solutions need be prepared only when making the pads. If the pads are stable for longer than a month, the necessity for preparing solutions of these usually expensive chemicals is greatly reduced. Thus this method is more economical than other techniques. When any of the enzyme or substrate solutions are used, only small volumes are necessary, thus providing another economy.

Another advantage of this method is the time involved. The complete analysis, from the measurement of a volume of the buffer solution to the end of recording the rate, usually takes from 3 to 5 min.

4. Other Systems

The described solid-surface method should be applicable to any solution method used previously. It could be used, for example, to assay for metal ions like Mg^{2+} in blood by placing an appropriate organic reagent on the pad surface. The ion would then react with the chelator to give a fluorescence, which is measured. Studies along these lines are currently in progress in our laboratory.

REFERENCES

1. Manual of Electrophoresis Procedures for the Model 310 System, G. K. Turner Associates, 12C, January 1965.

2. H. T. Gordon, J. Chromatogr., 22, 60 (1966).

3. E. Sawicki, T. W. Stanley, and W. C. Elbert, J. Chromatogr., 20, 348 (1965).

4. Y. Takemoto, Nature, 190, 1044 (1961).

5. J. Eisenbrand, Pharm. Acta Helv. , 39, 232 (1964).

6. W. Wadman, G. Thomas, and A.Pardee, Anal. Chem. , 26, 1192 (1954).

7. B. Hamman and M. Martin, Anal. Biochem. , 15, 305 (1966).

8. S. H. Van Horst, H. Tang, and V. Jurkovich, Anal. Chem. , 31, 135 (1959).

9. V. G. Shore and A. B. Pardee, Anal. Chem. , 28, 1479 (1956).

10. K. Semm and R. Fried, Naturwissenschaften, 38, 326 (1952).

11. F. Korte and H. Weitkamp, Angew Chem. , 70, 434 (1958).

12. D. Abelson, Nature, 188, 850 (1960).

13. E. Saeicki, T. W. Stanley, S. McPherson, and M. Morgan, Talanta, 13, 619 (1966).

14. J. W. Sease, J. Amer. Chem. Soc. , 69, 2242 (1947).

15. J. W. Sease, J. Amer. Chem. Soc. , 70, 3630 (1948).

16. J. G. Kirchner, J. M. Miller, and G. J. Keller, Anal. Chem. , 23, 420 (1951).

17. E. Stahl, Chem. Ztg. , 82, 323 (1958).

18. R. H. Reitsema, Anal. Chem. , 26, 960 (1954).

19. J. Avigan, D. S. Goodman, and D. Steinberg, J. Lipid Res. , 4, 100 (1963).

20. J. L. Brown and J. M. Johnson, J. Lipid Res. , 3, 480 (1962).

21. H. K. Mangold, J. Amer. Oil Chem. Soc. , 38, 708 (1961).

22. G. Manchata, Mikrochim. Acta, 79 (1960).

23. S. Huneck, J. Chromatogr. , 7, 561 (1962).

24. P. Schellenberg, Angew. Chem. , 84, 118 (1962).

25. J. A. Brown and M. M. March, Anal. Chem. , 25, 1865 (1953).

26. D. Jänchen and G. Pataki, J. Chromatogr. , 33, 391 (1968).

27. G. Pataki and A. Kunz, J. Chromatogr. , 23, 465 (1966).

28. G. Pataki and E. Stradky, Chimia, 20, 361 (1966).

29. W. M. Connors and W. K. Boak, J. Chromatogr. , 16, 243 (1964).

30. M. Beroza, K. R. Hill, and K. H. Norris, Anal. Chem. , 40, 1608 (1968).

31. B. L. Van Duuren and C. Bardi, Anal. Chem. , 35, 2198 (1963).

32. E. Sawicki, T. W. Stanley, and H. Johnson, Microchem. J. , 8, 257 (1964).

33. R. Passwater, American Instrument Co. , Silver Spring, Md. , private communication.

34. G. G. Guilbault and R. Zimmerman, Anal. Letters, 3, 145 (1970).

35. G. G. Guilbault and A. Vaughn, Anal. Chim. Acta, 55, 107 (1971).

36. G. G. Guilbault and R. Zimmerman, Anal. Chim. Acta, 58, 75 (1972).

Chapter 15

FLUORESCENT INDICATORS

I. INTRODUCTION

Fluorescent indicators are used in analysis in much the same way that colored indicators are--that is, to indicate the end point of a titration. However, instead of a color change, there is a change in fluorescence intensity as the indicator changes from one form to another. Such indicators are generally used in turbid, dark, or highly colored solutions where colored indicators are difficult to observe or lack sensitivity, and also in precipitation titrations.

There are four classes of fluorescent indicators: (a) acid-base indicators, where fluorescence intensity changes with pH; (b) oxidation-reduction indicators, whose fluorescence changes with the oxidation state of the molecule; (c) adsorption indicators, whose adsorption effects a change in fluorescence intensity; and (d) metallochromic indicators, where a change in the fluorescence occurs due to a change in the concentration of an ion involved in chelate formation.

II. APPARATUS

Assays with fluorescent indicators can employ a small ultraviolet headlamp (short or long wavelength) in a dark room or a home-made view box equipped with an ultraviolet-light source.

Alternatively any fluorometer is easily adapted to this assay. An inexpensive light source (generally $>$ 300 nm; i. e. , a 360 BL lamp) can be used as the

source. Diehl [1], White [2], and Wilkins [3] have described titration view
boxes for use in this application.

III. ACID-BASE INDICATORS

The use of fluorescent indicators in titrations of colored solutions
represents the most prominent of the indicator applications. A good fluorescent
acid-base indicator is one whose fluorescence is very intense and highly dependent on pH. The indicator should fluoresce in the visible region. 2-Naphthol
would be a good fluorescent acid-base indicator for the 8 to 9 pH range
(Fig. 143). The indicator would be useful for titrations in highly colored
samples, such as red wine or animal tissues and fluids, where regular colored
indicators would be difficult to use.

Either the molecular (HInd) or the ionized form (Ind$^-$) forms of the
indicator may be the fluorophor:

$$HInd \;\; \rightleftharpoons \;\; H^+ + Ind^-, \tag{217}$$

$$Ind^- + h\nu \longrightarrow Ind^* \longrightarrow Ind^- + h\nu \; (fluorescence), \tag{218}$$

or

$$HInd + h\nu \longrightarrow HInd^* \longrightarrow HInd + h\nu \; (fluorescence). \tag{219}$$

Chen [5] has shown that 4-methylumbelliferone can be used as a fluorescent
indicator for the titration of acids (λ_{ex} 360 nm) or bases (λ_{ex} 300-320 nm).
A list of all the readily available indicators covering the pH range 0 to 14

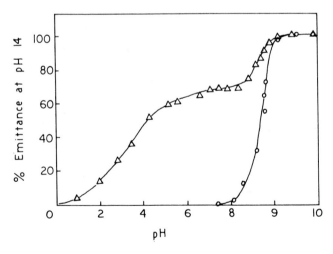

FIG. 143. Variation in the fluorescence intensity of 2-naphthol with pH:
o — o, 365-nm excitation; △ — △, 313-nm excitation. After Hercules and
Rogers, Ref. [4].

is given in Table 91.   Additional references can be found in chapters on this
subject in books or leaflets [6-8] and in papers by Dement [9, 10], who listed
60 fluorescent indicators and their pH values.

TABLE 91

Fluorescent Indicators

Indicator	Color change[a]	pH Range
4-Methylumbelliferone (first transition)	G-WB	0. 0-2. 0
3, 6-Dihydroxyphthalimide	G-YG	0. 0-2. 5
Benzoflavin	Y-G	0. 3-1. 7
Ethoxyacridone	G-B	1. 2-3. 2
3, 6-Tetramethyldiaminoxanthone	G-B	1. 2-3. 4
Esculin	C-B	1. 5-2. 0
Eosin yellowish	C-Y	2. 0-3. 5
7-Amino-1, 3-naphthalenedisulfonic acid	WB-B	2. 0-4. 0
1-Hydroxy-2-naphthoic acid	C-B	2. 5-3. 5
Salicylic acid	C-B	2. 5-4. 0
2', 4', 5', 7'-Tetrabromofluorescein sodium salt	C-G	2. 5-4. 5
2-Naphthylamine	C-V	2. 8-4. 4
Erythrosin	C-B	3. 0-4. 0
1-Naphthylamine	C-B	3. 0-4. 5
3-Hydroxy-2-naphthoic acid	B-G	3. 0-6. 8
o-Phenylenediamine	G-C	3. 1-4. 4
p-Phenylenediamine	C-OY	3. 1-4. 4
5-Aminosalicylic acid	C-G	3. 1-4. 4
o-Methoxybenzaldehyde	C-G	3. 1-4. 4
Chromotropic acid	C-B	3. 5-4. 5
Phloxine	C-Y	3. 4-5. 0
Fluorescein	WG-G	4. 0-5. 0
Quininic acid	Y-B	4. 0-5. 0
4, 5-Dihydroxy-2, 7-naphthalenedisulfonic acid disodium salt	WB-B	4. 0-6. 0
2', 7'-Dichlorofluorescein	WG-G	4. 0-6. 0
Resorufin	C-O	4. 0-6. 0
ß-Methyl esculetin	C-B	4. 0-6. 2
Acridine	G-B	4. 5-6. 0
3, 6-Dihydroxyxanthone	C-BG	5. 4-7. 6
3, 6-Dihydroxyphthalic acid	B-G	5. 8-8. 2

TABLE 91 --Continued

Indicator	Color change[a]	pH Range
Quinine (first transition)	B-V	5. 9-6. 1
5-Amino-2, 3-dihydro-1, 4-phthalazinedione	B-WB	6. 0-7. 0
3, 6-Dihydroxyphthalonitrile	B-G	6. 0-8. 0
2-Naphthol-6-sulfonic acid sodium salt	C-B	6. 0-8. 0
Brilliant Diazo Yellow	C-B	6. 5-7. 5
Thioflavin	C-G	6. 5-7. 6
4-Methylumbelliferone (second transition)	WB-B	6. 5-8. 0
Umbelliferone	O-B	6. 5-8. 0
Orcinaurine	C-G	6. 5-8. 0
Magnesium-8-quinolinol	C-Y	7. 0-7. 2
2-Naphthol	WB-B	7. 0-8. 5
2-Naphthol-6, 8-disulfonic acid dipotassium salt	WB-B	7. 0-8. 5
Morin	WG-G	7. 0-8. 5
1-Naphthol	C-BG	7. 0-9. 0
2-Naphthol-3, 6-disulfonic acid disodium salt	WB-B	7. 0-9. 0
2-Naphthol-6, 8-disulfonic acid dipotassium salt	C-B	7. 5-9. 0
trans-o-Hydroxycinnamic acid	C-G	7. 2-9. 0
1-Naphthol-4-sulfonic acid sodium salt	DB-LB	8. 0-9. 0
1-Naphthol-2-sulfonic acid sodium salt	DB-LB	8. 0-9. 0
Coumarin	WG-G	8. 0-9. 5
Acridine Orange	WYG-Y	8. 0-10
Naphthol AS	C-YG	8. 2-10. 3
2-Naphthol-sulfonic acid	B-V	9-10
Ethoxyphenylnaphthastilbazonium chloride	G-O	9. 0-11. 0
6, 7-Dimethoxyisoquinoline-1-carboxylic acid	Y-B	9. 5-11. 0
Quinine sulfate (second transition)	V-C	9. 5-10. 0
Eosin BN	C-Y	10. 5-14. 0
1-Naphthylamine	B-WB	12. 0-13. 0
7-Amino-1, 3-naphthalenedisulfonic acid	B-WO	12. 0-13. 0
Cotarmine	Y-C	12. 0-13. 0
1-Naphthoic acid	LB-G	12. 0-13. 0
2-Naphthoic acid	B-V	12. 0-13. 0
4-Amino-1-naphthalenesulfonic acid	B-G	12. 0-14. 0

[a]Key: C, colorless; B, blue; G, green; V, violet; O, orange; Y, yellow, D, dark; L, light; W, weak.

Care should be exercised not to pick an indicator whose emission is absorbed by the solution. Acridine, for example, has a green fluorescence in acid, a violet fluorescence in base. If the titration is performed in a yellow solution (one that absorbs in the violet), then the absorption spectrum of the medium will superimpose itself on the fluorescence spectrum of the indicator in alkaline solution. Thus acridine will have a green fluorescence in acid but no fluorescence in basic solutions.

To eliminate this problem we should select fluorescent indicators in accordance with the color of the solution studied. Thus two sets of indicators should be chosen, each with a different fluorescent color at each pH: one set could be used for blue and green solutions (with green or blue fluorescence), the other for red, yellow, and brown solutions (with red, yellow, and brown fluorescence).

IV. OXIDATION-REDUCTION INDICATORS

Oxidation-reduction indicators change fluorescence with change in oxidation state--that is, the oxidized state is nonfluorescent, the reduced state is fluorescent, or vice versa.

Some typical oxidation-reduction indicators and their applications are listed in Table 92.

TABLE 92

Oxidation-Reduction Indicators

Indicator	Application	Color of fluorescence	Refs.
Acriflavine	Iodometry, bromometry, permanganate titrations	Green	11, 12
α-Naphthoflavone	Iodometry, bromometry	Blue	13
Rivanol	Iodometry, bromometry	Green	14
Rhodamine C	Permanganate titrations, iodometry and bromometry	Orange	15, 16
Fluorescein	Iodometry, bromometry	Green	15, 16
Carmine	Iodometry	Blue	14

V. ADSORPTION INDICATORS

The best known example of the use of a fluorescent adsorption indicator is Fajan's method, which was introduced in the 1920s. This method is based on

Before end point   At end point   After end point

the adsorption of the Ind⁻ form of the indicator (such as dichlorofluorescein)
onto the precipitate of AgX charged positively immediately after the end point,
due to adsorption of the excess Ag^+ onto the AgX precipitate. This $(AgX)^+$ +
Ind⁻ adsorption complex, green fluorescent, then indicates the end point of the
Ag^+ titration of halides.

In choosing an adsorption indicator one must take into account the relative
ease of adsorption of the dye and the pK_a of the fluorescent indicator. Thus
eosin, which is a strongly adsorbing dye, is not used for Cl⁻ titrations, but is
used for titrations of Br⁻ and I⁻. In Cl⁻ titrations the less highly adsorbing
fluorescein or dichlorofluorescein is used. Also one must work in a pH region
above the pK_a of the indicator so that the indicator will exist in the adsorbing
anionic form.

Some fluorescent adsorption indicators, useful for titrations, are listed
in Table 93. These indicators and their uses are fluorescein (green) for halides

TABLE 93

Fluorescent Adsorption Indicators

Indicator	Fluorescence color	Ions assayed	Refs.
Acriflavine	Yellow-green	Cl^-, Br^-, I^-, CrO_4^{2-}, $Fe(CN)_6^{2-}$, Ag^+	17
Fluorescein	Green	Cl^-, Br^-, I^-, CrO_4^{2-}, Ag^+, Pb^{2+}	17-19
Dichlorofluorescein	Green	Cl^-	18
Eosin	Green	Cl^-, Br^-, I^-, CrO_4^{2-}, Ag^+, Pb^{2+}	17-19

TABLE 93--Continued

Indicator	Fluorescence color	Ions assayed	Refs.
Erythrosin	Green	Br^-, I^-, CrO_4^{2-}, Ag^+, Pb^{2+}	17, 19
4-Methylumbelli-ferone	Blue	Cl^-, Br^-, I^-, Ag^+, CrO_4^{2-}, Hg_2^{2+}	17, 19
Morin	Green	Pb^{2+}	17
2-Naphtholsulfonic acid	Blue-violet	Cl^-	18
Phloxin	Yellow-green	Cl^-, Br^-, I^-, Ag^+	19
Primulin	Blue	Cl^-, Br^-, I^-, Pb^{2+}, Ag^+	17, 19
Quinine sulfate	Blue	Hg_2^{2+}	17
Rhodamine 6C	Yellow-green	Cl^-, Br^-, I^-, CrO_4^{2-}, Ag^+	17
Thioflavin S	Blue	Cl^-, Br^-, I^-, Ag^+, Pb^{2+}	18, 20
Umbelliferone	Blue	Cl^-, Br^-, I^-, Ag^+, Pb^{2+}, Hg_2^{2+}, WO_4^{2-}	17, 19, 21

or chromate with silver and Pb^{2+} with chromate; eosin (green) for lead with phosphate and halides or chromate with silver; erythrosin (green) for lead with oxalate and halides or chromate with silver; phloxin (yellow-green) for halides with silver; dichlorofluorescein (green) for chloride with silver; Rhodamine 6C (yellow-green) for halides or chromate with silver; 4-methylumbelliferone or umbelliferone (blue) for halides with silver, lead with chromate or oxalate, mercurous ions with thiocyanate or chloride, and tungstate with lead; 2-naphtholsulfonic acid (blue-violet) for chloride with silver; primulin (blue) or thioflavin S (blue) for lead with phosphate and halides with silver; quinine sulfate (blue) for mercury(I) with chromate or ferrous hexocyanate, chloride with silver, and tungstate with lead; morin (green) for lead with ferrous hexocyanate; and acriflavine (yellow-green) for halides or chromate with silver and ferrous hexacyanate with silver.

VI. FLUOROCHROMIC INDICATORS FOR METALS

The G. Frederick Smith Chemical Co. lists several fluorochromic
indicators in their 1967 catalog [22]. One of these is calcein (fluorescein
methyleneiminodiacetic acid, or fluoxerone (134)). This compound has the
acid-base fluorescent properties of fluorescein and the chelating properties of
EDTA.

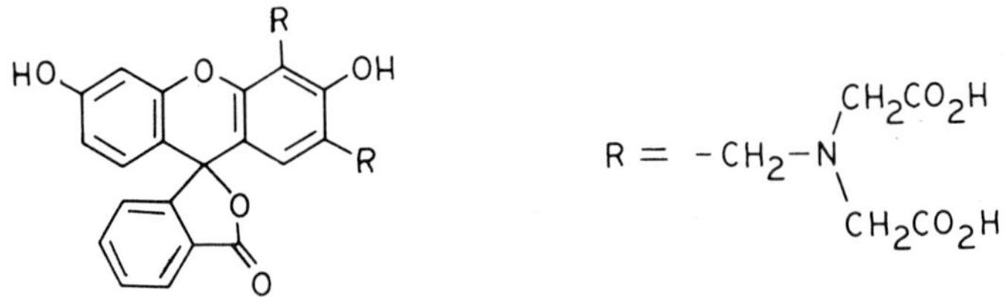

$$R = -CH_2-N \begin{matrix} CH_2CO_2H \\ \\ CH_2CO_2H \end{matrix}$$

(134)

Calcein has a strong adsorption from about 440 to 500 nm at pH 5 to 10 and
gives a rather narrow, bright yellow-green fluorescence emission band from
500 to 575 nm. The fluorescence is pH sensitive and very intense from pH 5 to
10 but drops sharply below pH 5 and above 10. The calcium salt retains a
strong fluorescence at pH values above 12, but the fluorescence of the magnesium salt is greatly diminished at pH values over 11. Calcein has been used for
the determination of calcium in blood serum [22, 23] and as an indicator for
the determination of copper [24, 25]. In an extensive study of calcein with
other elements Wallach and Steck [26] have shown that copper, nickel, and
cobalt destroy the fluoresceince of calcein, but aluminum increases it below
pH 2.5, and at a pH 2 the fluorescence is sufficiently strong to provide a method
for the determination of aluminum. Zinc displaces cobalt from the calcein
complex and results in a fluorescence that is sufficient for the determination of
zinc [26].

Wilkins [27] prepared a similar derivative of calcein, 4-methylumbelli-
ferone methyleneiminodiacetic acid, and called it Calcein Blue (135). By adding
N-methylglycine to this mixture and to the calcein preparation he also prepared

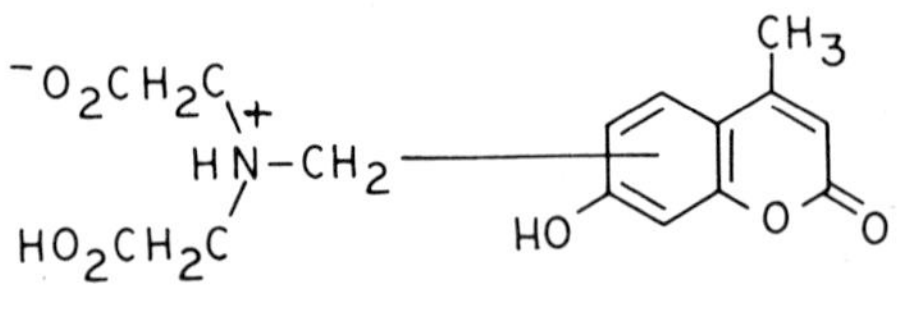

(135)

Methyl Calcein Blue and methyl calcein.   Calcein Blue retains the pH indicator properties of 4-methylumbelliferone and gives a brilliant blue fluorescence (415 nm) at pH 11.   Calcein Blue has an advantage over calcein for fluorometric titrations since its maximum fluorescence is produced with excitation at 330 nm in acid and at 370 nm in base.   Since 370 nm is a major component of most long-wavelength sources of ultraviolet radiation, this reagent is more sensitive than calcein (λ_{ex} 488 nm).

At pH 12 or above Calcein Blue does not fluoresce but does so on the addition of calcium, barium, or strontium.   At pH values of 4 to 10 ions such as those of copper and nickel quench the fluorescence of calcein and Calcein Blue.   In order to explain these phenomena Wilkins [28] suggested that metals form two types of complex with these indicators.   One type is illustrated by the quenching of the fluorescence with copper.   Here the indicator-Cu(II) combination takes the form of the indicator ion and like the indicator ion is not fluorescent above pH 12.   The other form, which Wilkins calls the "reversal type," is illustrated by the reappearance of the fluorescence above pH 12 on the addition of the calcium ion, which forms a Ca-indicator complex similar to the molecular form.   The use of methyl calcein and Methyl Calcein Blue as indicators in the HEDTA titration of aluminum, nickel, and manganese [29] and the use of metal fluorochromic indicators in the separation of nickel, chromium, cobalt, iron, and titanium by ion exchange and titration with EDTA [3] have been proposed.

New metallofluorometric indicators are continually being developed.   Bis-(carboxymethylaminomethyl)dichlorofluorescein can be used as a metallofluorochromic indicator in the titration of copper, cobalt, and nickel with EDTA [30].   Both 7-(2-hydroxy-4-sulfonaphthylazo)-8-quinolinol with its analogs and bisglycine-2,3-dichlorofluorescein are also proposed as metallofluorochromic indicators for the chelometric titrations of magnesium, calcium, copper, cobalt, nickel, iron, chromium, zinc, cadmium, and vanadium [31].

The preparation of two stilbene derivatives, 4,4'-diaminostilbene-N,N,N,',N'-tetraacetic acid and its 2,2'-disulfonic acid, has been described, and these compounds are recommended as metallofluorochromic indicators for the titration of magnesium, calcium, copper, cobalt, iron, nickel, chromium, zinc, cadmium, and vanadium(IV)[32].   Zincon, o-{2-[a-(2-hydroxy-5-sulfophenylazo)benzylidene]hydrazino} benzoic acid, sodium salt, is proposed as a metallofluorochromic end-point indicator for the direct determination of zinc and mercury by titration with EDTA [33].   Eggers [34] has described the preparation of a xanthoncomplexone as well as an umbellicomplexone metallofluorometric indicators that are similar in response to calcein.   A thesis of

194 pages on the preparation, properties, and uses of five new substituted coumarins as metallofluorescence indicators has been produced by Haitink [35] under the direction of H. Diehl. The five compounds are umbelliferone-8-methyleneiminodiacetic acid, 4-methylesculetinmethyleneiminodiacetic acid· 1/2H$_2$O, 4-methylumbelliferone-8-methyleneglycine, umbelliferone-8-methylene-glycine· 1/2H$_2$O, and 4-methylumbelliferone-8-methylenesarcosine· 1/2H$_2$O.

REFERENCES

1. H. Diehl, Fluorometric Reagents for Calcium and Magnesium, G. F. Smith Chemical Co., Columbus, 1964.

2. C. E. White, Practical Fluorescence, Dekker, New York, 1971, Chapter 2.

3. D. H. Wilkins, Talanta, 2, 355 (1959).

4. D. Hercules and L. Rogers, Anal. Chem., 30, 96 (1958).

5. R. F. Chen, Anal. Letters, 1, 423 (1968).

6. J. A. Radley and J. Grant, Fluorescence Analysis in UV Light, Van Nostrand, Princeton, N. J., 1954, pp. 419-429.

7. M. A. Konstantinova-Shlezinger, Fluorometric Analysis, Davey, New York, 1961, pp. 108-120.

8. Fluorescent Indicators, Eastman Organic Chemicals Bulletin, 29, No. 4.

9. J. Dement, in Handbook of Chemistry and Physics, 48th ed., Chemical Rubber Co., Cleveland, 1967-1968, p. 43.

10. J. Dement, J. Chem. Educ., 30, 145 (1953).

11. H. Goto and J. Kakita, J. Chem. Soc. Japan, 64, 515 (1943).

12. H. Goto and J. Kakita, Chem. Anal., 47, 4240 (1953).

13. H. Goto, Chem. Zbl., 2, 3519 (1940).

14. H. Goto and J. Kakita, J. Chem. Soc. Japan, 63, 470 (1942).

15. H. Goto, J. Chem. Soc Japan, 59, 1357 (1938).

16. H. Goto, Sci. Rept. Tohoko Imp. Univ., 29, 446 (1940); Chem. Zbl., 1, 1847 (1941).

17. E. Kocsis, J. Kallos, and G. Zador, Z. Anal. Chem., 126, 452 (1943).

18. H. Goto, Sci. Rept. Tohoko Imp. Univ., 28, 513 (1940); Chem. Zbl., 2, 2926 (1940).

19. E. Kocsis, G. Zador, and J. Kallos, Z. Anal. Chem., 126, 138 (1943).

20. E. Kocsis and G. Zador, Z. Anal. Chem., 124, 274 (1942).

21. A. del Compo and F. Sierra, Z. Anal. Chem., 104, 279 (1937).

22. H. Diehl, Calcein, Calmagnite and o,o'-Dihydroxyazobenzene Colorimetric Fluorometric Reagents for Calcium and Magnesium, G. Frederick Smith Chemical Co., Columbus, Ohio, 43223, 1964.

23. B. L. Kepner and D. M. Hercules, Anal. Chem., 35, 1238 (1963).

24. L. E. Gibbs and D. H. Wilkins, Talanta, 2, 161 (1959).

25. D. H. Wilkins and L. E. Gibbs, Talanta, 2, 201 (1959).

26. D. F. Wallach and T. L. Steck, Anal. Chem., 35, 1238 (1963).

27. D. H. Wilkins, Talanta, 4, 182 (1960).

28. D. H. Wilkins, Talanta, 2, 277 (1959); 4, 80 (1960).

29. D. H. Wilkins, Anal. Chim. Acta, 23, 309 (1960).

30. F. Bermejo-Martinez and A. Margolet, Chemist-Analyst, 53, 45 (1964).

31. F. Bermejo-Martinez, A. Badrinas, and A. P. Bouza, Inf. Quim. Anal. (Madrid), 14, 151 (1964).

32. G. F. Kirkbright, D. I. Rees, and W. I. Stephen, Anal. Chim. Acta, 27, 558 (1962).

33. G. K. Singhal and K. N. Tandon, Talanta, 14, 1351 (1967).

34. J. H. Eggers, Talanta, 4, 38 (1960).

35. G. M. Haitink, Dissertation Abstr., B28, No. 4, 1386-7 (1967); University Microfilms, Ann Arbor, Mich., No. 67-12969.

Chapter 16

FORENSIC AND ENVIRONMENTAL ANALYSIS

I. INTRODUCTION

Fluorescence has been used as a qualitative and quantitative analytical technique in forensic chemistry and analysis (i. e. , in crime detection) and also for the assay of toxic substances in air- and water-pollution research. In this chapter we consider the usefulness of fluorescence as an analytical method in these areas.

II. FORENSIC ANALYSIS

A. General Remarks

Fluorescence has proved to be a valuable technique in forensic analysis because of its high sensitivity and selectivity. Typical applications include the following: detection of drugs and poisons in tissues; determination of the age of a body from the fluorescence of organs and tissues; and detection of blood; sperm, and gun grease.

Luminescence analysis, however, has not been as widely accepted in criminology as other techniques. Because of the conservative tendency of the law courts in accepting evidence, expert opinions must be rigorously substantiated so as to exclude all possibility of error. The findings must be theoretically sound and interpreted exactly. Some of the uncertainty that formerly existed in the use of fluorescence has been eliminated due to better

instrumentation and more exacting procedures. It can now be used as a powerful tool for ascertaining the chemical nature of substances at a submicrogram
level.

A good chapter on fluorometric methods in forensic chemistry appears in a
book by Konstantinova-Shlezinger [1].

B. Selected Examples

Skin scars of different ages exhibit different fluorescences. Those 1 to 2
months old look dark and velvety, those 4 to 6 months old exhibit a bluish-white
fluorescence. In scars over 1 year old the fluorescence depends on the degree
of pigmentation, ranging from bluish white to dark. Hence the age of injuries
can be ascertained by ultraviolet-light examination and fluorescence. This same
age dependence exists for cartilage and can be used in the examination of
dismembered corpses [2].

Fluorometric examination of organs is very useful in the assay of poisoning
with intensely fluorescing substances. For instance, in cases of quinacrine
poisoning the tissues of the liver, lungs, spleen, and other internal organs
assume a bright yellowish green fluorescence [3].

The bright fluorescence of some petroleum fractions has been used for the
detection of traces of lubricants (used in vehicles) that stain the body and clothes
in traffic accidents. The fluorescence-microscopic investigation of skin
fluorescence in victims of traffic accidents may detect the presence of yellowish
and whitish-bluish surface layers that are easily removed with ether. In such
layers the epidermal cells sometimes acquire bright whitish or bluish fluorescence, probably due to the penetration of oils into the epidermis.

Wide use is being made of fluorometric analysis in the examination of hair
and traces of human blood and sperm. The fluorescence of hair is weak.
According to Boller [4], fluorescence microscopy is useful in the detection of
dyed hair. Such investigations are most conveniently carried out on cross
sections of hairs. This method may be applied not only to human but also to
animal hair in the examination of furs. The specific structure of the hair
cuticle can be rapidly examined by fluorochroming with a berberine sulfate
solution (1:1000).

A conclusive test for blood is the reaction of blood with concentrated
sulfuric acid to produce microaggregates of hematoporphyrin which emit bright
orange-red fluorescence (λ_{em} 597, 619, and 653 nm) detectable with the
fluorescence microscope [5]. Alternatively blood can be detected by the
fluorophor found by the peroxidative activity of blood on hydrogen peroxide and
hemoglobin [6].

Kostyakova [7] presented a method for the assay of quinine, quinacrine, and ethacridine in biological materials, by using the fluorescence of these compounds in aqueous solution. Quinine possesses a blue fluorescence, but on treatment with bromine water and ammonia shows a yellow-green fluorescence. Ethacridine and quinacrine have a similar yellow-green fluorescence, but their fluorescence is observed in an acid and an alkaline medium, respectively; as little as 10 ppb is detectable.

Pontus and Husson [8] proposed a method for the detection of sperm stains on the basis of their fluorescence in the 400- to 490-nm region (λ_{em} 420 nm). The detection of spermatozoa in stains is often difficult, yet it can be performed fluorometrically by reaction with berberine sulfate [9] or with Acridine Orange and auramine [10]. Furthermore live spermatozoa can be distinguished from dead ones by their fluorescence; the former are green, while the latter are red [11].

One of the important subdivisions of forensic medicine is the examination of bullet injuries in order to determine the point of entry and the signs of discharge of the weapon at point-blank range. Fluorochroming with acriflavine of Acridine Orange of celloidin sections (after the removal of celloidin with oil of cloves) distinctly shows the grains of gunpowder on the skin around the point of entry. The fluorescence of gunpowder grains is very bright, and they can be readily identified by their characteristic shape and the presence of a dark nonfluorescent fringe of graphite [12].

Another technique for examination of bullet injuries, described by Rozanov [13], involves the detection of the gun grease in the entry wound by its bluish fluorescence. Kozlov [1] proposed an ether extraction for the determination of traces of gun grease. If several shots were fired from the same gun, a comparison of the fluorescence brightness of the extracts can establish their sequence.

In forensic toxicological examinations it is very important to differentiate between ptomaines and alkaloids. Boshnzakov [14] described a fluorometric test for this purpose. Solutions of morphine salts appear dirty green, solutions of Pantopon brownish green, solutions of opium greenish brown, and strychnine solutions light blue when exposed to ultraviolet light. Ptomaine solutions appear turbid, with a pale-blue tint under ultraviolet illumination.

Another excellent application of fluorescence is its ability to detect drugs such as lysergic acid diethylamide (LSD). The fluorescence of LSD was first pointed out by Udenfriend et al. [15] and was applied to detection in tissues by Axelrod et al. [16] at levels of 6 ng/g of tissue.

Genest and Farmilio [17] described the application of fluorescence to the identification of LSD in seized material. It is possible to positively identify

LSD by its spectral characteristics (excitation and emission) together with the products of hydrolysis. Fluorescence on TLC plates containing LSD can also be measured directly without elution, thereby simplifying the method and preserving the laboratory evidence.

In contrast to practical forensic chemistry, in which fluorometric analysis alone can be used, criminology makes extensive use of "grading" fluorescence tests, mostly in their simplest form, such as visual examination of documents in the ultraviolet [18].

In spite of its simplicity, this technique will detect traces of etching, remnants of erased text, and starch particles in places of transferred seal imprints; it reveals invisible inks, traces of glue, and postmarks that are not readily noticeable in the daylight. Obviously failure to detect traces of forgery or other traces in such a test is not a convincing proof of their absence. Its advantage is that it does not require prolonged time, does not destroy the document, and does not require elaborate apparatus. Similar tests are used to detect adulteration of food products (margarin in butter, mineral oil in vegetable oil, etc.).

The fluorescence of paintings in the ultraviolet sometimes makes it possible to establish the authorship of a painting or traces of restoration, which may be important in expert examination of paintings. For instance, in a fluorescence examination of the painting "Rough Sea" alleged to be by Aivazovskii, Sal'kov [19] found a signature of another painter under the signature claimed to be Aivazovskii's, which was not visible in ordinary daylight. Another interesting case was the detection of a fluorescent signature of Rubens on a painting whose authenticity had been disputed [20].

III. AIR- AND WATER-POLLUTION ANALYSIS

One class of air pollutants that constitutes a very serious threat to health is that of the aromatic hydrocarbons, many of which have been characterized as carcinogenic. By far the most active research group in the area of detection of such pollutants is that of Sawicki and his colleagues at the Environmental Protection Agency. In one recent paper [21] Sawicki presented the fluorescence spectra and relative fluorescence values (K_Q) for a number of hydrocarbons (Table 94).

As is evident, most of the compounds are so fluorescent that less than 1 ppb can be easily detected. Thus fluorescence appears to be an ideal method for the assay of this large class of pollutants.

By a proper control of the excitation and emission wavelengths Sawicki found it was possible to separate the carcinogenic hydrocarbons from each

other. For example, benzo[a]pyrene is the only compound that has a fluorescence maximum at 546 nm. As little as 0.04 μg of benzo[a]pyrene can be
assayed in a mixture of 1 μg each of all the other hydrocarbons. Similarly as
little as 0.1% of perylene can be detected in pure benzopyrene.

Similarly each of the aromatic hydrocarbons can be assayed in mixtures by
using a chromatographic separation. The technique most commonly used is
thin-layer chromatography.

Carcinogenic hydrocarbons have also been reported in inhaled cigarette
smoke. Van Duuren [22-24] has described methods for the assay of hydrocarbons in cigarette smoke by TLC and fluorescence spectroscopy. He was
able to identify benzo[a]pyrene in cigarette smoke by its characteristic fluorescence spectrum. He also found dibenz[a, h]anthracene, both strong carcinogens,
in cigarette smoke. Muel [25] was able to show that 130 μg of benzo[a]pyrene
was released in the exhaust fumes of an automobile in 90 min. He also showed
that drinking water in Paris is free of this carcinogen.

TABLE 94

Fluorescence Intensities of Polynuclear Hydrocarbons in Pentane[a]

Compound	Wavelength maxima (nm)		
	λ_{ex}	λ_{em}	K_Q[b]
Quinine			1.0
Anthanthrene	420	430	45
Naphtho[2, 3-a]pyrene	457	458	∼33
Perylene	430	438	32
Dibenzo[b, k]chrysene	308	428	17[c]
7-Methyldibenzo[a, h]pyrene	460	467	13[c]
Benzo[k]fluoranthene	302	400	13
Benzo[a]pyrene	381	403	6
1, 4-Diphenylbutadiene	328	370	6
p-Terphenyl	284	338	6
Benzo[k, l]xanthene	363	418	5
11H-Benzo[a]fluorene	317	340	4
11H-Benzo[b]fluorene	312	340	4
7H-Benzo[c]fluorene	334	337	4
Benzo[b]naphtho[2, 3]furan	320	350	4
Anthracene	350	398	3
Benzo[b]chrysene	283	398	2

TABLE 94--Continued

| Compound | Wavelength maxima (nm) | | |
	λ_{ex}	λ_{em}	K_Q[b]
9-Methylanthracene	382	410	2
3-Methylcholanthrene	297	392	2
Benzo[b]fluoranthene	300	428	2
Dibenzo[a, e]pyrene	370	401	2
Tribenzo[a, e, i]pyrene	384	448	2
Dibenz[a, h]anthracene	292	394	1
Fluorene	300	321	1
Benz[a]anthracene	284	382	0.9
7,12-Dimethylbenz[a]anthracene	293	427	0.8
Fluoranthene	354	464	0.8
Dibenz[a, j]anthracene	300	410	0.7
Dibenz[a, c]anthracene	280	381	0.7
Picene	281	398	0.7
4-Methylpyrene	338	386	0.7
o-Phenylenepyrene	360	506	0.6
1-Methylpyrene	336	394	0.6
Chrysene	264	381	0.6
Coronene	337	450	0.6
Pyrene	330	382	0.5
3-Methylphenenthrene	292	368	0.5
Benzo[e]pyrene	329	389	0.4
Benzo[g, h, i]perylene	280	419	0.4
Acenaphthene	291	341	0.4
Phenanthrene	252	362	0.2
2-Methylphenanthrene	257	357	0.2
4-Cyclopenta[d, e, f]phenanthrene	294	362	0.2
Triphenylene	288	357	0.1
5,12-Dihydronaphthacene	282	340	0.1

[a]Measurements were made with an Aminco-Bowman spectrophotofluorometer and are reported as uncorrected instrumental readings. (After Sawicki et al. [21].

[b]Fluorescence relative to quinine.

[c]In chloroform.

Sawicki et al. [26, 27] have investigated the fluorescence characteristics of carbazole and polynuclear carbazoles present in tar. The addition of alkali forms the carbazole anion, which absorbs at longer wavelengths. The anion is unstable in the presence of water; base is therefore added in methanolic solution. Substitution of additional benzene rings onto the carbazole nucleus yields many derivatives, each with characteristic excitation and fluorescence properties. The maxima for a number of carbazole derivatives in neutral and alkaline dimethylformamide (DMF) are listed in Table 95. The combination of several separations on thin-layer chromatograms and spectrofluorometric analysis

TABLE 95

Fluorescence Intensities of Carbazoles and Polynuclear Carbazoles in
DMF and in Alkaline DMF[a]

Compound	DMF wavelength maxima (nm)			Alkaline DMF[b] wavelength maxima (nm)		
	λ_{ex}	λ_{em}	K_Q[c]	λ_{ex}	λ_{em}	K_Q[c]
Carbazole	291	359	10	288	422	6
11H-Benzo[a]carbazole	282	377	11	312	441	6
5H-Benzo[b]carbazole	284	407	10.5	300	521	3
7H-Benzo[c]carbazole	329	388	5.5	308	450	5
4H-Benzo[d,e,f]carbazole	350	378	7	410	462	5
7H-Dibenzo[c,g]carbazole	363	392	13	333	442	4
1-Azacarbazole	298	375	3	288	465	2
2-Hydroxycarbazole	310	350	6	350	405	3

[a]After Bender, Sawicki, and Wilson [27].

[b]DMF - 29% methanolic tetraethylammonium hydroxide (5:1).

[c]Relative fluorescence intensity.

permits assay and identification of the individual carbazoles in air samples. The presence of free carbazole in ordinary urban air was demonstrated in this way. The carcinogen 11H-benzo[a]carbazole was identified in air polluted by coal-tar-pitch fumes.

Oxidized organic fragments produced during combustion are a major class of atmospheric irritants. In this category aldehydes are among the most irritating to living systems. Formaldehyde, acetaldehyde, and other simple aliphatic aldehydes are found in air. The acroleins are another major class of irritants produced during combustion. They are highly reactive and can be made to condense with many reagents to yield highly colored and/or fluorescent derivatives.

Sawicki et al. [28] have described several colorimetric and fluorometric methods for the detection and assay of ß-hydroxyacrolein, which is a tautomer of malonaldehyde. This dialdehyde is highly reactive and condenses with many amines to yield intensely fluorescent derivatives. The products formed with ethyl p-aminobenzoate and 4,4'-sulfonyldianiline are the most intense and stable, and thus very useful for analysis. With ethyl p-aminobenzoic acid the λ_{ex} is 500 and λ_{em} is 550 nm. With 4,4'-sulfonyldianiline λ_{ex} is 475 and λ_{em} is 545 nm. From 0.3 to 150 ng is detectable.

Bredereck et al. [29] investigated the fluorescence of aromatic aldehydes. Apparently most aromatic aldehydes are nonfluorescent in hydrocarbon solution, but fluorescence activation frequently occurs on addition of polar solvents or directly in polar solvents. The activation appears to result from solvent inter-action with the excited molecule. Pyrene-3-aldehyde, naphthalene-2-aldehyde, anthracene-9-aldehyde, and acenaphthene-3-aldehyde all fluoresce in hydroxylic solvents, but not in pure hydrocarbon solvents. Tetracene-9-aldehyde and pyrene-4-aldehyde fluoresce in pure heptane as well as in hydroxylic solvents.

MacDougall published detailed reviews [30, 31] on the application of fluorescence assay to the determination of pesticide residues in animal and plant tissues. Details were presented concerning instrumentation, structure-fluorescence relationships, and chemical mechanisms of some of the available procedures. Analytical methods for the assay of pesticides are presented in Chapter 7 of this volume.

REFERENCES

1. M. A. Konstantinova-Shlezinger, Fluorometric Analysis, Israel Program for Scientific Translations, 1965, Chapter 19.

2. Kh. Takho-Godi, In Materialy III vsesoyuznogo soveshchaniza sudebno-meditsinkikh ekspertov, Riga, 1957, p. 27.

3. M. Khait and A. Starchevskaya, in Materialy III rasshirennoi nauchnoi konferentsii Kievskogo, Kiev Gosmedizdat, 1958, pp. 80 and 81.

4. W. Boller, Archiv Kriminol., 100, 207, 264 (1937).

5. V. N. Vinogradov, in Referaty dokladov IX rasshirennoi konferentsii Leningradskogo, Leningrad, 1955, p. 14.

6. W. Sphecht, Angew Chem., 50, 8 (1937).

7. L. Kostyakova, Zh. Anal. Khim. , 2, 27 (1947); 6, 251 (1951).

8. P. Pontus and Husson, Compt. Rend. Biol. , CXVI (21), 538 (1937).

9. V. N. Vinogradov and A. Tumanov, in Sbornik referatov, dokladov rasshirennoi nauchnoi konferentsii, Kharkov, 1956, p. 109.

10. Kh. Takho-Godi, Sudebno-meditsinskaya ekspertiza, 3, 27 (1958).

11. L. Stockinger, Mikroskopie, 4, 53 (1949).

12. V. N. Vinogradov, in Materialy III, vsesoyuznogo soveshchaniza sudebno-meditsinskikh ekspertov, Riga, 1957, p. 188.

13. B. M. Rozanov, Tr. Voenno-meditsinskoi Akad. im. S. M. Kirova, 53, 219 (1952).

14. A. N. Boshnyakov, Voenno-meditsinskii Zh. , 1, 75 (1953).

15. S. Udenfriend, D. Duggan, B. Vasta, and B. Brodie, J. Pharmacol. Exptl. Therap. , 120, 26 (1957).

16. J. Axelrod, R. Brady, B. Witkop, and E. Evarts, Ann. N. Y. Acad. Sci. , 66, 435 (1957).

17. K. Genest and C. G. Farmilio, J. Pharm. Pharmacol. , 16, 250 (1964).

18. N. Terziev, B. Kirichniskii, A. Eisman, and E. Gerken, Physical Investigations in Criminology, Moscow, 1948, p. 111.

19. A. Sal'kov, Sudebno-meditsinskaya Ekspertiza, No. 7, 1928.

20. E. Beil , R. Fabr, and A. Zhoryn, Sudebno-meditsinskaya Ekspertiza, No. 8.

21. E. Sawicki, T. Hauser, and T. Stanley, Intern. J. Air Pollution, 2, 253 (1960).

22. B. Van Duuren, J. Natl. Cancer Inst. , 21, 623 (1958).

23. B. L. Van Duuren, J. Natl. Cancer Inst. , 21, 1 (1958).

24. B. L. Van Duuren and A. Kosak, J. Org. Chem. , 23, 473 (1958).

25. B. Muel, Studies on the Luminescence of 3, 4-Benzopyrene, Ph. d. thesis, University of Paris, 1962.

26. D. F. Bender, E. Sawicki, and R. Wilson, Intern, J. Air and Water Pollution, 8, 633 (1964).

27. D. F. Bender, E. Sawicki, and R. Wilson, Anal. Chem. , 36, 1011 (1964).

28. E. Sawicki, T. Stanley, and H. Johnson, Anal. Chem. , 35, 199 (1963).

29. K. Bredereck, T. Förster, and H. Oesterlin, in Luminescence of Organic and Inorganic Materials, Wiley, New York, 1962, p. 161.

30. D. MacDougall, Residue Rev. , 1, 24 (1962).

31. D. MacDougall, Residue Rev. , 5, 119 (1964).

AUTHOR INDEX

Underlined numbers give the page on which the complete reference is listed.

A

Abbott, S. R. , 116(85), 134, 301, 339
Abelson, D. , 331(271), 347, 582, 595
Abou-Donia, M. B. , 386, 394
Aboul-Saad, W. G. , 490(164), 530
Abrams, A. , 520(415), 540
Ackerman, H. , 387, 395
Adachi, S. , 312(146), 342
Adams, J. , 325(239,240), 326(239,240), 346
Adams, R. , 291(31), 338
Adler, A. , 329, 346
Adler, E. , 381, 393
Aguila, J. F. , 223(54), 235, 266
Ainsworth, S. , 487, 529
Aksentsev, S. L. , 209(62), 212
Alarion, R. , 301, 339
Albers, R. , 300, 331(274), 339, 347
Albert, A. , 479, 528
Alberts, R. W. , 147(12), 159
Alder, J. F. , 432(40), 444
Aldrich, R. , 328, 346
Aledort, L. , 315(157), 342
Alexander, B. , 486(125), 529
Alexander, N. , 16(23), 29, 501(259), 534
Alford, W. C. , 226(173), 230(173), 248(173), 263(173), 264(173), 271
Algar, B. E. , 119(95), 135
Alkemade, C. T. J. , 411, 438, 440(49), 441(49), 442(5), 443, 445
Alkon, D. , 305, 341
Allaway, W. H. , 228(240), 259, 273
Allen, F. L. , 563, 575
Allen, M. B. , 549(37), 568(37), 571
Alonso, R. , 474(43,44), 526
Alpers, H. , 324(214), 345
Alvarez, C. , 325(238), 326(238), 345
Amann, H. , 558(96), 560(96), 561(96), 573
Ambrose, J. , 309(129), 341
Amesz, J. , 552(60), 553(60), 572
Amos, D. , 228(218), 255, 273
Anastasina, G. V. , 225(131), 230(267), 244(131), 263(267), 269, 275
Andersen, N. R. , 122(109), 135
Anderson, B. M. , 473(35), 525
Anderson, C. , 325(239,240,242), 326(239,240), 346

Anderson, J. M. , 552, 553(60), 554, 568(59), 572
Anderson, R. , 325(242), 346, 432(39), 444
Anderson, S. R. , 484, 488, 489, 528, 529, 530
Anderson, W. , 489, 530
Ando, K. , 473(32), 525
Andreae, W. A. , 228(221), 256, 273, 370, 393
Andrejack, J. , 223(51), 266
Andrushko, G. , 227(202), 252, 272
Anfinsen, C. B. , 477, 513(356), 514(356,358,359), 526, 538
Angell, E. , 323(193), 344
Antonis, A. , 299(65), 339
Antonovich, V. , 228(234), 258(234), 273
Anufrieva, E. V. , 477, 527
Apella, E. , 494(197), 532
Appel, W. , 558(96), 560(96), 561(96), 573
Aprison, M. H. , 368, 392
Arapova, E. , 216(13), 265
Archibald, R. , 335, 348
Argauer, R. , 2, 28, 153(25), 154(25), 155(25), 156(25), 157(25), 159, 279, 325(238), 326(238), 337, 345
Argerich, T. , 328, 346
Arian, S. , 504, 535
Arkhangelskaya, V. , 215(12), 265
Armentrout, D. N. , 411, 423(12), 426(12), 428, 431(12), 432(12), 434(12), 443
Armstrong, W. , 402, 407
Arnold, W. , 545, 564(129), 570, 575
Arnon, D. , 560, 574
Arnon, R. , 495(211), 518, 532, 539
Arnone, A. , 514, 538
Arquilla, E. R. , 479(96), 528
Arrio, B. , 519, 540
Arrio-Dupont, M. , 491(171), 530
Asahara, M. , 361, 391
Attalah, N. A. , 517, 539
Auerbach, M. E. , 323(193,197), 344
Augenstein, L. , 208, 209(63), 212, 498(246), 506(302,305), 534, 536
Auld, D. S. , 523(442), 541
Avigan, J. , 584(19), 595
Avioli, L. , 332, 347
Avron, M. , 543(1), 569

619

SUBJECT INDEX

A

Absorption spectrum, 2, 3, 4
Acenaphthene, fluorescence of, 614
 phosphorescence of, 190
Acetaldehyde, fluorescence assay, 300
 phosphorescence of, 190
Acetic acid, assay of, 296, 377, 378
Acetol, assay of, 299
Acetonaphthone, intersystem crossing,
 97
Acetone, fluorescence assay, 300
 phosphorescence of, 190
Acetophenone, intersystem crossing
 efficiency, 96, 97
Acetylcholine, assay of, 368, 369
N-Acetyl-ß-D-Glucosaminidase, assay
 of, 358, 359
N-Acetyl-L-tyrosine ethyl ester,
 phosphorescence, 163, 196
Acid-Base, indicators, 598-601
 table of, 599, 600
 reactions, in excited state, 105-107
Acidity, of excited molecules, 9, 105-
 107
Acids, organic, assay of, 295-298,
 376-378
Acridine, chemiluminescence of, 403
 fluorescence of, 286, 292
 lifetime, 16
Acridinium ion, excited state acidity
 of, 106
Acridone, lifetime, 16
Acriflavin, fluorescence of, 12, 292
 use in assay of O_2, 254
Acrolein, assay of, 301, 616
Actinomycin, fluorescence assay of,
 323
 phosphorescence of, 209
Activators, assay of, 380-384
Adenine, fluorescence of, 313
 phosphorescence of, 209, 210
Adenine nucleotides, fluorescence of,
 316
Adenosine, fluorescence of, 312-314
 phosphorescence of, 209, 210
Adenosine diphosphate, assay of, 315
 fluorescence of, 312-314
 phosphorescence of, 209

Adenosine monophosphate, fluorescence assay of, 312-315
 phosphorescence, 209
Adenosine triphosphate, assay of,
 315, 320, 381, 382, 405, 406
 fluorescence of, 312-314
 phosphorescence of, 209
 in photosynthesis, 544, 545
Adenylic acid, assay of, 296, 313
 phosphorescence of, 210
Adsorption indicators, 601-604
 table of, 602, 603
Air pollutants, fluorescence assay
 of, 612-616
 on solid surfaces, 583
 phosphorescence assay of, 200-201
Albumin, bovine serum,
 ANS conjugates, 484, 488
 DNS conjugates, 470, 471, 474,
  480-483, 485, 486
 fluorescence of, 512, 513, 516,
  517
 labeling with other dyes, 470,
  484, 487, 493
 lifetime of, 16
 phosphorescence, 209
 other species, fluorescence of,
  470, 475, 489, 516
Alcohol dehydrogenase, assay of,
 209, 379
 fluorescence labeling of, 484, 510
Alcohols, assay of, 298-300, 379, 380
Aldehydes, fluorescence of, 81, 88,
 96-99, 300-301, 616
 phosphorescence of, 201
Aldosterone, assay of, 331
Aldrin, assay of, 385, 386, 388, 389
Aliphatic compounds, luminescence
 of, 80-82, 98, 99, 616
Alizarin Garnet R, reagent for Al,
 8, 9
Alkaloids, fluorescence of, 611
 phosphorescence assay of, 202, 203
Alkyl carbazoles, fluorescence of,
 289, 290
Alkyl halides, effect on ECL, 460-461
Alkyl thiohydantoins, assay of, 323

645

Hydroxyquinoline, fluorescence of,
   effect of H bonding, 108-110
   reagent for inorganic ions, 214, 234,
      250, 251
Hydroxytryptamine, fluorescence of,
   288, 306
Hydroxytryptophan, phosphorescence
   of, 192
Hypochlorite peroxide reaction, 405

I

Imidan, phosphorescence of, 192, 205
Imidazole acetic acid, assay of, 306
Immunoglobulins, labeling of, 477, 478,
   495
Impurities, effect on luminescence, 27
Incandescent lamps, 33
Indicators
   chemiluminescence, 399-401
   fluorescence, 597-607
      acid-base, 598-601
         table of, 599, 600
      adsorption, 601-604
         table of, 602, 603
      apparatus for, 597, 598
      metal ions, 604-606
      redox, 601
Indigo white, fluorescence of, 356-358
Indium, assay of, 227, 249, 385
   atomic fluorescence of, 411, 430
Indole, chemiluminescence of, 402
   lifetime of, 17
   quantum yield, 12
   phosphorescence of, 209
Indoleacetic acid, assay of, 296
   lifetime of, 17
   phosphorescence of, 192
   temperature effect on fluorescence, 24
Indole acetonitrile, phosphorescence of,
   192
Indole alkaloids, assay of, 203
Indole butyric acid, phosphorescence
   of, 192
Indole carboxylic acid, phosphorescence
   of, 192
Indole propionic acid, phosphorescence
   of, 192
Indoxyl esters, substrate for cholin-
      esterase, 356-358, 591
   derivatives, fluorescence of, 288
Inhibitors, assay of, 384-389
   of cholinesterase, 385-388
   of hexokinase, 388, 389
Inner filter effect, 114
Inorganic ions, assay of, 213-264
Inorganic reagents, assay of inorganic
   ions, 219-221

Instrumentation,
   atomic fluorescence, 423-425
   fluorescence, 19-20, 31-78
      basic instruments, 56-78
      corrected spectra, 71-74
      decay times, 74-76, 77
      cell compartment, 50, 51
      detectors, 53-56
      light sources, 33-38
      monochromators, 38-50
      practical considerations, 137, 138
      slits, 52, 53
   phosphorescence, 162, 163, 174-
      185, 198-200
Insulin, fluorescence labeling of, 479
   phosphorescence of, 209
Intensity, atomic fluorescence, 418-
   419
   fluorescence, 18
   phosphorescence, 167-168
Interferences, atomic fluorescence,
   442, 443
Intersystem crossing, in carbonyls,
   97-99
   in nitrogen heterycyclics, 91-96
Intramolecular energy transfer, 97-99
Intrinsic fluorescence, proteins, 497-
   521
Iodide, assay of, 220, 222, 226, 249
Iridium, asay of, 227, 249
   atomic fluorescence, 430
Iron, assay of, 220, 222, 226, 246,
   385
   atomic fluorescence of, 429
Isocitric acid, assay of, 297, 377, 378
Isoguanine, fluorescence of, 313
Isolan, phosphorescence of, 192, 204
Isoniazid, assay of, 324
Isoquinoline alkaloids, assay of, 203

K

Kautsky effect, 558, 559
Kelthane, phosphorescence, 189, 192,
   204
Kepone, phosphorescence of, 193, 204
α-Keto acids, assay of, 297, 298
Ketoglutaric acid, assay of, 297
Ketone bodies, assay of, 368, 369
Ketones, fluorescence of, 96-99, 300,
   301
   phosphorescence of, 201
Ketoses, assay of, 312, 373
Kinetic methods, 349-395
Klett fluorometer, 57-59
Kynurenic acid, assay of, 296
Kynureninase, assay of, 373